Higher
GCSE Mathematics:
Revision and Practice

Higher GCSE Mathematics: Revision and Practice

D. Rayner

Oxford University Press

Oxford University Press, Great Clarendon Street, Oxford OX2 6DP

Oxford New York
Athens Auckland Bangkok Bogota Buenos Aires
Calcutta Cape Town Chennai Dar es Salaam
Delhi Florence Hong Kong Istanbul Karachi
Kuala Lumpur Madrid Melbourne Mexico City
Mumbai Nairobi Paris São Paulo Singapore
Taipei Tokyo Toronto Warsaw

and associated companies in
Berlin Ibadan

Oxford is a trade mark of Oxford University Press

First published 1994
Reprinted 1994, 1995 (twice), 1996 (twice), 1997, 1998

ISBN 0 19 914574 1 (School edition)
0 19 914535 0 (Bookshop edition)

Artwork by Nick Hawken and Julian Page

Typeset and illustrated by Tech-Set, Gateshead, Tyne and Wear
Printed and bound in Great Britain by
Butler & Tanner Ltd, Frome and London

Preface

This book is for candidates working through Key Stage 4 towards a GCSE in Mathematics: it covers the National Curriculum for the Higher Tier. The book can be used both in the classroom and by students working on their own. There are explanations, worked examples and numerous exercises which, it is hoped, will help students to build up confidence. The author believes that people learn mathematics by *doing* mathematics. The questions are graded in difficulty throughout the exercises. Questions marked † are more difficult and are included to provide a stimulus for the most able students.

The book can be used either as a course book over the last two or three years before the Key Stage 4 examinations or as a revision text in the final year. The contents list shows where all the topics appear and an index at the back of the book provides further reference.

The work is collected into sections on Number, Algebra, Shape and Space, Handling Data, Probability, and Using and Applying Mathematics. This is done for ease of reference but the material can be taught in any convenient order. Most teachers and students will prefer to alternate freely between topics from each of the sections.

Each part concludes with a selection of past GCSE and sample Key Stage 4 questions from all the main boards. At the end of the book, there are several revision exercises which provide mixed questions across the curriculum. There are also multiple choice questions for variety.

The section on Using and Applying Mathematics begins with a discussion of conjectures, implication and proof. There follows a selection of starters for coursework projects which have been tried and tested. They can be used to provide practice in the strategies involved in attempting 'open-ended' problems. Outline solutions have been provided for these.

The author is indebted to the many pupils and colleagues who have assisted him in this work. He is particularly grateful to Philip Cutts, Julie Anderson and Micheline Rayner for their invaluable work. Thanks are also due to the following examination boards for kindly allowing the use of questions from their past mathematics papers and sample Key Stage 4 questions:

University of London Examinations and Assessment Council	[L]
Midlands Examinations Group	[M]
Northern Examinations and Assessment Board	[N]
Southern Examining Group	[S]
Welsh Joint Education Committee	[W]

The sample questions from these boards represent one stage in the development of nationally approved material and do not necessarily reflect the style currently accepted.

D. Rayner 1993

CONTENTS

1 Number

1.1 Arithmetic review

Exercise 1

Evaluate the following without a calculator:

1. $7{\cdot}6 + 0{\cdot}31$
2. $15 + 7{\cdot}22$
3. $7{\cdot}004 + 0{\cdot}368$
4. $0{\cdot}06 + 0{\cdot}006$
5. $4{\cdot}2 + 42 + 420$
6. $3{\cdot}84 - 2{\cdot}62$
7. $11{\cdot}4 - 9{\cdot}73$
8. $4{\cdot}61 - 3$
9. $17 - 0{\cdot}37$
10. $8{\cdot}7 + 19{\cdot}2 - 3{\cdot}8$
11. $25 - 7{\cdot}8 + 9{\cdot}5$
12. $3{\cdot}6 - 8{\cdot}74 + 9$
13. $20{\cdot}4 - 20{\cdot}399$
14. $2{\cdot}6 \times 0{\cdot}6$
15. $0{\cdot}72 \times 0{\cdot}04$
16. $27{\cdot}2 \times 0{\cdot}08$
17. $0{\cdot}1 \times 0{\cdot}2$
18. $(0{\cdot}01)^2$
19. $2{\cdot}1 \times 3{\cdot}6$
20. $2{\cdot}31 \times 0{\cdot}34$
21. $0{\cdot}36 \times 1000$
22. $0{\cdot}34 \times 100\,000$
23. $3{\cdot}6 \div 0{\cdot}2$
24. $0{\cdot}592 \div 0{\cdot}8$
25. $0{\cdot}1404 \div 0{\cdot}06$
26. $3{\cdot}24 \div 0{\cdot}002$
27. $0{\cdot}968 \div 0{\cdot}11$
28. $600 \div 0{\cdot}5$
29. $0{\cdot}007 \div 4$
30. $2640 \div 200$
31. $1100 \div 5{\cdot}5$
32. $(11 + 2{\cdot}4) \times 0{\cdot}06$
33. $(0{\cdot}4)^2 \div 0{\cdot}2$
34. $77 \div 1000$
35. $(0{\cdot}3)^2 \div 100$
36. $(0{\cdot}1)^4 \div 0{\cdot}01$
37. $\dfrac{92 \times 4{\cdot}6}{2{\cdot}3}$
38. $\dfrac{180 \times 4}{36}$
39. $\dfrac{0{\cdot}55 \times 0{\cdot}81}{4{\cdot}5}$
40. $\dfrac{63 \times 600 \times 0{\cdot}2}{360 \times 7}$
41. $479 \div 15$
42. $874 \div 21$
43. $8735 \div 25$
44. $1924 \div 31$
45. $4260 \div 51$
46. $51063 \div 19$

Exercise 2

1. A maths teacher bought 40 calculators at £8·20 each and a number of other calculators costing £2·95 each. In all she spent £387. How many of the cheaper calculators did she buy?

2. At a temperature of 20°C the common amoeba reproduces by splitting in half every 24 hours. If we start with a single amoeba, how many will there be after (a) 8 days, (b) 16 days?

3. Find a pair of positive integers a and b for which
$18a + 65b = 1865$

4. Find all the missing digits in these multiplications.

(a) $\begin{array}{r} 5* \\ 9\times \\ \hline **6 \\ \hline \end{array}$ (b) $\begin{array}{r} *7 \\ *\times \\ \hline 4*6 \\ \hline \end{array}$ (c) $\begin{array}{r} 5* \\ *\times \\ \hline 1*4 \\ \hline \end{array}$

5. Find the least positive whole number n for which $582416035 + n$ is exactly divisible by 11.

6. Red sticks are 5 cm long and blue sticks are 8 cm.

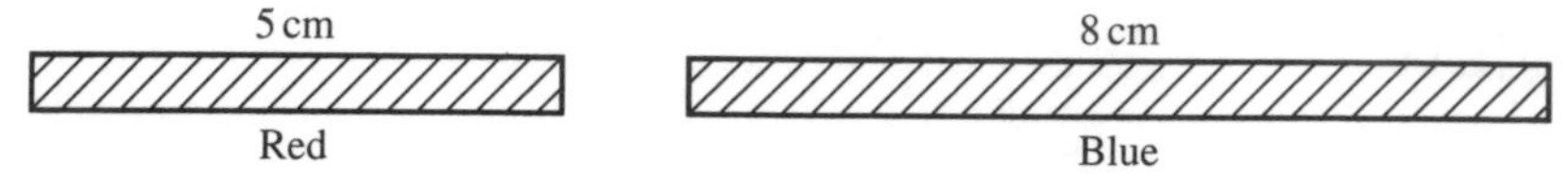

(a) How can you measure 31 cm using these sticks?
(b) How can you measure 17 cm using these sticks?
(c) How can you measure 1 cm using these sticks?

7. A group of friends share a bill for £13·69 equally between them. How many were in the group?

8. Put three different numbers in the circles so that when you add the numbers at the end of each line you always get a square number.

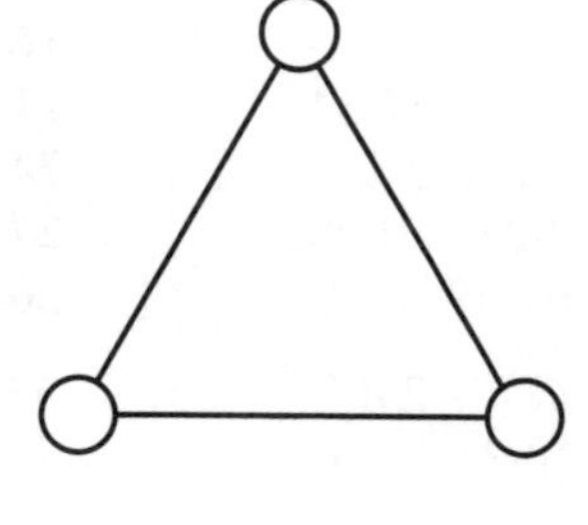

9. Put four different numbers in the circles so that when you add the numbers at the end of each line you always get a square number.

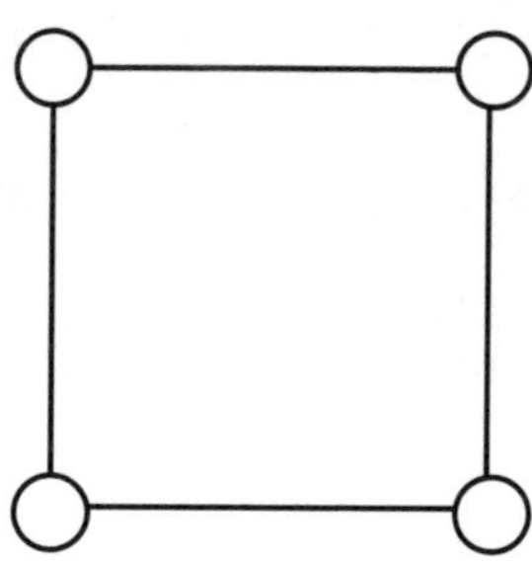

10. You are given that $41 \times 271 = 11111$. Work out the following *in your head*.
(a) 246×271
(b) $22222 \div 271$
(c) This time you can write down a (little!) working. Work out $41^2 \times 271$.

11. S_1 is the sum of all the even numbers from 2 to 1000 inclusive. S_2 is the sum of all the odd numbers from 1 to 999 inclusive. Work out $S_1 - S_2$.

Prime numbers

A prime number is divisible only by itself and by one.
e.g. 2, 3, 5, 7, 11, 13
NB One is *not* a prime number

Example 1

Express 6930 as a product of primes.

$2\overline{)69^13^10}$	(divide by 2)
$3\overline{)34^16^15}$	(divide by 3)
$3\overline{)11^25^15}$	(divide by 3)
$5\overline{)\ 38^35}$	(divide by 5)
$7\overline{)\ \ 77}$	(divide by 7)
11	(stop because 11 is prime)

$\therefore\ 6930 = 2 \times 3 \times 3 \times 5 \times 7 \times 11$

Exercise 3

1. Which of the following are prime numbers?

3, 11, 15, 19, 21, 23, 27, 29, 31, 37, 39, 47, 51, 59, 61, 67, 72, 73, 87, 99.

2. Express each of the following numbers as a product of primes.
(a) 600 (b) 693 (c) 2464
(d) 3510 (e) 4000 (f) 22 540

3. The first of three consecutive numbers is n. Find the smallest value of n if the three numbers are all non-prime.

4. (a) Express 1008 and 840 as products of their prime factors.
(b) Find the H.C.F. (highest common factor) of 1008 and 840.
(c) Find the smallest number which can be multiplied by 1008 to give a square number.

5. (a) Express 19 800 and 12 870 as products of their prime factors.
(b) Find the H.C.F. of 19 800 and 12 870.
(c) Find the smallest number which can be multiplied by 19 800 to give a square number.

6. (a) Is 263 a prime number?
By how many numbers do you need to divide 263 so that you can find out?
(b) Is 527 a prime number?
(c) Suppose you used a computer to find out if 1147 was a prime number. Which numbers would you tell the computer to divide by?

Estimation

It is always sensible to check that the answer to a calculation is 'about the right size'.

> *Example 2*
>
> Estimate the value of $\dfrac{57{\cdot}2 \times 110}{2{\cdot}146 \times 46{\cdot}9}$, correct to one significant figure.
>
> We have approximately, $\dfrac{60 \times 100}{2 \times 50} = 60$
>
> On a calculator the value is 62·52 (to 4 significant figures).

Exercise 4

In this exercise there are 25 questions, each followed by three possible answers. In each case only one answer is correct.
Write down each question and decide (by estimating) which answer is correct.

1. $7{\cdot}2 \times 9{\cdot}8$ [52·16, 98·36, 70·56]
2. $2{\cdot}03 \times 58{\cdot}6$ [118·958, 87·848, 141·116]
3. $23{\cdot}4 \times 19{\cdot}3$ [213·32, 301·52, 451·62]
4. $313 \times 107{\cdot}6$ [3642·8, 4281·8, 33678·8]
5. $6{\cdot}3 \times 0{\cdot}098$ [0·6174, 0·0622, 5·98]
6. $1200 \times 0{\cdot}89$ [722, 1068, 131]
7. $0{\cdot}21 \times 93$ [41·23, 9·03, 19·53]
8. $88{\cdot}8 \times 213$ [18914·4, 1693·4, 1965·4]
9. $0{\cdot}04 \times 968$ [38·72, 18·52, 95·12]
10. $0{\cdot}11 \times 0{\cdot}089$ [0·1069, 0·0959, 0·00979]
11. $13{\cdot}92 \div 5{\cdot}8$ [0·52, 4·2, 2·4]
12. $105{\cdot}6 \div 9{\cdot}6$ [8·9, 11, 15]
13. $8405 \div 205$ [4·6, 402, 41]
14. $881{\cdot}1 \div 99$ [4·5, 8·9, 88]
15. $4{\cdot}183 \div 0{\cdot}89$ [4·7, 48, 51]
16. $6{\cdot}72 \div 0{\cdot}12$ [6·32, 21·2, 56]
17. $20{\cdot}301 \div 1010$ [0·0201, 0·211, 0·0021]
18. $0{\cdot}28896 \div 0{\cdot}0096$ [312, 102·1, 30·1]
19. $0{\cdot}143 \div 0{\cdot}11$ [2·3, 1·3, 11·4]
20. $159{\cdot}65 \div 515$ [0·11, 3·61, 0·31]
21. $(5{\cdot}6 - 0{\cdot}21) \times 39$ [389·21, 210·21, 20·51]
22. $\dfrac{17{\cdot}5 \times 42}{2{\cdot}5}$ [294, 504, 86]
23. $(906 + 4{\cdot}1) \times 0{\cdot}31$ [473·21, 282·131, 29·561]
24. $\dfrac{543 + 472}{18{\cdot}1 + 10{\cdot}9}$ [65, 35, 85]
25. $\dfrac{112{\cdot}2 \times 75{\cdot}9}{6{\cdot}9 \times 5{\cdot}1}$ [242, 20·4, 25·2]

26. There are about 7000 cinemas in the U.K. and every day about 300 people visit each one.
The population of the U.K. is about 60 million.

A film magazine report said:

'Over 3% of British people go to the cinema everyday'.

Is the magazine report fair?
Show the working you did to decide.

27. The petrol consumption of a large car is 22 miles per gallon and petrol costs £2·45 per gallon.
Jasper estimates that the petrol costs of a round trip of about 1200 miles will be £130.
Is this a reasonable estimate?

28. The 44 teachers in a rather difficult school decide to buy 190 canes at £2·42 each.
They share the cost equally between them.
The headmaster used a calculator to work out the cost per teacher and got an answer of £1·05 to the nearest penny.
Without using a calculator, work out an estimate for the answer to check whether or not he got it right. Show your working.

Ratio

The word 'ratio' is used to describe a fraction. If the *ratio* of a boy's height to his father's height is 4 : 5, then he is $\frac{4}{5}$ as tall as his father.

Example 3

Change the ratio 2 : 5 into the form

(a) $1 : n$ (b) $m : 1$

(a) $2 : 5 = 1 : \frac{5}{2}$
$= 1 : 2{\cdot}5$

(b) $2 : 5 = \frac{2}{5} : 1$
(b) $= 0{\cdot}4 : 1$

Example 4

Divide £60 between two people A and B in the ratio 5 : 7.

Consider £60 as 12 equal parts (i.e. $5 + 7$). Then A receives 5 parts and B receives 7 parts.

$\therefore$ A receives $\frac{5}{12}$ of £60 = £25

B receives $\frac{7}{12}$ of £60 = £35

Exercise 5

Express the following ratios in the form $1:n$.

1. 2:6 **2.** 5:30 **3.** 2:100

4. 5:8 **5.** 4:3 **6.** 8:3

Express the following ratios in the form $n:1$.

7. 12:5 **8.** 5:2 **9.** 4:5

In Questions **10** to **13**, divide the quantity in the ratio given.

10. £40; (3:5) **11.** £120, (3:7)

12. 180 kg; (1:5:6) **13.** 184 minutes; (2:3:3)

14. When £143 is divided in the ratio 2:4:5, what is the difference between the largest share and the smallest share?

15. If $\frac{5}{8}$ of the children in a school are boys, what is the ratio of boys to girls?

16. A man and a woman share a bingo prize of £1000 between them in the ratio 1:4. The woman shares her part between herself, her mother and her daughter in the ratio 2:1:1. How much does her daughter receive?

17. A man and his wife share a sum of money in the ratio 3:2. If the sum of money is doubled, in what ratio should they divide it so that the man still receives the same amount?

18. In a herd of x cattle, the ratio of the number of bulls to cows is 1:6. Find the number of bulls in the herd in terms of x.

19. If $x:3 = 12:x$, calculate the positive value of x.

20. £400 is divided between Ann, Brian and Carol so that Ann has twice as much as Brian and Brian has three times as much as Carol. How much does Brian receive?

21. A cake weighing 550 g has three ingredients: flour, sugar and raisins. There is twice as much flour as sugar and one and a half times as much sugar as raisins. How much flour is here?

22. A brother and sister share out their collection of 5000 stamps in the ratio 5:3. The brother then shares his stamps with two friends in the ratio 3:1:1, keeping most for himself. How many stamps do each of his friends receive?

1.2 Percentages

Percentages are simply a convenient way of expressing fractions or decimals.
'50% of £60' means $\frac{50}{100}$ of £60, or more simply $\frac{1}{2}$ of £60. Percentages are used very frequently in everyday life and are misunderstood by a large number of people.
What are the implications if 'inflation falls from 10% to 8%'? Does this mean prices will fall?

Example 1

A car costing £2400 is reduced in price by 10%.
Find the new price.

$$10\% \text{ of } £2400 = \frac{10}{100} \times \frac{2400}{1}$$
$$= £240$$

$$\text{New price of car} = £(2400 - 240)$$
$$= £(2160)$$

Example 2

After a price increase of 10% a television set costs £286. What was the price before the increase?

The price before the increase is 100%

$$\therefore \quad 110\% \text{ of old price} = £286$$

$$\therefore \quad 1\% \text{ of old price} = £\frac{286}{110}$$

$$\therefore \quad 100\% \text{ of old price} = £\frac{286}{110} \times \frac{100}{1}$$

$$\text{Old price of TV} = £260$$

Exercise 6

1. Calculate
(a) 30% of £50 (b) 45% of 2000 kg
(c) 4% of $70 (d) 2·5% of 5000 people

2. In a sale, a jacket costing £40 is reduced by 20%. What is the sale price?

3. The charge for a telephone call costing 12p is increased by 10%. What is the new charge?

4. In peeling potatoes 4% of the mass of the potatoes is lost as 'peel'. How much is *left* for use from a bag containing 55 kg?

5. Work out, to the nearest penny

(a) 6·4% of £15·95 (b) 11·2% of £192·66

(c) 8·6% of £25·84 (d) 2·9% of £18·18

6. Find the total bill:

5 golf clubs at £18·65 each

60 golf balls at £16·50 per dozen

1 bag at £35·80

V.A.T. at $17\frac{1}{2}$% is added to the total cost.

7. In 1994 a club has 250 members who each pay £95 annual subscription. In 1995 the membership increases by 4% and the annual subscription is increased by 6%. What is the total income from subscriptions in 1995?

8. In 1993 the prison population was 48 700 men and 1600 women. What percentage of the total prison population were men?

9. In 1991 there were 21 280 000 licensed vehicles on the road. Of these, 16 486 000 were private cars. What percentage of the licensed vehicles were private cars?

10. A quarterly telephone bill consists of £19·15 rental plus 4·7p for each dialled unit. V.A.T. is added at $17\frac{1}{2}$%. What is the total bill for Mrs Jones who used 915 dialled units?

11. Johnny thinks his goldfish got chickenpox. He lost 70% of his collection of goldfish. If he has 60 survivors, how many did he have originally?

12. The average attendance at Everton football club fell by 7% in 1994. If 2030 fewer people went to matches in 1994, how many went in 1993?

13. When heated an iron bar expands by 0·2%. If the increase in length is 1 cm, what is the original length of the bar?

14. In the last two weeks of a sale, prices are reduced first by 30% and then by a *further* 40% of the new price. What is the final sale price of a shirt which originally cost £15?

15. During a Grand Prix car race, the tyres on a car are reduced in weight by 3%. If they weigh 388 kg at the end of the race, how much did they weigh at the start?

16. Over a period of 6 months, a colony of rabbits increases in number by 25% and then by a further 30%. If there were originally 200 rabbits in the colony how many were there at the end?

17. A television costs £376 including $17\frac{1}{2}$% V.A.T. How much of the cost is tax?

18. The cash price for a car was £7640. Mr Elder bought the car on the following hire purchase terms: 'A deposit of 20% of the cash price and 36 monthly payments of £191·60'. Calculate the total amount Mr Elder paid.

Exercise 7

British Gas

Share Offer

In 1986 the Government sold shares in British Gas to members of the public. Because the shares were oversubscribed, most people did not receive all the shares they applied for. A simple formula was devised to work out the allocation of shares. People received 10% of the number they applied for plus another 300 shares. So if someone applied for 4000 shares, she received (10% of 4000) + 300 shares (i.e. 700).

People had to pay 50p for each share they received. When dealing in the shares opened on the Stock market the price of shares went up to a value between 60p and 75p (share prices vary from day to day).

Answer the following questions.

1. Mr Jones applied for 500 shares. How much profit did he make when he sold his entire allocation at 64p per share?

2. Mrs Miller applied for 5000 shares and sold her entire allocation at a price of 67p per share. How much profit did she make?

3. Mr Gandee sold his allocation when the price was 65p. He received a total of £1105. How many shares did he apply for?

4. Ms Ludwell sold her allocation when the price was 69p. She received a total of £586·50. How many shares did she apply for?

5. Mr Wyatt sold his allocation when the price was 66p and he made a *profit* of £400. How many shares did he apply for?

6. Mrs Morgan sold her allocation when the price was 68p and she made a profit of £270. How many shares did she apply for?

7. Miss Green applied for 2600 shares and then sold her allocation to make a profit of £67·20. At what price per share did she sell the shares?

8. Mr Singh applied for 60 000 shares and then sold his allocation to make a profit of £1386. At what price per share did he sell the shares?

In the next exercise use the formulae:

$$\text{Percentage profit} = \frac{\text{actual profit}}{\text{original price}} \times \frac{100}{1}$$

$$\text{Percentage loss} = \frac{\text{actual loss}}{\text{original price}} \times \frac{100}{1}$$

Example 3

A radio is bought for £16 and sold for £20.
What is the percentage profit?

Actual profit = £4

$$\therefore \text{ Percentage profit} = \frac{4}{16} \times \frac{100}{1} = 25\%$$

The radio is sold at a 25% profit.

Example 4

A car is sold for £2280, at a loss of 5% on the cost price. Find the cost price.

Do *not* calculate 5% of £2280!
The loss is 5% of the cost price.

$$\therefore \quad 95\% \text{ of cost price} = £2280$$

$$1\% \text{ of cost price} = £\frac{2280}{95}$$

$$\therefore \quad 100\% \text{ of cost price} = £\frac{2280}{95} \times \frac{100}{1}$$

$$\text{Cost price} = £2400$$

Exercise 8

1. The first figure is the cost price and the second figure is the selling price. Calculate the percentage profit or loss in each case.
   (a) £20, £25 (b) £400, £500
   (c) £60, £54 (d) £9000, £10 800
   (e) £460, £598 (f) £512, £550·40
   (g) £45, £39·60 (h) 50p, 23p

2. A car dealer buys a car for £500, gives it a clean, and then sells it for £640. What is the percentage profit?

3. A damaged carpet which cost £180 when new, is sold for £100. What is the percentage loss?

4. During the first four weeks of her life, a baby girl increases her weight from 3·2 kg to 4·7 kg. What percentage increase does this represent? (Give your answer to 3 sig. fig.)

5. When V.A.T. is added to the cost of a car tyre, its price increases from £16·50 to £18·48. What is the rate at which V.A.T. is charged?

6. In order to increase sales, the price of a Concorde airliner is reduced from £30 000 000 to £28 400 000. What percentage reduction is this?

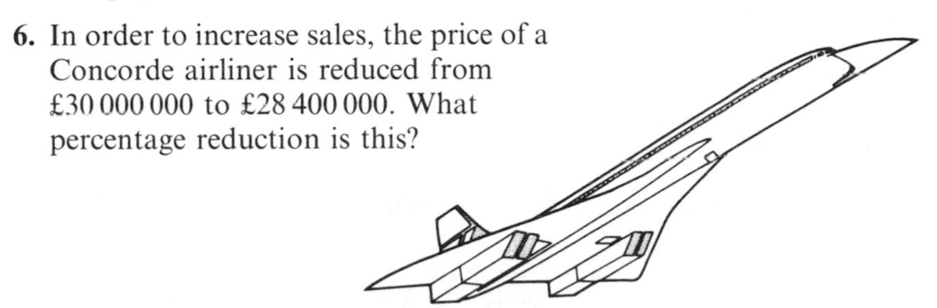

7. Find the *cost* price of the following:
   (a) selling price £55, profit 10%
   (b) selling price £558, profit 24%
   (c) selling price £680, loss 15%
   (d) selling price £11·78, loss 5%

8. An oven is sold for £600, thereby making a profit of 20%, on the cost price. What was the cost price?

9. A pair of jeans is sold for £15, thereby making a profit of 25% on the cost price. What was the cost price?

10. A book is sold for £5·40, at a profit of 8% on the cost price. What was the cost price?

11. An obsolete can of worms is sold for 48p, incurring a loss of 20%. What was the cost price?

12. A car, which failed its MOT test, was sold for £143, thereby making a loss of 35% on the cost price. What was the cost price?

13. If an employer reduces the working week from 40 hours to 35 hours, with no loss of weekly pay, calculate the percentage increase in the hourly rate of pay.

14. The rental for a television set changed from £80 per year to £8 per month. What is the percentage increase in the yearly rental?

15. The label on a carton of yoghurt is shown.
The figure on the right is smudged out. Work out what that figure should be.

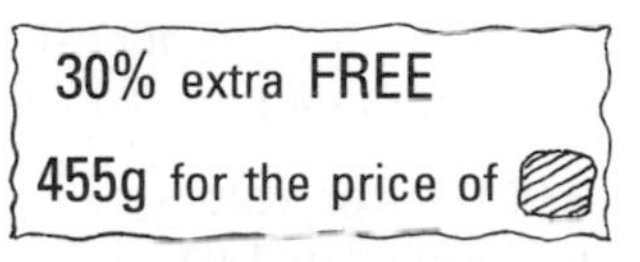

16. Given that $G = ab$, find the percentage increase in G when both a and b increase by 10%.

Compound interest

Suppose a bank pays a fixed interest of 10% on money in deposit accounts. A man puts £500 in the bank.

After one year he has
$500 + 10\%$ of $500 = £550$

After two years he has
$550 + 10\%$ of $550 = £605$

[Check that this is $1{\cdot}10^2 \times 500$]

After three years he has
$605 + 10\%$ of $605 = £665{\cdot}50$

[Check that this is $1{\cdot}10^3 \times 500$]

In general after n years the money in the bank will be £$(1{\cdot}10^n \times 500)$

Exercise 9

These questions are easier if you use a calculator with a $\boxed{y^x}$ button.

1. A bank pays interest of 9% on money in deposit accounts. Mrs Wells puts £2000 in the bank. How much has she after (a) one year, (b) two years, (c) three years?

2. A bank pays interest of 11%. Mr Olsen puts £5000 in the bank. How much has he after (a) one year, (b) three years, (c) five years?

3. A computer operator is paid £10 000 a year. Assuming her pay is increased by 7% each year, what will her salary be in four years time?

4. A new car is valued at £15 000. At the end of each year its value is reduced by 15% of its value at the start of the year. What will it be worth after 6 years?

5. Twenty years ago a bus driver was paid £50 a week. He is now paid £185 a week. Assuming an average rate of inflation of 7%, has his pay kept up with inflation?

6. Assuming an average inflation rate of 8%, work out the probable cost of the following items in 10 years:
(a) car £6500 (b) T.V. £340
(c) house £50 000

7. The population of an island increases by 10% each year. After how many years will the original population be doubled?

8. A bank pays interest of 11% on money in deposit accounts. After how many years will a sum of money have trebled?

9. (a) Draw the graph of $y = 1{\cdot}08^x$ for values of x from 0 to 10.
(b) Solve approximately the equation $1{\cdot}08^x = 2$.
(c) Money is invested at 8% interest. After how many years will the money have doubled?

1.3 Calculator

To use a calculator efficiently you sometimes have to think ahead and make use of the memory, inverse [$1/x$] and [$+/-$] buttons.

In the example below the buttons are:

[$\sqrt{}$]	square root	[y^x]	raises number y to the power x
[x^2]	square	[M in]	puts number in memory
[$1/x$]	reciprocal	[MR]	recalls number from memory

Example 1

Evaluate the following to 4 significant figures:

(a) $\dfrac{2\cdot3}{4\cdot7 + 3\cdot61}$ Find the denominator first.

[4·7] [+] [3·61] [=] [M in]

[2·3] [÷] [MR] [=] Answer 0·2768 (to four sig. fig.)

(b) $\left(\dfrac{1}{0\cdot084}\right)^4$

[0·084] [$1/x$] [y^x] [4] [=] Answer 20 090 (to four sig. fig.)

(c) $\sqrt[3]{[3\cdot2 \times (1\cdot7 - 1\cdot64)]}$

[1·7] [−] [1·64] [=] [×] [3·2] [=]

[y^x] [0·333333] [=] Answer 0·5769 (to four sig. fig.)

Note: To find a cube root, raise to the power $\frac{1}{3}$.
or as a decimal 0·333 . . .
Of course, if your calculator has a '$\sqrt[3]{}$' button, use that.

Exercise 10

Use a calculator to evaluate the following, giving the answers to 4 significant figures:

1. $\dfrac{7\cdot351 \times 0\cdot764}{1\cdot847}$ **2.** $\dfrac{0\cdot0741 \times 14700}{0\cdot746}$ **3.** $\dfrac{0\cdot0741 \times 9\cdot61}{23\cdot1}$

4. $\dfrac{417\cdot8 \times 0\cdot00841}{0\cdot07324}$ **5.** $\dfrac{8\cdot41}{7\cdot601 \times 0\cdot00847}$ **6.** $\dfrac{4\cdot22}{1\cdot701 \times 5\cdot2}$

7. $\dfrac{9{\cdot}61}{17{\cdot}4 \times 1{\cdot}51}$ **8.** $\dfrac{8{\cdot}71 \times 3{\cdot}62}{0{\cdot}84}$ **9.** $\dfrac{0{\cdot}76}{0{\cdot}412 - 0{\cdot}317}$

10. $\dfrac{81{\cdot}4}{72{\cdot}6 + 51{\cdot}92}$ **11.** $\dfrac{111}{27{\cdot}4 + 2960}$ **12.** $\dfrac{27{\cdot}4 + 11{\cdot}61}{5{\cdot}9 - 4{\cdot}763}$

13. $\dfrac{6{\cdot}51 - 0{\cdot}1114}{7{\cdot}24 + 1{\cdot}653}$ **14.** $\dfrac{5{\cdot}71 + 6{\cdot}093}{9{\cdot}05 - 5{\cdot}77}$ **15.** $\dfrac{0{\cdot}943 - 0{\cdot}788}{1{\cdot}4 - 0{\cdot}766}$

16. $\dfrac{2{\cdot}6}{1{\cdot}7} + \dfrac{1{\cdot}9}{3{\cdot}7}$ **17.** $\dfrac{8{\cdot}06}{5{\cdot}91} - \dfrac{1{\cdot}594}{1{\cdot}62}$ **18.** $\dfrac{4{\cdot}7}{11{\cdot}4 - 3{\cdot}61} + \dfrac{1{\cdot}6}{9{\cdot}7}$

19. $\dfrac{3{\cdot}74}{1{\cdot}6 \times 2{\cdot}89} - \dfrac{1}{0{\cdot}741}$ **20.** $\dfrac{1}{7{\cdot}2} - \dfrac{1}{14{\cdot}6}$ **21.** $\dfrac{1}{0{\cdot}961} \times \dfrac{1}{0{\cdot}412}$

22. $\dfrac{1}{7} + \dfrac{1}{13} - \dfrac{1}{8}$ **23.** $4{\cdot}2\left(\dfrac{1}{5{\cdot}5} - \dfrac{1}{7{\cdot}6}\right)$ **24.** $\sqrt{}(9{\cdot}61 + 0{\cdot}1412)$

25. $\sqrt{\left(\dfrac{8{\cdot}007}{1{\cdot}61}\right)}$ **26.** $(1{\cdot}74 + 9{\cdot}611)^2$ **27.** $\left(\dfrac{1{\cdot}63}{1{\cdot}7 - 0{\cdot}911}\right)^2$

28. $\left(\dfrac{9{\cdot}6}{2{\cdot}4} - \dfrac{1{\cdot}5}{0{\cdot}74}\right)^2$ **29.** $\sqrt{\left(\dfrac{4{\cdot}2 \times 1{\cdot}611}{9{\cdot}83 \times 1{\cdot}74}\right)}$ **30.** $(0{\cdot}741)^3$

31. $(1{\cdot}562)^5$ **32.** $(0{\cdot}32)^3 + (0{\cdot}511)^4$ **33.** $(1{\cdot}71 - 0{\cdot}863)^6$

34. $\left(\dfrac{1}{0{\cdot}971}\right)^4$ **35.** $\sqrt[3]{}(4{\cdot}714)$ **36.** $\sqrt[3]{}(0{\cdot}9316)$

37. $\sqrt[3]{\left(\dfrac{4{\cdot}114}{7{\cdot}93}\right)}$ **38.** $\sqrt[4]{}(0{\cdot}8145 - 0{\cdot}799)$ **39.** $\sqrt[5]{}(8{\cdot}6 \times 9{\cdot}71)$

40. $\sqrt[3]{\left(\dfrac{1{\cdot}91}{4{\cdot}2 - 3{\cdot}766}\right)}$ **41.** $\left(\dfrac{1}{7{\cdot}6} - \dfrac{1}{18{\cdot}5}\right)^3$ **42.** $\dfrac{\sqrt{}(4{\cdot}79) + 1{\cdot}6}{9{\cdot}63}$

43. $\dfrac{(0{\cdot}761)^2 - \sqrt{}(4{\cdot}22)}{1{\cdot}96}$ **44.** $\sqrt[3]{\left(\dfrac{1{\cdot}74 \times 0{\cdot}761}{0{\cdot}0896}\right)}$ **45.** $\left(\dfrac{8{\cdot}6 \times 1{\cdot}71}{0{\cdot}43}\right)^3$

46. $\dfrac{9{\cdot}61 - \sqrt{}(9{\cdot}61)}{9{\cdot}61^2}$ **47.** $\dfrac{9{\cdot}6 \times 10^4 \times 3{\cdot}75 \times 10^7}{8{\cdot}88 \times 10^6}$

48. $\dfrac{8{\cdot}06 \times 10^{-4}}{1{\cdot}71 \times 10^{-6}}$ **49.** $\dfrac{3{\cdot}92 \times 10^{-7}}{1{\cdot}884 \times 10^{-11}}$ **50.** $\left(\dfrac{1{\cdot}31 \times 2{\cdot}71 \times 10^5}{1{\cdot}91 \times 10^4}\right)^5$

51. $\left(\dfrac{1}{9{\cdot}6} - \dfrac{1}{9{\cdot}99}\right)^{10}$ **52.** $\dfrac{\sqrt[3]{}(86{\cdot}6)}{\sqrt[4]{}(4{\cdot}71)}$ **53.** $\dfrac{23{\cdot}7 \times 0{\cdot}0042}{12{\cdot}48 - 9{\cdot}7}$

54. $\dfrac{0{\cdot}482 + 1{\cdot}6}{0{\cdot}024 \times 1{\cdot}83}$ **55.** $\dfrac{8{\cdot}52 - 1{\cdot}004}{0{\cdot}004 - 0{\cdot}0083}$ **56.** $\dfrac{1{\cdot}6 - 0{\cdot}476}{2{\cdot}398 \times 41{\cdot}2}$

57. $\left(\frac{2{\cdot}3}{0{\cdot}791}\right)^7$ **58.** $\left(\frac{8{\cdot}4}{28{\cdot}7 - 0{\cdot}47}\right)^3$ **59.** $\left(\frac{5{\cdot}114}{7{\cdot}332}\right)^5$

60. $\left(\frac{4{\cdot}2}{2{\cdot}3} + \frac{8{\cdot}2}{0{\cdot}52}\right)^3$ **61.** $\frac{1}{8{\cdot}2^2} - \frac{3}{19^2}$ **62.** $\frac{100}{11^3} + \frac{100}{12^3}$

63. $\frac{7{\cdot}3 - 4{\cdot}291}{2{\cdot}6^2}$ **64.** $\frac{9{\cdot}001 - 8{\cdot}97}{0{\cdot}95^3}$ **65.** $\frac{10{\cdot}1^2 + 9{\cdot}4^2}{9{\cdot}8}$

1.4 Standard form

When dealing with either very large or very small numbers, it is not convenient to write them out in full in the normal way. It is better to use standard form. Most calculators represent large and small numbers in this way

The number $a \times 10^n$ is in standard form when $1 \leqslant a < 10$ and n is a positive or negative integer.

e.g. This calculator shows $2{\cdot}3 \times 10^8$.

2.3 08

Example 1

Write the following numbers in standard form:

(a) $2000 = 2 \times 1000 = 2 \times 10^3$

(b) $150 = 1{\cdot}5 \times 100 = 1{\cdot}5 \times 10^2$

(c) $0{\cdot}0004 = 4 \times \frac{1}{10\,000} = 4 \times 10^{-4}$

Exercise 11

Write the following numbers in standard form:

1. 4000 **2.** 500 **3.** 70 000
4. 60 **5.** 2400 **6.** 380
7. 46 000 **8.** 46 **9.** 900 000
10. 2560 **11.** 0·007 **12.** 0·0004
13. 0·0035 **14.** 0·421 **15.** 0·000 055
16. 0·01 **17.** 564 000 **18.** 19 million

19. The population of China is estimated at 1100 000 000. Write this in standard form.

20. A hydrogen atom has a mass of 0·000 000 000 000 000 000 000 001 67 grams. Write this mass in standard form.

21. The area of the surface of the Earth is about 510 000 000 km^2. Express this in standard form.

22. A certain virus is 0·000 000 000 25 cm in diameter. Write this in standard form.

23. Avogadro's number is 602 300 000 000 000 000 000 000. Express this in standard form.

24. The speed of light is 300 000 km/s. Express this speed in cm/s in standard form.

25. A very rich oil sheikh leaves his fortune of £$3{\cdot}6 \times 10^8$ to be divided between 100 relatives. How much does each relative receive? Give the answer in standard form.

Many calculators have an [EXP] button which is used for standard form.

To enter $1{\cdot}6 \times 10^7$ into the calculator:

press [1·6] [EXP] [7]

To enter $3{\cdot}8 \times 10^{-3}$

press [3·8] [EXP] [3] [+/−]

Example 2

Calculate $(4{\cdot}9 \times 10^{11}) \div (3{\cdot}5 \times 10^{-4})$.

[4·9] [EXP] [11] [÷] [3·5] [EXP] [4] [+/−] [=]

The answer is $1{\cdot}4 \times 10^{15}$.

Exercise 12

In Questions **1** to **12**, give the answer in standard form.

1. $5\,000 \times 3\,000$
2. $60\,000 \times 5\,000$
3. $0{\cdot}000\,07 \times 400$
4. $0{\cdot}0007 \times 0{\cdot}000\,01$
5. $8\,000 \div 0{\cdot}004$
6. $(0{\cdot}002)^2$
7. $150 \times 0{\cdot}000\,6$
8. $0{\cdot}000\,033 \div 500$
9. $0{\cdot}007 \div 20\,000$
10. $(0{\cdot}0001)^4$
11. $(2\,000)^3$
12. $0{\cdot}00592 \div 8\,000$

13. If $a = 512 \times 10^2$
$b = 0{\cdot}478 \times 10^6$
$c = 0{\cdot}0049 \times 10^7$
arrange a, b and c in order of size (smallest first).

14. If the number $2{\cdot}74 \times 10^{15}$ is written out in full, how many zeros follow the 4?

15. If the number $7{\cdot}31 \times 10^{-17}$ is written out in full, how many zeros would there be between the decimal point and the first significant figure?

16. If $x = 2 \times 10^5$ and $y = 3 \times 10^{-3}$ correct to one significant figure, find the greatest and least possible values of

(i) xy (ii) $\dfrac{x}{y}$

17. Oil flows through a pipe at a rate of $40\,\text{m}^3/\text{s}$. How long will it take to fill a tank of volume $1{\cdot}2 \times 10^5\,\text{m}^3$?

18. Given that $L = 2\sqrt{\dfrac{a}{k}}$, find the value of L in standard form when $a = 4{\cdot}5 \times 10^{12}$ and $k = 5 \times 10^7$.

19. The mean weight of all the men, women, children and babies in the UK is 42·1 kg. The population of the UK is 56 million. Work out the total weight of the entire population giving your answer in kg in standard form. Give your answer to a sensible degree of accuracy.

20. Percy, a rather mean gardener, is trying to estimate the number of seeds in a 50 gram packet. He counts 30 seeds from the packet and finds their weight is 6×10^{-2} g.

Use his sample to estimate the total number of seeds in the packet.

21. A light year is the distance travelled by a beam of light in a year. Light travels at a speed of approximately 3×10^5 km/s.
(a) Work out the length of a light year in km.
(b) Light takes about 8 minutes to reach the Earth from the Sun. How far is the Earth from the Sun in km?

22.† (a) The number 10 to the power 100 (10 000 sexdecillion) is called a 'Googol'! If it takes $\frac{1}{5}$ second to write a zero and $\frac{1}{10}$ second to write a 'one', how long would it take to write the number 100 'Googols' in full?
(b) The number 10 to the power of a 'Googol' is called a 'Googolplex'. Using the same speed of writing, how long in years would it take to write 1 'Googolplex' in full? You may assume that your pen has enough ink.

1.5 Negative numbers

For adding and subtracting, use the number line.

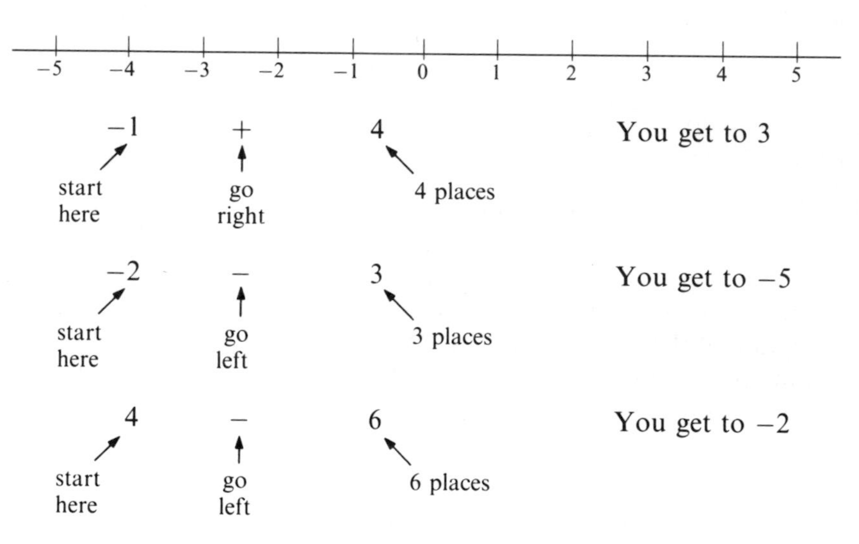

When you have two (+) or (−) signs together, use this rule.

$++ = +$ $\quad$ $+- = -$
$-- = +$ $\quad$ $-+ = -$

Example 1

(a) $3 - (-6) = 3 + 6 = 9$

(b) $-4 + (-5) = -4 - 5 = -9$

(c) $-5 - (+7) = -5 - 7 = -12$

Exercise 13

1. $+7 + (+6)$
2. $+11 + (+200)$
3. $-3 + (-9)$
4. $-7 + (-24)$
5. $-5 + (-61)$
6. $+0{\cdot}2 + (+5{\cdot}9)$
7. $+5 + (+4{\cdot}1)$
8. $-8 + (-27)$
9. $+17 + (+1{\cdot}7)$
10. $-2 + (-3) + (-4)$
11. $-7 + (+4)$
12. $+7 + (-4)$
13. $-9 + (+7)$
14. $+16 + (-30)$
15. $+14 + (-21)$
16. $-7 + (+10)$
17. $-19 + (+200)$
18. $+7{\cdot}6 + (-9{\cdot}8)$
19. $-1{\cdot}8 + (+10)$
20. $-7 + (+24)$
21. $+7 - (+5)$
22. $+9 - (+15)$
23. $-6 - (+9)$
24. $-9 - (+5)$
25. $+8 - (+10)$
26. $-19 - (-7)$
27. $-10 - (+70)$
28. $-5{\cdot}1 - (+8)$
29. $-0{\cdot}2 - (+4)$
30. $+5{\cdot}2 - (-7{\cdot}2)$
31. $-4 + (-3)$
32. $+6 - (-2)$
33. $+8 + (-4)$
34. $-4 - (+6)$
35. $+7 - (-4)$
36. $+6 + (-2)$
37. $+10 - (+30)$
38. $+19 - (+11)$
39. $+4 + (-7) + (-2)$

Example 2

When two numbers with the same sign are multiplied together, the answer is positive.

(a) $+7 \times (+3) = +21$ (b) $-6 \times (-4) = +24$

Example 3

When two numbers with different signs are multiplied together, the answer is negative.

(a) $-8 \times (+4) = -32$ (b) $+7 \times (-5) = -35$

When dividing numbers, the rules are the same as in multiplication.

Exercise 14

1. $+2 \times (-4)$
2. $+7 \times (+4)$
3. $-4 \times (-3)$
4. $-6 \times (-4)$
5. $-6 \times (-3)$
6. $+5 \times (-7)$
7. $-7 \times (-7)$
8. $-4 \times (+3)$
9. $+0{\cdot}5 \times (-4)$
10. $-1\frac{1}{2} \times (-6)$
11. $-8 \div (+2)$
12. $+12 \div (+3)$
13. $+36 \div (-9)$
14. $-40 \div (-5)$
15. $-70 \div (-1)$
16. $-56 \div (+8)$
17. $-\frac{1}{2} \div (-2)$
18. $-3 \div (+5)$
19. $+0{\cdot}1 \div (-10)$
20. $-0{\cdot}02 \div (-100)$
21. $-11 \times (-11)$
22. $-6 \times (-1)$
23. $+12 \times (-50)$
24. $-\frac{1}{2} \div (+\frac{1}{2})$
25. $-600 \div (+30)$
26. $-5{\cdot}2 \div (+2)$
27. $+7 \times (-100)$
28. $-6 \div (-\frac{1}{3})$
29. $100 \div (-0{\cdot}1)$
30. -8×-80
31. $-3 \times (-2) \times (-1)$
32. $+3 \times (-7) \times (+2)$
33. $+0{\cdot}4 \div (-1)$
34. $-16 \div (+40)$
35. $+0{\cdot}2 \times (-1000)$
36. $-7 \times (-5) \times (-1)$
37. $-14 \div (+7)$
38. $-7 \div (-14)$
39. $+1\frac{1}{4} \div (-5)$
40. $-6 \times (-\frac{1}{2}) \times (-30)$

Exercise 15

1. $-7+(-3)$
2. $-6-(-7)$
3. $-4\times(-3)$
4. $-4\times(+7)$
5. $4-(+6)$
6. $-4\times(-4)$
7. $+6\div(-2)$
8. $+8-(-6)$
9. $-7\times(+4)$
10. $-8\div(-2)$
11. $+10\div(-60)$
12. $(-3)^2$
13. $40-(+70)$
14. $-6\times(-4)$
15. $(-1)^5$
16. $-8\div(+4)$
17. $+10\times(-3)$
18. $-7\times(-1)$
19. $+10+(-7)$
20. $+12-(-4)$
21. $+100+(-7)$
22. $-60\times(-40)$
23. $-20\div(-2)$
24. $(-1)^{20}$

1.6 Fractions

Common fractions are added or subtracted from one another directly only when they have a common denominator.

Example 1

(a) $\frac{3}{4}+\frac{2}{5} = \frac{15}{20}+\frac{8}{20} = \frac{23}{20} = 1\frac{3}{20}$

(b) $2\frac{3}{8}-1\frac{5}{12} = \frac{19}{8}-\frac{17}{12} = \frac{57}{24}-\frac{34}{24} = \frac{23}{24}$

(c) $\frac{2}{5}\times\frac{6}{7} = \frac{12}{35}$

(d) $2\frac{2}{5}\div 6 = \frac{12}{5}\div 6 = \frac{\cancel{12}^{2}}{5}\times\frac{1}{\cancel{6}_{1}} = \frac{2}{5}$

Exercise 16

Evaluate and simplify your answer.

1. $\frac{3}{4}+\frac{4}{5}$
2. $\frac{1}{3}+\frac{1}{8}$
3. $\frac{5}{6}+\frac{6}{9}$
4. $\frac{3}{4}-\frac{1}{3}$
5. $\frac{3}{5}-\frac{1}{3}$
6. $\frac{1}{2}-\frac{2}{5}$
7. $\frac{2}{3}\times\frac{4}{5}$
8. $\frac{1}{7}\times\frac{5}{6}$
9. $\frac{5}{8}\times\frac{12}{13}$
10. $\frac{1}{3}\div\frac{4}{5}$
11. $\frac{3}{4}\div\frac{1}{6}$
12. $\frac{5}{6}\div\frac{1}{2}$
13. $\frac{3}{8}+\frac{1}{5}$
14. $\frac{3}{8}\times\frac{1}{5}$
15. $\frac{3}{8}\div\frac{1}{5}$
16. $1\frac{3}{4}-\frac{2}{3}$
17. $1\frac{3}{4}\times\frac{2}{3}$
18. $1\frac{3}{4}\div\frac{2}{3}$
19. $3\frac{1}{2}+2\frac{3}{5}$
20. $3\frac{1}{2}\times 2\frac{3}{5}$
21. $3\frac{1}{2}\div 2\frac{3}{5}$
22. $(\frac{3}{4}-\frac{2}{3})\div\frac{3}{4}$
23. $(\frac{3}{5}+\frac{1}{3})\times\frac{5}{7}$
24. $\dfrac{\frac{3}{8}-\frac{1}{5}}{\frac{7}{10}-\frac{2}{3}}$
25. $\dfrac{\frac{2}{3}+\frac{1}{5}}{\frac{3}{4}-\frac{1}{3}}$

26. Arrange the fractions in order of size:

(a) $\frac{7}{12}, \frac{1}{2}, \frac{2}{3}$ (b) $\frac{3}{4}, \frac{2}{3}, \frac{5}{6}$ (c) $\frac{1}{3}, \frac{17}{24}, \frac{5}{8}, \frac{3}{4}$ (d) $\frac{5}{6}, \frac{8}{9}, \frac{11}{12}$

27. Find the fraction which is mid-way between the two fractions given:

(a) $\frac{2}{5}, \frac{3}{5}$ (b) $\frac{5}{8}, \frac{7}{8}$ (c) $\frac{2}{3}, \frac{3}{4}$

(d) $\frac{1}{3}, \frac{4}{9}$ (e) $\frac{4}{15}, \frac{1}{3}$ (f) $\frac{3}{8}, \frac{11}{24}$

28. In the equation below all the asterisks stand for the same number. What is the number?

$$\left[\frac{*}{*} - \frac{*}{6} = \frac{*}{30}\right]$$

29. Work out one half of one third of 65% of £360.

30. Find the value of n if

$$\left(1\tfrac{1}{3}\right)^n - \left(1\tfrac{1}{3}\right) = \frac{28}{27}$$

31. A rubber ball is dropped from a height of 300 m. After each bounce the ball rises to $\frac{4}{5}$ of its previous height. How high, to the nearest cm, will it rise after the fourth bounce?

32. Steve Braindead spends his income as follows:

(a) $\frac{2}{5}$ of his income goes in tax,

(b) $\frac{2}{3}$ of what is left goes on food, rent and transport,

(c) he spends the rest on cigarettes, beer and betting.

What fraction of his income is spent on cigarettes, beer and betting?

33. A formula used by opticians is

$$\frac{1}{f} = \frac{1}{u} + \frac{1}{v}$$

Given that $u = 3$ and $v = 5\frac{1}{2}$ find the exact value of f.

34. When it hatches from its egg, the shell of a certain crab is 1 cm across. When fully grown the shell is approximately 10 cm across. Each new shell is one-third bigger than the previous one. How many shells does a fully grown crab have during its life?

35. Figs. 1 and 2 show an equilateral triangle divided into thirds and quarters. They are combined in Fig. 3. Calculate the fraction of Fig. 3 that is shaded.

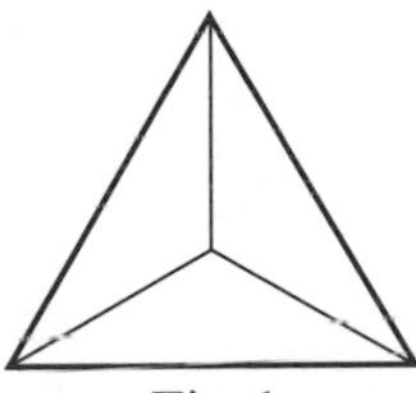
Fig. 1

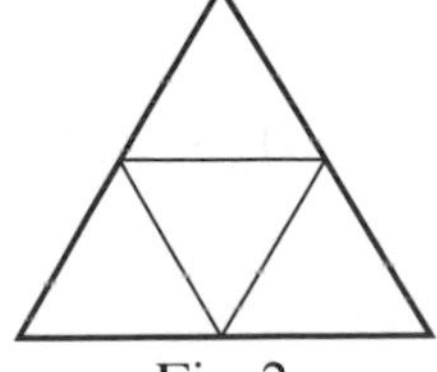
Fig. 2

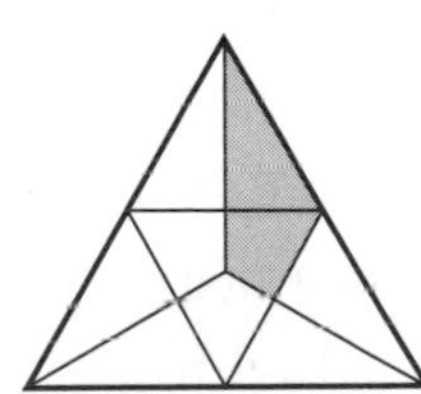
Fig. 3

36. Glass A contains 100 ml of water and glass B contains 100 ml of wine.

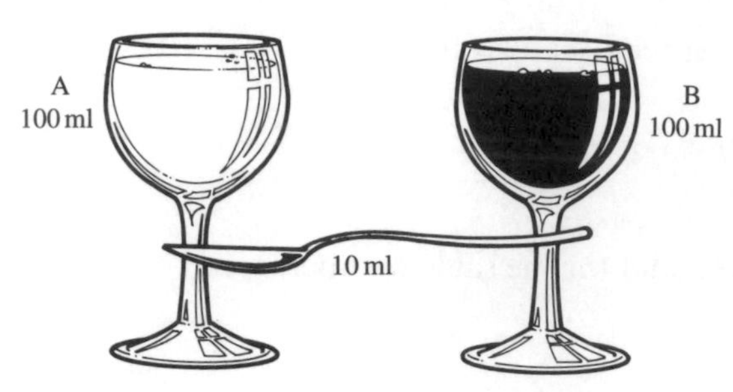

A 10 ml spoonful of wine is taken from glass B and mixed thoroughly with the water in glass A. A 10 ml spoonful of the mixture from A is returned to B. Is there now more wine in the water or more water in the wine?

Fractions, decimals and percentages

A decimal fraction is simply a fraction expressed in tenths, hundredths etc. Percentages are a convenient way of writing hundredths.

$$\frac{1}{4} = \frac{25}{100} = 25\% = 0{\cdot}25$$

Example 2

(a) Change $\frac{7}{8}$ to a decimal fraction.

(b) Change 0·35 to a vulgar fraction.

(c) Change $\frac{3}{8}$ to a percentage

(a) $\frac{7}{8}$ means divide 8 into 7

$\frac{7}{8} = 0{\cdot}875$ $\qquad 8\overline{)7{\cdot}000}$ giving $0{\cdot}875$

(b) $0{\cdot}35 = \frac{35}{100} = \frac{7}{20}$ $\qquad$ (c) $\frac{3}{8} = (\frac{3}{8} \times \frac{100}{1})\% = 37\frac{1}{2}\%$

Exercise 17

1. Change the fractions to decimals.

(a) $\frac{1}{4}$ (b) $\frac{2}{5}$ (c) $\frac{3}{8}$ (d) $\frac{5}{12}$ (e) $\frac{1}{6}$ (f) $\frac{2}{7}$

2. Change the decimals to fractions and simplify.

(a) 0·2 (b) 0·45 (c) 0·36 (d) 0·125 (e) 1·05 (f) 0·007

3. Change to percentages.

(a) $\frac{1}{4}$ (b) $\frac{1}{10}$ (c) 0·72 (d) 0·075 (e) 0·02 (f) $\frac{1}{3}$

4. Arrange in order of size (smallest first)

(a) $\frac{1}{2}$; 45%; 0·6 (b) 0·38; $\frac{6}{16}$; 4%

(c) 0·111; 11%; $\frac{1}{9}$ (d) 32%; 0·3; $\frac{1}{3}$

Evaluate, giving the answer to 2 decimal places:

5. $\frac{1}{4} + \frac{1}{3}$
6. $\frac{2}{3} + 0{\cdot}75$
7. $\frac{8}{9} - 0{\cdot}24$
8. $\frac{7}{8} + \frac{5}{9} + \frac{2}{11}$
9. $\frac{1}{3} \times 0{\cdot}2$
10. $\frac{5}{8} \times \frac{1}{4}$
11. $\frac{8}{11} \div 0{\cdot}2$
12. $(\frac{4}{7} - \frac{1}{3}) \div 0{\cdot}4$

Arrange the numbers in order of size (smallest first)

13. $\frac{1}{3}$, 0·33, $\frac{4}{15}$
14. $\frac{2}{7}$, 0·3, $\frac{4}{9}$
15. 0·71, $\frac{7}{11}$, 0·705
16. $\frac{4}{13}$, 0·3, $\frac{5}{18}$

1.7 Formulas

When a calculation is repeated many times it is often helpful to use a formula.
When a building society offers a mortgage it may use a formula like '$2\frac{1}{2}$ times the main salary plus the second salary'.
Publishers use a formula to work out the selling price of a book based on the production costs and the expected sales of the book.

Exercise 18

1. The final speed v of a car is given by the formula $v = u + at$. [u = initial speed, a = acceleration, t = time taken]. Find v when $u = 15$, $a = 0{\cdot}2$, $t = 30$.

2. The time period T of a simple pendulum is given by the formula
$$T = 2\pi\sqrt{\left(\frac{\ell}{g}\right)},$$
where ℓ is the length of the pendulum and g is the gravitational acceleration.
Find T when $\ell = 0{\cdot}65$, $g = 9{\cdot}81$ and $\pi = 3{\cdot}142$.

3. The total surface area A of a cone is related to the radius r and the slant height ℓ by the formula $A = \pi r(r + \ell)$.
Find A when $r = 7$ and $\ell = 11$.

4. The sum S of the squares of the integers from 1 to n is given by $S = \frac{1}{6}n(n + 1)(2n + 1)$. Find S when $n = 12$.

5. The acceleration a of a train is found using the formula
$$a = \frac{v^2 - u^2}{2s}.$$
Find a when $v = 20$, $u = 9$ and $s = 2{\cdot}5$.

6. Einstein's famous equation relating energy, mass and the speed of light is $E = mc^2$. Find E when $m = 0{\cdot}0001$ and $c = 3 \times 10^8$.

7. The area A of a parallelogram with sides a and b is given by $A = ab \sin\theta$, where θ is the angle between the sides. Find A when $a = 7$, $b = 3$ and $\theta = 30°$.

8. The distance s travelled by an accelerating rocket is given by $s = ut + \frac{1}{2}at^2$. Find s when $u = 3$, $t = 100$ and $a = 0{\cdot}1$.

9. The formula for the velocity of sound in air is $V = 72\sqrt{T + 273}$ where V is the velocity in km/h and T is the temperature of the air in °C.
(a) Find the velocity of sound where the temperature is 26°C.
(b) Find the temperature if the velocity of sound is 1200 km/h.
(c) Find the velocity of sound where the temperature is −77°C.

10. Find a formula for the area of the shape below, in terms of a, b and c.

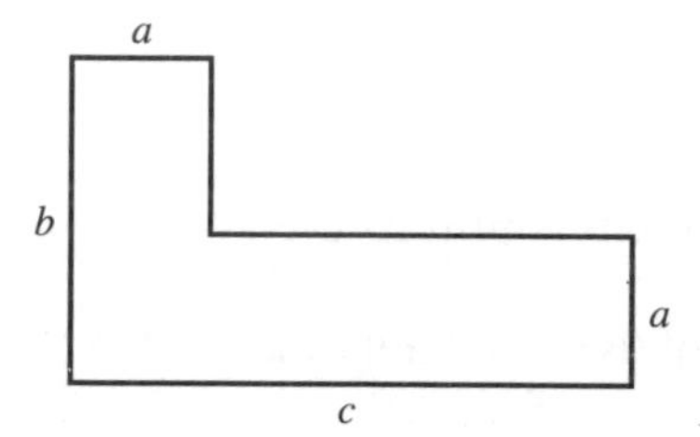

11. Find a formula for the shaded part below, in terms of p, q and r.

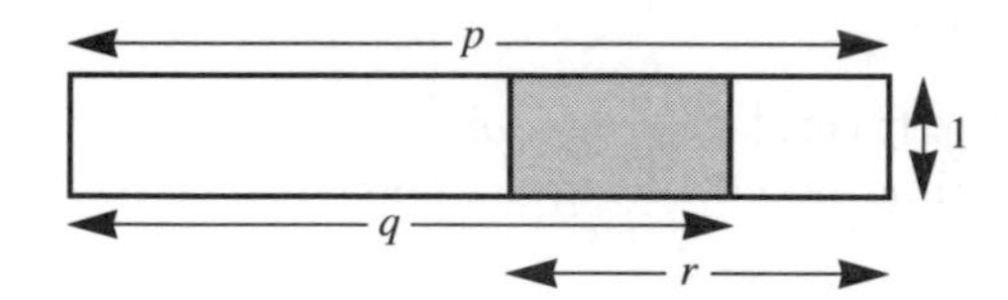

12. An intelligent fish lays brown eggs or white eggs and it likes to lay them in a certain pattern. Each brown egg is surrounded by six white eggs.

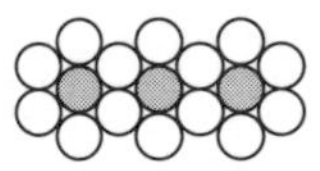

Here there are 3 brown eggs and 14 white eggs.
(a) How many eggs does it lay altogether if it lays 200 brown eggs?
(b) How many eggs does it lay altogether if it lays n brown eggs?

13. In the diagrams below, the rows of black tiles are surrounded by white tiles.

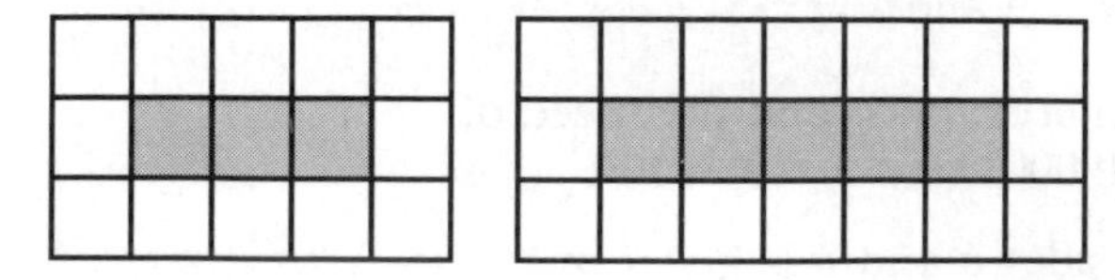

Find a formula for the number of white tiles w which would be needed to surround a row of n black tiles.

Substitution in a formula

Example 1

When $a = 3$, $b = -2$, $c = 5$, find the value of:

(a) $3a + b = (3 \times 3) + (-2)$
$= 9 - 2$
$= 7$

(b) $ac + b^2 = (3 \times 5) + (-2)^2$
$= 15 + 4$
$= 19$

(c) $a(c - b)$
$= 3[5 - (-2)]$
$= 3[7]$
$= 21$

Notice that working *down* the page is often easier to follow.

Exercise 19

Evaluate the following:
For Questions **1** to **12** $a = 3$, $c = 2$, $e = 5$.

1. $3a - 2$
2. $4c + e$
3. $2c + 3a$
4. $5e - a$
5. $e - 2c$
6. $e - 2a$
7. $4c + 2e$
8. $7a - 5e$
9. $c - e$
10. $10a + c + e$
11. $a + c - e$
12. $a - c - e$

For Questions **13** to **24** $h = 3$, $m = -2$, $t = -3$.

13. $2m - 3$
14. $4t + 10$
15. $3h - 12$
16. $6m + 4$
17. $9t - 3$
18. $4h + 4$
19. $2m - 6$
20. $m + 2$
21. $3h + m$
22. $t - h$
23. $4m + 2h$
24. $3t - m$

For Questions **25** to **36** $x = -2$, $y = -1$, $k = 0$.

25. $3x + 1$
26. $2y + 5$
27. $6k + 4$
28. $3x + 2y$
29. $2k + x$
30. xy
31. xk
32. $2xy$
33. $2(x + k)$
34. $3(k + y)$
35. $5x - y$
36. $3k - 2x$

$2x^2$ means $2(x^2)$
$(2x)^2$ means 'work out $2x$ and *then* square it'
$-7x$ means $-7(x)$
$-x^2$ means $-1(x^2)$

Example 2

When $x = -2$, find the value of

(a) $2x^2 - 5x = 2(-2)^2 - 5(-2)$
$= 2(4) + 10$
$= 18$

(b) $(3x)^2 - x^2 = (3 \times -2)^2 - 1(-2)^2$
$= (-6)^2 - 1(4)$
$= 36 - 4$
$= 32.$

Exercise 20

If $x = -3$ and $y = 2$, evaluate the following:

1. x^2 **2.** $3x^2$ **3.** y^2 **4.** $4y^2$
5. $(2x)^2$ **6.** $2x^2$ **7.** $10 - x^2$ **8.** $10 - y^2$
9. $20 - 2x^2$ **10.** $20 - 3y^2$ **11.** $5 + 4x$ **12.** $x^2 - 2x$
13. $y^2 - 3x^2$ **14.** $x^2 - 3y$ **15.** $(2x)^2 - y^2$ **16.** $4x^2$
17. $(4x)^2$ **18.** $1 - x^2$ **19.** $y - x^2$ **20.** $x^2 + y^2$
21. $x^2 - y^2$ **22.** $2 - 2x^2$ **23.** $(3x)^2 + 3$ **24.** $11 - xy$
25. $12 + xy$ **26.** $(2x)^2 - (3y)^2$ **27.** $2 - 3x^2$ **28.** $y^2 - x^2$

Example 3

When $a = -2$, $b = 3$, $c = -3$, evaluate (a) $\frac{2a(b^2 - a)}{c}$ (b) $\sqrt{(a^2 + b^2)}$

(a) $(b^2 - a) = 9 - (-2)$
$= 11$

$\therefore \frac{2a(b^2 - a)}{c} = \frac{2 \times (-2) \times (11)}{-3}$
$= 14\frac{2}{3}$

(b) $a^2 + b^2 = (-2)^2 + (3)^2$
$= 4 + 9$
$= 13$

$\therefore \sqrt{(a^2 + b^2)} = \sqrt{13}$

Exercise 21

Evaluate the following:
In Questions **1** to **16**, $a = 4$, $b = -2$, $c = -3$.

1. $a(b + c)$ **2.** $a^2(b - c)$ **3.** $2c(a - c)$ **4.** $b^2(2a + 3c)$
5. $c^2(b - 2a)$ **6.** $2a^2(b + c)$ **7.** $2(a + b + c)$ **8.** $3c(a - b - c)$
9. $b^2 + 2b + a$ **10.** $c^2 - 3c + a$ **11.** $2b^2 - 3b$ **12.** $\sqrt{(a^2 + c^2)}$
13. $\sqrt{(ab + c^2)}$ **14.** $\sqrt{(c^2 - b^2)}$ **15.** $\frac{b^2}{a} + \frac{2c}{b}$ **16.** $\frac{c^2}{b} + \frac{4b}{a}$

In Questions **17** to **32**, $k = -3$, $m = 1$, $n = -4$.

17. $k^2(2m - n)$ **18.** $5m\sqrt{(k^2 + n^2)}$ **19.** $\sqrt{(kn + 4m)}$ **20.** $kmn(k^2 + m^2 + n^2)$
21. $k^2m^2(m - n)$ **22.** $k^2 - 3k + 4$ **23.** $m^3 + m^2 + n^2 + n$ **24.** $k^3 + 3k$
25. $m(k^2 - n^2)$ **26.** $m\sqrt{(k - n)}$ **27.** $100k^2 + m$ **28.** $m^2(2k^2 - 3n^2)$
29. $\frac{2k + m}{k - n}$ **30.** $\frac{kn - k}{2m}$ **31.** $\frac{3k + 2m}{2n - 3k}$ **32.** $\frac{k + m + n}{k^2 + m^2 + n^2}$

33. Find $K = \sqrt{\left(\frac{a^2 + b^2 + c^2 - 2c}{a^2 + b^2 + 4c}\right)}$
if $a = 3$, $b = -2$, $c = -1$.

34. Find $W = \frac{kmn(k + m + n)}{(k + m)(k + n)}$
if $k = \frac{1}{2}$, $m = -\frac{1}{3}$, $n = \frac{1}{4}$.

1.8 Numerical problems

Foreign exchange

Money is changed from one currency into another using the method of proportion.

Exchange rates

Country	Rate of exchange
Belgium (franc)	BF 87·0 = £1
France (franc)	Fr 10·9 = £1
Germany (mark)	Dm 4·2 = £1
Italy (lire)	lire 2280 = £1
Spain (peseta)	Ptas 182 = £1
United States (dollar)	$1·74 = £1

Example 1

If a bottle of wine costs 8 francs in France, what is the cost in British money?

$$10{\cdot}9 \text{ francs} = £1$$

$$\therefore \quad 1 \text{ franc} = £\frac{1}{10{\cdot}9}$$

$$8 \text{ francs} = £\frac{1}{10{\cdot}9} \times 8 = £0{\cdot}73$$

(to the nearest penny)

The bottle costs approximately 73p in British money.

Exercise 22

Give your answers correct to two decimal places. Use the exchange rates given in the table.

1. Change the amount of British money into the foreign currency stated.
(a) £20 [French francs] (b) £70 [dollars]
(c) £200 [pesetas] (d) £1·50 [marks]
(e) £2·30 [lire] (f) 90p [dollars]

2. Change the amount of foreign currency into British money.
(a) Fr. 500 (b) $2500
(c) DM 7·5 (d) BF 900
(c) Lire 500,000 (f) Pts 950

3. An L.P. costs £4·50 in Britain and $4·70 in the United States. How much cheaper, in British money, is the record when bought in the USA?

4. A bottle of Cointreau costs 582 pesetas in Spain and Fr 48 in France. Which is the cheaper in British money, and by how much?

5. The EEC 'Butter Mountain' was estimated in 1988 to be costing Fr 218 000 per day to maintain the storage facilities. How much is this in pounds?

6. A Jaguar XJS is sold in several countries at the prices given below.

Britain	£15,000
Belgium	BF 1 496 400
France	Fr 194 020
Germany	DM 52 080
USA	$24 882

Write out in order a list of the prices converted into pounds.

7. A traveller in Switzerland exchanges 1300 Swiss francs for £400. What is the exchange rate?

8. An Irish gentleman on holiday in Germany finds that his wallet contains $700. If he changes the money at a bank how many marks will he receive?

9. An English soccer fan is arrested in France and has to pay a fine of 2 000 francs. He has 10 000 German marks in his wallet. How much has he left in British money, after paying the fine?

10. In Britain, a pint of beer cost 65p. In France a third of a litre of the same beer cost 4 francs. If 1 pint is approximately 0·568 litre, calculate the cost in pence of a pint of beer bought in France.

11. The rate of exchange between the pound and foreign currencies is constantly changing. The figures given at the start of this section are now out of date. Look up the current figures in a newspaper. Which currency has made the biggest change against the pound in percentage terms?

Map scales

Example 2

A map is drawn to a scale of 1 to 50 000.
Calculate the length of a road which appears as 3 cm long on the map.

1 cm on the map is equivalent to 50 000 cm on the Earth

$$\therefore \quad 3\text{ cm} \equiv 3 \times 50\,000 = 150\,000\text{ cm} = 1500\text{ m} = 1{\cdot}5\text{ km}$$

The road is 1·5 km long.

Exercise 23

1. On a map of scale 1 : 100 000, the distance between Tower Bridge and Hammersmith Bridge is 12·3 cm. What is the actual distance in km?
2. On a map of scale 1 : 15 000, the distance between Buckingham Palace and Brixton Underground Station is 31·4 cm. What is the actual distance in km?
3. If the scale of a map is 1 : 10 000, what will be the length on this map of a road which is 5 km long?
4. The distance from Hertford to St Albans is 32 km. How far apart will they be on a map of scale 1 : 50 000?
5. The 17th hole at the famous St Andrews golf course is 420 m in length. How long will it appear on a plan of the course of scale 1 : 8000?
6. The scale of a map is 1 : 1000. What are the actual dimensions of a rectangle which appears as 4 cm by 3 cm on the map? What is the area on the map in cm^2? What is the actual area in m^2?
7. The scale of a map is 1 : 100. What area does 1 cm^2 on the map represent? What area does 6 cm^2 represent?
8. The scale of a map is 1 : 20 000. What area does 8 cm^2 represent?

Speed, distance and time

Calculations involving these three quantities are simpler when the speed is *constant*. The formulae connecting the quantities are as follows:

(a) distance = speed × time

(b) $\text{speed} = \dfrac{\text{distance}}{\text{time}}$

(c) $\text{time} = \dfrac{\text{distance}}{\text{speed}}$

A helpful way of remembering these formulae is to write the letters D, S and T in a triangle,

thus:

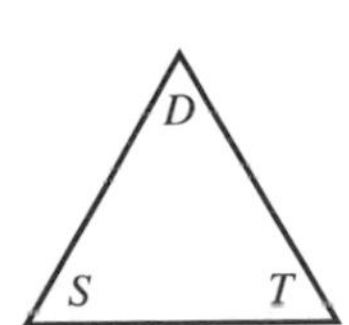

to find D, cover D and we have ST.

to find S, cover S and we have $\dfrac{D}{T}$

to find T, cover T and we have $\dfrac{D}{S}$

Great care must be taken with the units in these questions.

Example 3

A man is running at a speed of 8 km/h for a distance of 5200 metres. Find the time taken in minutes.

$$5200 \text{ metres} = 5{\cdot}2 \text{ km}$$

$$\text{time taken in hours} = \left(\frac{D}{S}\right) = \frac{5{\cdot}2}{8}$$

$$= 0{\cdot}65 \text{ hours}$$

$$\text{time taken in minutes} = 0{\cdot}65 \times 60$$

$$= 39 \text{ minutes}$$

Example 4

Change the units of a speed of 54 km/h into metres per second.

$$54 \text{ km/hour} = 54\,000 \text{ metres/hour}$$

$$= \frac{54\,000}{60} \text{ metres/minute}$$

$$= \frac{54\,000}{60 \times 60} \text{ metres/second}$$

$$= 15 \text{ m/s.}$$

Exercise 24

1. Find the time taken for the following journeys:
(a) 100 km at a speed of 40 km/h
(b) 250 miles at a speed of 80 miles per hour
(c) 15 metres at a speed of 20 cm/s. (answer in seconds)
(d) 10^4 metres at a speed of 2·5 km/h

2. Change the units of the following speeds as indicated:
(a) 72 km/h into m/s
(b) 30 m/s into km/h
(c) 0·012 m/s into cm/s
(d) 9000 cm/s into m/s

3. Find the speeds of the bodies which move as follows:
(a) a distance of 600 km in 8 hours
(b) a distance of 4×10^4 m in 10^{-2} seconds
(c) a distance of 500 m in 10 minutes (in km/h)

4. Find the distance travelled (in metres) in the following:
(a) at a speed of 40 km/h for $\frac{1}{4}$ hour
(b) at a speed of 338·4 km/h for 10 minutes
(c) at a speed of 15 m/s for 5 minutes
(d) at a speed of 14 m/s for 1 hour

5. A car travels 60 km at 30 km/h and then a further 180 km at 160 km/h. Find
(a) the total time taken
(b) the average speed for the whole journey

6. A cyclist travels 25 kilometres at 20 km/h and then a further 80 kilometres at 25 km/h.

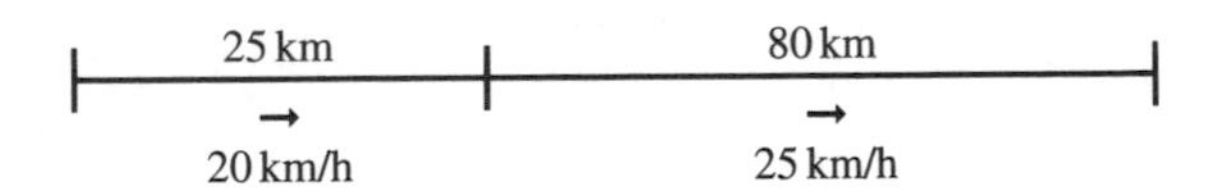

Find
(a) the total time taken
(b) the average speed for the whole journey

7. Sebastian Coe ran two laps around a 400 m track. He completed the first lap in 50 seconds and then decreased his speed by 5% for the second lap.
Find
(a) his speed on the first lap
(b) his speed on the second lap
(c) his total time for the two laps
(d) his average speed for the two laps.

8. The airliner Concorde flies 2000 km at a speed of 1600 km/h and then returns due to bad weather at a speed of 1000 km/h. Find the average speed for the whole trip.

9. A train travels from A to B, a distance of 100 km, at a speed of 20 km/h. If it had gone two and a half times as fast, how much earlier would it have arrived at B?

10. Two men running towards each other at 4 m/s and 6 m/s respectively are one kilometre apart. How long will it take before they meet?

11. A car travelling at 90 km/h is 500 m behind another car travelling at 70 km/h in the same direction. How long will it take the first car to catch the second?

12. How long is a train which passes a signal in twenty seconds at a speed of 108 km/h?

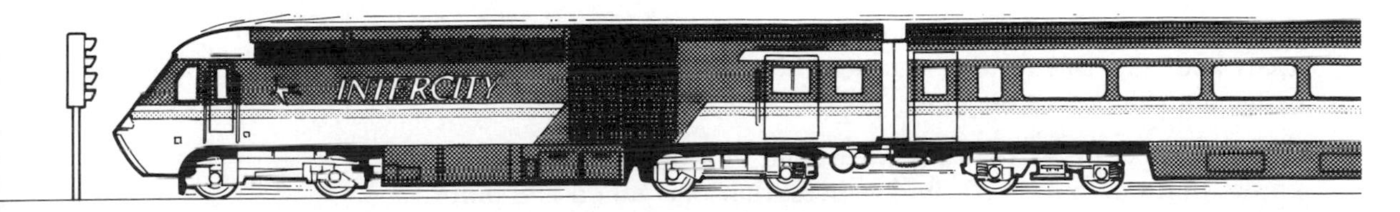

13. A train of length 180 m approaches a tunnel of length 620 m. How long will it take the train to pass completely through the tunnel at a speed of 54 km/h?

14. An earthworm of length 15 cm is crawling along at 2 cm/s. An ant overtakes the worm in 5 seconds. How fast is the ant walking?

15. A train of length 100 m is moving at a speed of 50 km/h. A horse is running alongside the train at a speed of 56 km/h. How long will it take the horse to overtake the train?

16. A car completes a journey at an average speed of 40 m.p.h. At what speed must it travel on the return journey if the average speed for the complete journey (out and back) is 60 m.p.h.?

Mixed questions

Exercise 25

1. 8% of 2500 + 37% of $P = 348$. Find the value of P.

2. Express 419 965 in terms of its prime factors.

3. A map is 278 mm wide and 445 mm long. When reduced on a photocopier, the copy is 360 mm long. What is the width of the copy, to the nearest millimetre?

4. How many prime numbers are there between 120 and 130?

5. Booklets have a mass of 19 g each and they are posted in an envelope of mass 38 g. Postage charges are shown in the table below.

Mass (in grams) not more than	60	100	150	200	250	300	350	600
Postage (in pence)	24	30	37	44	51	59	67	110

(a) A package consists of 15 booklets in an envelope. What is the total mass of the package?
(b) The mass of a second package is 475 g. How many booklets does it contain?
(c) What is the postage charge on a package of mass 320 g?
(d) The postage on a third package was £1·10. What is the largest number of booklets it could contain?

6. A rabbit runs at $7\,\text{m s}^{-1}$ and a hedgehog at $\frac{1}{2}\,\text{m s}^{-1}$. They are 90 m apart and start to run towards each other. How far does the hedgehog run before they meet?

7. Mark's recipe for a cake calls for 6 fluid ounces of milk. Mark knows that one pint is 20 fluid ounces, that one gallon is 8 pints and that 5 litres is roughly one gallon. He only has a measuring jug marked in millilitres. How many millilitres of milk does he need for the cake? Give your answer to a sensible degree of accuracy.

8. What is the smallest number greater than 1000 that is exactly divisible by 13 and 17?

9. In January a car dealer increased the price of a Jaguar 'Sovereign' by 8%. During the year demand for the car went down and in July the dealer decided to reduce the price by 7·5% in an effort to sell more cars.
Was it cheaper to buy the car in February or in July? Explain your answer.

10. (a) The sum of the factors of n (including 1 and n) is 7. Find n.
(b) The sum of the factors of p (including 1 and p) is 6. Find p.

11. Find the smallest value of n for which

$$1^2 + 2^2 + 3^2 + 4^2 + 5^2 + \ldots + n^2 > 800$$

12. The 'reciprocal' of 2 is $\frac{1}{2}$. The reciprocal of 7 is $\frac{1}{7}$. The reciprocal of x is $\frac{1}{x}$.
Find the square root of the reciprocal of the square root of the reciprocal of ten thousand.

13. A wicked witch stole a newborn baby from its parents. On the baby's first birthday the witch sent the grief-stricken parents 1 penny. On the second birthday she sent 2 pence. On the third birthday she sent 4 pence and so on, doubling the amount each time. How much did the witch send the parents on the twenty-first birthday?

14. The total mass of a jar one quarter full of jam is 250 g. The total mass of the same jar three quarters full of jam is 350 g.

$\frac{1}{4}$
250 g

$\frac{3}{4}$
350 g

What is the mass of the empty jar?

15. On an aircraft a certain weight of luggage is carried free and a charge of £10 per kilogram is made for any excess luggage.
(a) If the charge for 60 kg of luggage is £400, find the charge for 35 kg.
(b) If the charge for 30 kg of luggage is £100, find the charge for 15 kg.
(c) If a fifth of the luggage is carried free, what is the average cost per kilogram of the luggage?

16. Evaluate

(a) $\frac{1}{3} \times \frac{2}{4} \times \frac{3}{5} \times \ldots \times \frac{9}{11} \times \frac{10}{12}$.

(b) $[(-2)^{-2}]^{-2}$

17. In France petrol for a car costs 5·54 francs per litre. The exchange rate is 9·60 francs to the pound and we can assume that one gallon equals 4·5 litres.
A family uses the car to travel a total distance of 1975 km and the owner of the car estimates the petrol consumption of the car to be 27 miles per gallon. Take 1 km equal to 0·62 miles. Calculate the cost of the petrol in pounds to the nearest pound.

Exercise 26

1. I have lots of 1p, 2p, 3p and 4p stamps. How many different combinations of stamps can I make which total 5p?

2. Find n if

$$8+9+10+\ldots+n = 5^3$$

3. Copy and complete.

$$3^2+4^2+12^2 = 13^2$$
$$5^2+6^2+30^2 = 31^2$$
$$6^2+7^2+\quad =$$
$$x^2+\quad+\quad =$$

4. You are told that 8 cakes and 6 biscuits cost 174 pence and 2 cakes and 4 biscuits cost 66 pence. Without using simultaneous equations, work out the cost of each of the following.
(a) 4 cakes and 3 biscuits
(b) 10 cakes and 10 biscuits
(c) 3 cakes and 3 biscuits
(d) 1 cake
(e) 1 biscuit

5. Pages 6 and 27 are on the same (double) sheet of a newspaper. What are the page numbers on the opposite side of the sheet? How many pages are there in the newspaper altogether?

6. Use the numbers 1, 2, 3, 4, 5, 6, 7, 8, 9 once each and in their natural order to obtain an answer of 100. You may use only the operations $+$, $-$, $\times$, $\div$.

7. The ruler below has eleven marks and can be used to measure lengths from one unit to twelve units.

Design a ruler which can be used to measure all the lengths from one unit to twelve units but this time put the minimum possible number of marks on the ruler.

8. Each packet of washing powder carries a token and four tokens can be exchanged for a free packet. How many free packets will I receive if I buy 64 packets?

9. Find three consecutive square numbers whose sum is 149.

10. The diagrams show magic squares in which the sum of the numbers in any row, column or diagonal is the same. Find the value of x in each square.

(a)

	x	6
3		7
		2

(b)

4		5	16
x		10	
	7	11	2
1			13

11. Work out $100 - 99 + 98 - 97 + 96 - \ldots + 4 - 3 + 2 - 1$

12. The smallest three-digit product of a one-digit prime and a two-digit prime is
(a) 102 (b) 103 (c) 104 (d) 105 (e) 106

13. Pythagoras, the Greek mathematician, was also a shrewd businessman. Suppose he deposited £1 in the Bank of Athens in the year 500 B.C. at 1% compound interest. What would the investment be worth to his descendants in the year 2000 A.D.?

14. Apart from 1, 3 and 5, all odd numbers less than 100 can be written in the form $p + 2^n$ where p is a prime number and n is greater than or equal to 2.

e.g. $43 = 11 + 2^5$
$27 = 23 + 2^2$

For the odd numbers 7, 9, 11, ... 99, write as many as you can in the form $p + 2^n$

15. The digits 1, 9, 9, 4 are used to form fractions less than 1. Each of the four digits must be used,

e.g. $\frac{9}{419}$ or $\frac{4}{919}$

(a) Write down, smallest first, the three smallest fractions that can be made.

(b) Write down the largest fraction less than one that can be made.

1.9 Irrational numbers

A rational number can always be written exactly in the form $\frac{a}{b}$ where a and b are whole numbers.

Example 1

$\frac{3}{7}$ $\quad 1\frac{1}{2} = \frac{3}{2}$ $\quad 5{\cdot}14 = \frac{257}{50}$ $\quad 0{\cdot}\dot{6} = \frac{2}{3}$

All these are rational numbers.

An irrational number cannot be written in the form $\frac{a}{b}$.
$\sqrt{2}$, $\sqrt{5}$, π, $\sqrt[3]{2}$ are all irrational numbers. The proof that $\sqrt{2}$ is irrational is given on page 379.

Example 2

The recurring decimal 0·631 631 631 ... can be written in the form $\frac{a}{b}$.

Let $\quad r = 0{\cdot}631\,631\,631\ldots$
Multiply by 1000, $\quad 1000r = 631{\cdot}631\,631\,631\ldots$
Subtract, $\quad 999r = 631$

$\therefore \quad r = \frac{631}{999}$ so the number is rational.

Numbers like $\sqrt{2}$, $\sqrt{3}$, $\sqrt{7}$ are called *surds*. The following rules apply.

$\sqrt{a} \times \sqrt{b} = \sqrt{ab}$ $\quad$ e.g. $\sqrt{2} \times \sqrt{3} = \sqrt{6}$

$\sqrt{a} \div \sqrt{b} = \sqrt{\left(\frac{a}{b}\right)}$ $\quad$ e.g. $\sqrt{7} \div \sqrt{2} = \sqrt{\frac{7}{2}}$

Exercise 27

1. Which of the following numbers are rational?

$\frac{\pi}{2}$	$\sqrt{5}$	$(\sqrt{17})^2$	$\sqrt{3}$
3·14	$\frac{\sqrt{12}}{\sqrt{3}}$	π^2	$3^{-1} + 3^{-2}$
$7^{-\frac{1}{2}}$	$\frac{22}{7}$	$\sqrt{2} + 1$	$\sqrt{2{\cdot}25}$

2. (a) Write down any rational number between 4 and 6.
(b) Write down any irrational number between 4 and 6.

3. Write in the form $\frac{a}{b}$.

(a) $0{\cdot}\dot{2}$ (b) $0{\cdot}\dot{2}\dot{9}$ (c) $0{\cdot}\dot{5}4\dot{1}$

4. Think of two *irrational* numbers x and y such that $\frac{x}{y}$ is a *rational* number.

5. Explain the difference between a rational number and an irrational number.

6. (a) Is it possible to multiply a rational number by an irrational number to give an answer which is rational?
(b) Is it possible to multiply two irrational numbers together to give a rational answer?
(c) If either or both are possible, give an example.

7. Without using a calculator, simplify the following.

(a) $\sqrt{20} + \sqrt{45}$ (b) $\frac{\sqrt{80}}{\sqrt{45}}$

(c) $\frac{2}{\sqrt{2}}$ (d) $\sqrt{2} \times \sqrt{3} \times \sqrt{6}$

1.10 Measurements and errors

Measurement is approximate

(a) A length of some cloth is measured for a dress. You might say the length is 145 cm to the nearest cm.
The actual length could be anything from 144·5 cm to 145·49999 . . . cm using the normal convention which is to round up a figure of 5 or more. Clearly 145·4999 . . . is effectively 145·5 and we could use this figure.

(b) When measuring the length of a page in a book, you might say the length is 437 mm to the nearest mm.
In this case the actual length could be anywhere from 436·5 mm to 437·5 mm. We write 'length is between 436·5 mm and 437·5 mm'.

> In both cases (a) and (b), the measurement expressed to a given unit is in possible error of half a unit.

(c) (i) Similarly if you say your weight is 57 kg to the nearest kg, you could actually weigh anything from 56·5 kg to 57·5 kg.
(ii) If your brother was weighed on more sensitive scales and the result was 57·2 kg, his actual weight could be from 57·15 kg to 57·25 kg.
(iii) The weight of a butterfly might be given as 0·032 g. The actual weight could be from 0·0315 g to 0·0325 g.

Common sense

When you find the answer to a problem, you should choose the degree of accuracy appropriate for that situation.

(a) Suppose you were calculating how much tax someone should pay in a year and your actual answer was £2153·6752. It would be sensible to give the answer as £2154 to the nearest pound.

(b) In calculating the average speed of a car journey from Bristol to Cardiff, it would not be realistic to give an answer of 38·241 km/h. A more sensible answer would be 38 km/h.

(c) Always check to see if your answers are sensible.
 (i) If you do a calculation using say, trigonometry, and you find the height of a child's 'Wendy House' is 636·475 m you should know that something has gone wrong.
 (ii) If you are calculating the speed of a train and find the answer to be 0·23 km/h, either your calculation is at fault or the train needs a new engine!

Approximations

(a) Reminders: 35·2 ¦ 6 = 35·3 to 3 sig. fig.
↑
'5 or more'

0·041 ¦ 2 = 0·041 to 2 sig. fig.
↑↑
Do not count zeros at the beginning.

15·26 ¦ 66 = 15·27 to 2 decimal places
↑
0·349 ¦ 7 = 0·350 to 3 decimal places
↑

(b) Suppose you timed a race with a stopwatch and got 13·2 seconds while your friend with an electronic watch got 13·20 seconds.
Is there any difference? Yes!
When you write 13·20, the figure is accurate to 2 decimal places even though the last figure is a zero.
The figure 13·2 is accurate to only one decimal place.

(c) Similarly a weight, given as 12·00 kg, is accurate to 4 significant figures while '12 kg' is accurate to only 2 significant figures.

(d) Suppose you measure the length of a line in cm and you want to show that it is accurate to one decimal place. The measured length using a ruler might be 7 cm. To show that it is accurate to one decimal place, you must write 'length = 7·0 cm'.

Exercise 28

Questions **1** to **14**, give a measurement. Write down the upper and lower bounds of the number.

1. mass = 17 kg
2. $d = 256$ km
3. length = 2·4 m
4. $m = 0{\cdot}34$ grams
5. $v = 2{\cdot}04$ m/s
6. $x = 12{\cdot}0$ cm [N.B. not 12 cm!]
7. $T = 81{\cdot}4$°C
8. $M = 0{\cdot}3$ kg
9. $d = 4{\cdot}00$ cm
10. $y = 0{\cdot}07$ m
11. mass = 0·1 tonne
12. $t = 615$ seconds
13. $d = 7{\cdot}13$ m
14. $n = 52$ million (nearest million)

Questions **15** to **20**, give the result of a calculation. Give the answer to an appropriate degree of accuracy.

15. Average speed of a train between London and Manchester = 98·2513 km/h.

16. Weekly expenditure on food by the average British family = £39·732.

17. Amount of rainfall in Oxford during a thunderstorm = 2·241 inches.

18. Number of new cars sold in Britain in August 1995 = 412, 618.

19. Average lifetime of an electric light bulb = 2162·23 hours.

20. Quantity of water in a swimming pool = 958 617·7 litres.

In Questions **21** to **28**, write 'Yes' if the answer is sensible and 'No' if the answer is not.

21. Total weight of apples off a large tree = 62 kg.

22. Cost of a school meal in U.S.A. = $35·80.

23. Time taken by an aircraft to fly non-stop from London to Hong Kong = 14·2 h.

24. Ratio of population of China to population of UK = 200 : 1.

25. Daily takings at a Tesco superstore = £42 400.

26. Number of bricks needed to build an 'average' size house = 3·2 million.

27. Time required for your maths teacher to run 100 m = 11·2 seconds.

28. Weight of a 'typical' saloon car = 384 600 kg.

1.11 Errors in calculations

This section is concerned with the inaccuracy of numbers used in a calculation. We are not trying to prevent students from making mistakes!

(a) Here is a rectangle with sides measuring 37 cm by 19 cm to the nearest cm.
What are the largest and smallest possible areas of the rectangle consistent with this data?

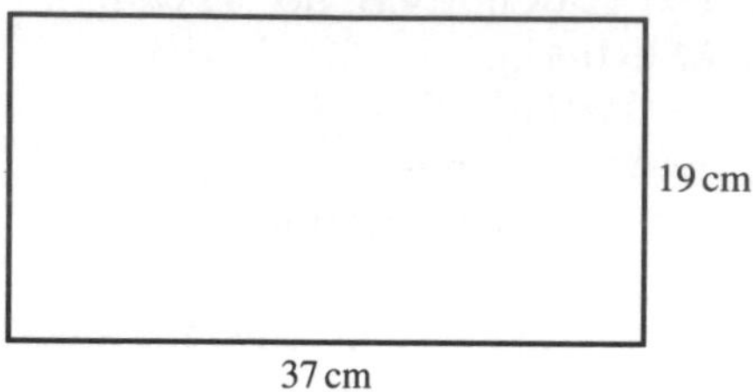

We know the length is between 36·5 cm and 37·5 cm
and the width is between 18·5 cm and 19·5 cm

largest possible area $= 37{\cdot}5 \times 19{\cdot}5$
$= 731{\cdot}25\ \text{cm}^2$

smallest possible area $= 36{\cdot}5 \times 18{\cdot}5$
$= 675{\cdot}25\ \text{cm}^2$

(b) A common problem occurs when a measured quantity is multiplied by a large number, e.g. a marble is weighed and found to be 12·3 grams. Find the weight of 100 identical marbles.

A simple answer is $100 \times 12{\cdot}3 = 1230$ g. We must realise that each marble could weigh from 12·25 g to 12·35 g.

$100 \times 12{\cdot}25 = 1225$ g
$100 \times 12{\cdot}35 = 1235$ g

So there is a possible error of up to 5 grams above or below our initial answer of 1230 g.

(c) Given that the numbers 8·6, 3·2 and 11·5 are accurate to 1 decimal place, calculate the upper and lower bounds for the calculation

$$\left(\frac{8{\cdot}6 - 3{\cdot}2}{11{\cdot}5}\right)$$

For the upper bound, make the top line of the fraction as large as possible and make the bottom line as small as possible.

$$\text{upper bound} = \left(\frac{8{\cdot}65 - 3{\cdot}15}{11{\cdot}45}\right)$$
$$= 0{\cdot}480\,3493$$
$$= 0{\cdot}480 \quad \text{(to 3 s.f.)}$$

$$\text{lower bound} = \left(\frac{8{\cdot}55 - 3{\cdot}25}{11{\cdot}55}\right)$$
$$= 0{\cdot}458\,8744$$
$$= 0{\cdot}459 \quad \text{(to 3 s.f.)}$$

Exercise 29

1. If $a = 3{\cdot}1$ and $b = 7{\cdot}3$, correct to 1 decimal place, find the largest possible value of
(i) $a + b$ (ii) $b - a$

2. If $x = 5$ and $y = 7$ to one significant figure, find the smallest possible values of

(i) $x + y$ (ii) $y - x$ (iii) $\frac{x}{y}$

3. In the diagram, ABCD and EFGH are rectangles with AB = 10 cm, BC = 7 cm, EF = 7 cm and FG = 4 cm, all figures accurate to the nearest cm.
Find the largest possible value of the shaded area.

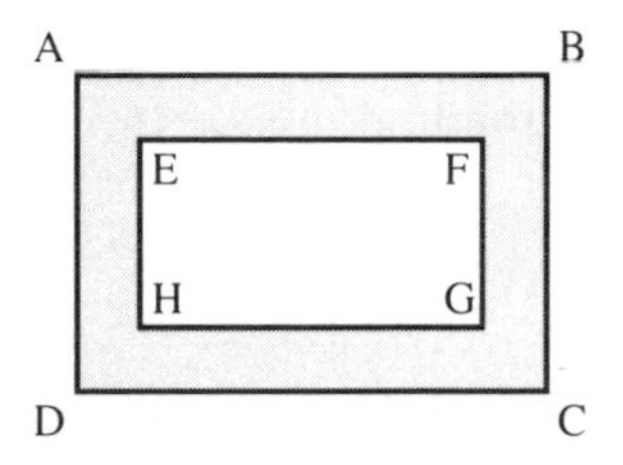

4. When a voltage V is applied to a resistance R the power consumed P is given by $P = \frac{V^2}{R}$.
If you measure V as 12·2 and R as 2·6, calculate the largest and smallest possible values of P.

5. A cyclist was timed along a straight piece of road. The time taken was 24·5 seconds, to the nearest 0·1 second, and the distance was 420 metres, to the nearest metre. Calculate the maximum and minimum values for the speed of the cyclist consistent with this data.

6. The velocity v of a body is calculated from the formula $v = \frac{2s}{t} - u$

where u, s and t are measured correct to 1 decimal place. Find the largest possible value for v when $u = 2{\cdot}1$, $s = 5{\cdot}7$ and $t = 2{\cdot}2$. Find also the smallest possible value for v consistent with these figures.

7. Use the formula $z = \frac{a - x^2}{2t}$ to find the largest value that z could have when $a = 71{\cdot}4$, $x = 5{\cdot}3$ and $t = 5{\cdot}4$, all correct to one decimal place.

8. A formula for velocity is $v = \sqrt{(u^2 + 2as)}$. Find the smallest possible value for v if $u = 11{\cdot}5$, $a = -9{\cdot}8$ and $s = 4{\cdot}0$, all correct to one decimal place.

9. The price of gold is \$22·65 per gram, and the density of gold is $19{\cdot}2\,\text{g/cm}^3$ (i.e. $1\,\text{cm}^3$ of gold weighs 19·2 g). These figures can be used as exact. A solid gold bar in the shape of a cuboid has dimensions $8{\cdot}6 \times 4{\cdot}1 \times 2{\cdot}4$, all measurements in cm correct to the nearest 0·1 cm.
   (a) What is the range of possible prices for this gold bar? Write '... < price < ...' giving your answers to the nearest \$10.
   (b) The gold bar was in fact weighed on some scales giving a weight of 1574 g to the nearest gram. Using this value, work out the range of possible prices for the gold bar. Again write '... < price < ...'.
   (c) Explain why the weighing method appears to give a smaller range of possible prices.

Percentage error

Suppose the answer to the calculation $4{\cdot}8 \times 33{\cdot}4$ is estimated to be 150. The exact answer to the calculation is 160·32.
The percentage error is

$$\left(\frac{160{\cdot}32 - 150}{160{\cdot}32}\right) \times \frac{100}{1} = 6{\cdot}4\% \qquad (2\,\text{s.f.})$$

In general,

$$\text{percentage error} = \left(\frac{\text{actual error}}{\text{exact value}}\right) \times 100$$

Exercise 30

Give the answers correct to 3 significant figures, unless told otherwise.

1. A man's salary is estimated to be £20 000. His actual salary is £21 540. Find the percentage error.
2. The speedometer of a car showed a speed of 68 m.p.h. The actual speed was 71 m.p.h. Find the percentage error.
3. The rainfall during a storm was estimated at 3·5 cm. The exact value was 3·1 cm. Find the percentage error.
4. The dimensions of a rectangle were taken as 80×30. The exact dimensions were $81{\cdot}5 \times 29{\cdot}5$. Find the percentage error in the area.
5. The answer to the calculation $3{\cdot}91 \times 21{\cdot}8$ is estimated to be 80. Find the percentage error.
6. An approximate value for π is $\frac{22}{7}$. A more accurate value is given on most calculators. Find the percentage error.

7. A wheel has a diameter of 58·2 cm. Its circumference is estimated at 3×60. Find the percentage error.

8. To the calculation $56{\cdot}2 \times 9{\cdot}1 - 4{\cdot}53$, Katy and Louise gave estimates of $50 \times 10 - 4$ and $60 \times 9 - 5$ respectively. Who made the more accurate estimate?

9. A speed of 60 km/h is taken to be roughly 18 m/s. Find the percentage error.

10. The fraction $\frac{19}{11}$ gives an approximate value for $\sqrt{3}$. Calculate the percentage error in using $\frac{19}{11}$ as the value of $\sqrt{3}$. Give the answer to 2 s.f.

11. Here are three approximate values using trigonometry.

$\sin 45° \approx \frac{7}{10}$ $\qquad$ $\cos 40° \approx \frac{3}{4}$ $\qquad$ $\tan 60° \approx 1{\cdot}7$

For which approximation is the percentage error the smallest? Give the percentage error for this one.

12. The most common approximate value for π is $\frac{22}{7}$.

Two better approximations are $\frac{355}{113}$ and $\sqrt{\left(\sqrt{\left(\frac{2143}{22}\right)}\right)}$

Which of these two values has the smaller percentage error?

13. It is given that $x = \dfrac{a}{b + c}$ and that $a = 58$, $b = 27{\cdot}5$ and $c = 19{\cdot}2$. The values for a, b and c have a maximum percentage error of 10%. Work out the minimum possible value of x correct to 3 s.f.

14.† The surface area S of a sphere is given by $S = 4\pi r^2$.

The exact value of r can be taken as 10 cm.

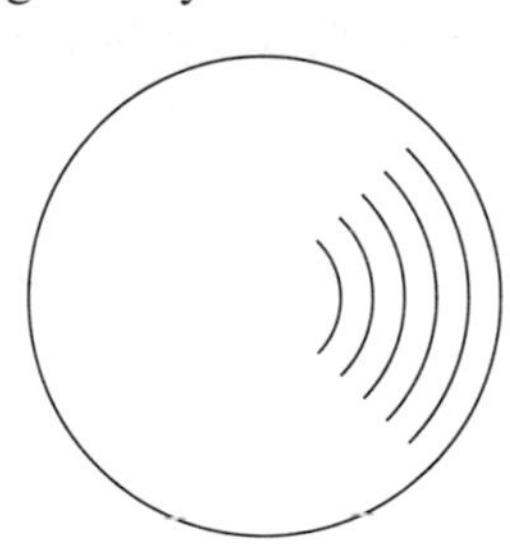

(a) Work out the value of S using $r = 10$ cm. Write down all the figures on your calculator.

(b) Suppose the radius is measured as 10·2 cm, i.e. an error of 2% of the exact value.

(i) Work out S_1, the surface area using $r = 10{\cdot}2$.

(ii) Write down $S_1 - S$.

(iii) Express the error as a percentage of the exact surface area S. Give this answer correct to 2 s.f.

Examination exercise 1

1. The table shows the mean number of copies sold per day by some daily papers in 1988.

Paper	Copies
Sun	4 146 644
Mirror	3 082 215
Daily Mail	1 792 701
Daily Express	1 679 438
Daily Telegraph	1 138 673
Star	1 013 688
Guardian	470 023
Times	450 626
Today	408 078
Independent	375 317
Financial Times	286 774
Total	14 844 177

(a) Which paper sold about half a million more copies than the Daily Telegraph?
(b) Which two papers each sold about four times as many copies as the Times?
(c) Write down two papers which sold equal numbers of copies, correct to the nearest 100 000.
(d) The Guardian sold 470 023 copies. What percentage of the total was this?
(e) The Sun sold 4 146 644 copies. Write this number in standard form correct to three significant figures. [S]

2. Jane is going on holiday to France and needs to buy some films for her camera. If she buys them in England before she goes, they will cost her £2·25 each. If she waits until she gets to France and buys them there, they will cost her 20·16 francs each.
If the exchange rate is £1 to 9·60 francs, find out how much, in English currency, she can save on each film by buying them in France. [M]

3. The approximate area of the World is 135 085 000 km^2.
(a) Express the number 135 085 000 in standard form.

The approximate area of Africa is 30 132 000 km^2.
(b) Calculate the area of Africa as an approximate percentage of the area of the World. (Give your answer correct to one decimal place.) [L]

4. In each of the following, put in a decimal point so that the measurement is reasonable.
(a) Diameter of a 2p coin. 2750 cm
(b) Weight of an average woman 5714 kg
(c) Capacity of a petrol tank on a moped 7698 litres
(d) Length of a football pitch 1076 m [M]

5. The scale of a map was 1 : 2000. The map is reduced in size so that it uses a quarter of the original area of paper. What is the scale of the new map? [S]

6. (a) Find the exact value of

$$\left[\left(\frac{2}{3}\right)^2 + \frac{5}{6}\right] \div 11\frac{1}{2}$$

(b) Given that $a = 64$, evaluate

$a^{\frac{1}{2}} + a^{\frac{2}{3}}$ [M]

7. A sum of money was invested at a fixed rate of compound interest added annually.
At the end of the first year it had amounted to £990 and at the end of the second year it had amounted to £1089.
(a) Calculate the yearly rate of interest.
(b) Calculate the original sum of money which was invested. [N]

8. Aziz wants to make a plastic mug in the shape of a cylinder. The volume, V, of the mug must be 510 cubic centimetres and the height, h, 9·6 centimetres. Aziz has to work out the radius, r, in centimetres. He knows that r is given by the formula

$$r = \sqrt{\frac{V}{\pi H}}.$$

(a) Taking $\pi = 3{\cdot}14$, calculate the value of r, correct to 1 decimal place.
(b) Explain clearly how Aziz could estimate the value of r, correct to 1 significant figure, if he did not have a calculator or mathematical tables. [M]

9. A metal rivet fits into a hole as shown.
The diameter of the rivet is $\frac{3}{5}''$.
The diameter of the hole is $\frac{7}{8}''$.
Calculate the distance a, which you should express as
(a) a fraction of an inch, (b) a decimal.

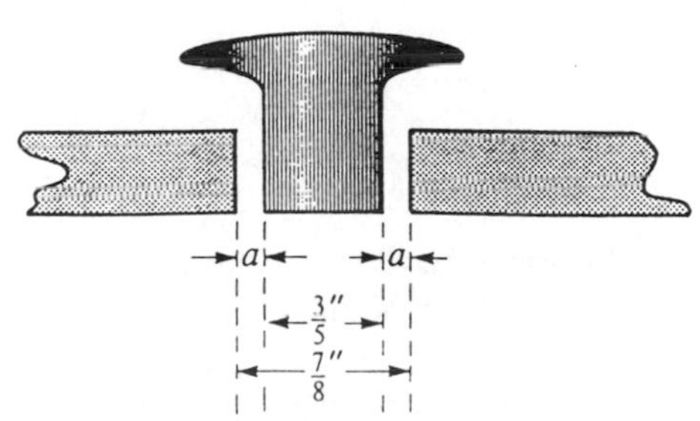

[N]

10. Supergrowth Unit Trust claims that the value of its units is likely to grow by 21% compound interest per annum. Assuming that this claim is true, calculate the value, after 5 years, of an investment of £1000 in Supergrowth Unit Trust. [M]

11. Write down the prime factors of
(a) 273,
(b) 2002.
Hence find the highest common factor of 273 and 2002. [S]

12. In a General Election a candidate loses his deposit if he does not obtain at least 5% of the total votes cast in the constituency for which he is seeking election.
(a) In a certain constituency, three candidates, A, B and C, had put up for election.
Out of the 42 560 votes cast, 21 523 people voted for candidate A, 18 862 voted for candidate B, and the rest voted for candidate C.
Decide whether, and by how many votes, candidate C lost or saved his deposit.
(b) In another constituency, there were just two candidates, R and S. The winner, who was candidate R, received 16 017 votes. By letting x be the number of people who voted for candidate S, or otherwise, calculate the least number of votes candidate S would have to obtain in order not to lose his deposit. [N]

13. This item appeared in a Sunday newspaper.

> Rubik's New Puzzle
>
> A new version of the Rubik Cube Puzzle has been invented. It has 256 times the number of arrangements of the original one. The original Rubik Cube had 43×10^{30} different arrangements.

Calculate the number of arrangements of the new puzzle, giving your answer in standard form, correct to one place of decimals. [N]

14.

> **It's a fact!**
>
> – SOMEONE IN THE UK IS BURGLED EVERY 66 SECONDS –

(a) Assuming that this statement is true, find correct to 2 significant figures, the number of burglaries in the UK in 1 year.
(b) Write your answer to part (a) in standard form. [M]

15. The time, T seconds, that an experiment takes depends on the temperature, t degrees, and is given by the formula

$$T = t^4 \quad t + 1$$

Calculate the time the experiment takes when the temperature is -3 degrees. [N]

16. The time, T minutes, taken by the moon to eclipse the sun totally is given by the formula

$$T = \frac{1}{v}\left(\frac{rD}{R} - d\right)$$

d and D are the diameters, in kilometres, of the moon and sun respectively;
r and R are the distances, in kilometres, of the moon and sun respectively from the earth;
v is the speed of the moon in kilometres per minute.
Given that

$d = 3.48 \times 10^3$, $D = 1.41 \times 10^6$, $r = 3.82 \times 10^5$,
$R = 1.48 \times 10^8$, $v = 59.5$

calculate the time taken for a total eclipse, giving your answer in minutes, correct to 2 significant figures. [M]

17. The world record for the 100 metres sprint is 9·83 seconds.
(a) Calculate the average speed of the athlete in km/h.
(b) A newspaper report said that the athlete was running at 40 km/h. calculate the percentage error of this estimate. [M]

18. The area of a rectangular field is 28 500 square metres and its length is 195 metres, both measurements being correct to three significant figures.
(a) Find
  (i) the greatest possible breadth of the field,
  (ii) the smallest possible breadth of the field.
(b) What is the breadth of the field correct to two significant figures? [N]

19. As part of a project, some children are timing cars as they travel along a section of road which they have measured as 100 metres. One car is timed at 6 seconds.
(a) Assuming these figures to be exact, find the average speed of this car, in metres per second.
(b) In fact, the length of road was measured correct to the nearest 10 metres and the car was timed correct to the nearest second. Find the maximum average speed of the car, correct to the nearest metre per second. [M]

2 Algebra

2.1 Sequences

Here is a sequence of 'houses' made from matches.

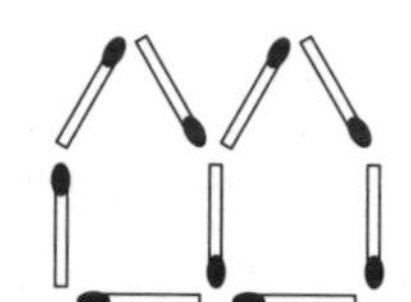
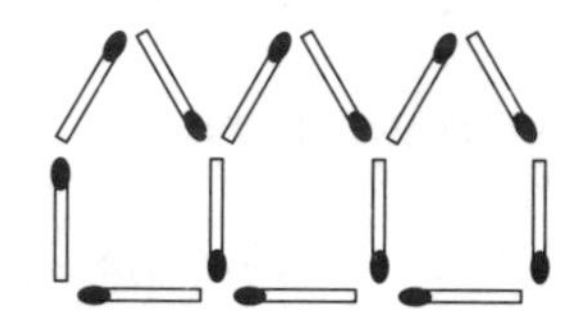
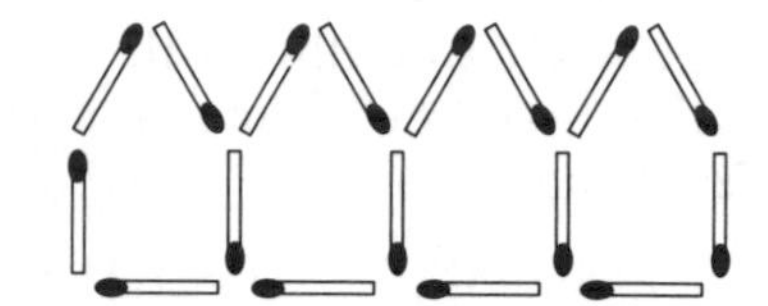

The table on the right records the number of houses h and the number of matches m.

h	m
1	5
2	9
3	13
4	17

If the number in the h column goes up one at a time, look at the number in the m column. If it goes up (or down) by the same number each time, the function connecting m and h is linear. This means that there are no terms in h^2 or anything more complicated.

In this case, the numbers in the m column go up by 4 each time. This suggests that a column for $4h$ might help.

h	m	$4h$
1	5	4
2	9	8
3	13	12
4	17	16

Now it is fairly clear that m is one more than $4h$.

So the formula linking m and h is $m = 4h + 1$

Example 1

The table shows how r changes with n. What is the formula linking r with n?

n	r
2	3
3	8
4	13
5	18

Because r goes up by 5 each time, try writing another column for $5n$.
The table shows that r is always 7 less than $5n$, so the formula linking r with n is
$r = 5n - 7$

n	r	$5n$
2	3	10
3	8	15
4	13	20
5	18	25

The method only works when the first set of numbers goes up by one each time. A more subtle approach has to be thought out if this is not the case!

Exercise 1

1. Below is a sequence of diagrams showing black tiles b and white tiles w with the related table.

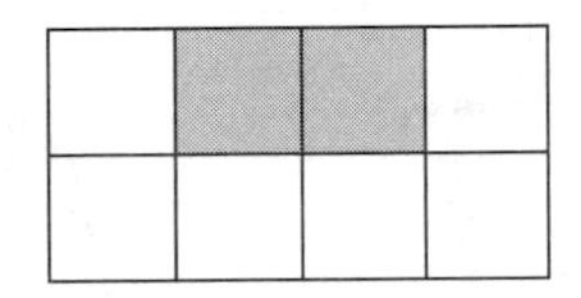
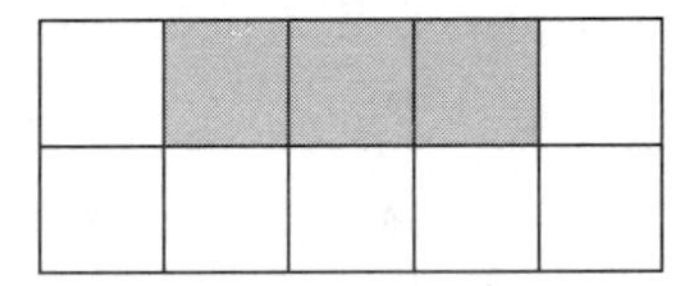

b	w
1	5
2	6
3	7
4	8

What is the formula for w in terms of b? Write it as $w = \ldots$

2. This is a different sequence with black tiles b and white tiles w and the related table.

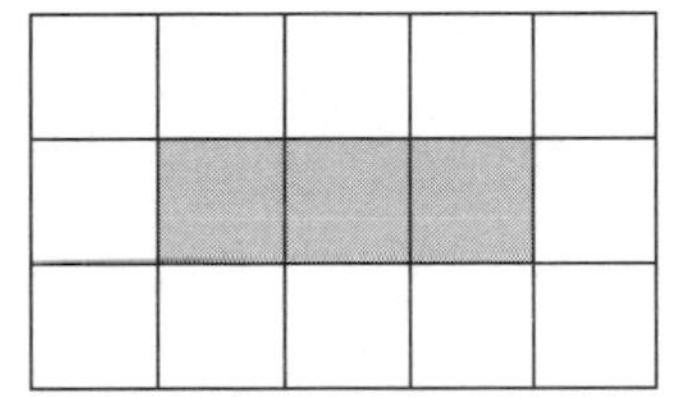
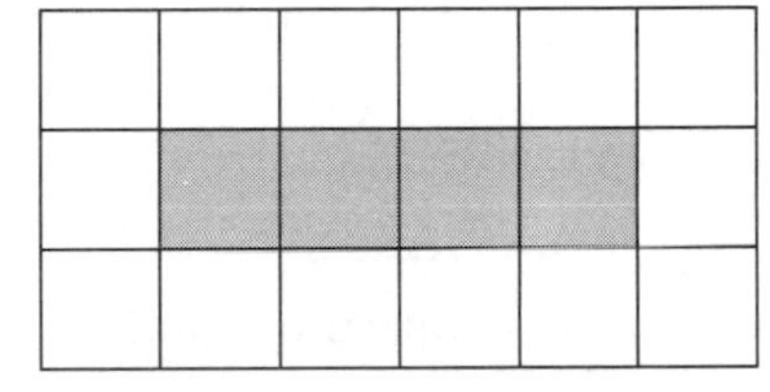

b	w
2	10
3	12
4	14
5	16

What is the formula? Write it as $w = \ldots$

3. Here is a sequence of I's.

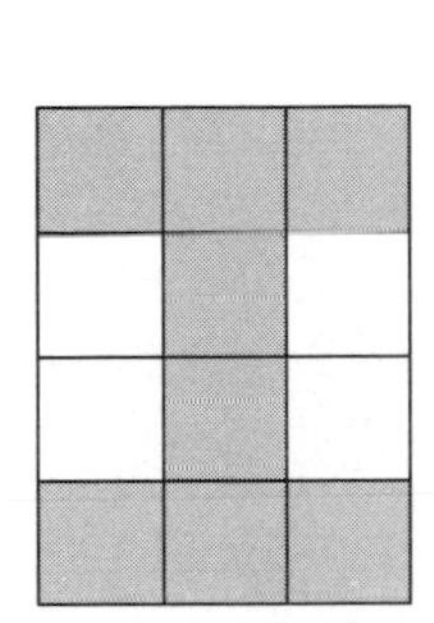
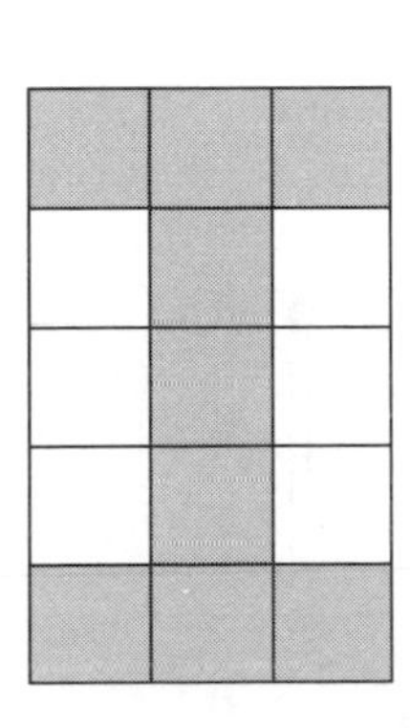
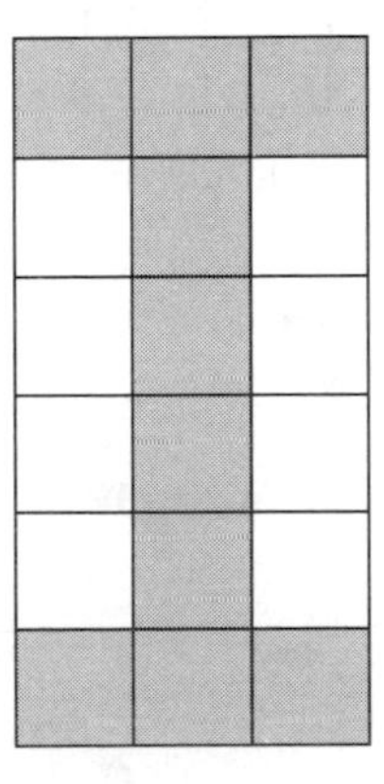

Make your own table for black tiles b and white tiles w. What is the formula for w in terms of b?

4. This sequence shows matches m arranged in triangles t.

  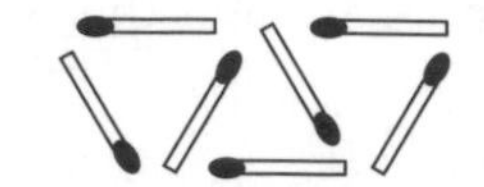 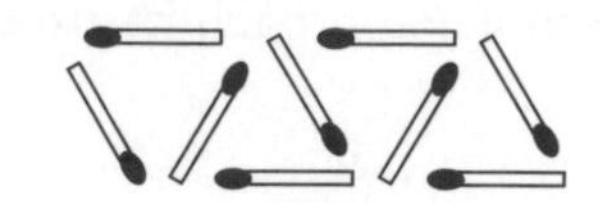

t	m
1	3
2	5
⋮	⋮

Make a table for t and m starting like this:
Continue the table and find a formula for m in terms of t. Write it as $m = \ldots$

5. Here is a different sequence of matches and triangles.

Make a table and find a formula connecting m and t.

6. In this sequence, there are triangles t and squares s around the outside.

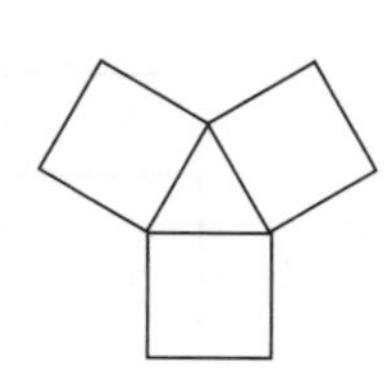 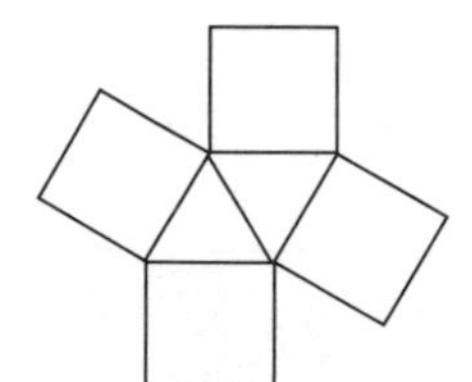 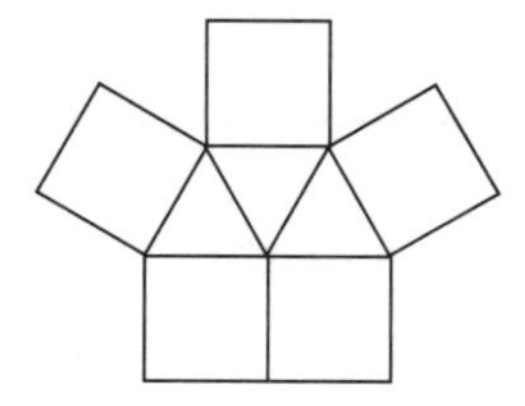 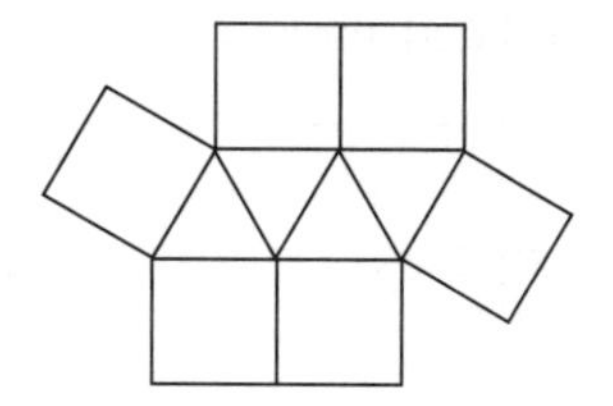

What is the formula connecting t and s?

7. Look at the tables below. In each case, find a formula connecting the two letters.

(a)

n	p
1	3
2	8
3	13
4	18

(b)

n	k
2	17
3	24
4	31
5	38

(c)

n	w
3	17
4	19
5	21
6	23

8. This is one member of a sequence of cubes c made from matches m.

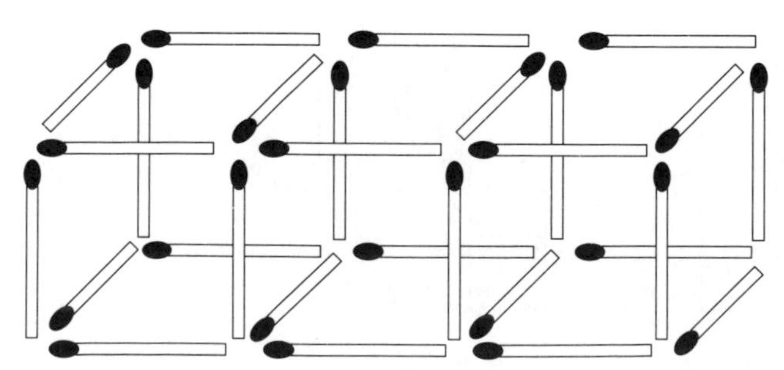

Find a formula connecting m and c.

9. In these tables, the numbers on the left do not go up by one each time. Try to find a formula in each case.

(a)

n	y
1	4
3	10
7	22
8	25

(b)

n	h
2	5
3	9
6	21
10	37

(c)

n	k
3	14
7	26
9	32
12	41

10. Some attractive loops can be made by fitting pentagon tiles together.

Diagram $n = 1$ Diagram $n = 2$ Diagram $n = 3$

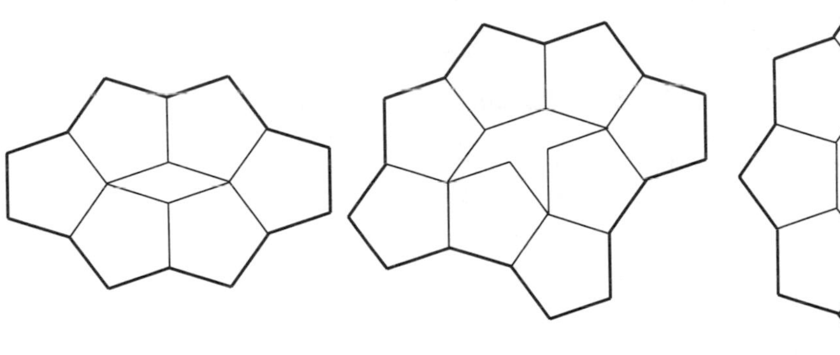

6 tiles t
14 external edges e

8 tiles t
17 external edges e

10 tiles t
20 external edges e

Make your own tables involving n, t and e.
(a) Find a formula connecting n and t.
(b) Find a formula connecting n and e.
(c) Find a formula connecting t and e.

11. When a group of workers were made redundant, the management used a formula to calculate the redundancy money paid.

Here are the details.

Number of years worked N	Redundancy money £R
1	£5400
2	£5800
3	£6200
4	£6600
10	£9000

(a) Sam has worked for 7 years. How much redundancy money will he receive?
(b) Write down the formula connecting the amount of redundancy paid R and the number of years worked N.

General term of a sequence

Here is a sequence. 1, 4, 9, 16, ...

The sequence can be rewritten. $1^2, 2^2, 3^2, 4^2, \ldots$

The 8th term is 8^2.

The 23rd term is 23^2.

The sequence can be written precisely by giving the *general* term, usually called the *n*th term.

The nth term is n^2.

Here is another sequence. 2, 4, 8, 16, ...

$2^1, 2^2, 2^3, 2^4, \ldots$

By rewriting the sequence, we can see that the nth term is 2^n.

Exercise 2

Look at each sequence and find a formula for the nth term.

Not all of the formulas are linear.

1. 3, 6, 9, 12, ...
2. 5, 10, 15, 20, ...
3. 3, 9, 27, 81, ...
4. 1, 3, 5, 7, ...
5. $1 \times 3, 2 \times 4, 3 \times 5, 4 \times 6, \ldots$
6. 3, 5, 7, 9, ...
7. 2, 5, 8, 11, ...
8. 0, 3, 8, 15, 24, ...
9. $2 \times 3, 3 \times 4, 4 \times 5, 5 \times 6, \ldots$
10. $\frac{1}{2}, \frac{2}{3}, \frac{3}{4}, \frac{4}{5}, \ldots$
11. $\frac{1}{3}, \frac{2}{4}, \frac{3}{5}, \frac{4}{6}, \ldots$
12. $1, \frac{1}{4}, \frac{1}{9}, \frac{1}{16}, \ldots$
13.† 2, 6, 12, 20, ...
14.† 4, 10, 18, 28, ...
15.† $3 \times 2, 3 \times 4, 3 \times 8, 3 \times 16, \ldots$
16.† 2, 8, 18, 32, 50, ...
17.† $2 \times 2, 4 \times 3, 6 \times 4, 8 \times 5, \ldots$
18.† 9, 99, 999, 9999, ...

Harder formulas

Most of the formulas up to now have been linear. Linear formulas have no terms with n^2, n^3, $\frac{1}{n}$, etc. This section looks at patterns which may give non-linear formulas.

Example 2

In the sequence of circle diagrams, each point is joined to every other point on the circle by a straight line. Find a formula connecting the number of lines l with the number of points p.

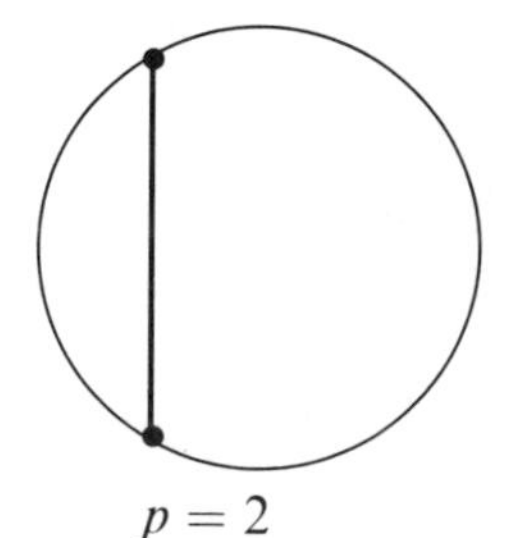

$p = 2$
$l = 1$

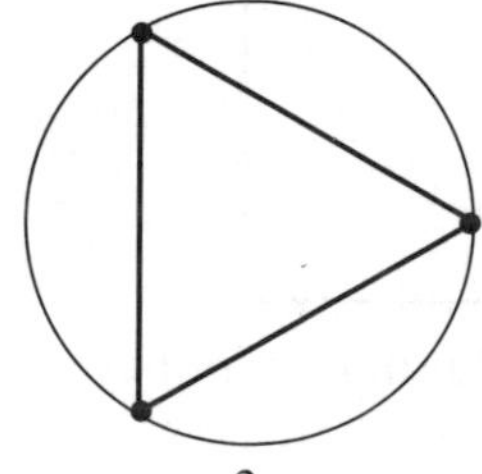

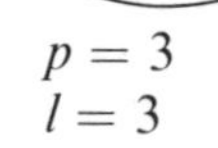

$p = 3$
$l = 3$

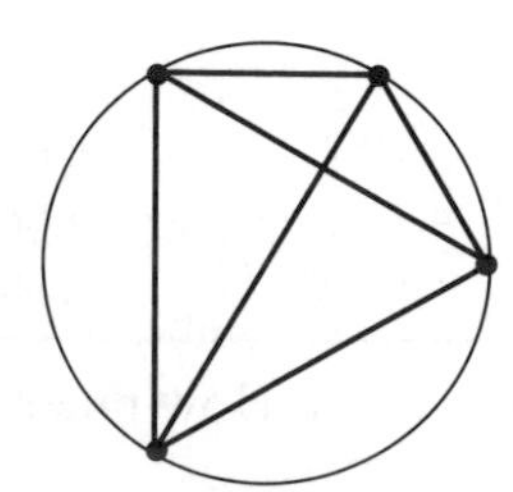

$p = 4$
$l = 6$

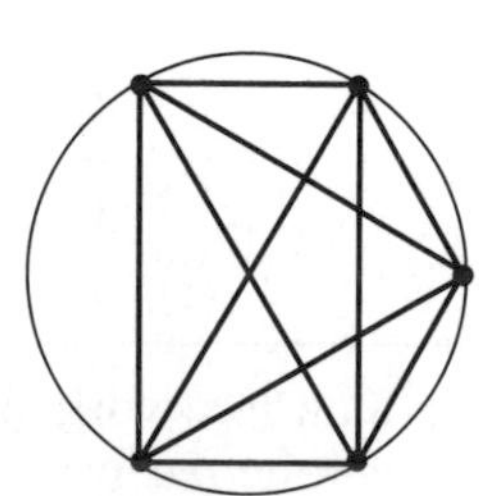

$p = 5$
$l = 10$

The table shows that p goes up one at a time, but l does not go up by the same number each time. So the formula is non-linear.

p	l
2	1
3	3
4	6
5	10
6	15

Often a difference table will reveal a pattern in the numbers. The second table shows the difference between one value and the next in the previous column. Such a table can be used to predict the next values in this case. The next two values of l are 21 $(= 15 + 6)$ for $p = 7$ and 28 $(= 21 + 7)$ for $p = 8$. This can be confirmed by drawing these two diagrams.

p	l	first difference	second difference
2	1		
		2	
3	3		1
		3	
4	6		1
		4	
5	10		1
		5	
6	15		

However, some more experimenting is needed to find the actual formula connecting p and l. Here, because the second difference column in the table is always the same number, it might help to work out columns for $p(p+1)$, $p(p-1)$, $p(p+2)$, or $p(p-2)$ etc. Here is the table.

p	l	$p(p+1)$	$p(p-1)$
2	1	6	2
3	3	12	6
4	6	20	12
5	10	30	20
6	15	42	30

The third column seems to be no help, but the fourth column shows a clear relationship to l giving the formula

$$l = \frac{1}{2}\,p(p-1)$$

This example shows that there is no set method which will always produce a formula. Working out other columns may or may not help. More detailed coverage of this is given on page 56.

Exercise 3

1. Here is a sequence of matchstick squares.

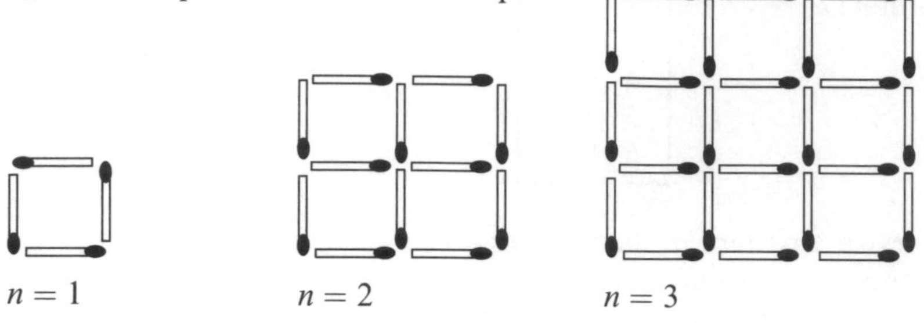

A formula is needed connecting the number of matches m and the diagram number n.

Here is the table.

n	m
1	4
2	12
3	24
4	40

(a) Without drawing the diagram, predict the number of matches in diagram number 5.

(b) Work out a column of values for $n(n+1)$ or $n(n-1)$ or $n(n+2)$ etc.
Find a formula for m in terms of n. Write it as $m = \ldots$

2. Count the number of sticks s and the *maximum* number of crossovers c in this sequence of diagrams.

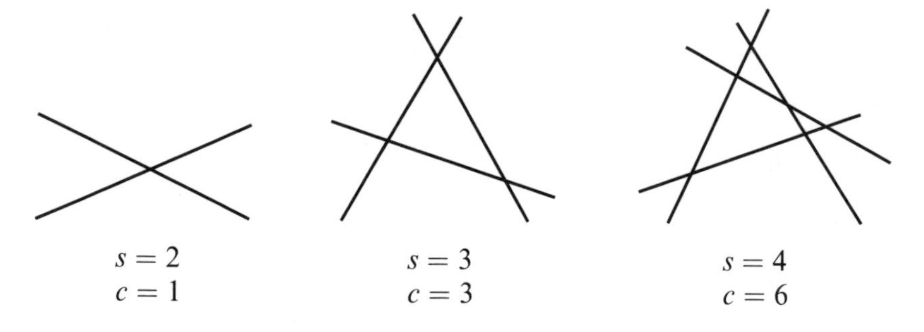

Here is a start on the table.

s	c
2	1
3	3
4	6
5	10

(a) Predict the maximum number of crossovers for $s = 6$.
Check your result by drawing.

(b) Find a formula for c in terms of s. Write it as $c = \ldots$

3. Look at the tables and find a formula connecting each pair of letters.

(a)

n	r
2	10
3	18
4	28
5	40

(b)

n	t
2	3
3	6
4	10
5	15

(c)

n	p
1	4
2	9
3	16
4	25

4. To form the next diagram, new squares are added round the outside of the previous diagram, giving this sequence.

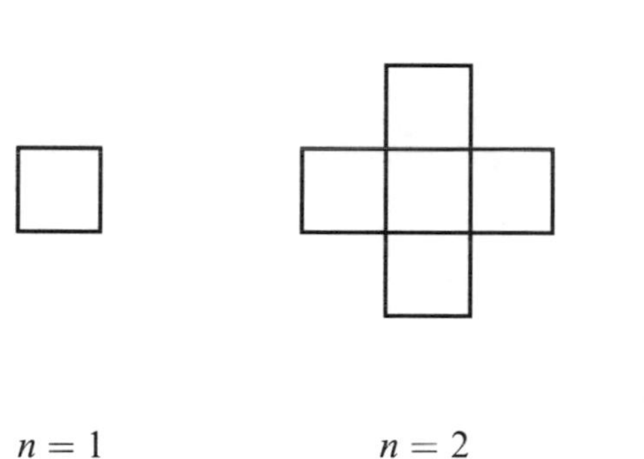
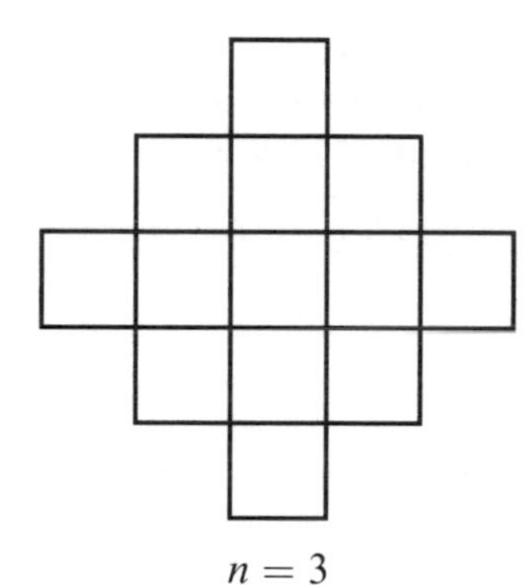

$n = 1$ $n = 2$ $n = 3$

(a) Draw the diagram $n = 4$. Count the squares s in each diagram and make a table.
(b) Without drawing the diagram for $n = 5$, can you predict the number of squares it will have?
Now check by drawing it.
(c) Find a formula for s in terms of n.

5. Two pupils, Anna and Dipika, did the same investigation and got the same results. The patterns they noticed were different.

Anna's table

n	t	pattern
1	3	1×3
2	8	2×4
3	15	3×5
4	24	4×6

Dipika's table

n	t	pattern
1	3	$1^2 + 2 \times 1$
2	8	$2^2 + 2 \times 2$
3	15	$3^2 + 2 \times 3$
4	24	$4^2 + 2 \times 4$

(a) What formula do you think Anna got ($t = \ldots$)?
(b) What formula do you think Dipika got ($t = \ldots$)?
(c) Are they in fact the same formula, written differently? Explain your answer.

6. In the famous *Towers of Hanoi* investigation, the number of moves needed to transfer discs from one peg to another is counted. One disc is moved at a time and no disc can ever be placed on top of a disc smaller than itself.

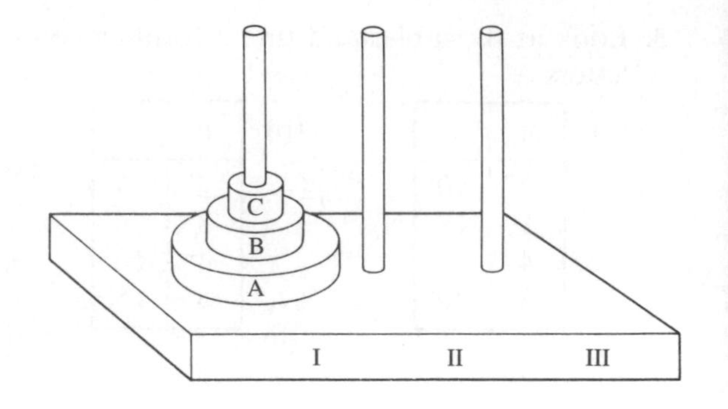

Here are the results for 2, 3, 4 and 5 discs.

number of discs d	minimum number of moves needed m
2	3
3	7
4	15
5	31

Find a formula for m in terms of d. Write it as $m = \dots$

Finding a function

Consider the function $f(x) = x^2 + 2x + 5$

x	$f(x)$	differences 1st	2nd
0	5		
		3	
1	8		2
		5	
2	13		2
		7	
3	20		2
		9	
4	29		

Consider the function $f(x) = 2x^2 - x + 7$

x	$f(x)$	differences 1st	2nd
0	7		
		1	
1	8		4
		5	
2	13		4
		9	
3	22		4
		13	
4	35		

In both the above cases, $f(x)$ is a quadratic function and the second differences are all the same. We can use this property of differences to find unknown functions.

Here are values of x and the corresponding values of an unknown function $f(x)$.

x	$f(x)$	differences	
0	1		
		4	
1	5		6
		10	
2	15		6
		16	
3	31		

Since the second differences are the same, try a quadratic function for $f(x)$.

Let $f(x) = ax^2 + bx + c$

$$
\begin{aligned}
\text{When } x = 0, f(x) &= 1. \\
1 &= 0 + 0 + c \ldots [1] \\
\text{When } x = 1, f(x) &= 5. \\
5 &= a + b + c \ldots [2] \\
\text{When } x = 2, f(x) &= 15. \\
15 &= 4a + 2b + c \ldots [3]
\end{aligned}
$$

Solving the three simultaneous equations [1], [2] and [3], we obtain $a = 3$, $b = 1$, $c = 1$.

$\therefore\ f(x) = 3x^2 + x + 1$

Confirm that when $x = 3$, $f(x) = 31$ which fits the given data.

Notes

If we make a difference table for an unknown function and we find that the *third* differences are constant then we try a cubic function for $f(x)$

i.e. let $f(x) = ax^3 + bx^2 + cx + d$

This topic is beyond the scope of most formal syllabuses, but it is an interesting and useful technique for dealing with investigations when you are trying to find a pattern or rule.

Exercise 4

Find the function $f(x)$.

1.

x	$f(x)$
0	4
1	7
2	12
3	19
4	28

2.

x	$f(x)$
1	8
2	15
3	24
4	35

3.

x	$f(x)$
0	3
1	12
2	25
3	42

4.

x	$f(x)$
1	2
2	10
3	24
4	44

5.

x	$f(x)$
0	0
1	1
2	3
3	6

6.

x	$f(x)$
0	5
1	6
2	13
3	32
4	69

7.

x	$f(x)$
0	0
1	6
2	24
3	60
4	120

8.

x	$f(x)$
0	−1
1	1
2	9
3	29
4	67

2.2 Solving linear equations

Many questions in mathematics are easier to answer if algebra is used. It is often best to let the unknown quantity be x and to try to translate the question into the form of an equation.

We will start by concentrating on solving the equation. Later we will look at questions where we will have to make up the equation first and then solve it.

Basic rules

Rule 1. Treat both sides the same.

Example 1

(a) We can add the same to both sides.

$3x - 1 = 4$

$3x - 1 + 1 = 4 + 1$ [add 1]

(b) We can subtract the same from both sides.

$5x + 7 = 10$

$5x + 7 - 7 = 10 - 7$ [take away 7]

(c) We can multiply both sides by the same factor.

$\frac{x}{4} = 3$

$4\left(\frac{x}{4}\right) = 4 \times 3$ [multiply by 4]

(d) We can divide both sides by the same factor.

$3x = 8$

$\frac{3x}{3} = \frac{8}{3}$ [divide by 3]

Rule 2. If the x term is negative, take it to the other side where it becomes positive.

Rule 3. If there are x terms on both sides, collect them on one side.

Example 2

$$\begin{aligned} 4 - 3x &= 2 \\ 4 &= 2 + 3x \\ 2 &= 3x \\ \frac{2}{3} &= x \end{aligned}$$

Example 3

$$\begin{aligned} 2x - 7 &= 5 - 3x \\ 2x + 3x &= 5 + 7 \\ 5x &= 12 \\ x &= \frac{12}{5} = 2\tfrac{2}{5} \end{aligned}$$

Exercise 5

Solve the following equations:

1. $2x - 5 = 11$
2. $3x - 7 = 20$
3. $2x + 6 = 20$
4. $5x + 10 = 60$
5. $8 = 7 + 3x$
6. $12 = 2x - 8$
7. $-7 = 2x - 10$
8. $3x - 7 = -10$
9. $12 = 15 + 2x$
10. $5 + 6x = 7$
11. $100x - 1 = 98$
12. $7 = 7 + 7x$
13. $\frac{x}{100} + 10 = 20$
14. $1000x - 5 = -6$
15. $-4 = -7 + 3x$
16. $2x + 4 = x - 3$
17. $x - 3 = 3x + 7$
18. $5x - 4 = 3 - x$
19. $4 - 3x = 1$
20. $5 - 4x = -3$
21. $7 = 2 - x$
22. $3 - 2x = x + 12$
23. $6 + 2a = 3$
24. $a - 3 = 3a - 7$
25. $2y - 1 = 4 - 3y$
26. $7 - 2x = 2x - 7$
27. $7 - 3x = 5 - 2x$
28. $8 - 2y = 5 - 5y$
29. $x - 16 = 16 - 2x$
30. $x + 2 = 3{\cdot}1$
31. $-x - 4 = -3$
32. $-3 - x = -5$

Rule 4. If there is a fraction in the x term, multiply out to simplify the equation.

Example 4a

$$\frac{2x}{3} = 10$$
$$2x = 30$$
$$x = \frac{30}{2} = 15$$

Example 4b

$$3 = \frac{x}{4} - 4$$
$$7 = \frac{x}{4}$$
$$28 = x$$

Exercise 6

1. $\frac{x}{5} = 7$
2. $\frac{x}{10} = 13$
3. $7 = \frac{x}{2}$
4. $\frac{x}{2} = \frac{1}{3}$
5. $\frac{3x}{2} = 5$
6. $\frac{4x}{5} = -2$
7. $7 = \frac{7x}{3}$
8. $\frac{3}{4} = \frac{2x}{3}$
9. $\frac{5x}{6} = \frac{1}{4}$
10. $-\frac{3}{4} = \frac{3x}{5}$
11. $\frac{x}{2} + 7 = 12$
12. $\frac{x}{3} - 7 = 2$
13. $\frac{x}{5} - 6 = -2$
14. $4 = \frac{x}{2} - 5$
15. $10 = 3 + \frac{x}{4}$
16. $\frac{a}{5} - 1 = -4$
17. $-\frac{x}{2} + 1 = -\frac{1}{4}$
18. $-\frac{3}{5} + \frac{x}{10} = -\frac{1}{5} - \frac{x}{5}$

Rule 5. When an equation has brackets, multiply out the brackets first.

Example 5

$$\begin{aligned} x - 2(x-1) &= 1 - 4(x+1) \\ x - 2x + 2 &= 1 - 4x - 4 \\ x - 2x + 4x &= 1 - 4 - 2 \\ 3x &= -5 \\ x &= -\frac{5}{3} \end{aligned}$$

Example 6

$$\begin{aligned} \frac{5}{x} &= 2 \\ 5 &= 2x \\ \frac{5}{2} &= x \end{aligned}$$

Exercise 7

Solve the following equations.

1. $x + 3(x+1) = 2x$
2. $1 + 3(x-1) = 4$
3. $2x - 2(x+1) = 5x$
4. $2(3x-1) = 3(x-1)$
5. $4(x-1) = 2(3-x)$
6. $4(x-1) - 2 = 3x$
7. $4(1-2x) = 3(2-x)$
8. $3 - 2(2x+1) = x + 17$
9. $4x = x - (x-2)$
10. $7x = 3x - (x+20)$
11. $5x - 3(x-1) = 39$
12. $3x + 2(x-5) = 15$
13. $7 - (x+1) = 9 - (2x-1)$
14. $10x - (2x+3) = 21$
15. $3(2x+1) + 2(x-1) = 23$
16. $5(1-2x) - 3(4+4x) = 0$
17. $7x - (2-x) = 0$
18. $3(x+1) = 4 - (x-3)$
19. $3y + 7 + 3(y-1) = 2(2y+6)$
20. $4(y-1) + 3(y+2) = 5(y-4)$

Exercise 8

Solve the following equations.

1. $\frac{7}{x} = 21$
2. $30 = \frac{6}{x}$
3. $\frac{5}{x} = 3$
4. $\frac{9}{x} = -3$
5. $11 = \frac{5}{x}$
6. $-2 = \frac{4}{x}$
7. $\frac{x+1}{3} = \frac{x-1}{4}$
8. $\frac{x+3}{2} = \frac{x-4}{5}$
9. $\frac{2x-1}{3} = \frac{x}{2}$
10. $\frac{3x+1}{5} = \frac{2x}{3}$
11. $\frac{5}{x-1} = \frac{10}{x}$
12. $\frac{12}{2x-3} = 4$
13. $2 = \frac{18}{x+4}$
14. $\frac{5}{x+5} = \frac{15}{x+7}$
15. $\frac{4}{x} + 2 = 3$
16. $\frac{6}{x} - 3 = 7$
17. $\frac{9}{x} - 7 = 1$
18. $-2 = 1 + \frac{3}{x}$
19. $4 - \frac{4}{x} = 0$
20. $5 - \frac{6}{x} = -1$
21. $\frac{x}{3} + \frac{x}{4} = 1$

Solving problems using linear equations

So far we have concentrated on solving given equations. Making up our own equations helps to solve problems which are difficult to solve. There are four steps.

(a) Let the unknown quantity be x (or any other suitable letter) and state the units where appropriate.
(b) Write the problem in the form of an equation.
(c) Solve the equation and give the answer in words.
(d) Check your solution using the problem and *not* your equation.

Example 7

Find three consecutive even numbers which add up to 792.

(a) Let the smallest number be x.
Then the other numbers are $(x+2)$ and $(x+4)$ because they are consecutive *even* numbers.

(b) Form an equation.
$$x+(x+2)+(x+4) = 792$$

(c) Solve:
$$3x+6 = 792$$
$$3x = 786$$
$$x = 262$$
The three numbers are 262, 264 and 266.

(d) Check. $262+264+266 = 792$ √

Exercise 9

Solve each problem by forming an equation. The first questions are easy but should still be solved using an equation, in order to practise the method.

1. The length of a rectangle is twice the width. If the perimeter is 20 cm, find the width.

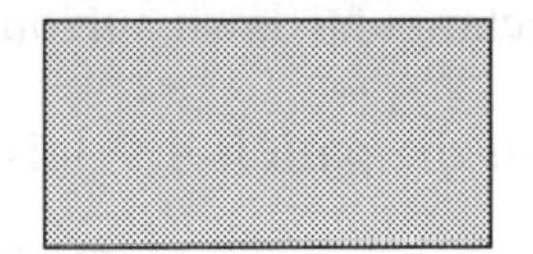

2. The width of a rectangle is one third of the length. If the perimeter is 96 cm, find the width.

3. The sum of three consecutive numbers is 276. Find the numbers. Let the first number be x.

4. The sum of four consecutive numbers is 90. Find the numbers.

5. The sum of three consecutive odd numbers is 177. Find the numbers.

6. Find three consecutive even numbers which add up to 1524.

7. When a number is doubled and then added to 13, the result is 38. Find the number.

8. If AB is a straight line, find x.

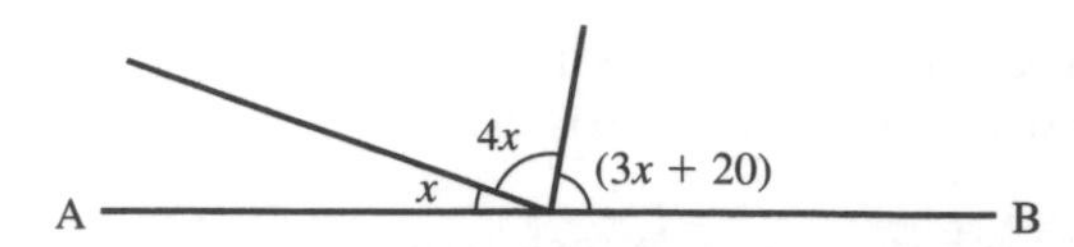

9. The difference between two numbers is 9. Find the numbers, if their sum is 46.

10. The three angles in a triangle are in the ratio 1 : 3 : 5. Find the angles.

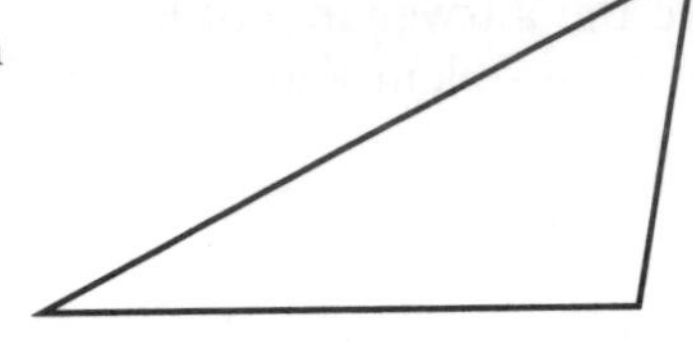

11. The sum of three numbers is 28. The second number is three times the first and the third is 7 less than the second. What are the numbers?

12. If the perimeter of the triangle is 22 cm, find the length of the shortest side.

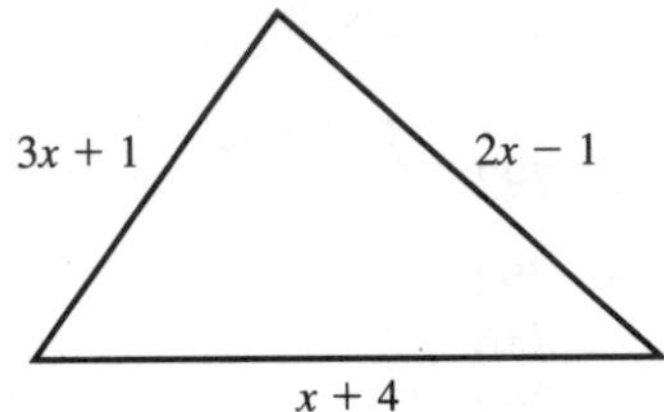

13. David weighs 5 kg less than John, who in turn is 8 kg lighter than Paul. If their total weight is 197 kg, how heavy is each person?

14. If the perimeter of the rectangle is 34 cm, find x.

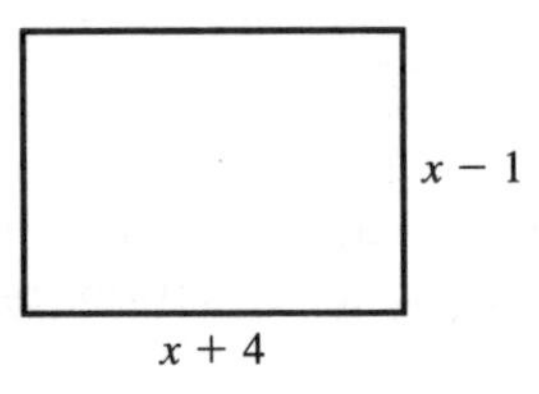

15. The diagram shows a rectangular lawn surrounded by a footpath x m wide.

(a) Show that the area of the path is $4x^2 + 14x$.

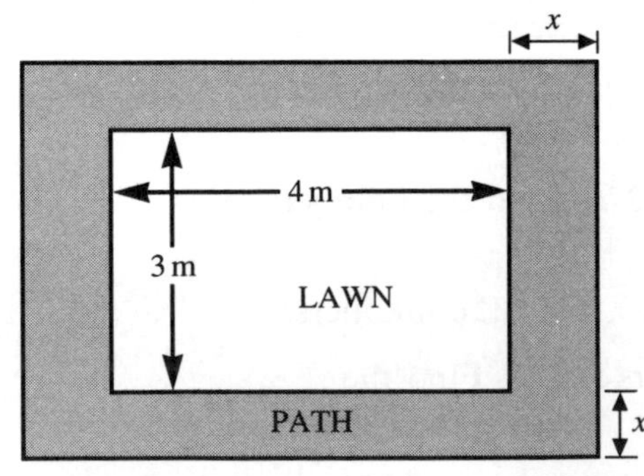

(b) Find an expression, in terms of x for the distance around the outside edge of the path.

(c) Find the value of x when this perimeter is 20 m.

Exercise 10

1. Shirin has two long planks of wood, both the same length. She needs to cut them into nine shelves each of the same length. She cuts six shelves from the first plank and has 15 cm of wood left over.

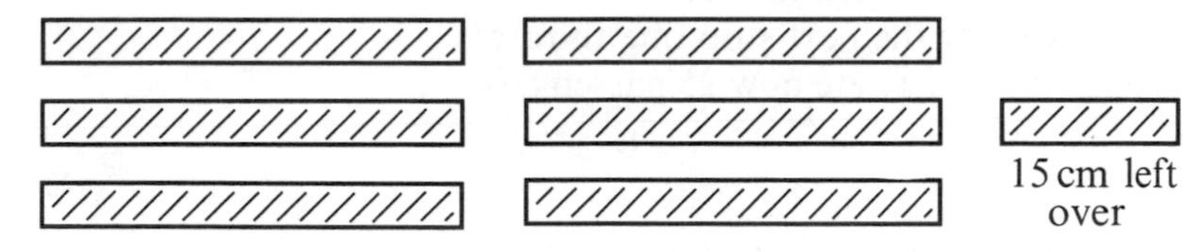

She then cuts three shelves from the second plank and has 135 cm left over.

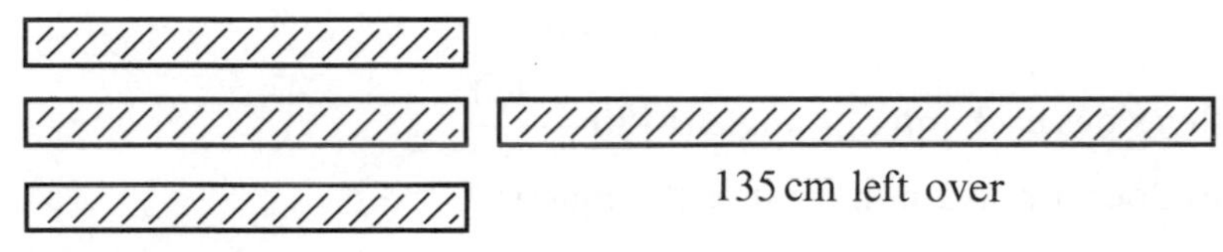

How long was each shelf?

2. Every year a man is paid £500 more than the previous year. If he receives £17 800 over four years, what was he paid in the first year?

3. A man buys x cans of beer at 30p each and $(x + 4)$ cans of lager at 35p each. The total cost was £3·35. Find x.

4. A man is 32 years older than his son. Ten years ago he was three times as old as his son was then. Find the present age of each.

5. A man runs to a telephone and back in 15 minutes. His speed on the way to the telephone is 5 m/s and his speed on the way back is 4 m/s. Find the distance to the telephone.

6. A car completes a journey in 10 minutes. For the first half of the distance the speed was 60 km/h and for the second half the speed was 40 km/h. How far is the journey?

7. A lemming runs from a point A to a cliff at 4 m/s, jumps over the edge at B and falls to C at an average speed of 25 m/s. If the total distance from A to C is 500 m and the time taken for the journey is 41 seconds, find the height BC of the cliff.

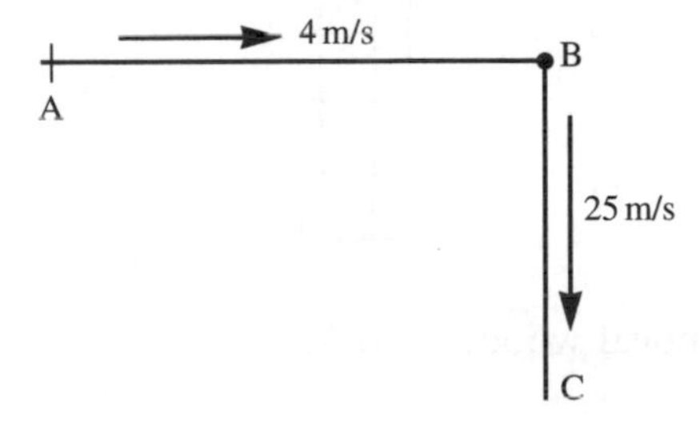

8. A bus is travelling with 48 passengers. When it arrives at a stop, x passengers get off and 3 get on. At the next stop half the passengers get off and 7 get on. There are now 22 passengers. Find x.

9. A bus is travelling with 52 passengers. When it arrives at a stop, y passengers get off and 4 get on. At the next stop one third of the passengers get off and 3 get on. There are now 25 passengers. Find y.

10. Mr Lee left his fortune to his 3 sons, 4 daughters and his wife. Each son received twice as much as each daughter and his wife received £6000, which was a quarter of the money. How much did each son receive?

11. In a regular polygon with n sides each interior angle is $180 - \frac{360}{n}$ degrees. How many sides does a polygon have if each angle is 156°?

12. A sparrow flies to see a friend at a speed of 4 km/h. His friend is out, so the sparrow immediately returns home at a speed of 5 km/h. The complete journey took 54 minutes. How far away does his friend live?

13. Consider the equation $an^2 = 182$ where a is any number between 2 and 5 and n is a positive integer. What are the possible values of n?

14. Consider the equation $\frac{k}{x} = 12$ where k is any number between 20 and 65 and x is a positive integer. What are the possible values of x?

15.†

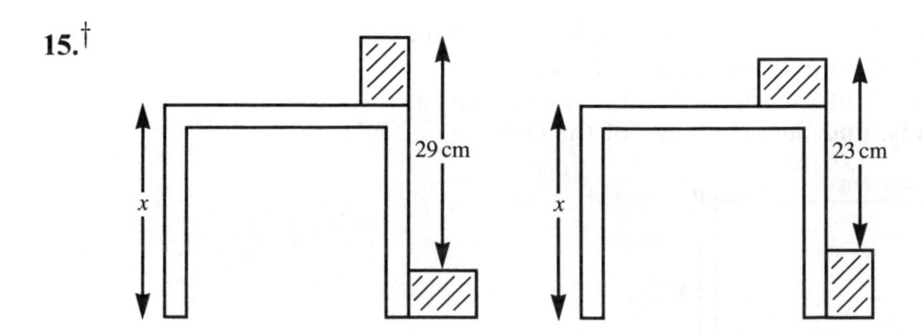

The diagrams show a table with two identical wooden blocks. Calculate the height of the table, x.

2.3 Trial and improvement

Some problems cannot be solved using linear equations. In such cases, the method of 'trial and improvement' is often a help.

Exercise 11

1. Think of a rectangle having an area of $72\,\text{cm}^2$ whose base is twice its height.

   $72\,\text{cm}^2$ height

   Write down the length of the base.

2. Find a rectangle of area $75\,\text{cm}^2$ so that its base is three times its height.

3. In each of the rectangles below, the base is twice the height. The area is shown inside the rectangle. Find the base and the height.

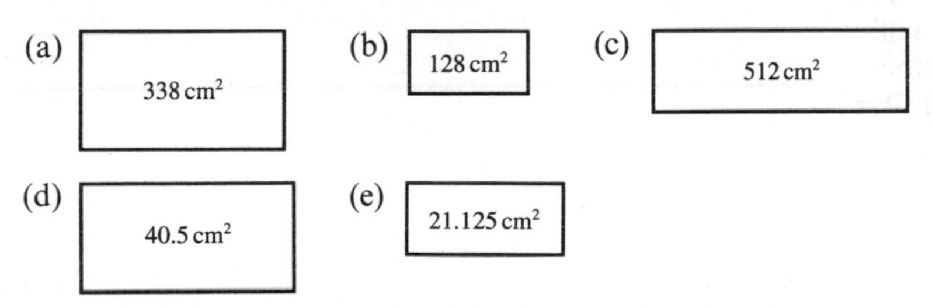

4. In this rectangle, the base is 1 cm more than the height and the area is $90\,\text{cm}^2$.

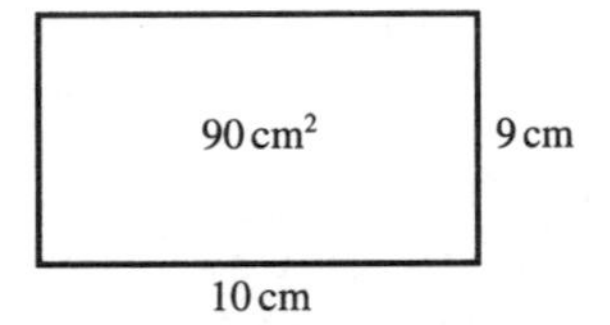

   In each of the rectangles below, the base is 1 cm more than the height. Find each base and height.

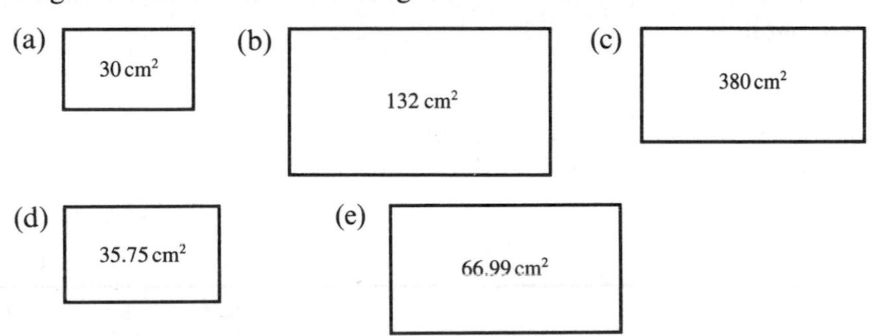

5. A problem! Once again the base of the rectangle is 1 cm more than the height.
   Try to find the base and the height of the rectangle.

   $75\,\text{cm}^2$

Inexact answers

In some questions it is not possible ever to find an answer which is precisely correct. However, we can find answers which are nearer and nearer to the exact one, perhaps to the nearest 0·1 cm or even to the nearest 0·01 cm.

In this rectangle, the base is 1 cm more than the height h cm. The area is 80 cm^2.
Find the height h.

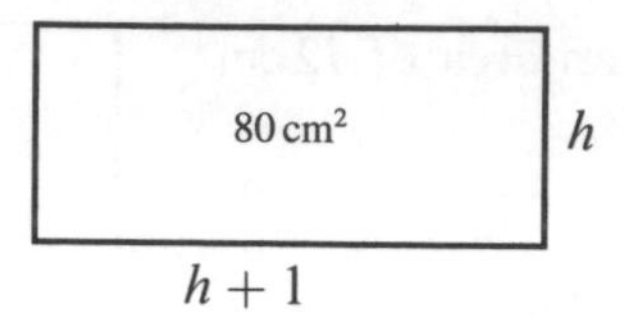

Try a special case and be systematic.

(a) Suppose $h = 8$.

$8(8+1) = 72 < 80$

This is too small.
Draw a line. Write 'too small' on the left-hand end and 'too big' on the right-hand end. Put 8 on the 'too small' end.

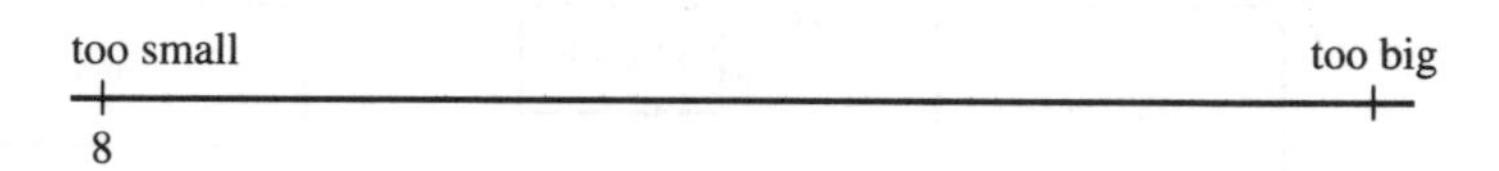

(b) Try $h = 9$.

$9(9+1) = 90 > 80$

This is too big, so put the result on the right-hand end.

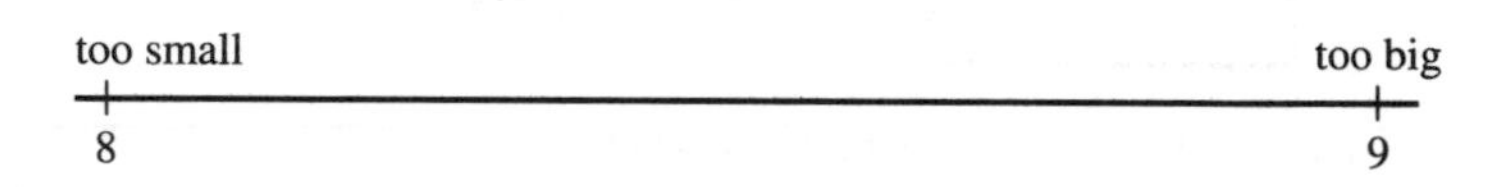

(c) Try an intermediate number, say $h = 8{\cdot}5$.

$8{\cdot}5(8{\cdot}5+1) = 80{\cdot}75 > 80$

This is slightly too big. Put the result on the right. Do *not* put it in the middle.

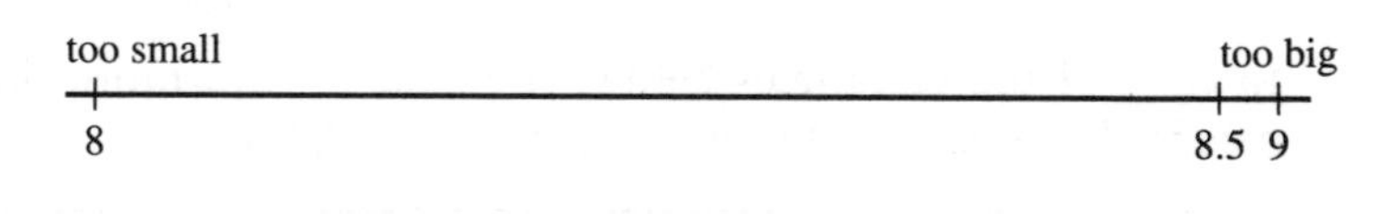

(d) Carry on adjusting the number you try, placing it suitably on the line. The number in brackets is the result of the calculation each time.

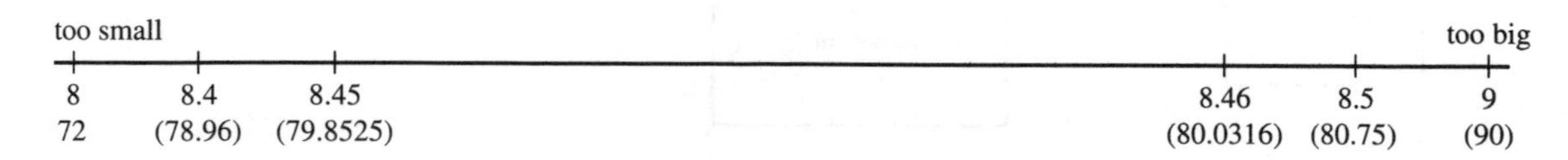

Clearly, the value for h is 'between 8·45 cm and 8·46 cm'.
For many problems, this will be accurate enough. Further trials will give further improvement in accuracy, e.g. h is 'between 8·458 cm and 8·459 cm'.

Exercise 12

1. In these two rectangles, the base is 1 cm more than the height.

(a)

100 cm^2

h

$h + 1$

(b)

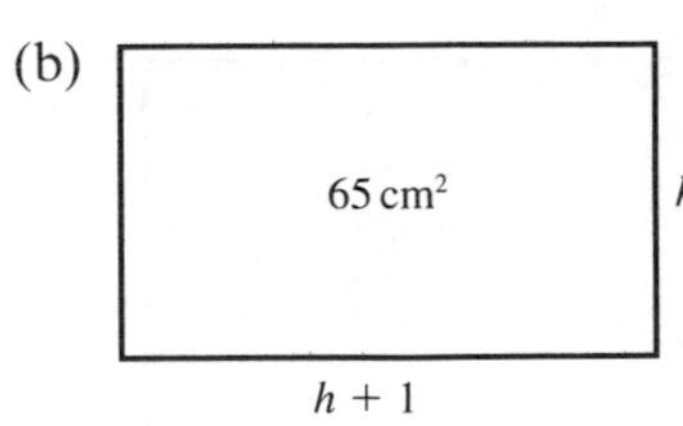

Find the height of each one, writing the answer as 'h is between _ cm and _ cm'. The two numbers should be 0·01 cm apart.

2. Find solutions to the following equations, giving the answer in the form 'x is between _ and _'. The two numbers should be 0·01 apart.

(a) $x(x - 3) = 11$ (b) $3x(x - 2) = 21$

(c) $x^3 = 300$ (d) $x^2(x + 1) = 50$

(e) $x + \dfrac{1}{x} = 6$ (f) $x^5 = 313$

(g) $x^x = 100$. Here you need a calculator with a $\boxed{x^y}$ button.

3. An engineer wants to make a solid metal cube of volume 526 cm^3.
Call the edge of the cube x and write down an equation.
Find x giving your answer in the form given in Question 2.

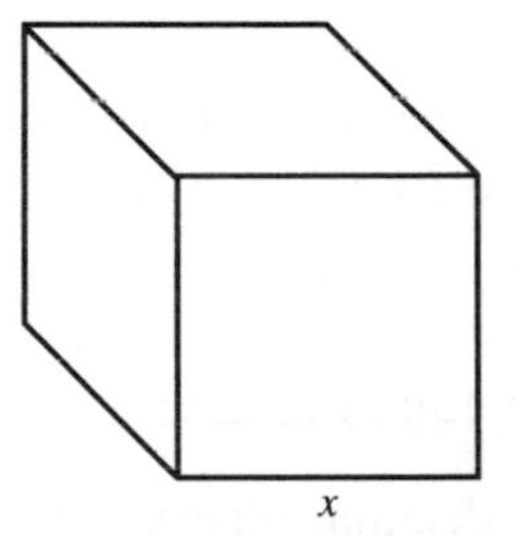

4. In this rectangle, the base is 2 cm more than the height.
Find h if the diagonal is 15 cm.

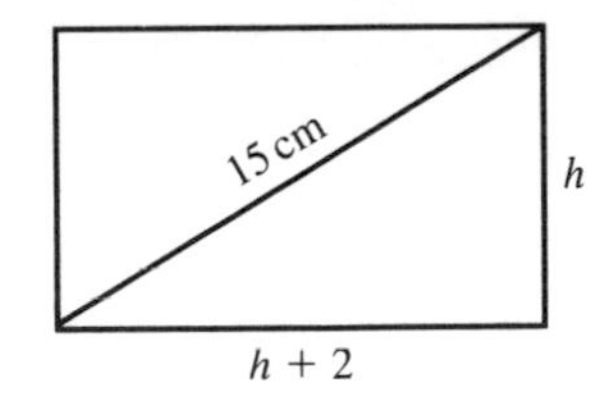

5. A designer for a supermarket chain wants to make a cardboard box of depth 6 cm. He has to make the length of the box 10 cm more than the width.

(a) What is the length of the box in terms of x?
The box is designed so that its volume is to be 9000 cm^3.

(b) Form an equation involving x.

(c) Solve the equation and hence give the dimensions of the box to the nearest cm.

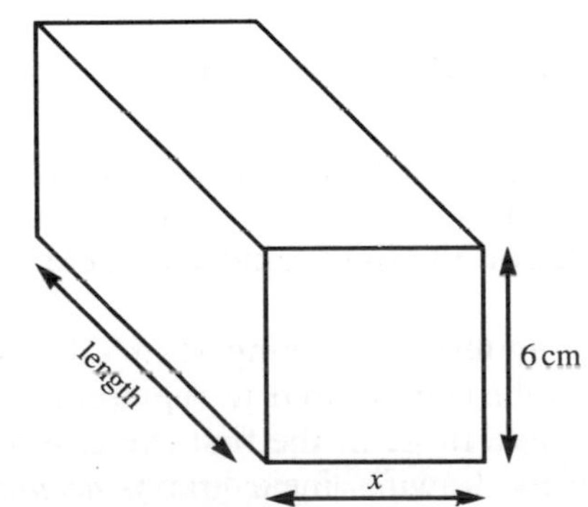

6. The diagram represents a rectangular piece of paper ABCD which has been folded along EF so that C has moved to G.

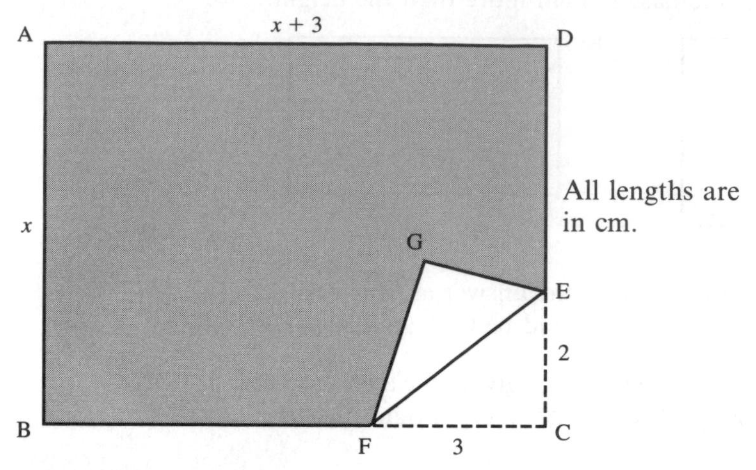

All lengths are in cm.

(a) Calculate the area of $\triangle$ ECF.
(b) Find an expression for the shaded area ABFGED in terms of x.

Given that the shaded area is $20\,\text{cm}^2$, show that $x(x+3) = 26$

Solve this equation, giving your answer correct to one decimal place.

7. In the rectangle PQRS, PQ $= x$ cm and QR $= 1$ cm. The line LM is drawn so that PLMS is a square.

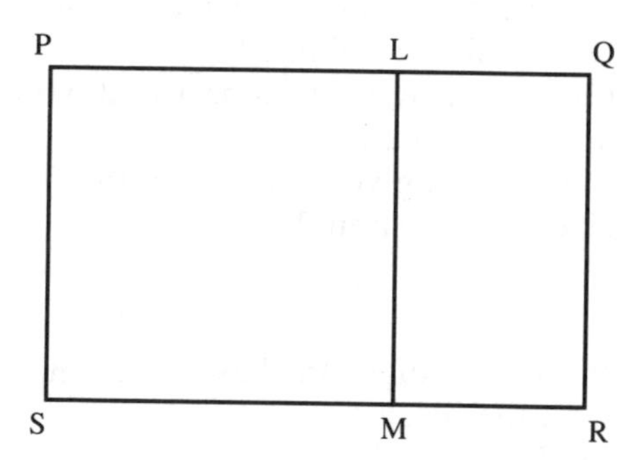

(a) Write down, in terms of x, the length LQ.

(b) If $\dfrac{PQ}{QR} = \dfrac{QR}{LQ}$, obtain an equation in x.

Hence find x correct to two decimal places.

2.4 Linear graphs

A *linear* graph is a graph which is a straight line. The graphs of $y = 3x - 1$, $y = 7 - 2x$, $y = \frac{1}{2}x + 100$ are all linear.
Notice that on the right hand side there are only x terms and number terms.
There are *no* terms involving x^2, x^3, $1/x$, etc.
Linear graphs can be used to represent a wide variety of practical situations like those in the first exercise below.
We begin by drawing linear graphs *accurately*.

Exercise 13

Draw the following graphs, using a scale of 2 cm to 1 unit on the x-axis and 1 cm to 1 unit on the y-axis.

1. $y = 2x + 1$ for $-3 \leqslant x \leqslant 3$

2. $y = 3x - 4$ for $-3 \leqslant x \leqslant 3$

3. $y = 8 - x$ for $-2 \leqslant x \leqslant 4$

4. $y = 10 - 2x$ for $-2 \leqslant x \leqslant 4$

5. $y = \dfrac{x+5}{2}$ for $-3 \leqslant x \leqslant 3$

6. $y = 3(x - 2)$ for $-3 \leqslant x \leqslant 3$

7. $y = \frac{1}{2}x + 4$ for $-3 \leqslant x \leqslant 3$

8. $v = 2t - 3$ for $-2 \leqslant t \leqslant 4$

9. $z = 12 - 3t$ for $-2 \leqslant t \leqslant 4$

10. Kendal Motors hires out vans.

Copy and complete the table where x is the number of miles travelled and C is the total cost in pounds.

x	0	50	100	150	200	250	300
C	35			65			95

Draw a graph of C against x, using scales of 2 cm for 50 miles on the x-axis and 1 cm for £10 on the C-axis.

(a) Use the graph to find the number of miles travelled when the total cost was £71.

(b) What is the formula connecting C and x?

11. Jeff sets up his own business as a plumber.

Copy and complete the table where C stands for his total charge and h stands for the number of hours he works.

h	0	1	2	3
C		33		

Draw a graph with h across the page and C up the page. Use scales of 2 cm to 1 hour for h and 2 cm to £10 for C.

(a) Use your graph to find how long he worked if his charge was £55.50.

(b) What is the equation connecting C and h?

12. The equation connecting the annual mileage, M miles, of a certain car and the annual running cost, £C is $C = \dfrac{M}{20} + 200$.

Draw the graph for $0 \leqslant M \leqslant 10\,000$ using scales of 1 cm for 1000 miles for M and 2 cm for £100 for C.

(a) From the graph find
  (i) the cost when the annual mileage is 7200 miles,
  (ii) the annual mileage corresponding to a cost of £320.

(b) There is a '200' in the formula for C. Explain where this figure might come from.

13. Some drivers try to estimate their annual cost of repairs £c in relation to their average speed of driving s km/h using the equation $c = 6s + 50$.

Draw the graph for $0 \leqslant s \leqslant 160$. From the graph find

(a) the estimated repair bill for a man who drives at an average speed of 23 km/h.

(b) the average speed at which a motorist drives if his annual repair bill is £1000.

2.5 Simultaneous linear equations

Graphical solution

Louise and Philip are two children; Louise is 5 years older than Philip. The sum of their ages is 12 years. How old is each child?

Let Louise be x years old and Philip be y years old.
The sum of their ages is 12,

so $x + y = 12$

The difference of their ages is 5,

so $x - y = 5$

Because both equations relate to the same information, both can be plotted to the same axes.

$x + y = 12$ goes through (0,12), (2,10), (6,6), (12,0).
$x - y = 5$ goes through (5,0), (7,2), (10,5).

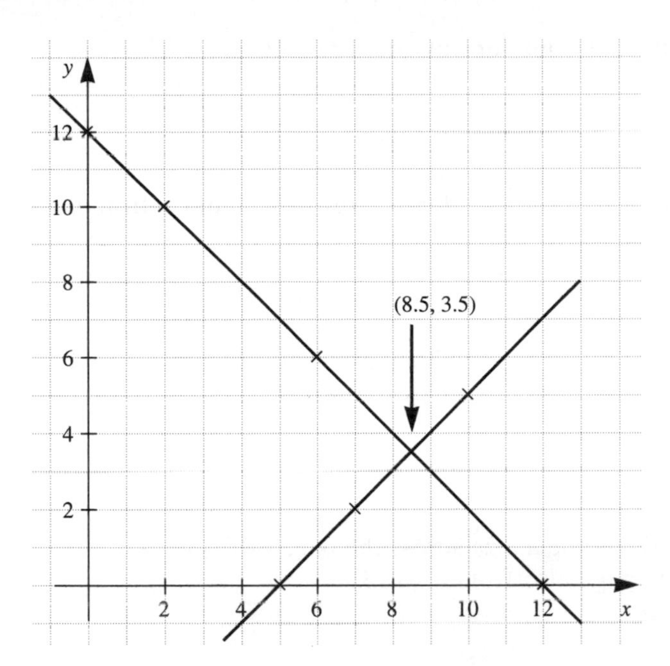

The values of x and y are found from the point where the two lines intersect. The point (8·5, 3·5) lies on both lines.

The solution is $x = 8{\cdot}5$, $y = 3{\cdot}5$

So Louise is $8\frac{1}{2}$ years old and Philip is $3\frac{1}{2}$ years old.

Note that only three points are required to define a straight line.

Exercise 14

1. Use the graphs below to solve the equations.

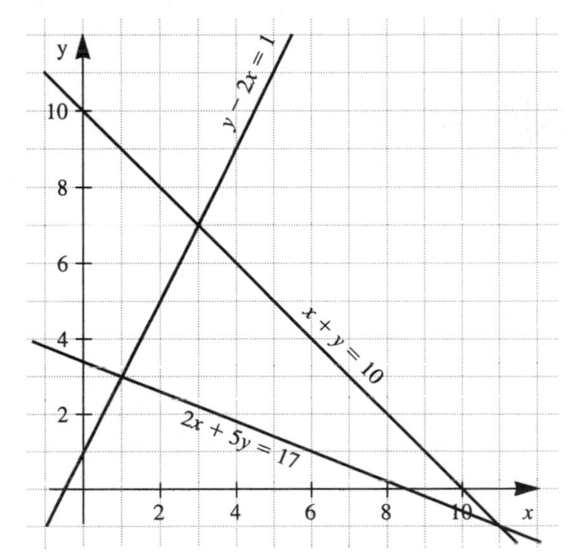

Solve (a) $x + y = 10$, $y - 2x = 1$ (b) $2x + 5y = 17$, $y - 2x = 1$ (c) $x + y = 10$, $2x + 5y = 17$

In Questions **2** to **6**, solve the simultaneous equations by first drawing graphs.

2. $x + y = 6$
$2x + y = 8$
Draw axes with x and y from 0 to 8.

3. $x + 2y = 8$
$3x + y = 9$
Draw axes with x and y from 0 to 9.

4. $x + 3y = 6$
$x - y = 2$
Draw axes with x from 0 to 8 and y from -2 to 4.

5. $5x + y = 10$
$x - y = -4$
Draw axes with x from -4 to 4 and y from 0 to 10.

6. $a + 2b = 11$
$2a + b = 13$
Here, the unknowns are a and b. Draw the a axis across the page from 0 to 13 and the b axis up the page also from 0 to 13.

7. There are four lines drawn here.

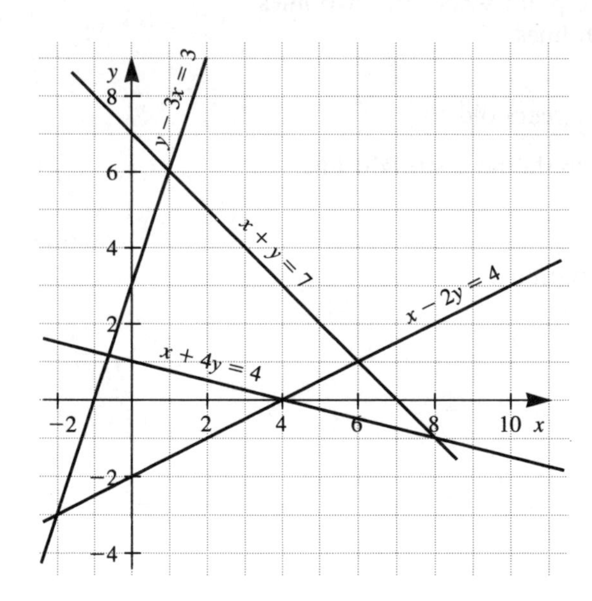

Write down the solutions to the following

(a) $x - 2y = 4$
$x + 4y = 4$

(b) $x + y = 7$
$y - 3x = 3$

(c) $y - 3x = 3$
$x - 2y = 4$

(d) $x + 4y = 4$
$x + y = 7$

(e) $x + 4y = 4$
$y - 3x = 3$

[Here, x and y should be given correct to 1 d.p.]

Algebraic solution

Simultaneous equations can be solved without drawing graphs. There are two methods: substitution and elimination. Choose the one which seems suitable.

(a) Substitution method

This method is used when one equation contains a single x or y as in equation [2] of Example 1 below.

Example 1

$3x - 2y = 0$ ... [1]
$2x + y = 7$ ... [2]

(a) Label the equations so that the working is made clear.
(b) In *this* case, write y in terms of x from equation [2].
(c) Substitute this expression for y in equation [1] and solve to find x.
(d) Find y from equation [2] using this value of x.

$$2x + y = 7$$
$$y = 7 - 2x$$

Substituting in [1]

$$3x - 2(7 - 2x) = 0$$
$$3x - 14 + 4x = 0$$
$$7x = 14$$
$$x = 2$$

Substituting in [2]

$$2 \times 2 + y = 7$$
$$y = 3$$

The solutions are $x = 2$, $y = 3$.

Exercise 15

Use the substitution method to solve the following:

1. $2x + y = 5$, $x + 3y = 5$
2. $x + 2y = 8$, $2x + 3y = 14$
3. $3x + y = 10$, $x - y = 2$
4. $2x + y = -3$, $x - y = 2$
5. $4x + y = 14$, $x + 5y = 13$
6. $x + 2y = 1$, $2x + 3y = 4$
7. $2x + y = 5$, $3x - 2y = 4$
8. $2x + y = 13$, $5x - 4y = 13$
9. $7x + 2y = 19$, $x - y = 4$
10. $b - a = -5$, $a + b = -1$
11. $a + 4b = 6$, $8b - a = -3$
12. $a + b = 4$, $2a + b = 5$
13. $3m = 2n - 6\frac{1}{2}$, $4m + n = 6$
14. $2w + 3x - 13 = 0$, $x + 5w - 13 = 0$
15. $x + 2(y - 6) = 0$, $3x + 4y = 30$
16. $2x = 4 + z$, $6x - 5z = 18$
17. $3m - n = 5$, $2m + 5n = 7$
18. $5c - d - 11 = 0$, $4d + 3c = -5$

(b) Elimination method

Use this method when the first method is unsuitable (some prefer to use it for every question).

Example 2

$$\begin{aligned} 2x + 3y &= 5 && \dots [1] \\ 5x - 2y &= -16 && \dots [2] \end{aligned}$$

[1] × 5	$10x + 15y = 25$	... [3]
[2] × 2	$10x - 4y = -32$	... [4]
[3] − [4]	$15y - (-4y) = 25 - (-32)$	
	$19y = 57$	
	$y = 3$	
Substitute in [1]	$2x + 3 \times 3 = 5$	
	$2x = 5 - 9 = -4$	
	$x = -2$	

The solutions are $x = -2$, $y = 3$.

Exercise 16

Use the elimination method to solve the following:

1. $2x + 5y = 24$, $4x + 3y = 20$

2. $5x + 2y = 13$, $2x + 6y = 26$

3. $3x + y = 11$, $9x + 2y = 28$

4. $x + 2y = 17$, $8x + 3y = 45$

5. $3x + 2y = 19$, $x + 8y = 21$

6. $2a + 3b = 9$, $4a + b = 13$

7. $2x + 7y = 17$, $5x + 3y = -1$

8. $5x + 3y = 23$, $2x + 4y = 12$

9. $3x + 2y = 11$, $2x - y = -3$

10. $3x + 2y = 7$, $2x - 3y = -4$

11. $x - 2y = -4$, $3x + y = 9$

12. $5x - 7y = 27$, $3x - 4y = 16$

13. $x + 3y - 7 = 0$, $2y - x - 3 = 0$

14. $3a - b = 9$, $2a + 2b = 14$

15. $2x - y = 5$, $\dfrac{x}{4} + \dfrac{y}{3} = 2$

16. $3x - y = 17$, $\dfrac{x}{5} + \dfrac{y}{2} = 0$

17. $4x - 0{\cdot}5y = 12{\cdot}5$, $3x + 0{\cdot}8y = 8{\cdot}2$

18. $0{\cdot}4x + 3y = 2{\cdot}6$, $x - 2y = 4{\cdot}6$

Solving problems using simultaneous equations

As with linear equations, solving problems involves four steps.

(a) Let the two unknown quantities be x and y.
(b) Write the problem in the form of two equations.
(c) Solve the equations and give the answers in words.
(d) Check your solution using the problem and not your equations.

Example 3

A motorist buys 24 litres of petrol and 5 litres of oil for £10·70, while another motorist buys 18 litres of petrol and 10 litres of oil for £12·40.
Find the cost of 1 litre of petrol and 1 litre of oil at this garage.

(a) Let the cost of 1 litre of petrol be x pence and the cost of 1 litre of oil be y pence.

(b) $24x + 5y = 1070$ ... [1]
$18x + 10y = 1240$ ... [2]

(c) Solve the equations: $x = 30$, $y = 70$

1 litre of petrol costs 30 pence.
1 litre of oil costs 70 pence.

(d) Check: $24 \times 30 + 5 \times 70 = 1070\text{p} = £10{\cdot}70$
$18 \times 30 + 10 \times 70 = 1240\text{p} = £12{\cdot}40$ ✓

Exercise 17

Solve each problem by forming a pair of simultaneous equations.

1. Find two numbers with a sum of 15 and a difference of 4.

2. Twice one number added to three times another gives 21. Find the numbers, if the difference between them is 3.

3. The average of two numbers is 7, and three times the difference between them is 18. Find the numbers.

4. Here is a puzzle from a newspaper. The ? and * stand for numbers which are to be found. The totals for the rows and columns are given.

Write down two equations involving ? and * and solve them to find the values of ? and *

?	*	?	*	36
?	*	*	?	36
*	?	*	*	33
?	*	?	*	36
39	33	36	33	

5. The line, with equation $y + ax = c$, passes through the points (1, 5) and (3, 1). Find a and c.
Hint: For the point (1, 5) put $x = 1$ and $y = 5$ into $y + ax = c$, etc.

6. The line $y = mx + c$ passes through (2, 5) and (4, 13). Find m and c.

7. A stone is thrown into the air and its height, h metres above the ground, is given by the equation

$h = at - bt^2$.

From an experiment we know that $h = 40$ when $t = 2$ and that $h = 45$ when $t = 3$.
Show that, $a - 2b = 20$
and $a - 3b = 15$.
Solve these equations to find a and b.

8. A television addict can buy either two televisions and three video-recorders for £1750 or four televisions and one video-recorder for £1250. Find the cost of one of each.

9. A pigeon can lay either white or brown eggs. Three white eggs and two brown eggs weigh 13 ounces, while five white eggs and four brown eggs weigh 24 ounces. Find the weight of a brown egg and of a white egg.

10. A tortoise makes a journey in two parts; it can either walk at 0·4 m/s or crawl at 0·3 m/s.

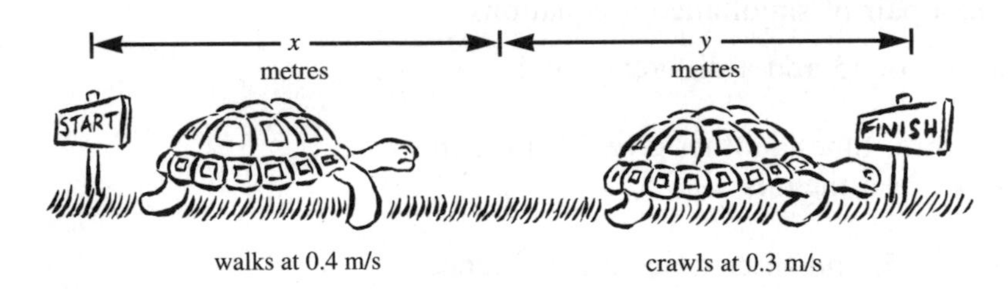

If the tortoise walks the first part and crawls the second, the journey takes 110 seconds.
If it crawls the first part and walks the second, the journey takes 100 seconds.
Let x metres be the length of the first part and y metres be the length of the second part.
Write down two simultaneous equations and solve them to find the lengths of the two parts of the journey.

11. A cyclist completes a journey of 500 m in 22 seconds, part of the way at 10 m/s and the remainder at 50 m/s. How far does she travel at each speed?

12. A bag contains forty coins, all of them either 2p or 5p coins. If the value of the money in the bag is £1·55, find the number of each kind.

13. A slot machine takes only 10p and 50p coins and contains a total of twenty-one coins altogether. If the value of the coins is £4·90, find the number of coins of each value.

14. Thirty tickets were sold for a concert, some at 60p and the rest at £1. If the total raised was £22, how many had the cheaper tickets?

15. A kipper can swim at 14 m/s with the current and at 6 m/s against it. Find the speed of the current and the speed of the kipper in still water.

16. If the numerator and denominator of a fraction are both decreased by one the fraction becomes $\frac{2}{3}$. If the numerator and denominator are both increased by one the fraction becomes $\frac{3}{4}$. Find the original fraction.

17. In three years time a pet mouse will be as old as his owner was four years ago. Their present ages total 13 years. Find the age of each now.

18. The diagram shows segments of three straight lines.

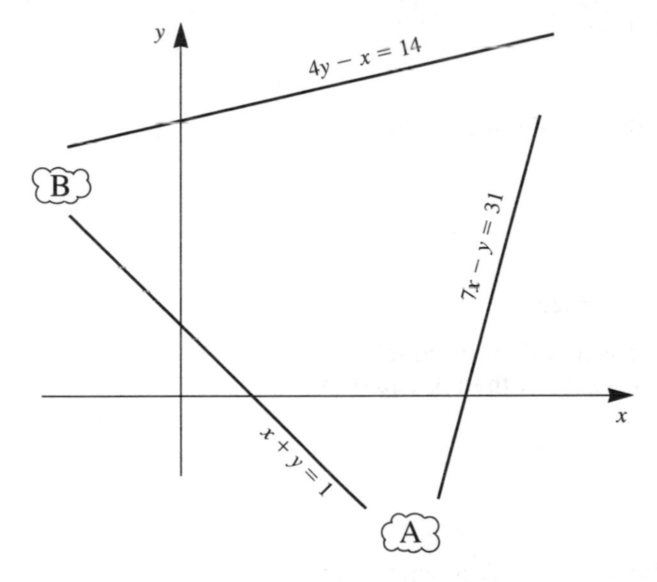

Find the coordinates of the points A and B.

19. A spider can walk at a certain speed and run at another speed. If she walks for 10 seconds and runs for 9 seconds she travels 85 m. If she walks for 30 seconds and runs for 2 seconds she travels 130 m. Find her speeds of walking and running.

20. A wallet containing £40 has three times as many £1 coins as £5 notes. Find the number of each kind.

21. The curve $y = ax^2 + bx + c$ passes through the points (1, 8), (0, 5) and (3, 20). Find the values of a, b and c and hence the equation of the curve.

22. The curve $y = ax^2 + bx + c$ passes through (1, 8), (−1, 2) and (2, 14). Find the equation of the curve.

2.6 Inequalities and regions

Example 1

A designer is asked to make a container for a new shampoo in the form of a cuboid with a square base. There are two constraints.
(a) Stability on the supermarket shelf. The stores insist that the area of the base of the box must exceed $20\,\text{cm}^2$.
(b) Artistry. The director of the project insists that the side of the square base must be less than 6 cm so that the picture on the box looks elegant.

Let the base of the box measure x cm by x cm.

For stability: $x^2 > 20$
$x > \sqrt{20}$
$x > 4{\cdot}5$ cm (to 1 d.p.)

For artistic reasons: $x < 6$

So the designer can make the box with the side of the square base any size between 4·5 cm and 6 cm.

$$4{\cdot}5 < x < 6$$

Here is the meaning of inequality symbols used.

$x < 4$	x is less than 4	$z \leqslant 10$	z is less than or equal to 10
$y > 7$	y is greater than 7	$t \geqslant -3$	t is greater than or equal to -3

Solving inequalities

Where a statement contains two symbols, then look at each part separately.
So, if n is an integer and $3 < n \leqslant 7$,
then n has to be greater than 3 but at the same time it has to be less than or equal to 7.

n could be 4, 5, 6 or 7 only.

Inequalities are solved in the same way as equations.
However, when the inequality is multiplied or divided by a negative number, the inequality is reversed.

e.g. $4 > -2$

When multiplied by -2, this becomes

$-8 < 4$

It is best to avoid this as shown in Example 2(b).

Example 2

Solve the inequalities (a) $2x - 1 > 5$ (b) $5 - 3x \leqslant 1$

(a) $2x - 1 > 5$

$2x > 5 + 1$

$x > \frac{6}{2}$

$x > 3$

(b) $5 - 3x \leqslant 1$

$5 \leqslant 1 + 3x$

$5 - 1 \leqslant 3x$

$\frac{4}{3} \leqslant x$

Exercise 18

In Questions **1** to **4**, x is an integer. List the values which x can take.

1. $1 < x < 4$
2. $1 \leqslant x < 6$
3. $\frac{1}{2} \leqslant x \leqslant 2$
4. $0 < x \leqslant 3\frac{1}{2}$

Solve the following inequalities:

5. $x - 3 > 10$
6. $x + 1 < 0$
7. $5 > x - 7$
8. $2x + 1 \leqslant 6$
9. $3x - 4 > 5$
10. $10 \leqslant 2x - 6$
11. $5x < x + 1$
12. $2x \geqslant x - 3$
13. $4 + x < -4$
14. $3x + 1 < 2x + 5$
15. $2(x + 1) > x - 7$
16. $7 < 15 - x$
17. $9 > 12 - x$
18. $4 - 2x \leqslant 2$
19. $3(x - 1) < 2(1 - x)$
20. $7 - 3x < 0$
21. $\frac{x}{3} < -1$
22. $\frac{2x}{5} > 3$
23. $2x > 0$
24. $\frac{x}{4} < 0$

Hint: in Questions **25** to **29**, solve the two inequalities separately.

25. $10 \leqslant 2x \leqslant x + 9$
26. $x < 3x + 2 < 2x + 6$
27. $10 \leqslant 2x - 1 \leqslant x + 5$
28. $x < 3x - 1 < 2x + 7$
29. $x - 10 < 2(x - 1) < x$

Squares and square roots in inequalities need care.

The equation $x^2 = 4$ becomes $x = \pm 2$ which is correct.

For the inequality $x^2 < 4$, we might wrongly write $x < \pm 2$.
Consider $x = -3$, say.
-3 is less than -2 and is also less than $+2$.
But $(-3)^2$ is not less than 4 and so
$x = -3$ does not satisfy the inequality $x^2 < 4$.
The correct solution for $x^2 < 4$
is $-2 < x < 2$

Example 3

Solve the inequality $2x^2 - 1 > 17$.

$$2x^2 - 1 > 17$$
$$2x^2 > 18$$
$$x^2 > 9$$
$$x > 3 \text{ or } x < -3$$

[Avoid the temptation to write $x > \pm 3$]!

Exercise 19

Solve the inequalities.

1. $x^2 < 25$
2. $x^2 \leqslant 16$
3. $x^2 > 1$
4. $2x^2 \geqslant 72$
5. $3x^2 + 5 > 5$
6. $5x^2 - 2 < 18$

For Questions **7** to **13**, list the solutions which satisfy the given condition.

7. $3a + 1 < 20$; a is a positive integer.
8. $b - 1 \geqslant 6$; b is a prime number less than 20.
9. $1 < z < 50$; z is a square number.
10. $2x > -10$; x is a negative integer.
11. $x + 1 < 2x < x + 13$; x is an integer.
12. $0 \leqslant 2z - 3 \leqslant z + 8$; z is a prime number.
13. $\frac{a}{2} + 10 > a$; a is a positive even number.
14. Given that $4x > 1$ and $\frac{x}{3} \leqslant 1\frac{1}{3}$, list the possible integer values of x.

15. State the smallest integer n for which $4n > 19$.

16. Given that $-4 \leqslant a \leqslant 3$ and $-5 \leqslant b \leqslant 4$, find
(a) the largest possible value of a^2
(b) the smallest possible value of ab
(c) the largest possible value of ab
(d) the value of b if $b^2 = 25$

17. For any shape of triangle ABC, complete the statement AB + BC □ AC, by writing $<$, $>$ or $=$ inside the box.

18. Find a simple fraction r such at $\frac{1}{3} < r < \frac{2}{3}$.

19. Find the largest prime number p such that $p^2 < 400$.

20. Find the values of x which satisfy each pair of simultaneous inequalities..

(a) $x < 6$
$-3 \leqslant x \leqslant 8$

(b) $x > -2$
$-4 < x < 2$

(c) $2x + 1 \leqslant 5$
$-12 \leqslant 3x - 3$

(d) $3x - 2 < 19$
$2x \geqslant -6$

21. Find the integer n such that $n < \sqrt{300} < n + 1$.

22. If $f(x) = 2x - 1$ and $g(x) = 10 - x$ for what values of x is $f(x) > g(x)$?

23. If $2^r > 100$, what is the smallest integer value of r?

24. Given $\left(\frac{1}{3}\right)^x < \frac{1}{200}$, what is the smallest integer value of x?

25.† Find the smallest integer value of x which satisfies $x^x > 10\,000$.

26.† What integer values of x satisfy
$100 < 5^x < 10\,000$?

27.† If x is an acute angle and $\sin x > \frac{1}{2}$, write down the range of values that x can take.

28.† If x is an acute angle and $\cos x > \frac{1}{4}$, write down the range of values that x can take.

29.† If a, c, d are positive and $ay - c < d$, what can you say about y?

30.† This is Fazleen's working to solve $\frac{1}{x} < \frac{3}{2}$.

$\frac{1}{x} < \frac{3}{2}$

$2 < 3x$ [cross multiply]

$\frac{2}{3} < x$ [divide by 3]

Her friend Shelina said it was wrong. She wrote:

Try $x = -1$: $\frac{1}{-1} < \frac{3}{2}$

$-1 < \frac{3}{2}$ which is correct.

So the answer $x > \frac{2}{3}$ must be wrong.

Who is correct? Where is the mistake?

Shading regions

It is useful to represent inequalities on a graph, particularly where two variables (x and y) are involved. The work on linear programming on page 363 requires a knowledge of shading regions.

Example 4

Draw a sketch graph and shade the area which represents the set of points that satisfy each of these inequalities.

(a) $x > 2$ (b) $1 \leqslant y \leqslant 5$ (c) $x + y \leqslant 8$

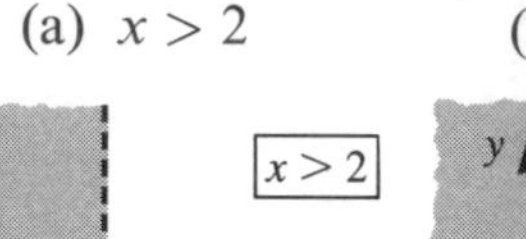
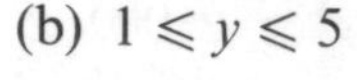
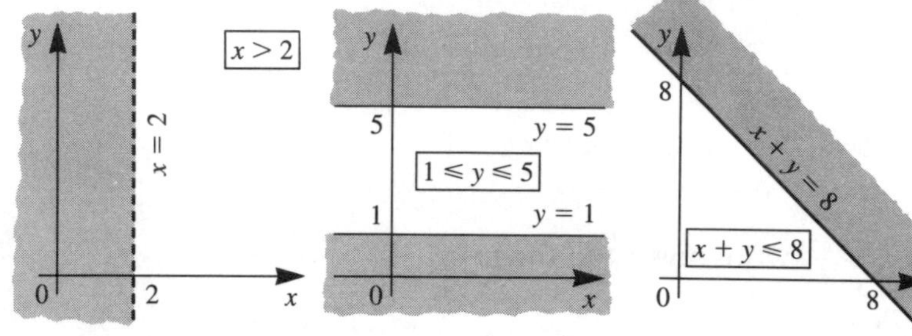

In each graph, the unwanted region is shaded. Then when several regions are shown on the same diagram, the required region shows up clearly.

In (a), the line $x = 2$ is shown as a broken line to indicate that the points on the line are *not* included.

In (b) and (c), the lines $y = 1, y = 5$ and $x + y = 8$ are shown as solid lines because points on the line *are* included in the region.

To decide which side to shade when the line is sloping, take a trial point. This can be any point which is not actually on the line.

In (c) above, the trial point could be (1, 1).
Is (1, 1) in the region $x + y \leqslant 8$?
It satisfies $x + y \leqslant 8$ because $1 + 1 = 2$ which is less than 8.
So, *below* the line is the required region for $x + y \leqslant 8$, and the unwanted region *above* the line is shaded.

Exercise 20

In Questions **1** to **6**, describe the region left unshaded.

1.

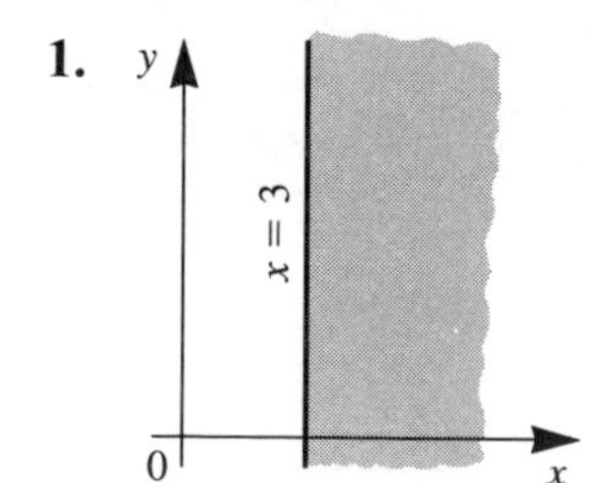

2.

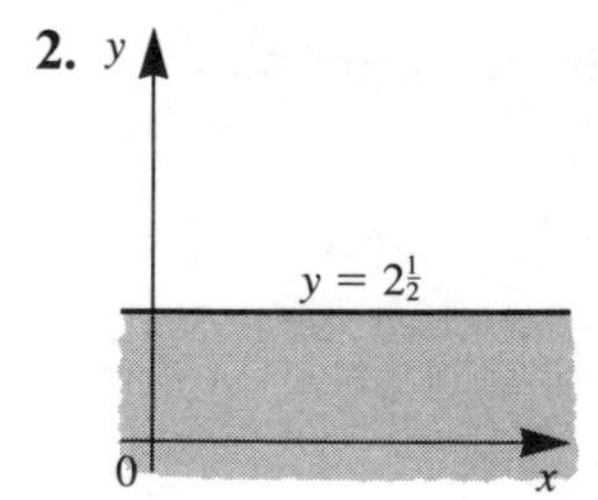

3.

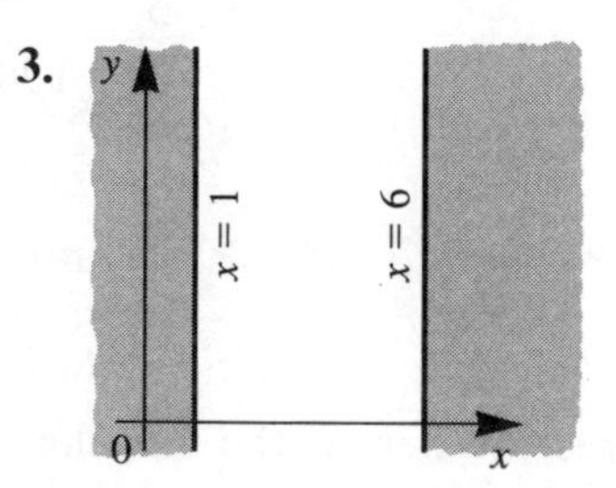

4. 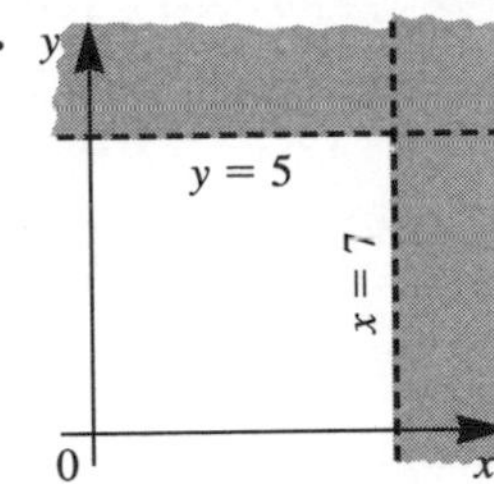

5. 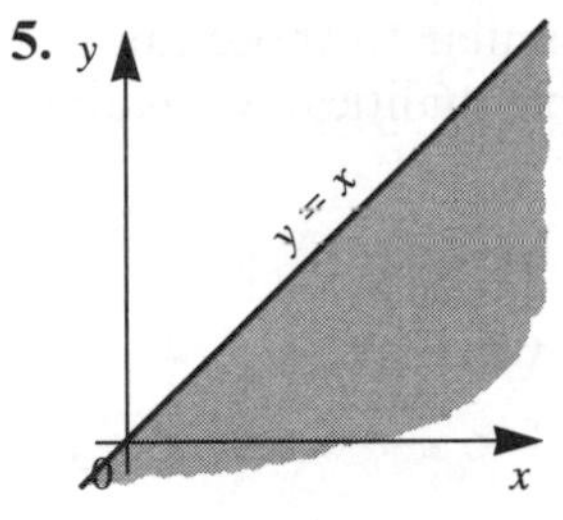

6. 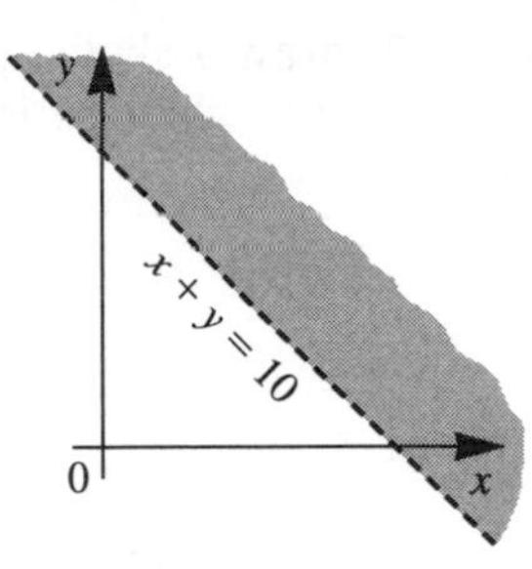

7. The point (1, 1), marked *, lies in the unshaded region. Use this as a trial point to describe the unshaded region.

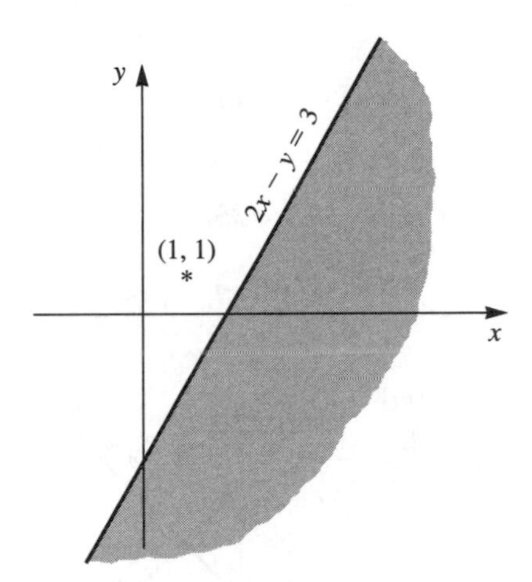

8. The point (3, 1), marked *, lies in the unshaded triangle. Use this as a trial point to write down the three inequalities which describe the unshaded region.

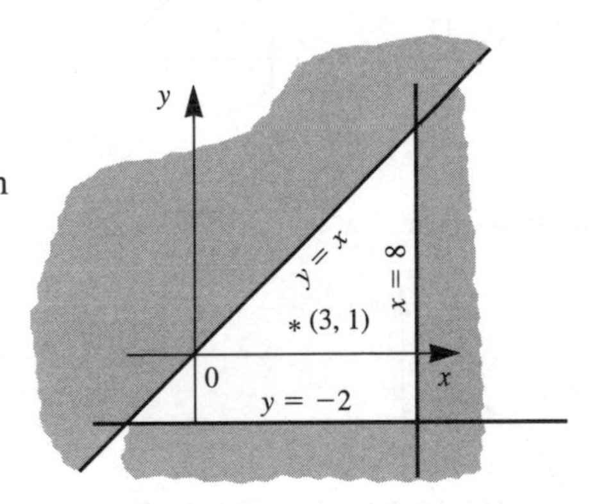

9. A trial point (1, 1) lies inside the unshaded triangles. Write down the three inequalities which describe each unshaded region.

(a)

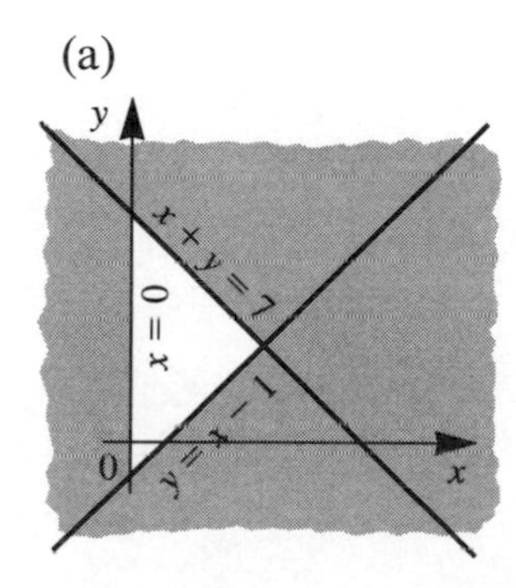

(b)

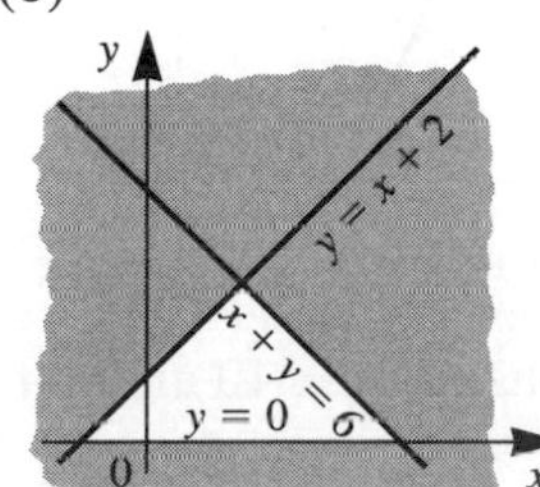

For Questions **10** to **27**, draw a sketch graph similar to those above and indicate the set of points which satisfy the inequalities by shading the unwanted region.

10. $2 < x < 7$

11. $0 < y < 3\frac{1}{2}$

12. $-2 < x < 2$

13. $x < 6$ and $y < 4$

14. $0 < x < 5$ and $y < 3$

15. $1 < x < 6$ and $2 < y < 8$

16. $-3 < x < 0$ and $-4 < y < 2$

17. $y < x$

18. $x + y < 5$

19. $y > x + 2$ and $y < 7$

20. $x > 0$ and $y > 0$ and $x + y < 7$

21. $x > 0$ and $x + y < 10$ and $y > x$

22. $8 > y > 0$ and $x + y > 3$

23. $x + 2y < 10$ and $x > 0$ and $y > 0$

24. $3x + 2y < 18$ and $x > 0$ and $y > 0$

25. $x > 0$, $y > x - 2$, $x + y < 10$

26. $3x + 5y < 30$ and $y > \frac{x}{2}$

27. $y > \frac{x}{2}$, $y < 2x$ and $x + y < 8$

28. The two lines $y = x + 1$ and $x + y = 5$ divide the graph into four regions A, B, C, D.

Write down the two inequalities which describe each of the regions A, B, C, D.

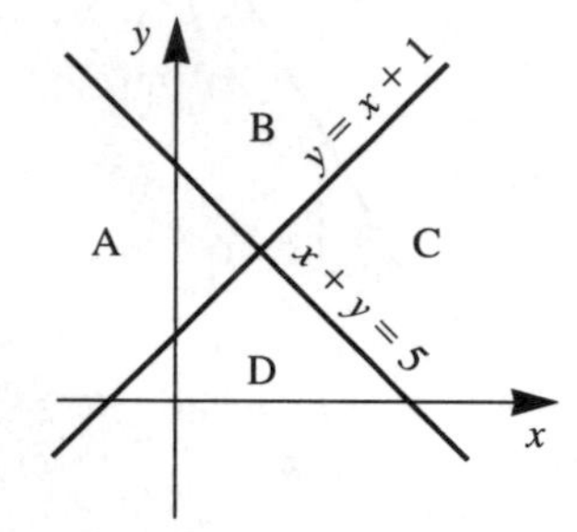

29.† Here is a sketch of $y = 2 - x - x^2$.

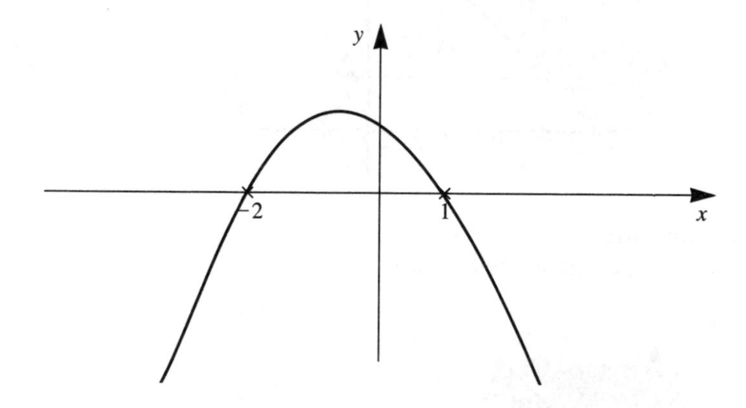

For what values of x is $(2 - x - x^2)$ positive?

30.† Using the same axes, draw the graphs of $xy = 10$ and $x + y = 9$ for values of x from 1 to 10.
Hence find all pairs of positive integers whose product is greater than 10 and whose sum is less than 9.

2.7 Brackets, factorising and equations

Expanding brackets

To work out the expression $(x+3)(x+2)$, the area of a rectangle can be a help.
From the diagram, the total area of the rectangle

$$= (x+3)(x+2)$$
$$= x^2+2x+3x+6$$
$$= x^2+5x+6$$

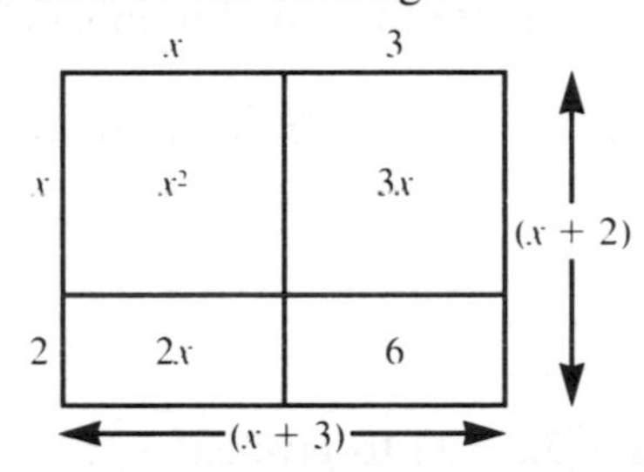

After a little practice, it is possible to do without the diagram.

Example 1

Remove the brackets and simplify $(3x-2)(2x-1)$.

$$\begin{aligned}(3x-2)(2x-1) &= 3x(2x-1)-2(2x-1)\\ &= 6x^2-3x-4x+2\\ &= 6x^2-7x+2\end{aligned}$$

Example 2

Remove the brackets and simplify $(x-3)^2$

Be careful with this expression. It is not x^2-9, nor even x^2+9.

$$\begin{aligned}(x-3)^2 &= (x-3)(x-3)\\ &= x(x-3)-3(x-3)\\ &= x^2-3x-3x+9\\ &= x^2-6x+9\end{aligned}$$

Exercise 21

Remove the brackets and simplify.

1. $(x+1)(x+3)$
2. $(x+3)(x+2)$
3. $(y+4)(y+5)$
4. $(x-3)(x+4)$
5. $(x+5)(x-2)$
6. $(x-3)(x-2)$
7. $(a-7)(a+5)$
8. $(z+9)(z-2)$
9. $(x-3)(x+3)$
10. $(k-11)(k+11)$
11. $(2x+1)(x-3)$
12. $(3x+4)(x-2)$
13. $(2y-3)(y+1)$
14. $(7y-1)(7y+1)$
15. $(x+4)^2$
16. $(x+2)^2$
17. $(x-2)^2$
18. $(2x+1)^2$
19. $(x+1)^2+(x+2)^2$
20. $(x-2)^2+(x+3)^2$

21. $(x+2)^2+(2x+1)^2$
22. $(y-3)^2+(y-4)^2$
23. $(x+2)^2-(x-3)^2$
24. $(x-3)^2-(x+1)^2$
25. (a) If n is an integer, explain why $(2n-1)$ is an odd number.
(b) Write down the next odd number after $(2n-1)$ and show that the sum of these two numbers is a multiple of 4.
(c) Write down the next odd number after these two and show that the sum of the three odd numbers is a multiple of 3.
(d) Show that the sum of the squares of these three odd numbers is $12n^2+12n+11$.
Is it possible for this expression to be a multiple of 11? If so give two possible values of n.

Factorising expressions

Earlier we expanded expressions such as $x(3x-1)$ to give $3x^2-x$. The reverse of this process is called factorising.

Example 3

Factorise: (a) x^2+7x (b) $3y^2-12y$ (c) $6a^2b-10ab^2$

(a) x is common to x^2 and $7x$. $\therefore\; x^2+7x = x(x+7)$.
The factors are x and $(x+7)$.

(b) $3y$ is common $\therefore\; 3y^2-12y = 3y(y-4)$

(c) $2ab$ is common $\therefore\; 6a^2b-10ab^2 = 2ab(3a-5b)$

Exercise 22

Factorise the following expressions completely.

1. x^2+5x
2. x^2-6x
3. $7x-x^2$
4. y^2+8y
5. $2y^2+3y$
6. $6y^2-4y$
7. $3x^2-21x$
8. $16a-2a^2$
9. $6c^2-21c$
10. $15x-9x^2$
11. $56y-21y^2$
12. $ax+bx+2cx$
13. $x^2+xy+3xz$
14. $x^2y+y^3+z^2y$
15. $3a^2b+2ab^2$
16. x^2y+xy^2
17. $6a^2+4ab+2ac$
18. $ma+2bm+m^2$
19. $2kx+6ky+4kz$
20. $ax^2+ay+2ab$
21. x^2k+xk^2
22. a^3b+2ab^2
23. $abc-3b^2c$
24. $2a^2e-5ae^2$
25. a^3b+ab^3
26. $x^3y+x^2y^2$
27. $6xy^2-4x^2y$
28. $3ab^3-3a^3b$
29. $2a^3b+5a^2b^2$
30. ax^2y-2ax^2z
31. $2abx+2ab^2+2a^2b$
32. $ayx+yx^3-2y^2x^2$

Solving equations

Example 4

Solve the equation $(x+3)^2 = (x+2)^2 + 3^2$

$$\begin{aligned} (x+3)^2 &= (x+2)^2 + 3^2 \\ (x+3)(x+3) &= (x+2)(x+2) + 9 \\ x^2 + 6x + 9 &= x^2 + 4x + 4 + 9 \\ 6x + 9 &= 4x + 13 \\ 2x &= 4 \\ x &= 2 \end{aligned}$$

Exercise 23

Solve the following equations:

1. $x^2 + 4 = (x+1)(x+3)$

2. $x^2 + 3x = (x+3)(x+1)$

3. $(x+3)(x-1) = x^2 + 5$

4. $(x+1)(x+4) = (x-7)(x+6)$

5. $(x-2)(x+3) = (x-7)(x+7)$

6. $(x-5)(x+4) = (x+7)(x-6)$

7. $2x^2 + 3x = (2x-1)(x+1)$

8. $(2x-1)(x-3) = (2x-3)(x-1)$

9. $x^2 + (x+1)^2 = (2x-1)(x+4)$

10. $x(2x+6) = 2(x^2 - 5)$

In Questions **11** and **12**, form an equation in x by means of Pythagoras' Theorem, and hence find the length of each side of the triangle.
(All the lengths are in cm.)

11.

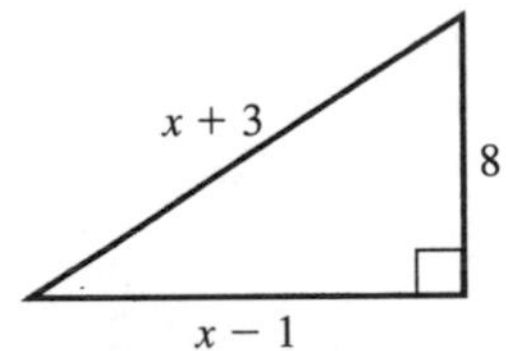

12.

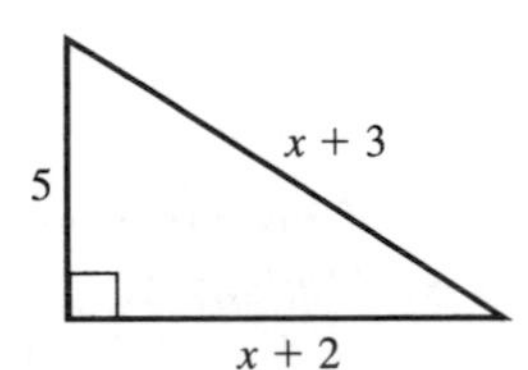

13. The area of the rectangle shown exceeds the area of the square by $2\,\text{cm}^2$. Find x.

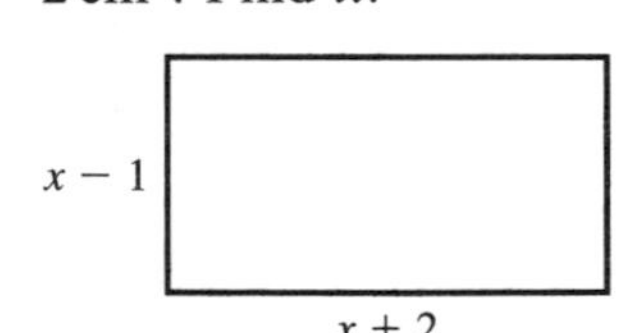

x

x

14. The area of the square exceeds the area of the rectangle by $13\,\text{m}^2$. Find y.

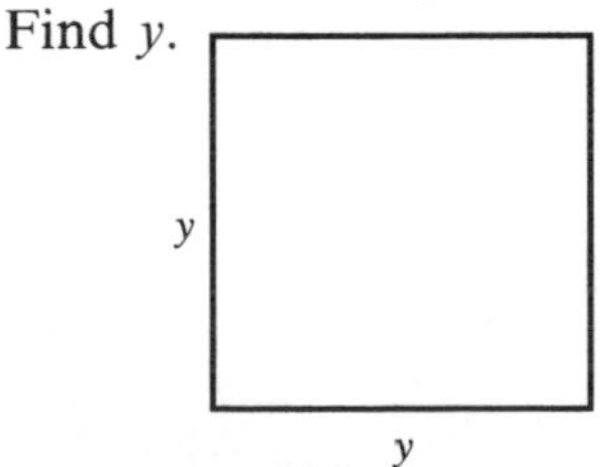

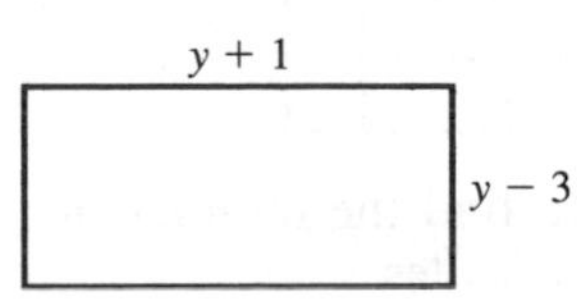

Examination exercise 2

1. (a) Solve the equations
(i) $7 - 2x = 9$
(ii) $\frac{9}{y} = 5$
(b) Factorise the expression $6z^2 - 9z$
(c) Multiply out the brackets giving your answer in its simplest form. $(2x + 5)(3x - 7)$ [S]

2. Solve the equation

$$\frac{(x+1)}{2} - \frac{(2x+1)}{3} = 1$$ [M]

3. The cost, £C, of making n articles is given by the formula
$C = a + bn$ where a and b are constants.
The cost of making 4 articles is £20 and the cost of making 7 articles is £29.
Write down two equations in a and b.
Solve these equations to find the values of a and b. [L]

4. A craftsman can be paid in one of two ways:
Method A A down payment of £300, then £4 an hour.
Method B £9 an hour.
(a) He works for n hours. Write down, in terms of n, an expression for his earnings using
(i) *Method A*,
(ii) *Method B*.
(b) Write down an inequality that will be true if his earnings by *Method B* are greater then his earnings by *Method A*.
(c) Solve your inequality for n. [M]

5. (a) Calculate
(i) $1^3 + 2^3 + 3^3 - (1 + 2 + 3)^2$
(ii) $1^3 + 2^3 + 3^3 + 4^3 - (1 + 2 + 3 + 4)^2$.
(b) Comment on your results for part (a). [M]

6. Sian is making a pattern of tiles by surrounding black tiles with white tiles.

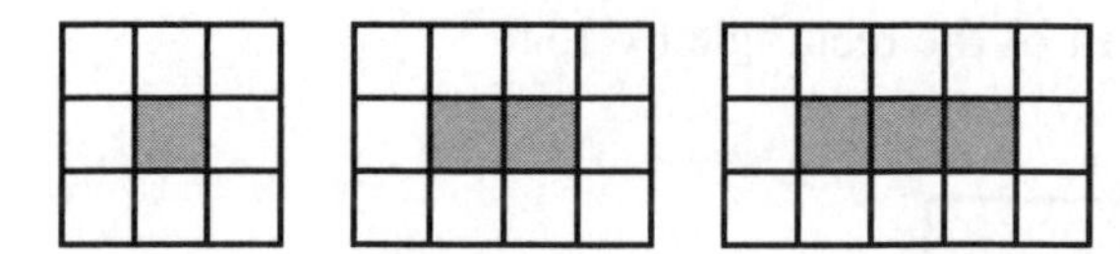

(a) Give a formula that Sian could use to find the number of white tiles required to surround n black tiles.
(b) If Sian had 81 white tiles, what is the greatest number of black tiles that can be surrounded? [N]

7. The diagram shows a number square with a 'T' marked on it. The number at the centre of the top row gives the T-number. The T-number of the 'T' shown is $T13$.

1	2	3	4	5	6	7	8	9	10
11	12	13	14	15	16	17	18	19	20
21	22	23	24	25	26	27	28	29	30
31	32	33	34	35	36	37	38	39	40
41	42	43	44	45	46	47	48	49	50
51	52	53	54	55	56	57	58	59	60
61	62	63	64	65	66	67	68	69	70

The 'T' can be TRANSLATED to other positions on the number square. In each position it must cover five numbers on the number square.

(a) What is the smallest T-number?
(b) Find the total of the numbers in $T15$.
(c) (i) Copy the diagram below and write on it, in terms of x, the numbers in Tx.

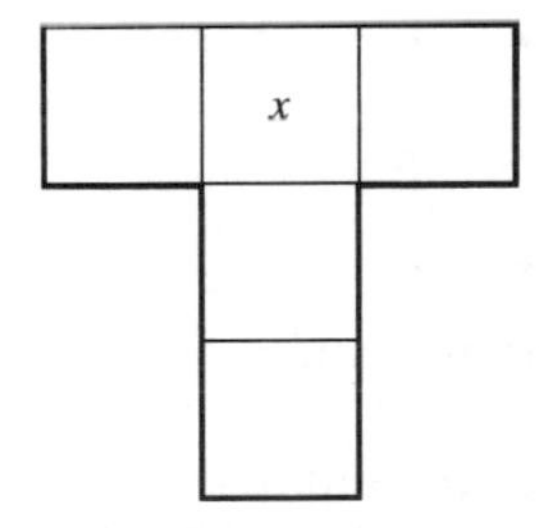

(ii) What is the total, in terms of x, of the numbers in Tx?
(iii) Which T-number has a total of 355?
(d) What is the difference between the totals of the numbers in Tx and $T(x+1)$? [S]

8. A scientist has a theory that the time taken for a comet to disintegrate in the Earth's atmosphere can be calculated from the equation

$100t = \sqrt{m}$ where t = time taken (seconds),
m = mass of comet (tonnes).

(a) A comet took 4·35 seconds to disintegrate.
Use the equation to calculate its mass.
(b) In November a comet of estimated mass $8{\cdot}5 \times 10^6$ tonnes was to enter the Earth's atmosphere.
How long would it take to disintegrate? [N]

9. PULL-UPS

At the end of a one minute 'pull-ups' contest the following facts were published.
The top five contestants completed a total of 163 pull-ups between them.
The winning contestant had completed only one more pull-up than the contestant who was second.
Three contestants tied for third place, each having completed only one fewer pull-up than the contestant who was second.
How many pull-ups were completed by the winning contestant? [S]

10. Shapes of area 5 cm^2 can be made using five one centimetre squares. Each square touches the next along an edge or at a corner.

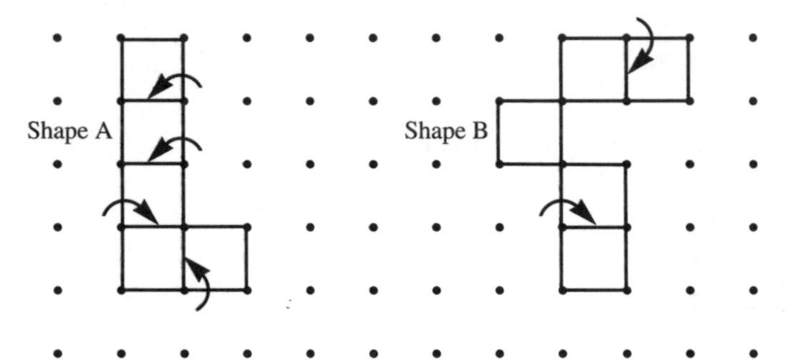

Shape A has four common edges, marked ↶ above, and a perimeter of 12 cm.
Shape B has two common edges, marked ⤵ above, and a perimeter of 16 cm.

(a) On dotted paper draw
  (i) a shape C with only one common edge,
  (ii) a shape D with perimeter 14 cm,
  (iii) the shape E with the smallest possible perimeter.

(b) Complete the table below.

Shape	Perimeter P cm	No. of common edges N
A	12	4
B	16	2
C		1
D	14	
E		

(c) Find a formula connecting P and N or express the relationship in words. [N]

11.

D	E	E	D
E	D	D	E
E	D	D	E
D	E	E	D

C			C
	M	M	
	M	M	
C			C

A "chessboard" consists of four different types of square. These are denoted by C, corner squares, of which there are four,
E, edge squares, which lie along the edge of the board, but are not corner squares,
M, middle squares, which are neither corner nor edge squares,
D, diagonal squares, which lie along both diagonals of the square.
(Some squares may be more than one type of square simultaneously.)
The diagrams above show the same 4×4 chessboard labelled in two different ways.
The table below shows the number of each type of square for various size boards.

Size of board	Type of square C	E	M	D
3×3	4	4	1	5
4×4	4	8	4	8
5×5	4	12	9	9
6×6				

(a) Complete the table on the right for the 6×6 board.

(b) How many middle squares will an 8×8 board have?
(c) A board has 48 edge squares. How many diagonal squares does it have?
(d) A board has 225 middle squares. How many edge squares does it have?
(e) Find the number of each type of square for a 100×100 board. [N]

12. Mary wants to find, correct to one decimal place, a solution of the equation

$x^2 + x = 1.$

By trying values of x, find a solution to Mary's equation.
Show your working clearly. [N]

13. Some pupils were making up number puzzles.
One pupil's number puzzle was written like this:

$2(3x - 1) < 18$

What is the largest whole number that this pupil could have thought of? [N]

14.

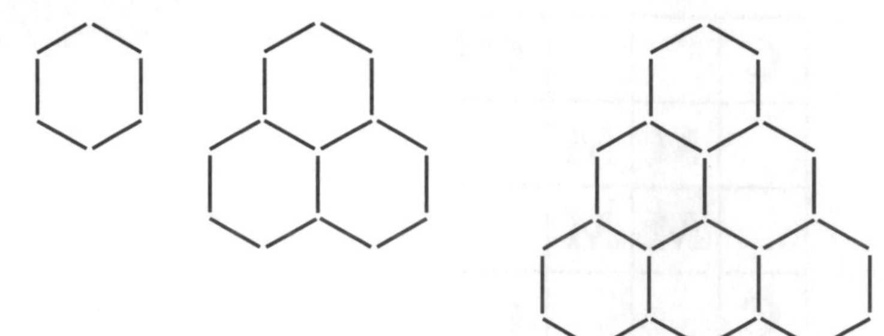

Bridget is doing an investigation which involves making patterns of hexagons with sticks. The first three of these patterns are shown above.

(a) Copy and complete this table.

number of rows, n	1	2	3	4	5
number of hexagons, h	1	3			
number of sticks, s	6				

(b) Write down a formula for h in terms of n.

(c) The formula for s in terms of n is of the form

$s = an^2 + bn,$

where a and b are constants.
Find the value of a and the value of b. [M]

15. The positive number n satisfies the equation

$n^2 + n = 80$

(a) Write down two consecutive whole numbers between which n must lie.

(b) Use trial and improvement and a calculator to find the value of n correct to 3 significant figures.

16. A motorist travels on country lanes on a journey of 80 miles at an average speed of x m.p.h.

(a) Write, in terms of x, an expression for the time taken in hours.

When he plans the same journey later a motorway has been opened. He is now able to travel the same distance at twice the speed that he could do along the country lanes.

(b) Write, in terms of x, an expression for the time taken along the motorway in hours.

The journey along the motorway takes 70 minutes less than the journey along the country lanes.

(c) Calculate the speed of the car for the journey along the country lanes. [S]

3 Shape and space 1

3.1 Three-dimensional coordinates

In three-dimensional space, we need three coordinates to describe how the position of a point relates to the common origin of three axes. In this diagram, O is the origin for the three axes. For A, $x = 5$, $y = 4$, $z = 2$, so A is written (5, 4, 2).

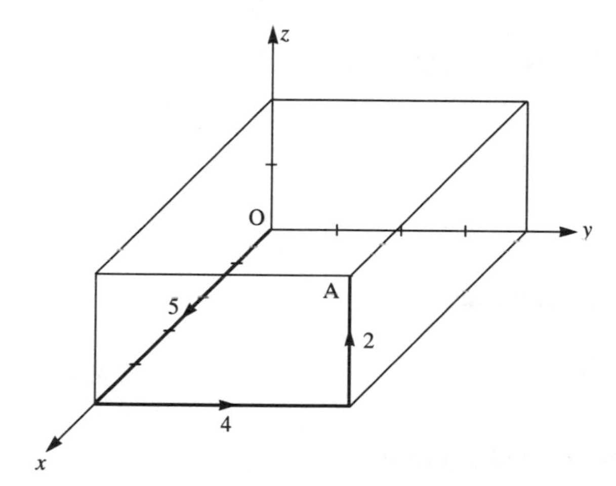

Exercise 1

1. Write down the coordinates of the points A, B, C, D.

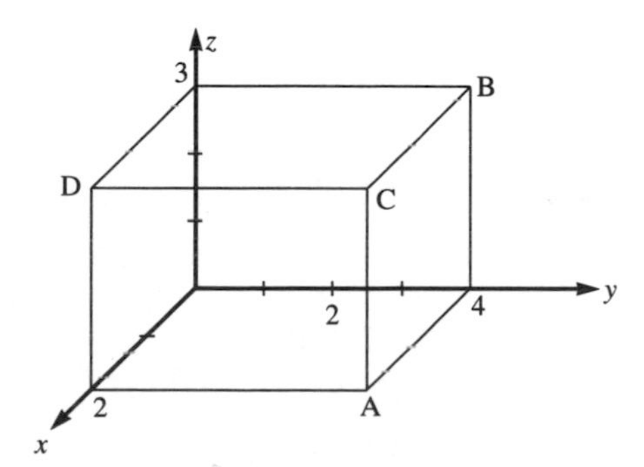

2. (a) Write down the coordinates of B, C, Q, R.

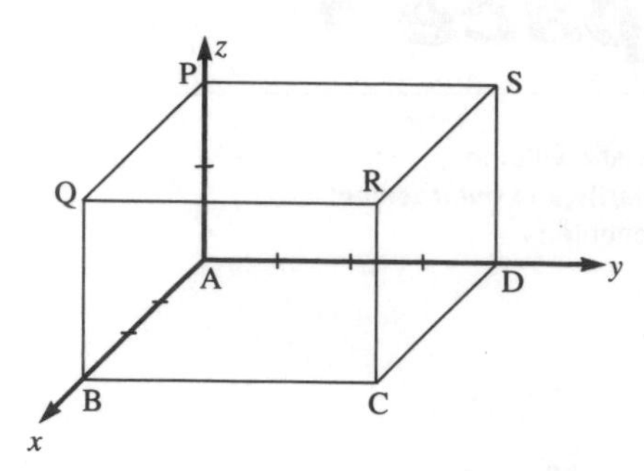

(b) Write down the coordinates of the mid-points of
(i) AD (ii) DS (iii) DC

(c) Write down the coordinates of the centre of the face
(i) ABCD (ii) PQRS (iii) RSDC

(d) Write down the coordinates of the centre of the box.

3. (a) Write down the coordinates of C, R, B, P, Q.

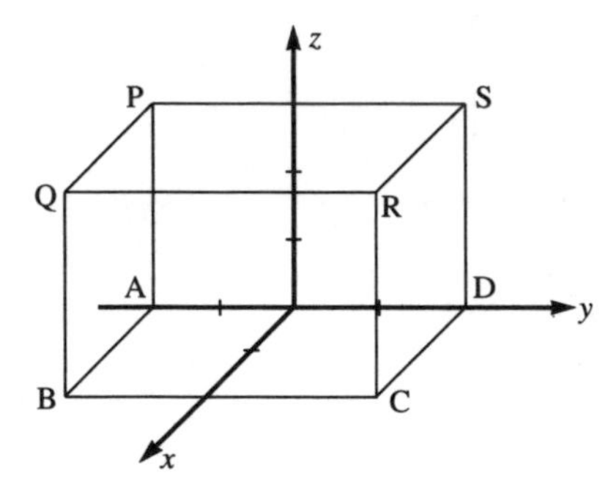

(b) Write down the coordinates of the mid-points of
(i) QB (ii) PQ

4. Use Pythagoras' theorem to calculate the lengths of
(i) LM (ii) LN (iii) NM

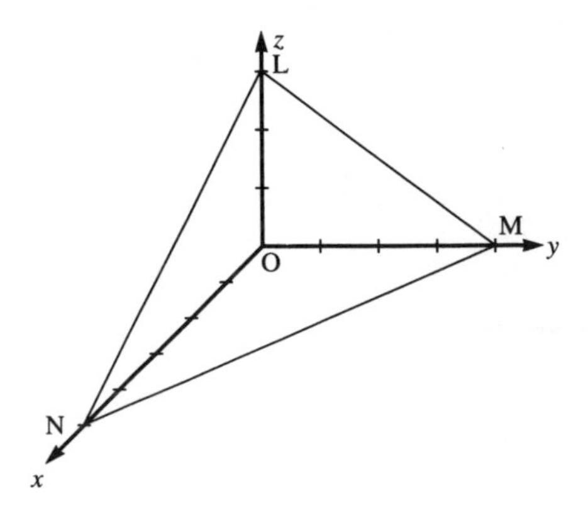

5. In the cuboid below, OP = 4, OQ = 7, OR = 5. Write down the coordinates of the centre of the face
(i) ABCR (ii) BCQD (iii) OPDQ

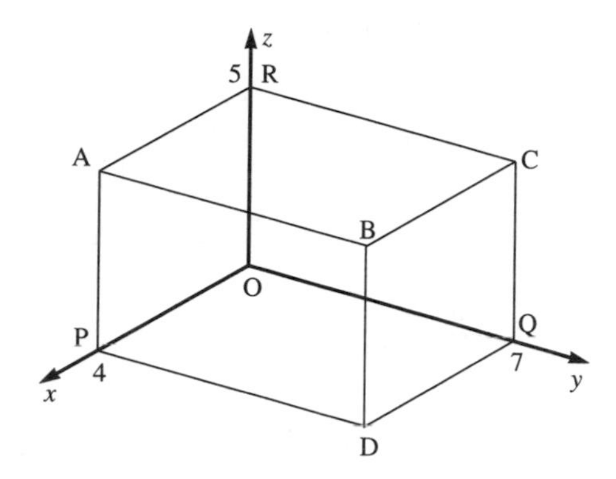

6. Measured from a control tower O, an aircraft is 20 km north, 30 km east and 5 km high. So with origin at O, its coordinates are (20, 30, 5).
It travels east at 3 km per minute. What are its coordinates after 5 minutes?

7. Draw your own axes and plot the points O(0,0,0), A(3,0,0), B(3,4,0), C(0,0,5)
Work out the length of the following lines.
(a) AB (b) OB (c) CB

8. A solid object has vertices at (0,0,0), (6,0,0), (6,6,0), (0,6,0) and (3,3,4).
Plot the points on a diagram and name the object.

9.† Plot the points O(0,0,0), A(6,8,0), B(6,8,10).
Work out (a) OA
(b) angle BOA

10.† A prisoner starts his escape tunnel at (0,0,0). He digs in a straight line to a sewer (5, 3, −2). He then crawls along the sewer to (5, 20, −2). Finally he digs in a straight line to escape at (0,24,0). Work out the total length of the three sections of his escape route. The units are given in metres.

3.2 Locus

In mathematics, the word *locus* describes the position of points which obey a certain rule. The locus can be the path traced out by a moving point.

Three important loci

(a) Circle

The locus of points which are equidistant from a fixed point O is shown. It is a **circle** with centre O.

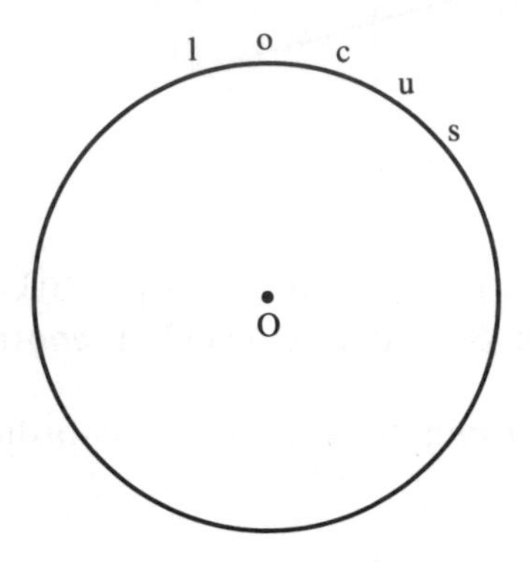

(b) Perpendicular bisector

The locus of points which are equidistant from two fixed points A and B is shown.

The locus is the **perpendicular bisector** of the line AB. Use compasses to draw arcs, as shown, or use a ruler and a protractor.

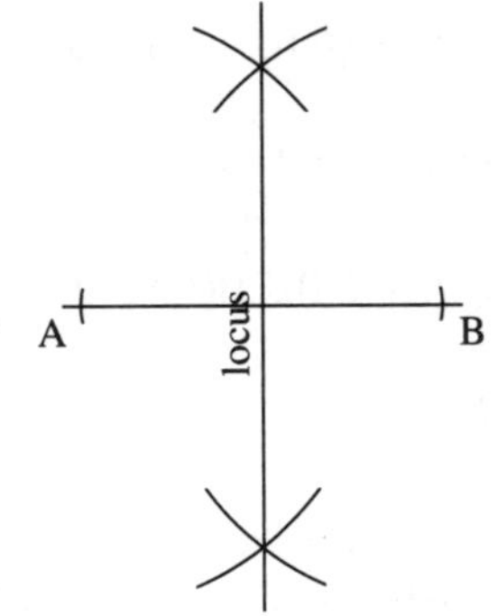

(c) Angle bisector

The locus of points which are equidistant from two fixed lines AB and AC is shown.

The locus is the line which bisects the angle BAC. Use compasses to draw arcs or use a protractor to construct the locus.

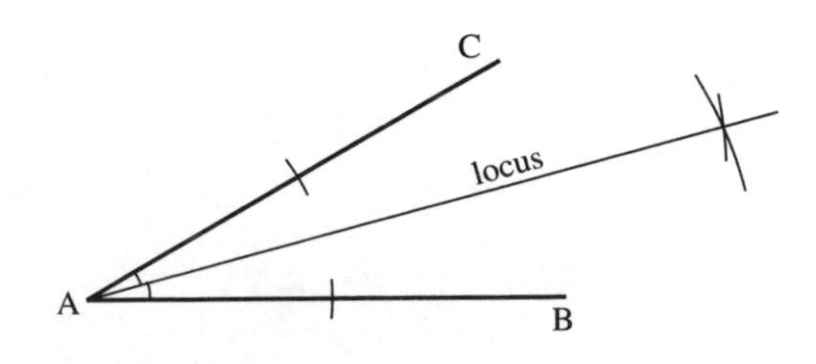

Exercise 2

1. Mark two points P and Q which are 10 cm apart. Draw the locus of points which are equidistant from P and Q.

2. Draw two lines AB and AC of length 8 cm, where $B\hat{A}C = 40°$. Draw the locus of points which are equidistant from AB and AC.

3. (a) Draw the triangle LMN full size.
   (b) Draw the locus of the points which are:
      (i) equidistant from L and N
      (ii) equidistant from LN and LM
      (iii) 4 cm from M
   [Draw the three loci in different colours].

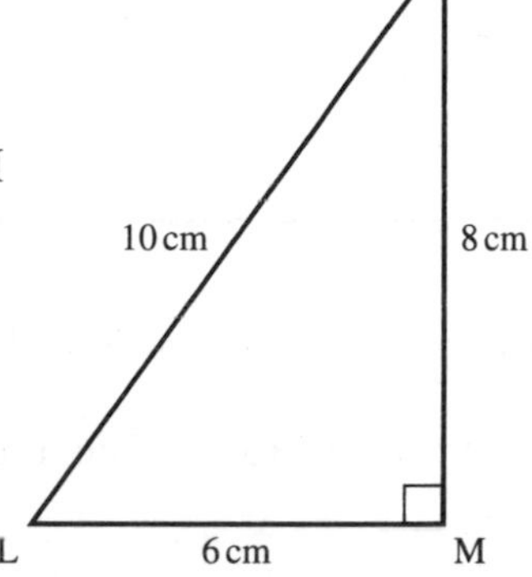

4. Draw a line AB of length 6 cm. Draw the locus of a point P so that angle ABP = 90°.

5. Channel 9 in Australia are planning the position of a new TV satellite to send pictures all over the country. The new satellite is to be placed:
   (a) an equal distance from Darwin and Adelaide.
   (b) vertically above a point not more than 2000 km from Perth.
   (c) vertically above a point not more than 1600 km from Brisbane.

Make a copy of the map on squared paper using the grid lines as reference.
Show clearly where the satellite could be placed so that it satisfies the conditions (a), (b), (c) above.

6. A and B are fixed pegs 10 cm apart. A piece of string of length 15 cm is tied to A and B and passes through a small ring R which can slide along the string.

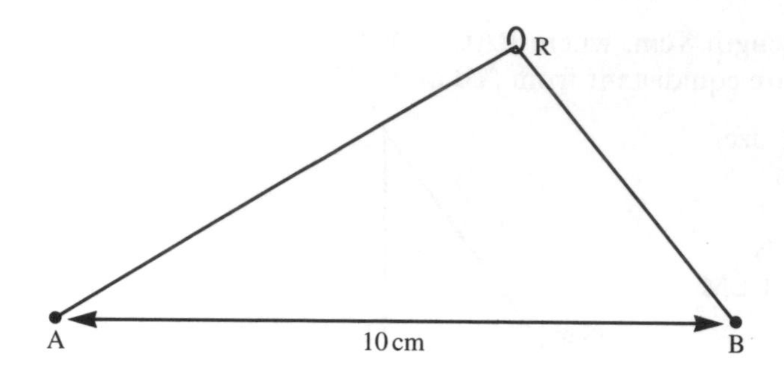

Draw, as accurately as you can, the locus of R if the string is always kept taut.
The locus of the point R where AR + BR = 15 cm is a curve called an ellipse.

7. The diagram shows the walls of a rectangular shed measuring 8 m by 5 m.
A goat is tied to the corner C by a rope 7 m long.
Make a scale drawing to show the boundary of the area which the goat can reach.

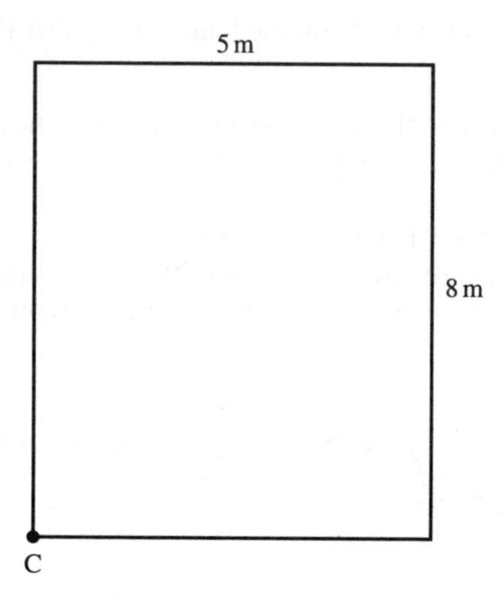

8. Rod AB rotates about the fixed point A. B is joined to C which slides along a fixed rod. AB = 15 cm and BC = 25 cm.

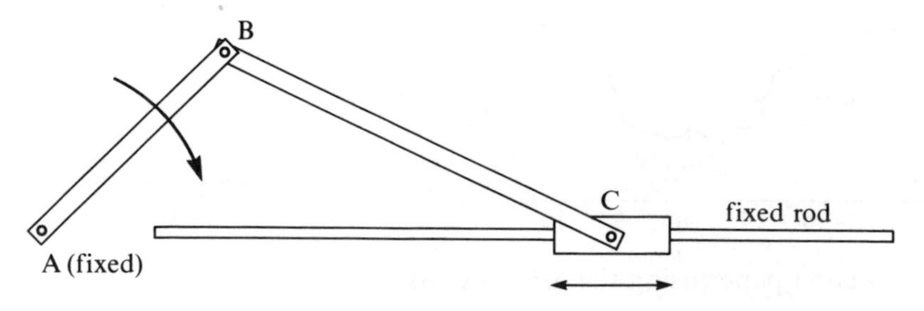

Describe the locus of C as AB rotates clockwise about A.
Find the smallest and the largest distance between C and A.

9. Cog A has 36 teeth and cog B has 24 teeth.

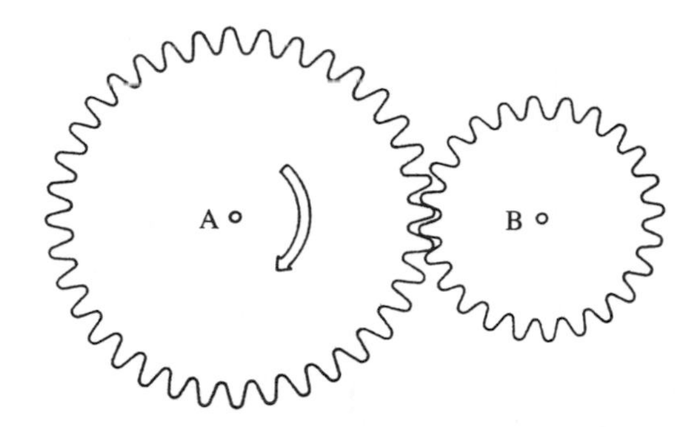

(a) When cog A turns clockwise in what direction does cog B turn?
(b) When cog A makes 10 complete revolutions how many times does cog B rotate?
(c) If cog B is rotating at a speed of 300 revolutions per minute at what speed is cog A rotating?

10. Draw two points M and N 16 cm apart. Draw the locus of a point P which moves so that the area of triangle MNP is 80 cm^2.

11. Describe the locus of a point which moves in three dimensional space and is equidistant from two fixed points.

12. A circle of radius 2 cm rolls around the perimeter of a square of side 8 cm.
Sketch the locus of the centre of the circle.

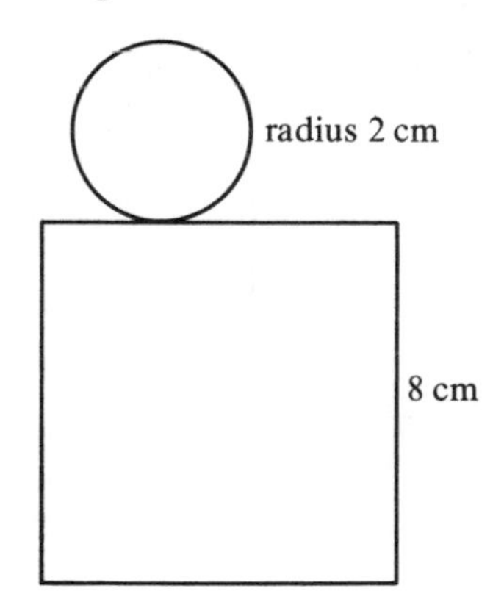

13. A circle, centre O, radius 5 cm, is inscribed inside a square ABCD. Point P moves so that OP $\leqslant$ 5 cm and BP$\leqslant$ 5 cm. Shade the set of points indicating where P can be.

14. Construct a triangle ABC where AB = 9 cm, BC = 7 cm and AC = 5 cm.
(a) Sketch and describe the locus of points within the triangle which are equidistant from AB and AC.
(b) Shade the set of points within the triangle which are less than 5 cm from B and are also nearer to AC than to AB.

15. A right-angled triangle KLM has hypotenuse KM of length 20 cm. Sketch the locus of L for all triangles KLM.

16. Copy the diagram shown.
$OC = 4$ cm.
Sketch the locus of P which moves so that PC is equal to the perpendicular distance from P to the line AB. This locus is called a *parabola*.

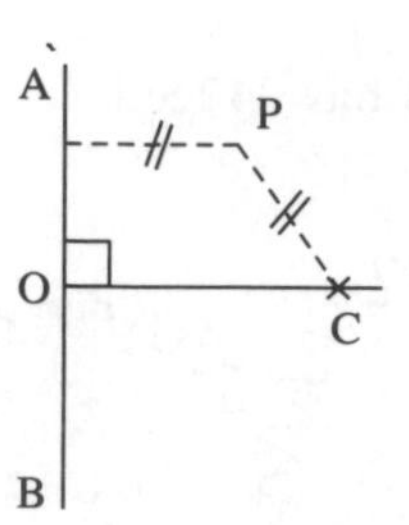

17. A rod OA of length 60 cm rotates about O at a constant rate of 1 revolution per minute. An ant, with good balance, walks along the rod at a speed of 1 cm per second. Sketch the locus of the ant for 1 minute after it leaves O.

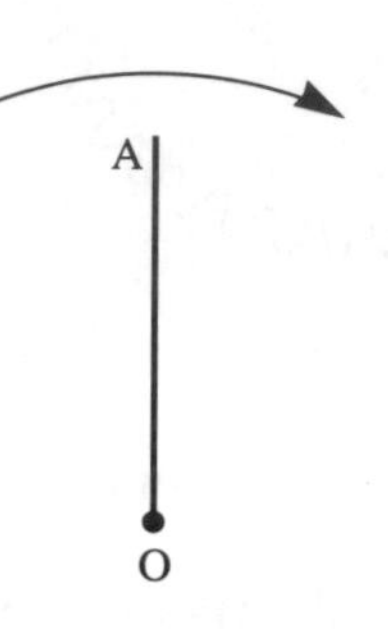

3.3 Pythagoras' Theorem

In a right-angled triangle, the square on the hypotenuse is equal to the sum of the squares on the other two sides.

$a^2 + b^2 = c^2$

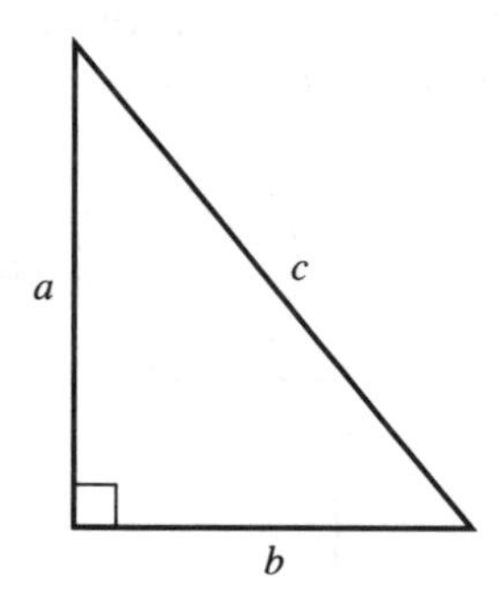

Example 1

Find the side marked *d*.

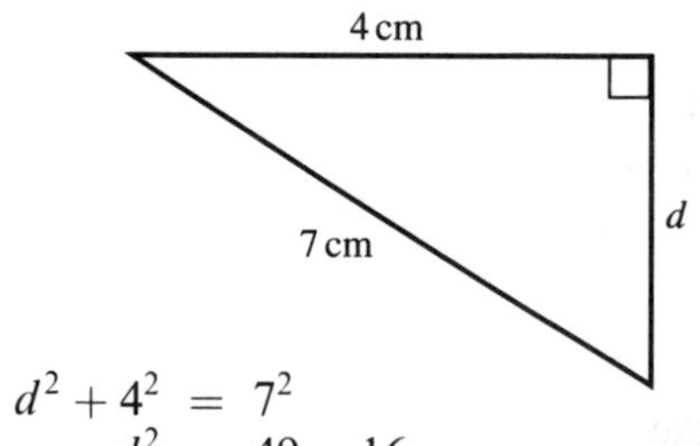

$$d^2 + 4^2 = 7^2$$
$$d^2 = 49 - 16$$
$$d = \sqrt{33} = 5{\cdot}74\text{ cm (3 S.F.)}$$

The *converse* is also true:
'If the square on one side of a triangle is equal to the sum of the squares on the other two sides, then the triangle is right-angled.'

Exercise 3

In Questions **1** to **4**, find x. All the lengths are in cm.

1.

x

8

6

2.

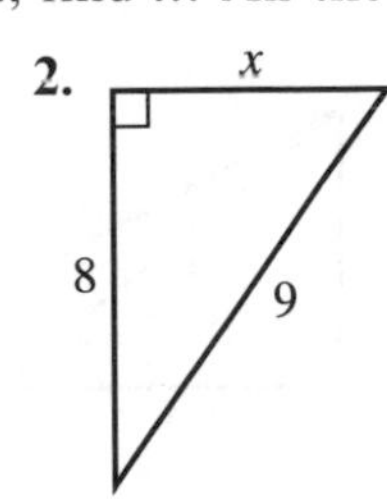

3.

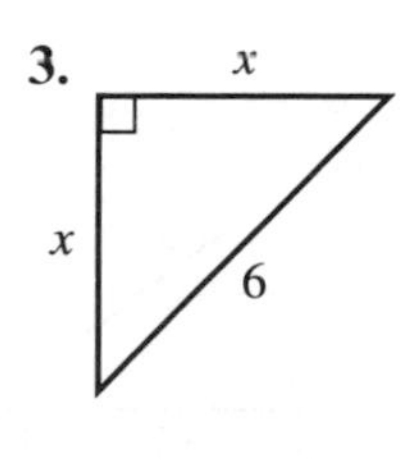

4.

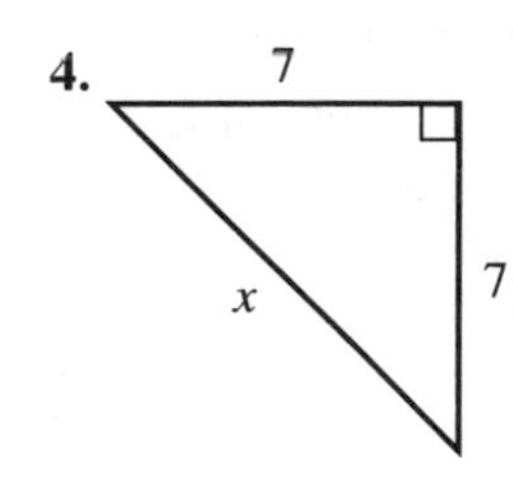

5. Find the length of a diagonal of a rectangle of length 9 cm and width 4 cm.

6. A square has diagonals of length 10 cm. Find the sides of the square.

7. A thin wire of length 18 cm is bent into the shape shown.

Calculate the length from A to B.

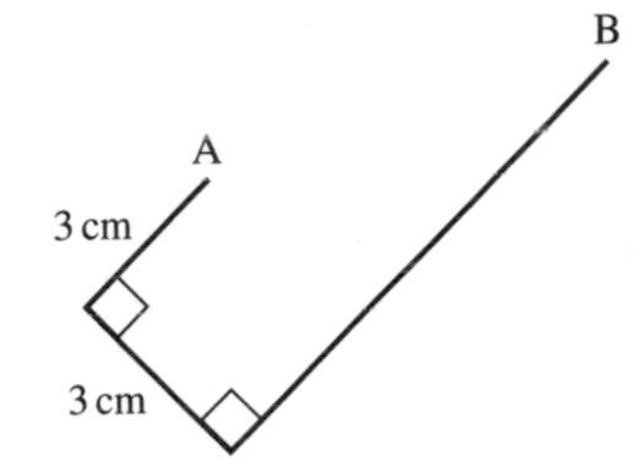

8. A 4 m ladder rests against a vertical wall with its foot 2 m from the wall. How far up the wall does the ladder reach?

9. A ship sails 20 km due North and then 35 km due East. How far is it from its starting point?

10. A paint tin is a cylinder of radius 12 cm and height 22 cm. Leonardo, the painter, drops his stirring stick into the tin and it disappears. Work out the maximum length of the stick.

11. In the diagram, A is at (1, 2) and B is at (6, 4)

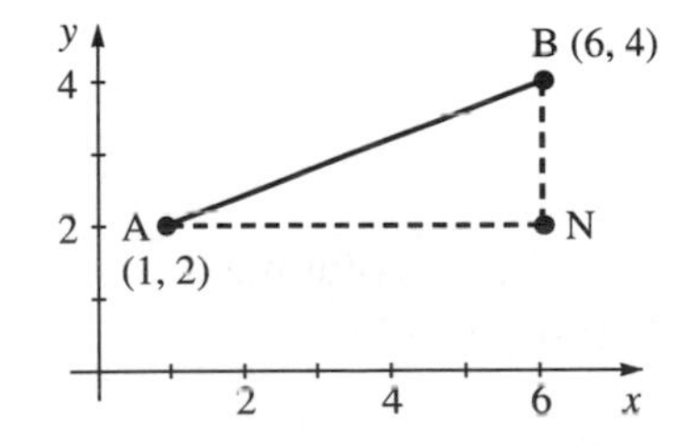

Work out the length AB. (First find the lengths of AN and BN).

12. On squared paper, plot P(1, 3), Q(6, 0), R(6, 6). Find the lengths of the sides of triangle PQR. Is the triangle isosceles?

In questions **13** to **16** find x.

13.

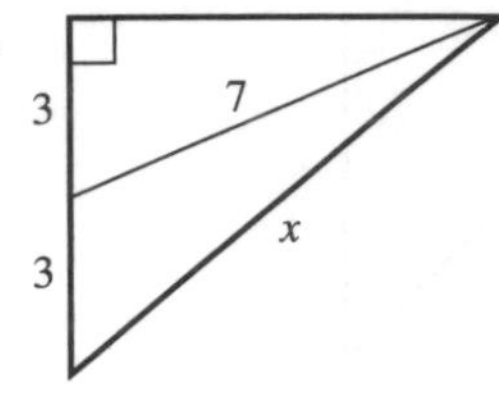

14.

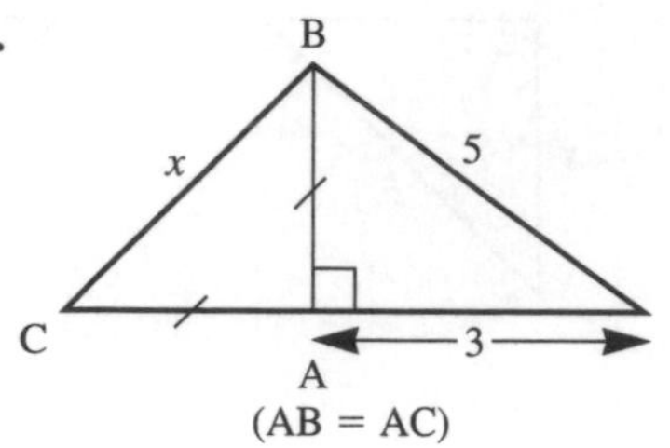

15.

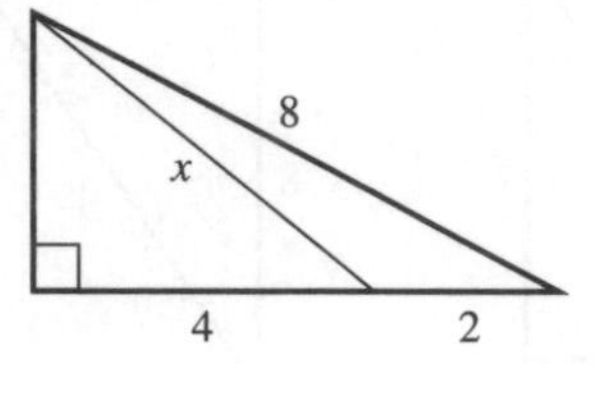

16.

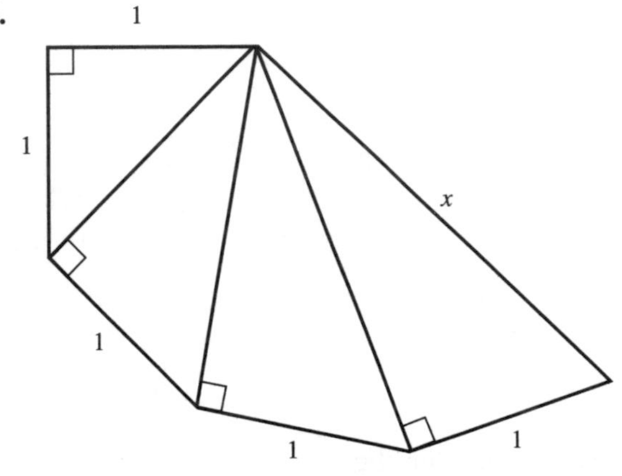

Exercise 4

1. The most well known right-angled triangle is the 3, 4, 5 triangle $[3^2 + 4^2 = 5^2]$.
It is interesting to look at other right-angled triangles where all the sides are whole numbers.

(a) (i) Find c if $a = 5$, $b = 12$
(ii) Find c if $a = 7$, $b = 24$
(iii) Find a if $c = 41$, $b = 40$

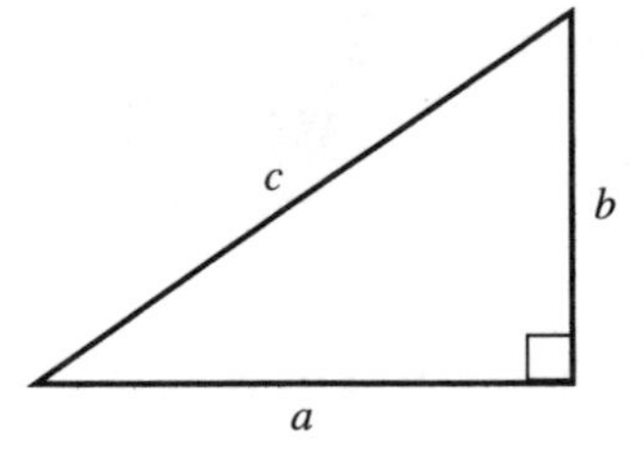

(b) Write the results in a table.

a	b	c
3	4	5
5	12	?
7	24	?
?	40	41

(c) Look at the sequences in the 'a' column and in the 'b' column. Also write down the connection between b and c for each triangle.

(d) Predict the next three sets of values of a, b, c.
Check to see if they really do form right-angled triangles.

2. The diagram shows a cuboid 5 cm by 4 cm by 12 cm.

Calculate (a) AC (b) AD.

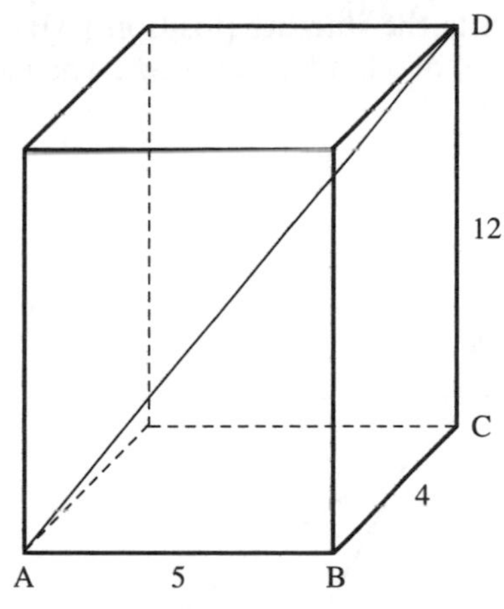

3. Find the length of a diagonal of a rectangular room of length 5 m, width 3 m and height 2·5 m.

4. Find the height of a rectangular box of length 8 cm, width 6 cm where the length of a diagonal is 11 cm.

5. TC is a vertical pole whose base lies at a corner of the horizontal rectangle ABCD.
The top of the pole T is connected by straight wires to points A, B and D.

Calculate (a) TC
(b) TD
(c) (harder) TA.

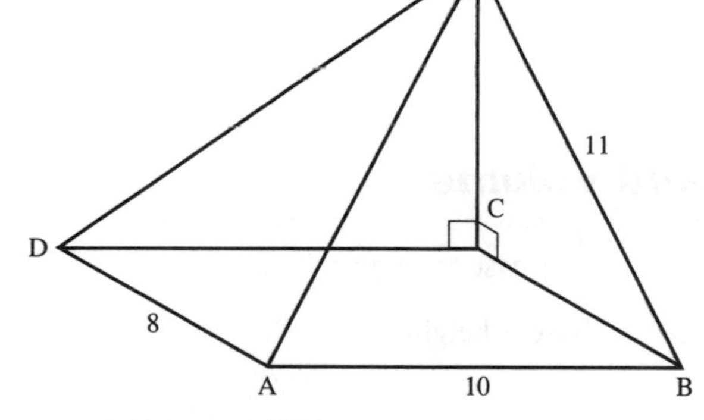

6. The diagonal of a rectangle exceeds the length by 2 cm. If the width of the rectangle is 10 cm, find the length.

7. A ladder reaches H when held vertically against a wall.
When the base is 6 feet from the wall, the top of the ladder is 2 feet lower than H.
How long is the ladder?

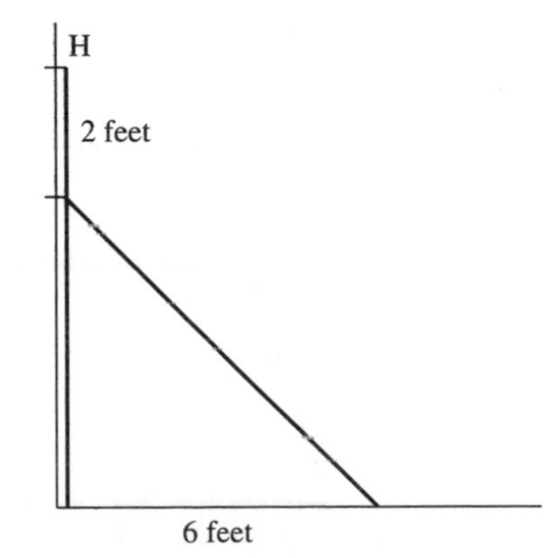

8. The diagram represents the starting position (AB) and the finishing position (CD) of a ladder as it slips. The ladder is leaning against a vertical wall.

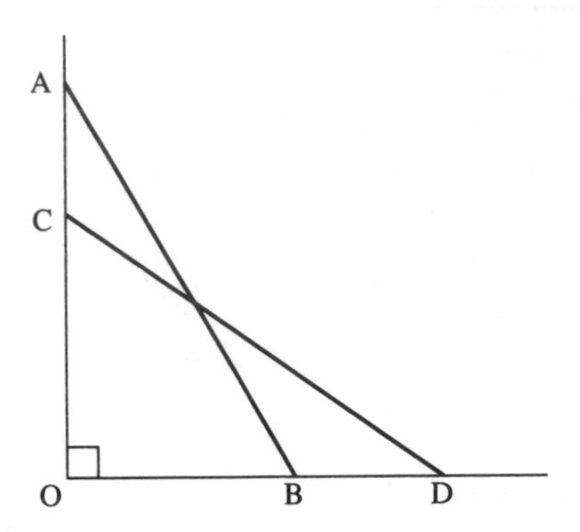

Given: AC $= x$, OC $=$ 4AC, BD $=$ 2AC and OB $= 5$ m.
Form an equation in x, find x and hence find the length of the ladder.

9. An aircraft is vertically above a point which is 10 km West and 15 km North of a control tower. If the aircraft is 4000 m above the ground, how far is it from the control tower?

3.4 Area and volume

For a **triangle,** area $= \frac{1}{2} \times$ base $\times$ height

For a **rectangle,** area $=$ base $\times$ height

Trapezium

A trapezium has two parallel sides.

area $= \frac{1}{2}(a+b) \times h$

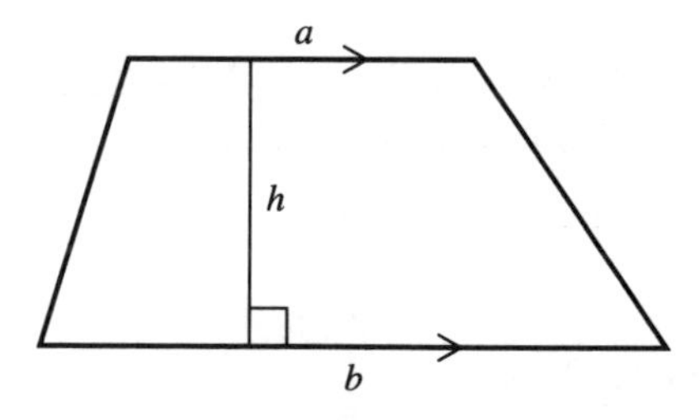

Parallelogram

area $= b \times h$

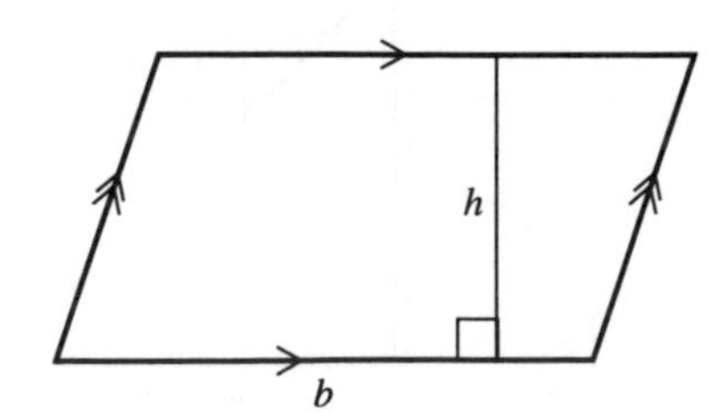

Exercise 5

For Questions **1** to **6**, find the area of each shape. All lengths are in cm.

1.

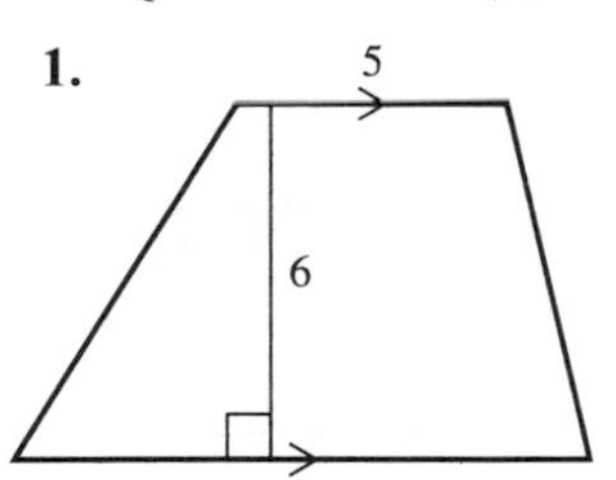

2.

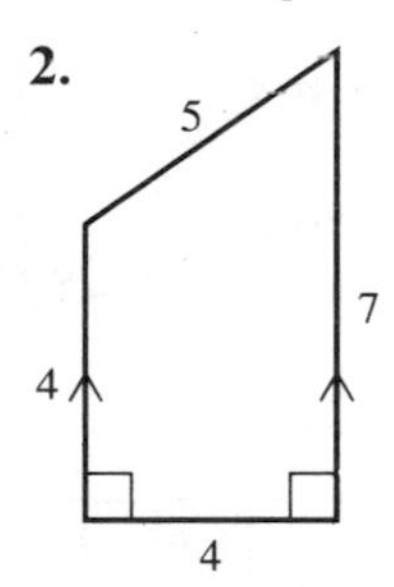

3.

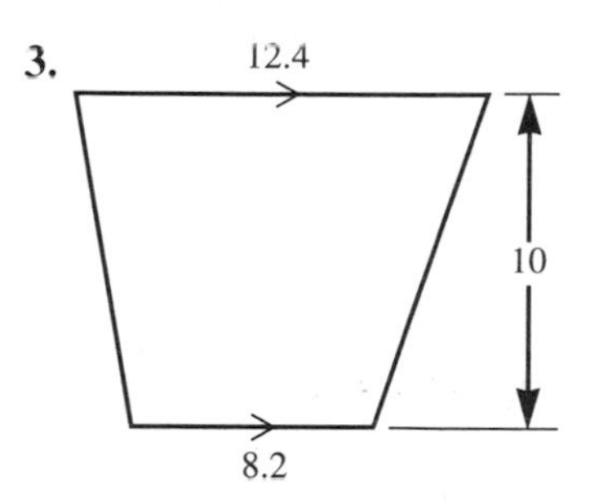

4.

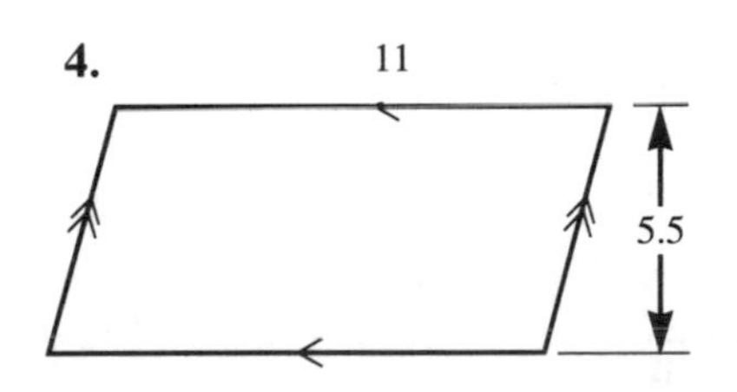

5.

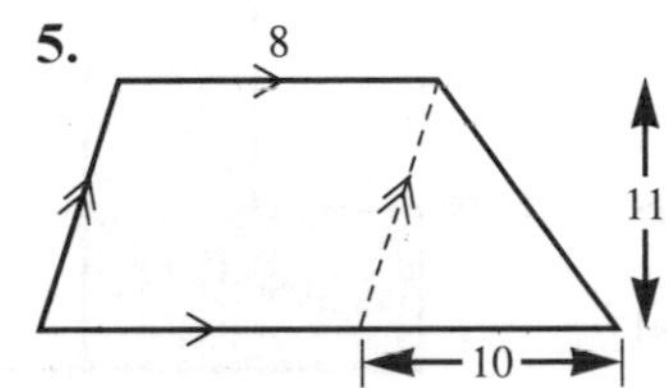

6.

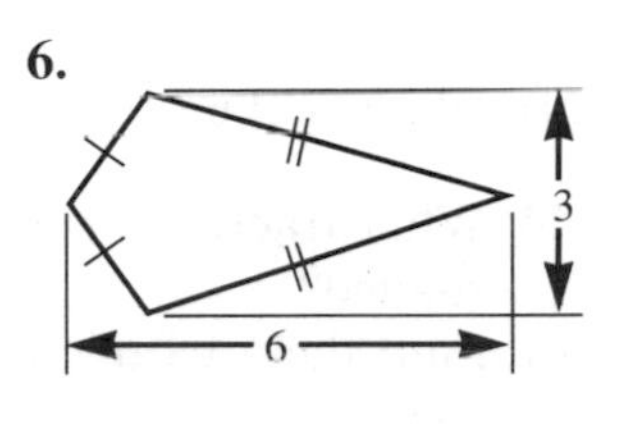

In Questions **7** to **10**, find the area shaded.

7.

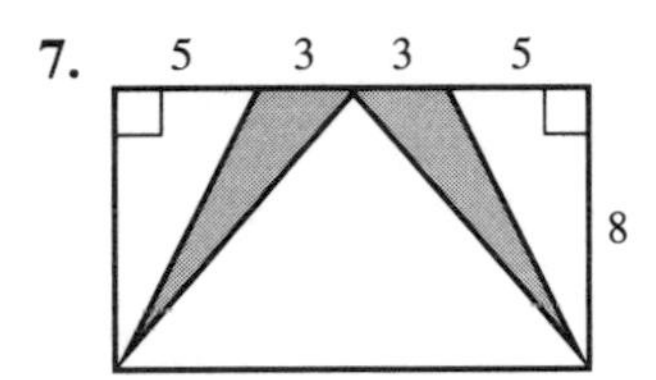

8.

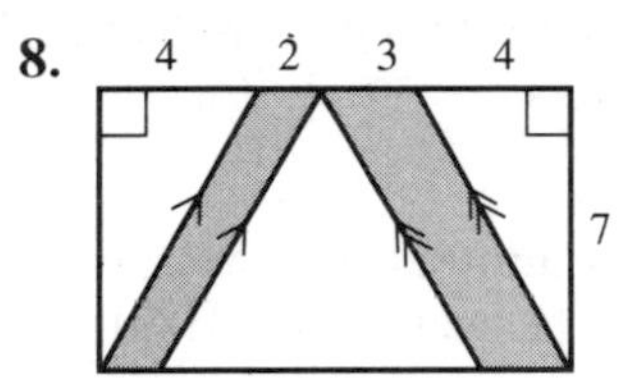

9.

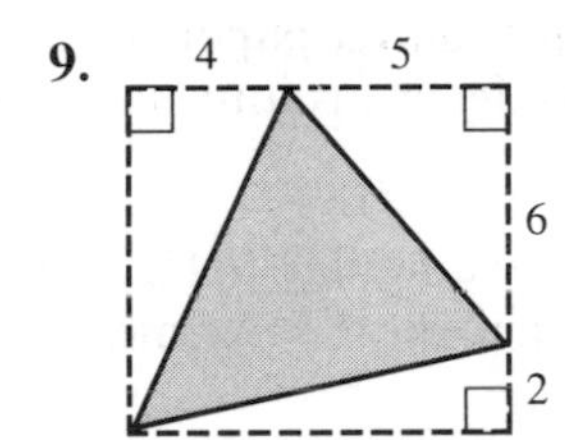

10.

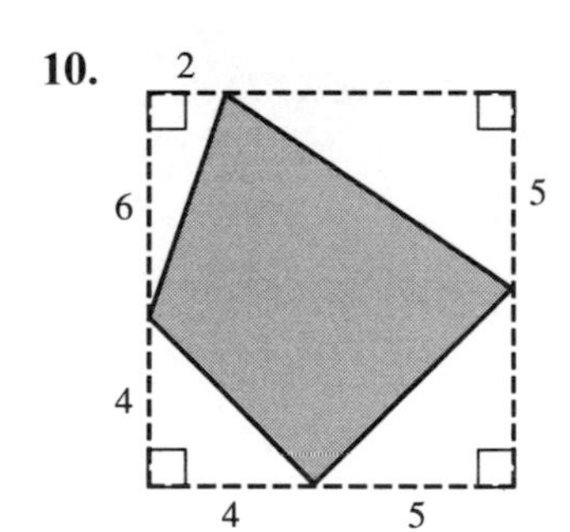

11.

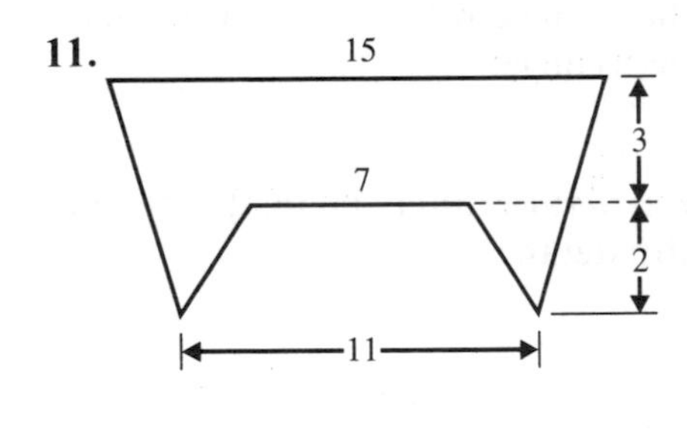

12.

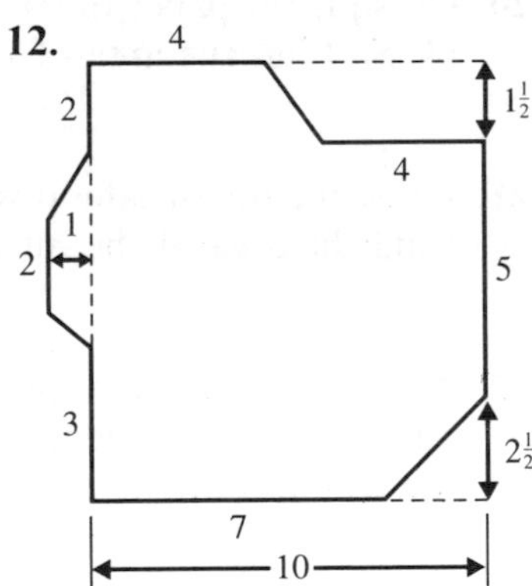

13. A rectangle has an area of $117\,m^2$ and a width of 9 m. Find its length.

14. A trapezium of area $105\,cm^2$ has parallel sides of length 5 cm and 9 cm. How far apart are the parallel sides?

15. A floor 5 m by 20 m is covered by square tiles of side 20 cm. How many tiles are needed?

16. The arrowhead has an area of $3{\cdot}6\,\text{cm}^2$. Find the length x.

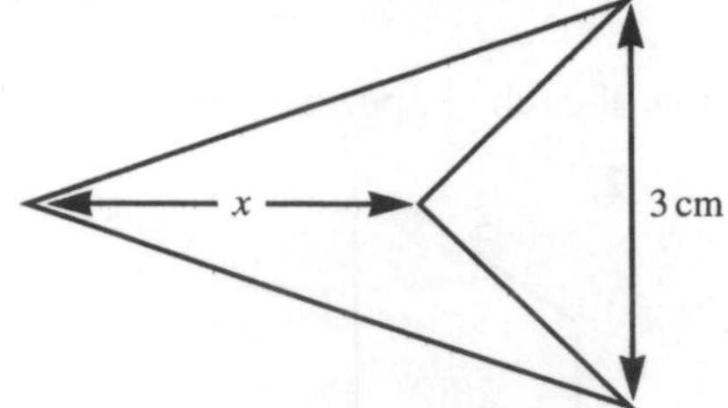

17. The diagram shows a square ABCD in which DX = XY = YC = AW. The area of the square is $45\,\text{cm}^2$.

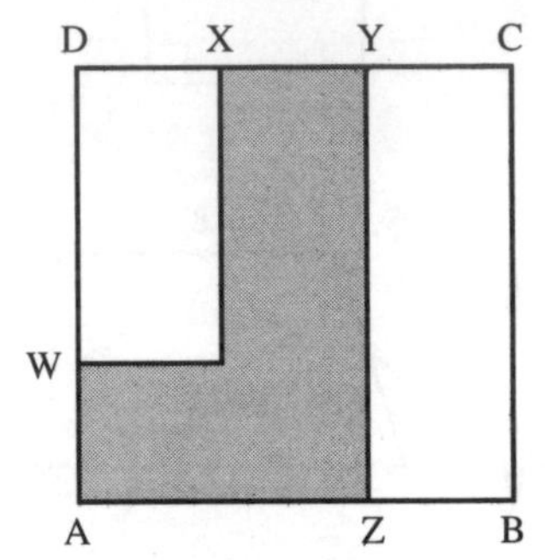

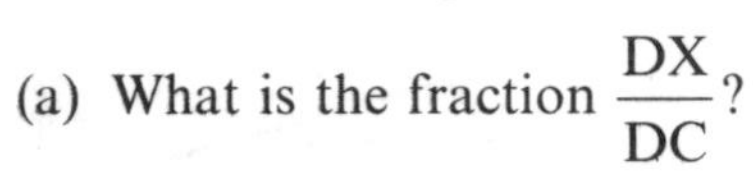

(a) What is the fraction $\frac{DX}{DC}$?

(b) What fraction of the square is shaded?

(c) Find the area of the unshaded part.

18. On squared paper draw a 7×7 square. Design a pattern which divides it up into nine smaller squares.

19. A rectangular field, 400 m long, has an area of 6 hectares. Calculate the perimeter of the field. [1 hectare = $10\,000\,\text{m}^2$].

20. On squared paper draw the triangle with vertices at (1, 1), (5, 3), (3, 5). Find the area of the triangle.

21. Draw the quadrilateral with vertices at (1, 1), (6, 2), (5, 5), (3, 6). Find the area of the quadrilateral.

22. A square wall is covered with square tiles. There are 85 tiles altogether along the two diagonals. How many tiles are there on the whole wall?

23. The side of the small square is half the length of the side of the large square. The L-shape has an area of $75\,\text{cm}^2$. Find the side length of the large square.

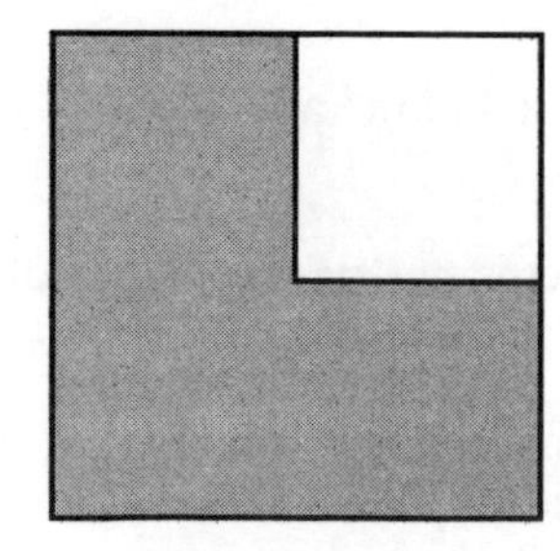

Circles

For any circle, the ratio $\left(\frac{\text{circumference}}{\text{diameter}}\right)$ is equal to π.

The value of π is usually taken to be 3·14, but this is not an exact value. Through the centuries, mathematicians have been trying to obtain a better value for π.

For example, in the third century A.D., the Chinese mathematician Liu Hui obtained the value 3·14159 by considering a regular polygon having 3072 sides! Ludolph van Ceulen (1540-1610) worked even harder to produce a value correct to 35 significant figures. He was so proud of his work that he had this value of π engraved on his tombstone.

Electronic computers are now able to calculate the value of π to many thousands of figures, but its value is still not exact. It was shown in 1761 that π is an *irrational number* which, like $\sqrt{2}$ or $\sqrt{3}$ cannot be expressed exactly as a fraction.

The first fifteen significant figures of π can be remembered from the number of letters in each word of the following sentence.

How I need a drink, cherryade of course, after the silly lectures involving Italian kangaroos.

There remain a lot of unanswered questions concerning π, and many mathematicians today are still working on them.

The following formulae should be memorised.

$$\text{circumference} = \pi d$$
$$= 2\pi r$$
$$\text{area} = \pi r^2$$

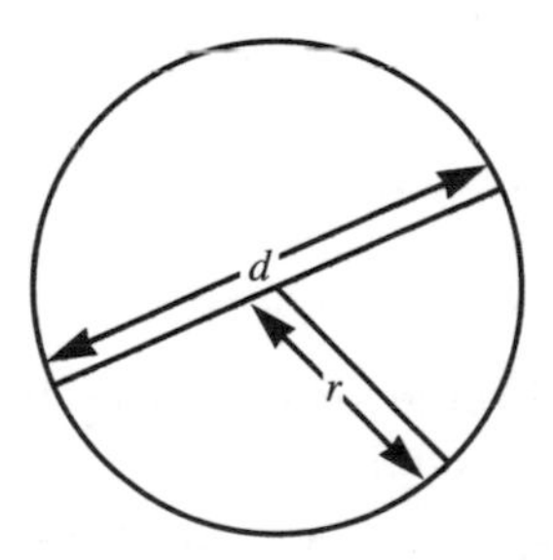

Example 1

Find the circumference and area of a circle of diameter 8 cm. Take π from a calculator.

Circumference $= \pi d$

$= \pi \times 8$

$= 25{\cdot}1$ cm (3 S.F.)

Area $= \pi r^2$

$= \pi \times 4^2$

$= 50{\cdot}3\ \text{cm}^2$ (3 S.F.)

Example 2

Find the perimeter and area of the shape below.

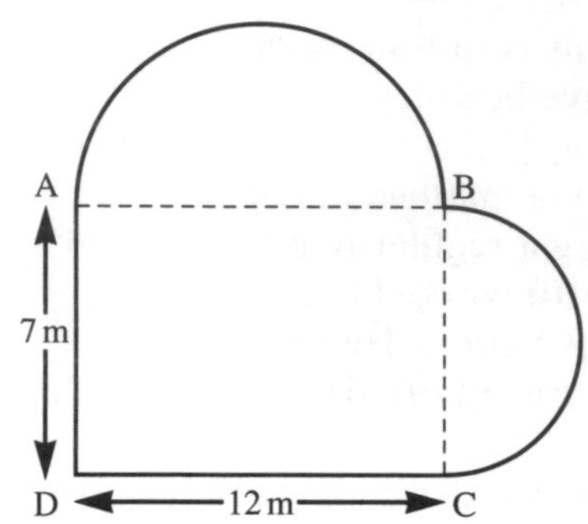

$$\begin{aligned} \text{Perimeter} &= \text{arc AB} + \text{arc BC} + \text{CD} + \text{DA} \\ &= \left(\frac{\pi \times 12}{2}\right) + \left(\frac{\pi \times 7}{2}\right) + 12 + 7 \\ &= 9\tfrac{1}{2}\pi + 19 \\ &= 48{\cdot}8\,\text{m (3 S.F.)} \end{aligned}$$

$$\begin{aligned} \text{Area} &= \text{large semicircle} + \text{small semicircle} + \text{rectangle} \\ &= \left(\frac{\pi \times 6^2}{2}\right) + \left(\frac{\pi \times 3{\cdot}5^2}{2}\right) + (12 \times 7) \\ &= 160\,\text{m}^2 \text{ (3 S.F.)} \end{aligned}$$

Exercise 6

For each shape find (a) the perimeter, (b) the area. All lengths are in cm unless otherwise stated. All the arcs are either semicircles or quarter circles.

1.

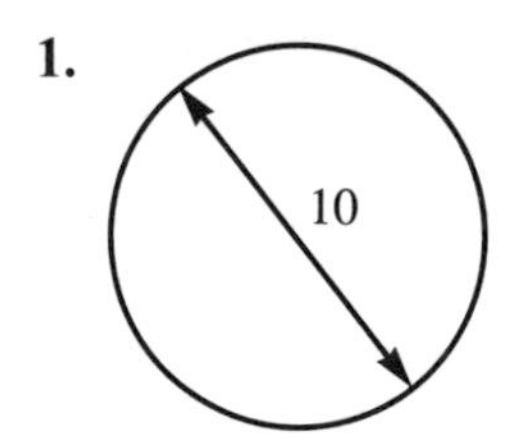

2.

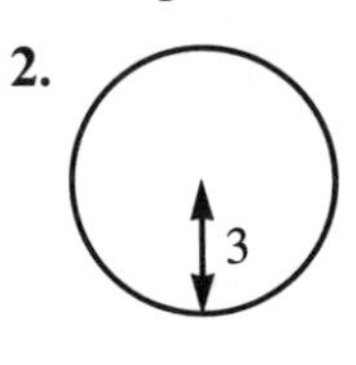

3.

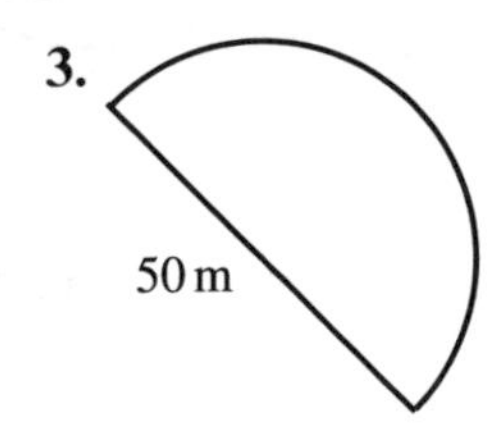

4.

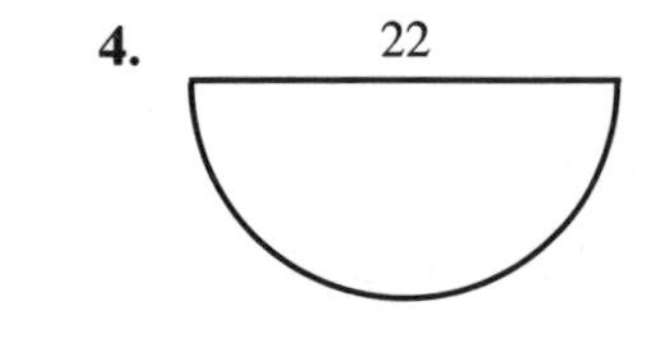

5.

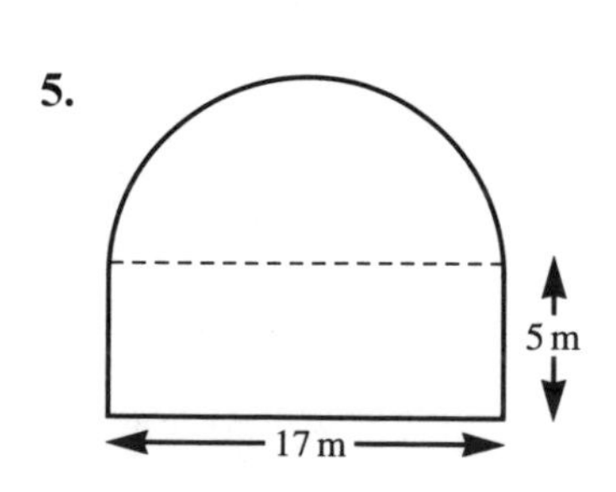

6.

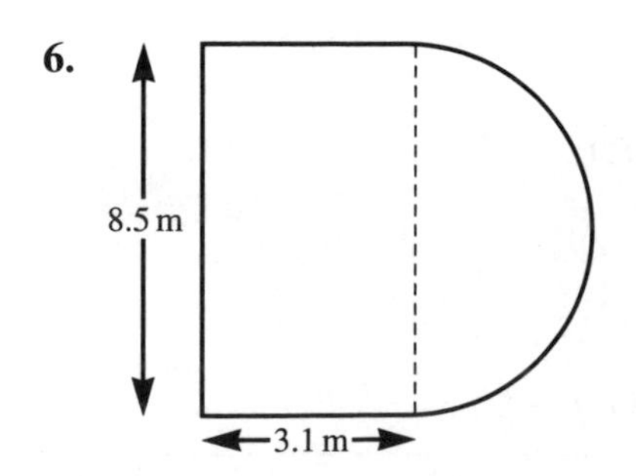

7.

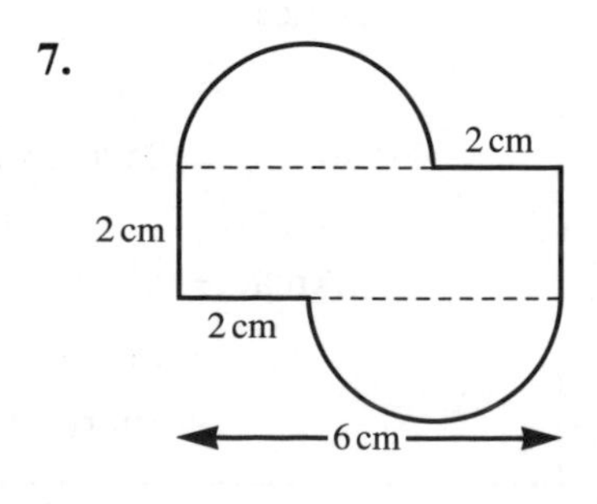

8.

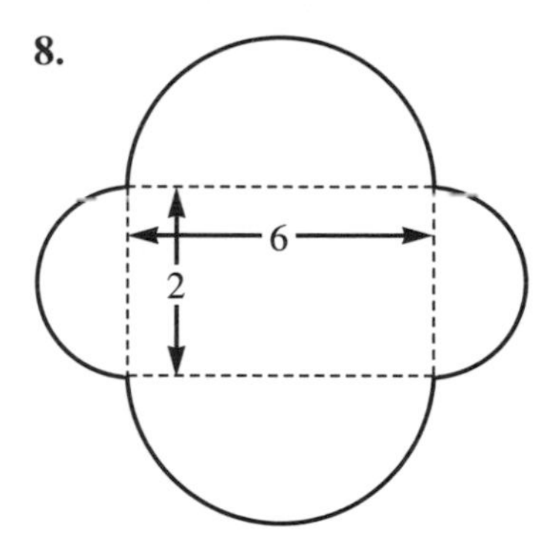

9.

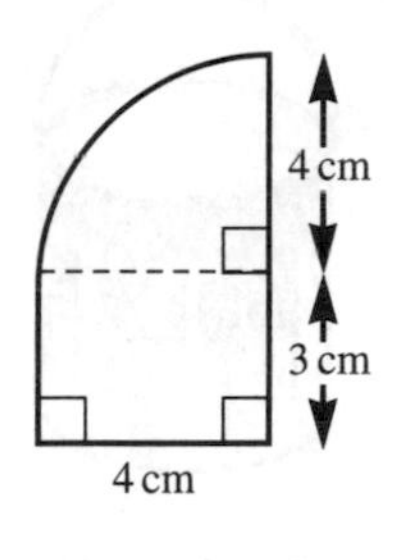

10.

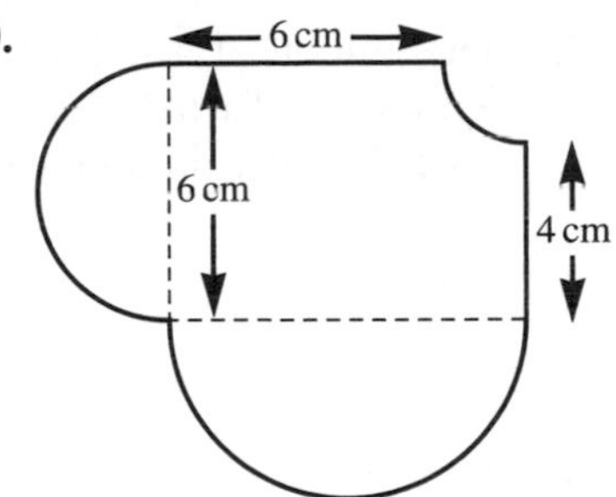

11.

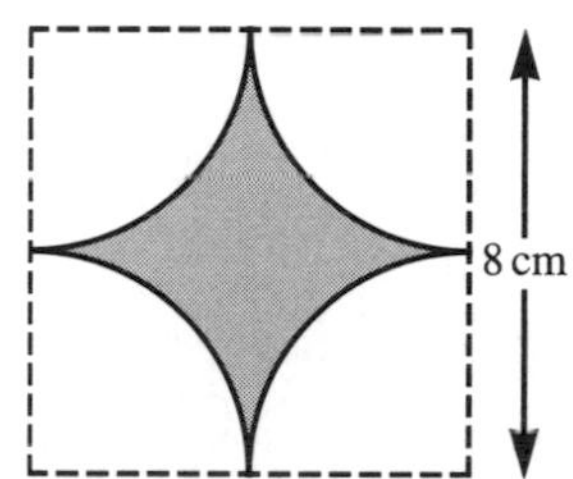

12.

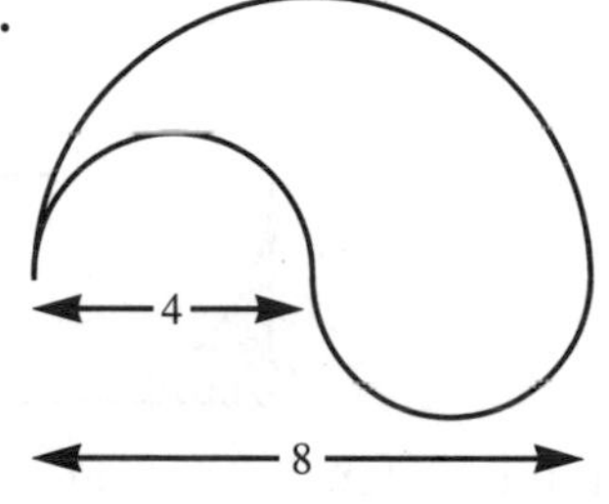

13.

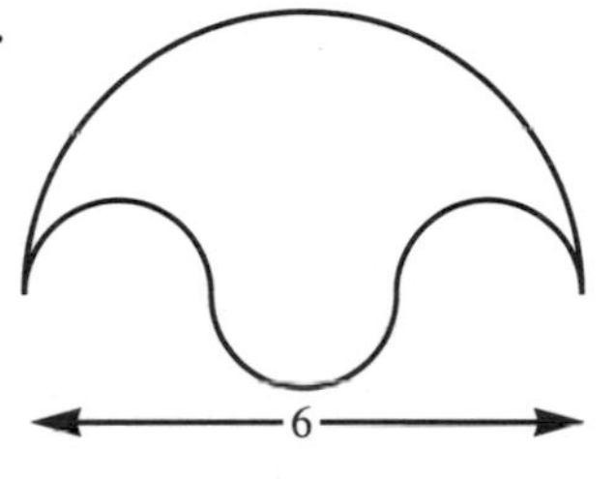

Example 3

A circle has a circumference of 20 m. Find the radius of the circle.

Let the radius of the circle be r m.

$$\text{Circumference} = 2\pi r$$

$$\therefore \quad 2\pi r = 20$$

$$\therefore \quad r = \frac{20}{2\pi}$$

$$r = 3{\cdot}18$$

The radius of the circle is 3·18 m (3 S.F.).

Example 4

A circle has an area of 45 cm^2. Find the radius of the circle.

Let the radius of the circle be r cm.

$$\pi r^2 = 45$$

$$r^2 = \frac{45}{\pi}$$

$$r = \sqrt{\left(\frac{45}{\pi}\right)} = 3{\cdot}78 \text{ (3 S.F.)}$$

The radius of the circle is 3·78 cm.

Exercise 7

1. A circle has an area of 15 cm^2. Find its radius.
2. An odometer is a wheel used by surveyors to measure distances along roads. The circumference of the wheel is one metre. Find the diameter of the wheel.
3. Find the radius of a circle of area 22 km^2.
4. Find the radius of a circle of circumference 58·6 cm.

5. The handle of a paint tin is a semicircle of wire which is 28 cm long. Calculate the diameter of the tin.

6. A circle has an area of $16\,\text{mm}^2$. Find its circumference.

7. A circle has a circumference of 2500 km. Find its area.

8. A circle of radius 5 cm is inscribed inside a square as shown. Find the area shaded.

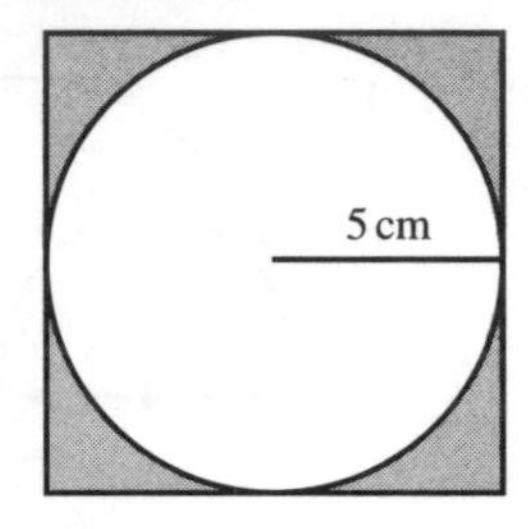

9. Discs of radius 4 cm are cut from a rectangular plastic sheet of length 84 cm and width 24 cm.

How many complete discs can be cut out? Find
(a) the total area of the discs cut
(b) the area of the sheet wasted.

10. The tyre of a car wheel has an outer diameter of 30 cm. How many times will the wheel rotate on a journey of 5 km?

11. A golf ball of diameter 1·68 inches rolls a distance of 4 m in a straight line. How many times does the ball rotate completely? (1 inch = 2·54 cm)

12. A circular pond of radius 6 m is surrounded by a path of width 1 m.
(a) Find the area of the path.
(b) The path is resurfaced with astroturf which is bought in packs each containing enough to cover an area of $7\,\text{m}^2$. How many packs are required?

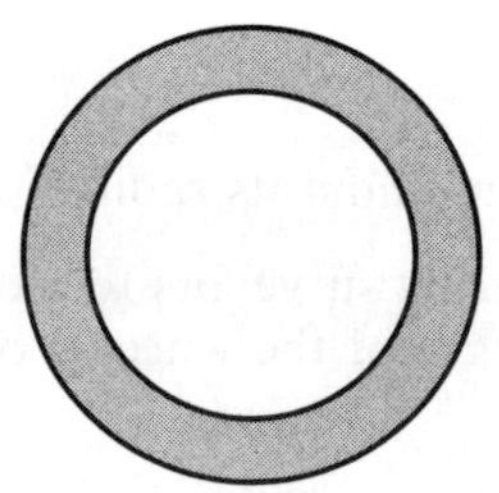

13. A rectangular metal plate has a length of 65 cm and a width of 35 cm. It is melted down and recast into circular discs of the same thickness. How many complete discs can be formed if
(a) the radius of each disc is 3 cm?
(b) the radius of each disc is 10 cm?

14. Calculate the radius of a circle whose area is equal to the sum of the areas of three circles of radii 2 cm, 3 cm and 4 cm respectively.

15. The diagram below shows a lawn (unshaded) surrounded by a path of uniform width (shaded). The curved end of the lawn is a semicircle of diameter 10 m.

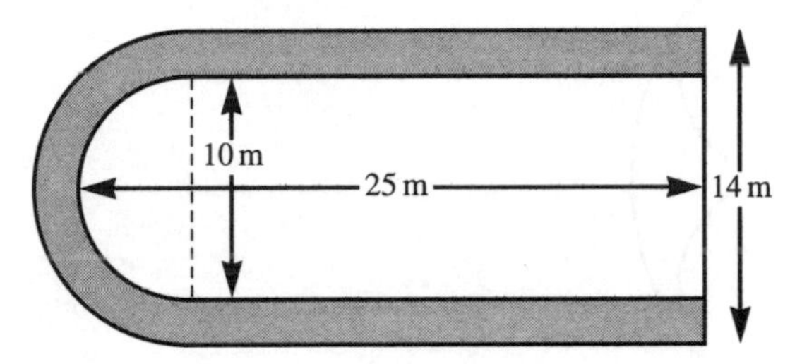

Calculate the total area of the path.

16. The diameter of a circle is given as 10 cm, correct to the nearest cm. Calculate
(a) the maximum possible circumference
(b) the minimum possible area of the circle consistent with this data.

17. A square is inscribed in a circle of radius 7 cm. Find
(a) the area of the square
(b) the area shaded.

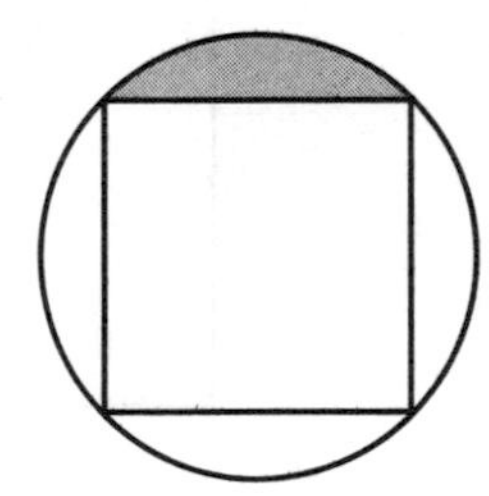

18. The governor of a prison has 100 m of wire fencing. What area can he enclose if he makes a circular compound?

19.† The semicircle and the isosceles triangle have the same base AB and the same area. Find the angle x.

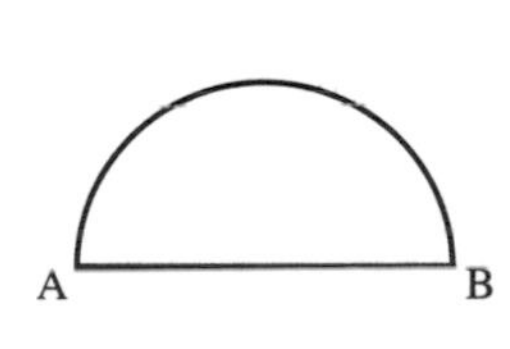

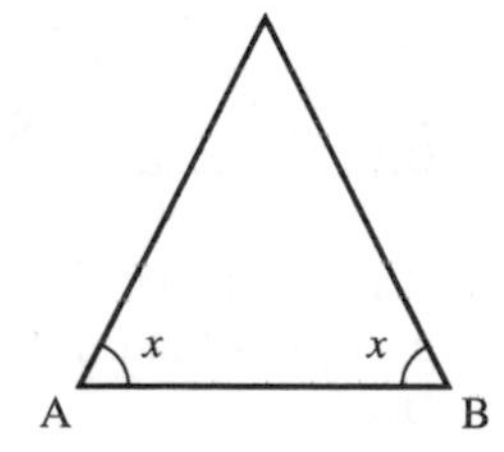

20.† Mr Gibson decided to measure the circumference of the Earth using a very long tape measure. For reasons best known to himself he held the tape measure 1 m from the surface of the (perfectly spherical) Earth all the way round. When he had finished Mrs Gibson told him that his measurement gave too large an answer. She suggested taking off 6 m. Was she correct? [Take the radius of the Earth to be 6400 km (if you need it)].

21.† The large circle has a radius of 10 cm. Find the radius of the largest circle which will fit in the middle.

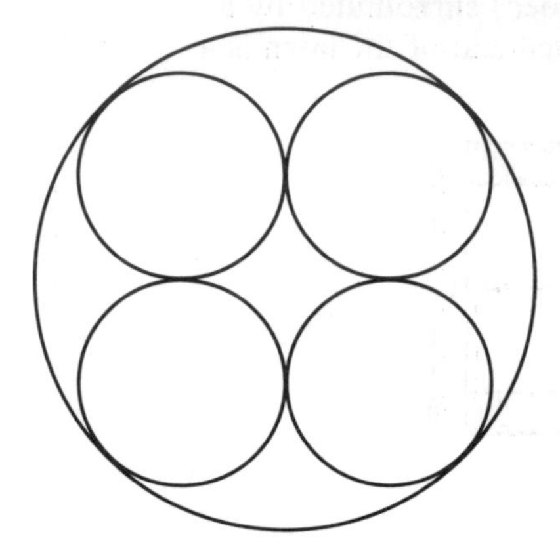

Mixed questions

Exercise 8

1. The sloping line divides the area of the square in the ratio 1 : 5. What is the ratio $a : b$?

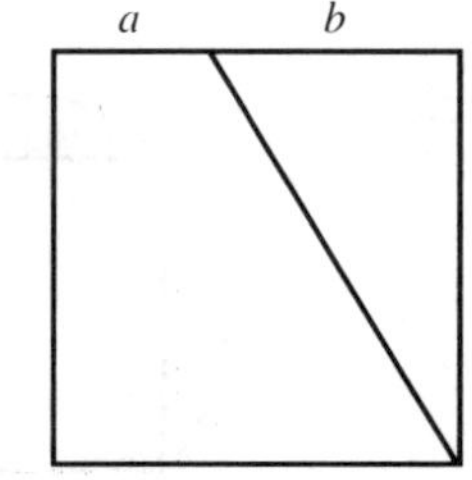

2. ABCD is a square of side 14 cm. P, Q, R, S are the mid-points of the sides. Semicircles are drawn with centres P, Q, R, S as shown on the diagram. Find the shaded area, using $\pi = \dfrac{22}{7}$.

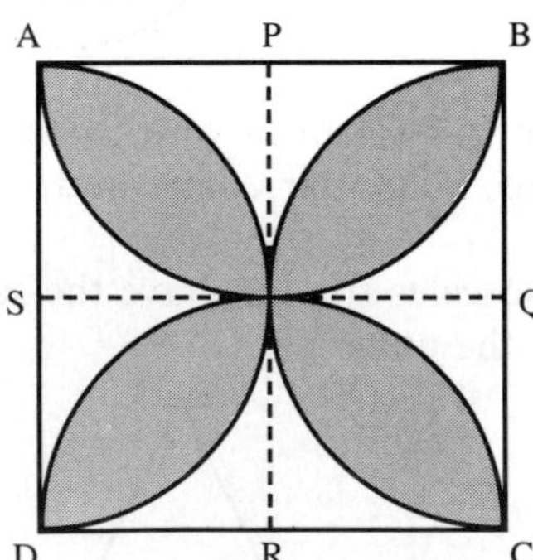

3. (Hard) Two equal circles are cut from a square piece of paper of side 2 m. Calculate the radius of the largest possible circles. Give your answer correct to 3 significant figures.

4. What fraction of the area of a circle of radius 4 cm is more than 3 cm from the centre?

5. These circles have radii 2, 3, 5 cm. Express the shaded area as a percentage of the area of the largest circle.

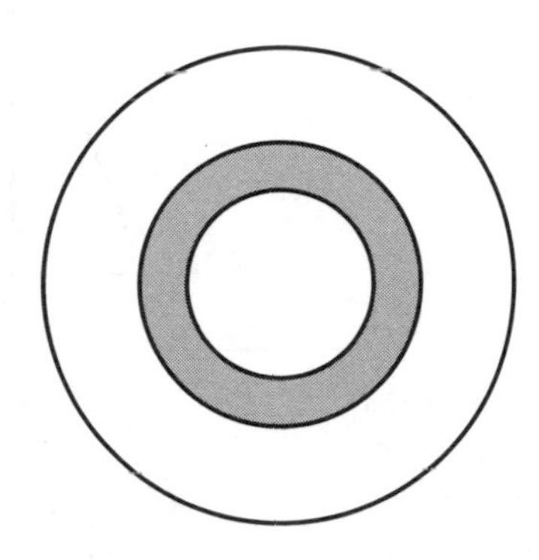

6. PQR is a right-angled triangle with vertices at P(2, 1), Q(4, 3), R(8, t). Given that angle PQR is 90°, find t.

7. A square of side x cm is drawn in the middle of a square of side 10 cm. The corners of the smaller square are 2 cm from the corners of the larger square and are on the diagonals of the larger square. Find x correct to 4 s.f.

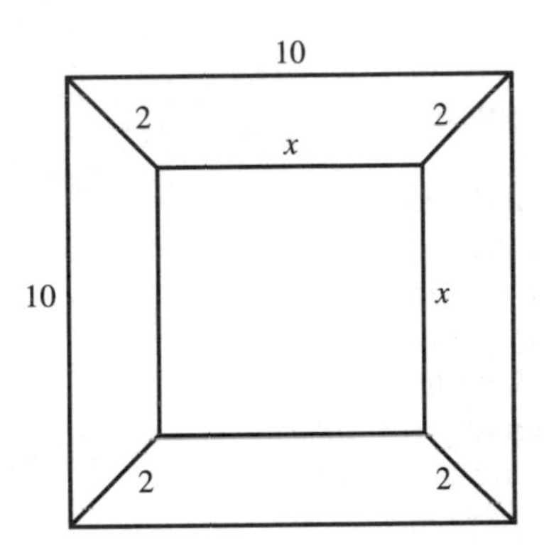

8. ABCDEFGH is a regular octagon. Calculate angle ACD.

9. The diagram shows two diagonals drawn on the faces of a cube. Calculate the angle between the diagonals.

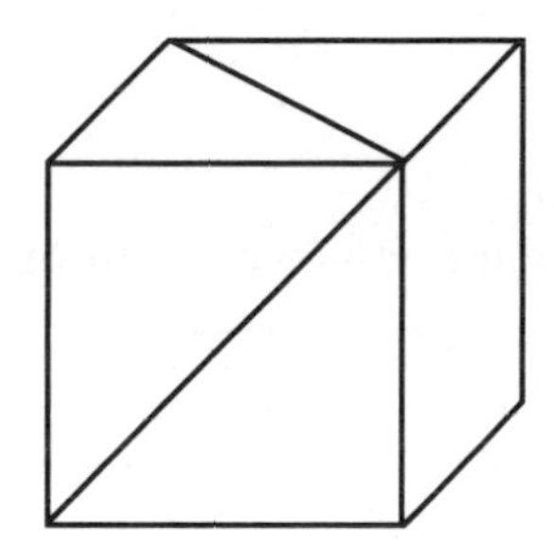

10.† The radius of the circle inscribed in a regular hexagon has length 4 cm. Find the area of the hexagon, correct to 3 s.f.

11.† In the diagram AB = BD = DC and $A\hat{D}C - D\hat{A}C = 70°$

Find $A\hat{D}B$.

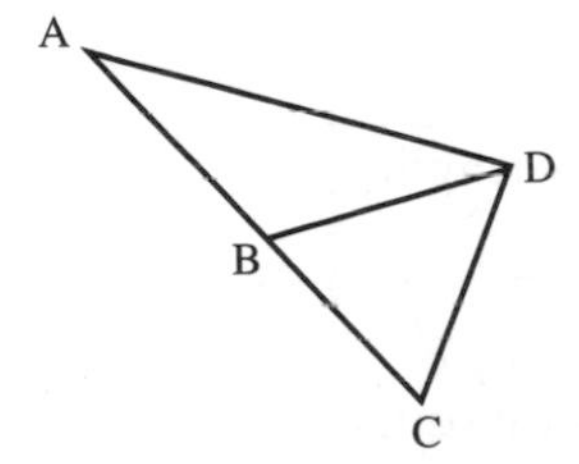

12.† In $\triangle ABC$, $AC = BC = 12$ cm.
CD is perpendicular to AB.
MD = 1 cm.
M is equidistant from A, B and C.
Calculate the length CM.

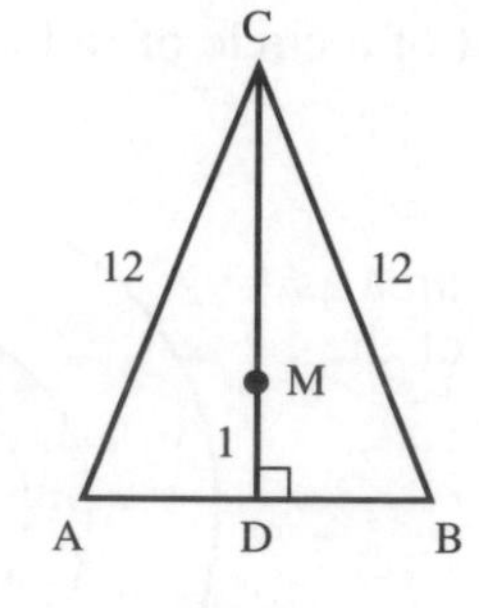

13.† The circumcircle and the inscribed circle of an equilateral triangle are shown. Calculate the ratio of the area of the circumcircle to the area of the inscribed circle.

14.† ABCDEFGH is a regular octagon of side 1 unit.
(a) Calculate the length AF
(b) Calculate the length AC
(c) (Much harder) You probably used a calculator to work out AF. Can you calculate the *exact* value of AF? Leave your answer in a form involving a square root.

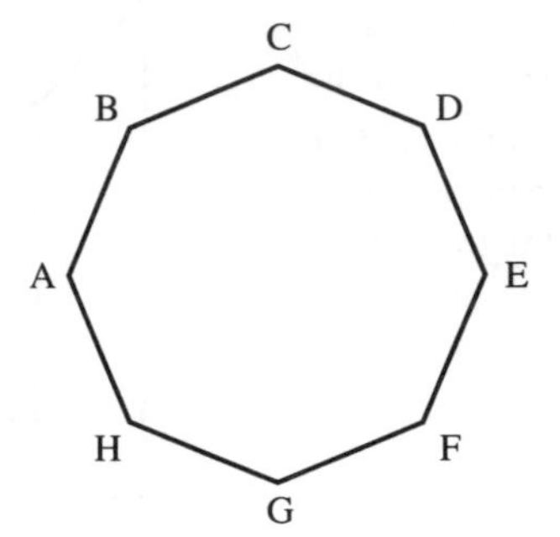

Volume

A **prism** is an object with the same cross section throughout its length.

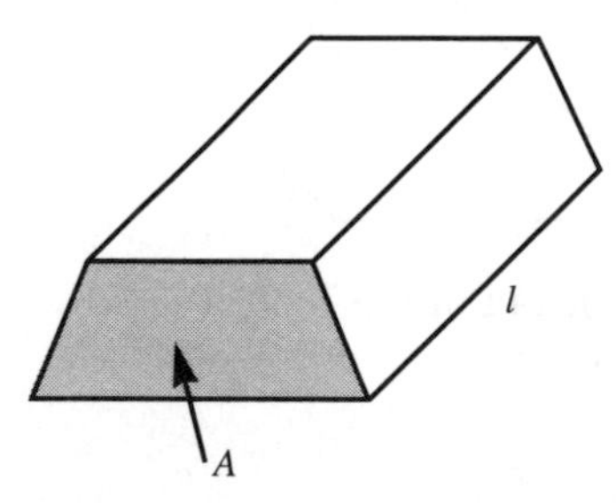

Volume of prism = (area of cross section) × length
$= A \times l$.

A **cuboid** is a prism whose six faces are all rectangles. A cube is a special case of a cuboid in which all six faces are squares.

A **cylinder** is a prism whose cross section is a circle.

radius $= r$
height $= h$

h

Volume of cylinder = area of cross section × length
Volume $= \pi r^2 h$

Example 5

Calculate the height of a cylinder of volume $500\,\text{cm}^3$ and base radius 8 cm.

Let the height of the cylinder be h cm.

$$\pi r^2 h = 500$$
$$3{\cdot}14 \times 8^2 \times h = 500$$
$$h = \frac{500}{3{\cdot}14 \times 64}$$
$$h = 2{\cdot}49 \text{ (3 S.F.)}$$

The height of the cylinder is 2·49 cm.

Exercise 9

1. Calculate the volume of the prisms. All lengths are in cm.

(a)

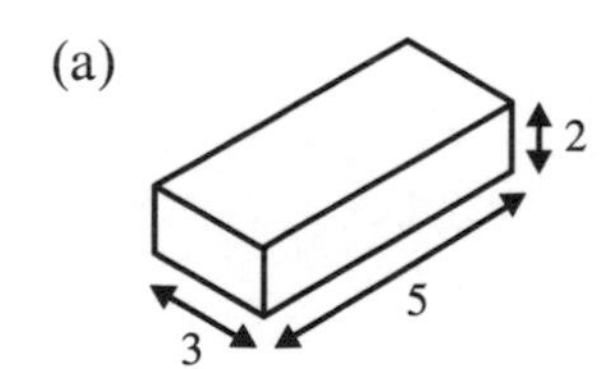

(b)

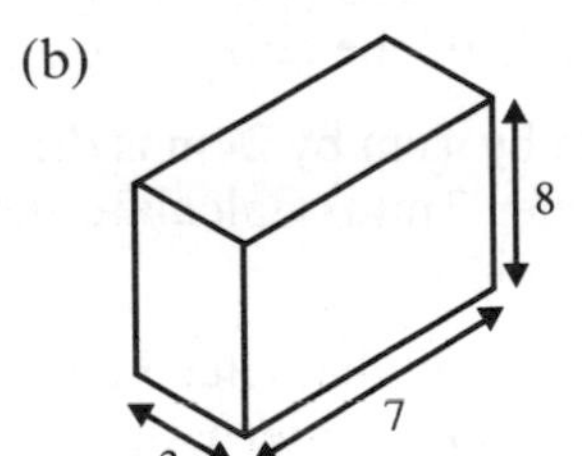

(c)

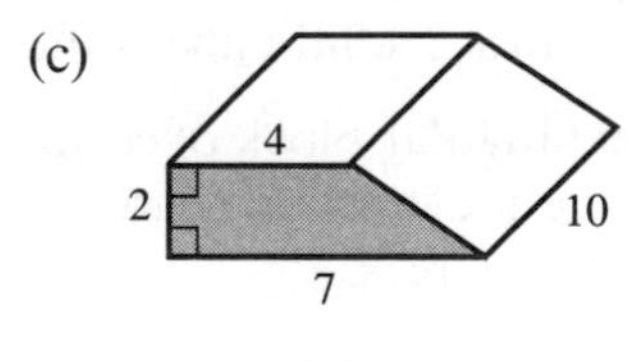

(d)

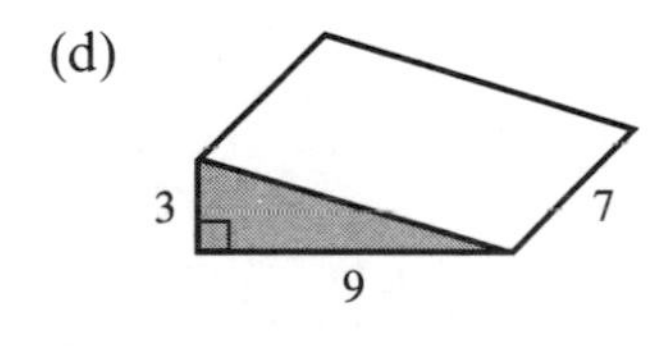

(e)

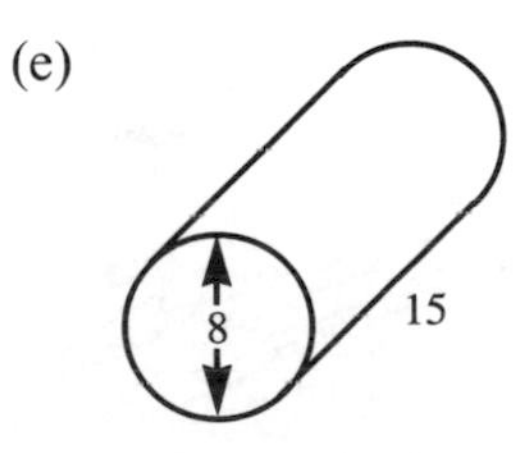

(f)

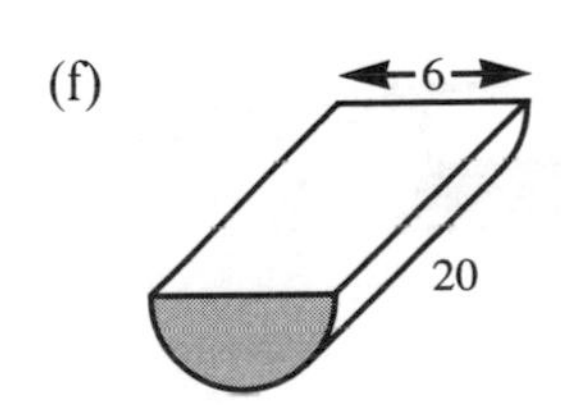

2. Calculate the volume of the following cylinders
(a) $r = 4\,\text{cm}$, $h = 10\,\text{cm}$
(b) $r = 11\,\text{m}$, $h = 2\,\text{m}$

3. The diagram shows a view of the water in a swimming pool.

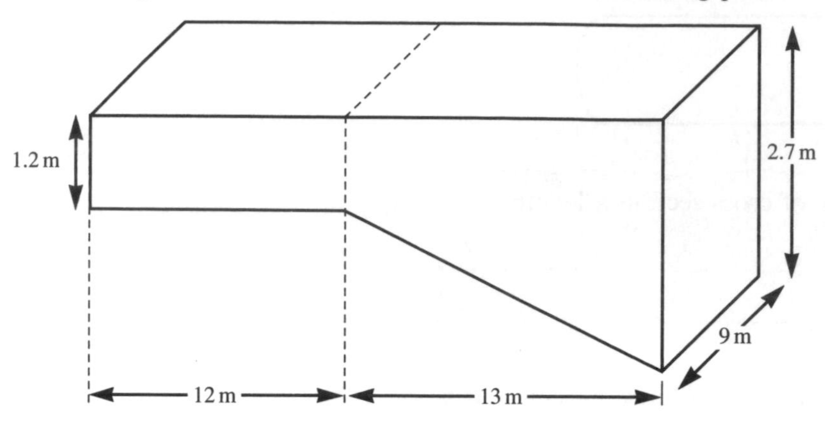

Calculate the volume of water in the pool.

4. Find the height of a cylinder of volume 200 cm^3 and radius 4 cm.

5. Find the length of a cylinder of volume 2 litres and radius 10 cm.

6. Find the radius of a cylinder of volume 45 cm^3 and length 4 cm.

7. When 3 litres of oil is removed from an upright cylindrical can, the level falls by 10 cm. Find the radius r of the can.

8. A solid cylinder of radius 4 cm and length 8 cm is melted down and recast into a solid cube. Find the side of the cube.

9. A solid rectangular block of copper 5 cm by 4 cm by 2 cm is drawn out to make a cylindrical wire of diameter 2 mm. Calculate the length of the wire.

10. Water flows through a circular pipe of internal diameter 3 cm at a speed of 10 cm/s. If the pipe is full, how much water issues from the pipe in one minute? (answer in litres)

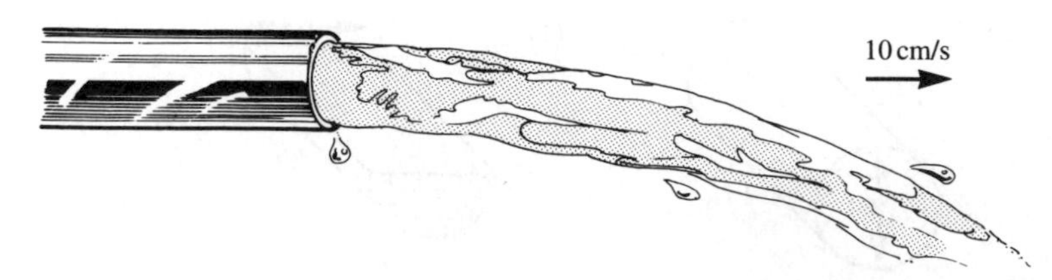

11. Water issues from a hose-pipe of internal diameter 1 cm at a rate of 5 litres per minute. At what speed is the water flowing through the pipe?

12. A cylindrical metal pipe has external diameter of 6 cm and internal diameter of 4 cm. Calculate the volume of metal in a pipe of length 1 m. If 1 cm^3 of the metal weighs 8 g, find the weight of the pipe.

13. Mr Gibson decided to build a garage and began by calculating the number of bricks required. The garage was to be 6 m by 4 m and 2·5 m in height. Each brick measures 22 cm by 10 cm by 7 cm. Mr Gibson estimated that he would need about 40 000 bricks. Is this a reasonable estimate?

14. A cylindrical can of internal radius 20 cm stands upright on a flat surface. It contains water to a depth of 20 cm. Calculate the rise h in the level of the water when a brick of volume 1500 cm^3 is immersed in the water.

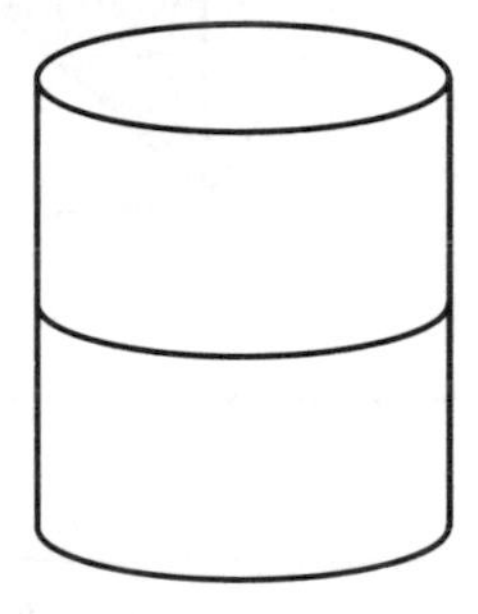

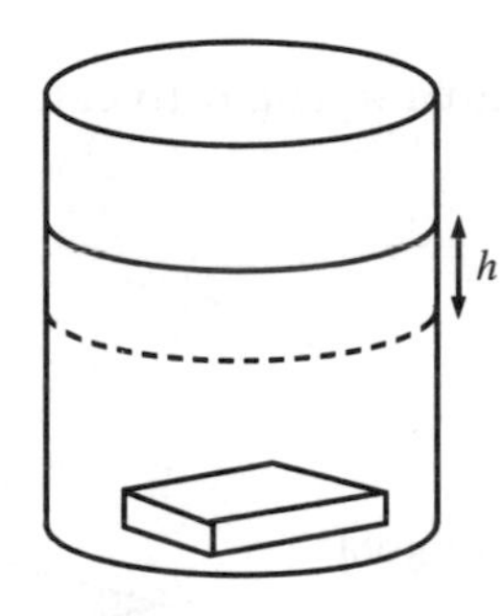

15. A cylindrical tin of height 15 cm and radius 4 cm is filled with sand from a rectangular box. How many times can the tin be filled if the dimensions of the box are 50 cm by 40 cm by 20 cm?

16. Rain which falls onto a flat rectangular surface of length 6 m and width 4 m is collected in a cylinder of internal radius 20 cm. What is the depth of water in the cylinder after a storm in which 1 cm of rain fell?

Nets

If the cube shown was made of cardboard, it could be cut along some of the edges and then laid out flat to form the *net* of the cube.

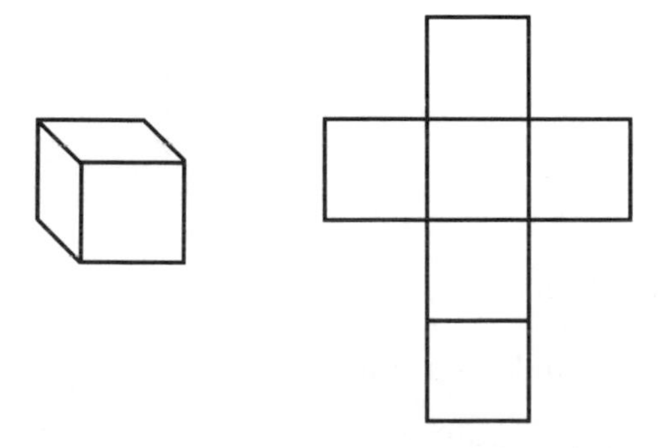

Exercise 10

1. Which of the nets below can be used to make a cube?

(a) 

(b) 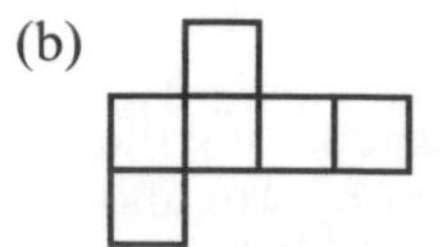

(c) 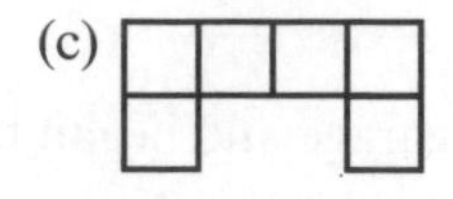

(d) 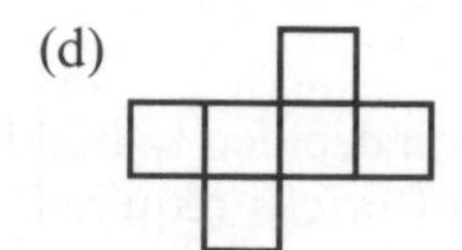

2. The diagram shows the net of a pyramid. The base is shaded. The lengths are in cm.

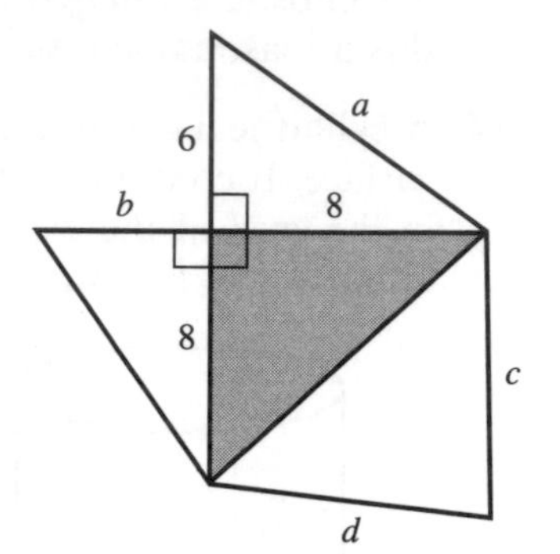

(a) How many edges will the pyramid have?
(b) How many vertices will it have?
(c) Find the lengths a, b, c, d.
(d) Use the formula $V = \frac{1}{3}$ base area $\times$ height to calculate the volume of the pyramid.

3. The diagram shows the net of a prism.

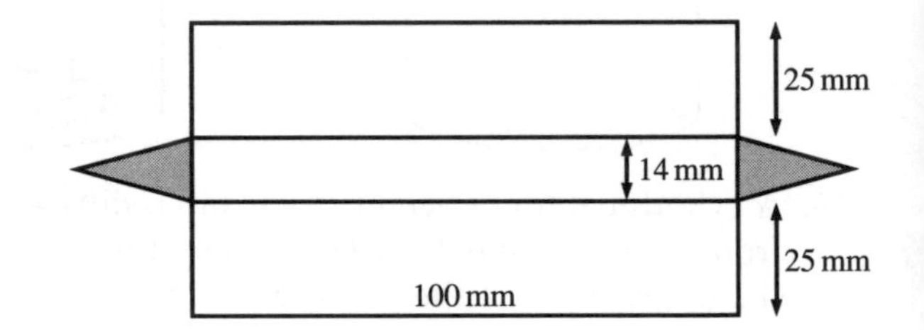

(a) Find the area of one of the triangular faces (shown shaded).
(b) Find the volume of the prism.

4. This is the net of a square-based pyramid.

What are the lengths a, b, c, x, y?

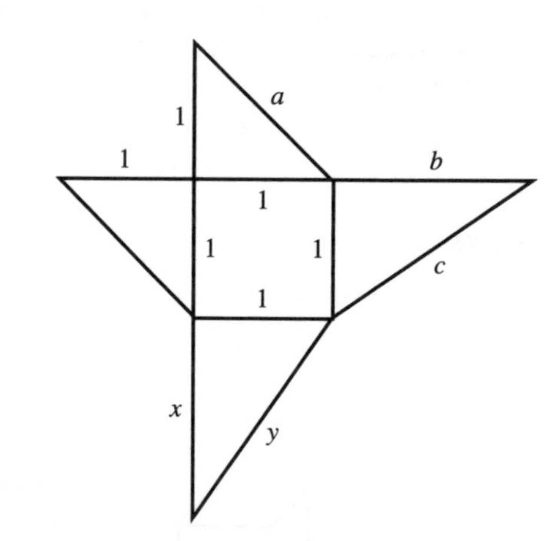

5. Sketch nets for the following
(a) Closed rectangular box 7 cm × 9 cm × 5 cm.
(b) Closed cylinder: length 10 cm, radius 6 cm.
(c) Prism of length 12 cm, cross section an equilateral triangle of side 4 cm.

3.5 Similarity and enlargement

If one shape is an enlargement of another, the two shapes are mathcmatically *similar*.

The two triangles A and B are similar if they have the same angles.

For other shapes to be similar, not only must corresponding angles be equal, but also corresponding edges must be in the same proportion.

The two quadrilaterals C and D are similar. All the edges of shape D are twice as long as the edges of shape C.

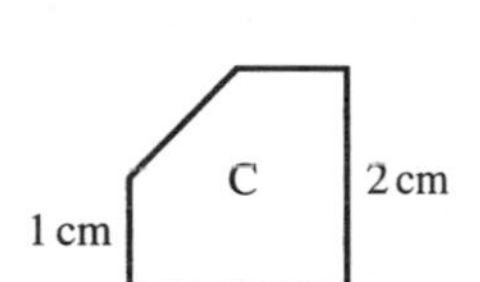

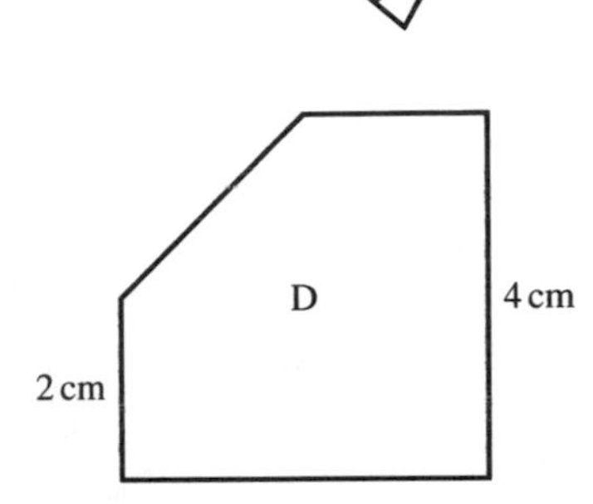

The two rectangles E and F are not similar even though they have the same angles.

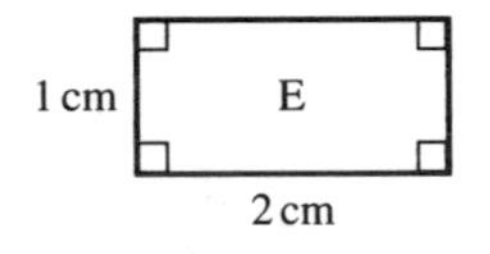

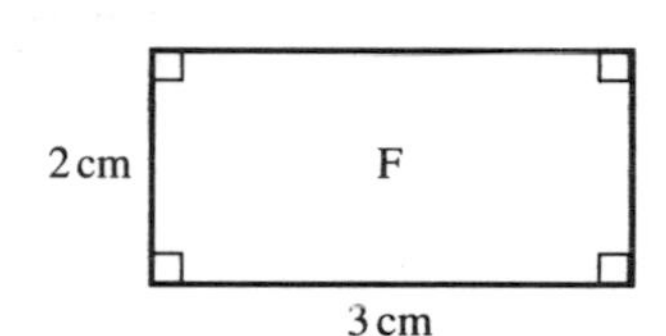

Example 1

In the triangles PQR and PST, QR is parallel to ST.

(a) Show that the triangles are similar.

(b) Given the lengths shown, calculate QS.

(a) In triangles PQR and PST,
angle P is common,
$P\hat{Q}R = P\hat{S}T$ (corresponding angles).
$\therefore$ triangles PQR and PST are similar.

(b) $\dfrac{ST}{QR} = \dfrac{PS}{PQ}$ (corresponding sides are proportional)

$\dfrac{7}{5} = \dfrac{PS}{4}$

$\therefore \quad PS = \frac{28}{5} = 5\frac{3}{5}$

$\therefore \quad QS = 5\frac{3}{5} - 4 = 1\frac{3}{5}$

Exercise 11

1. Which of the shapes B, C, D is similar to shape A?

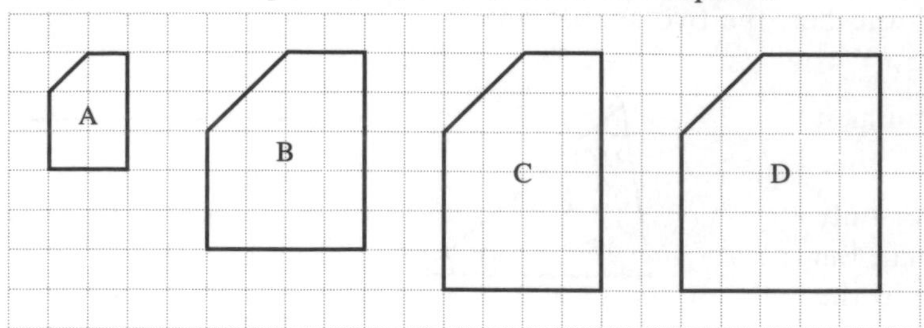

In Questions **2** to **12**, find the sides marked with letters; all lengths are given in cm.

2.

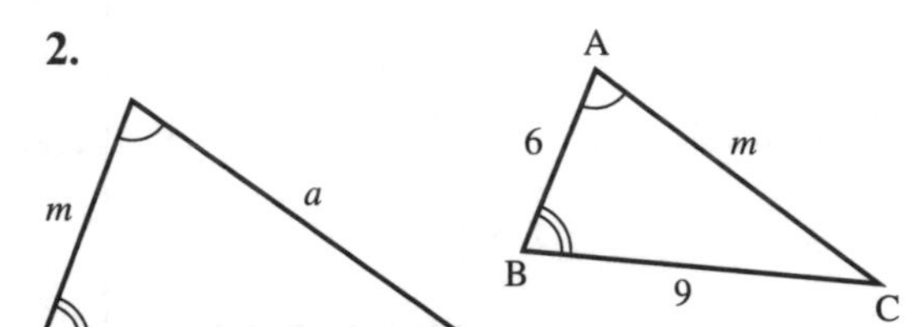

3.

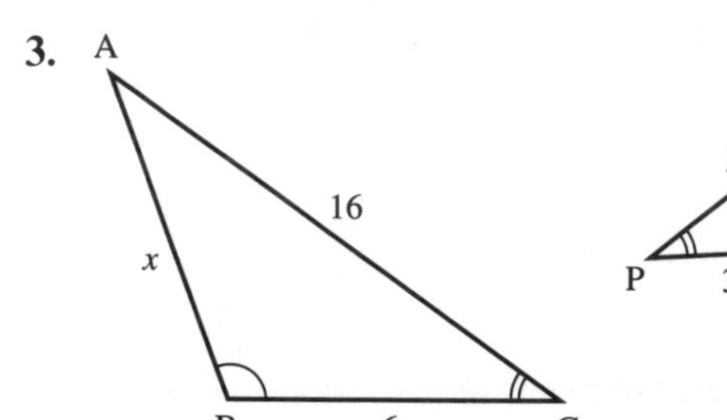

4.

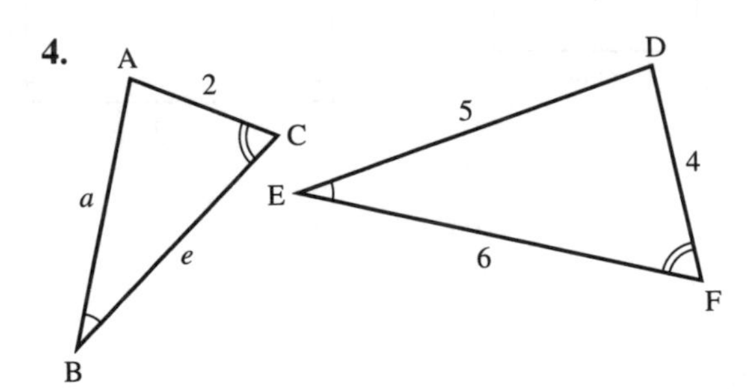

5.

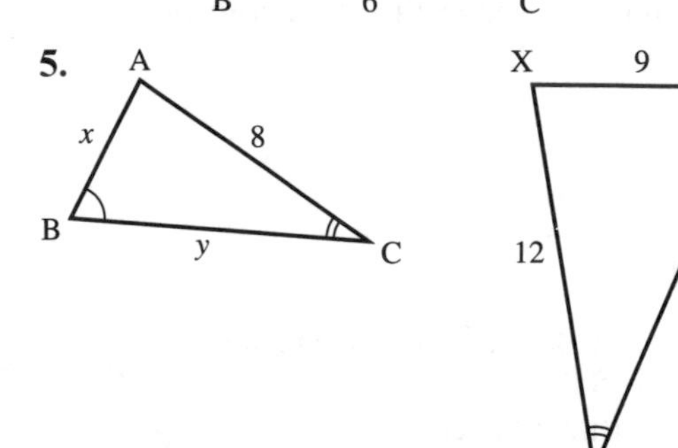

6.

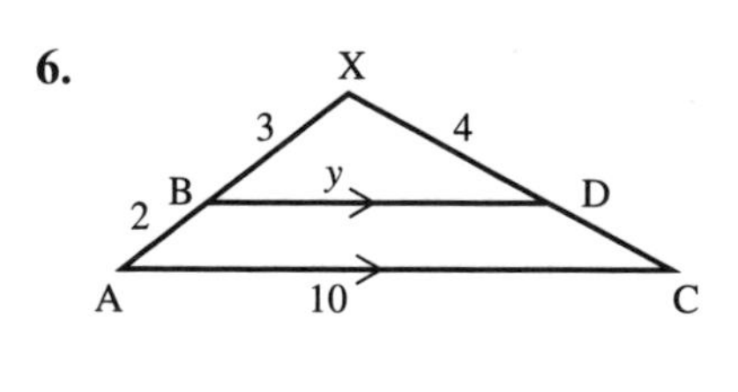

7.

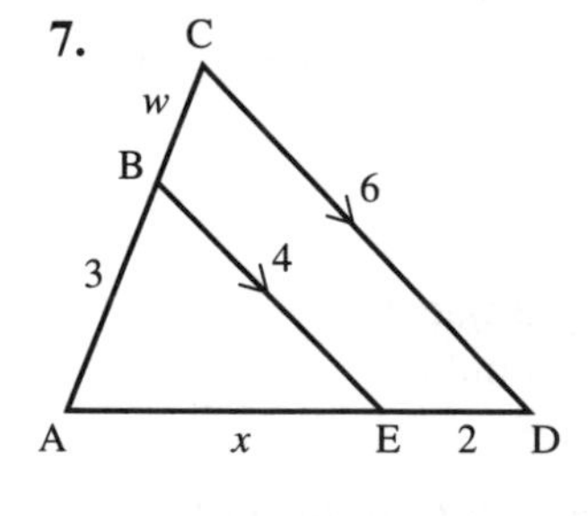

8.

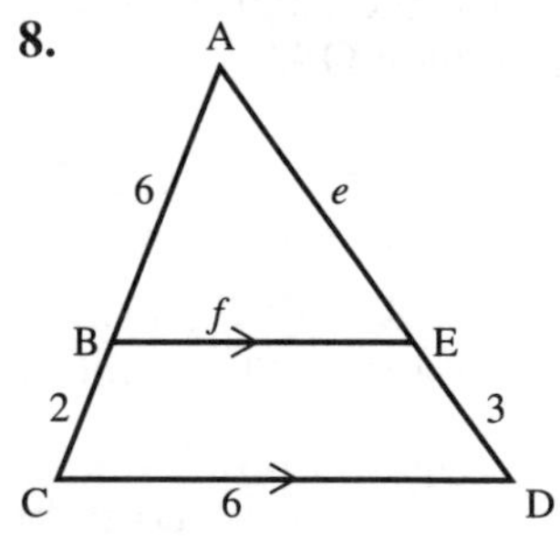

9.

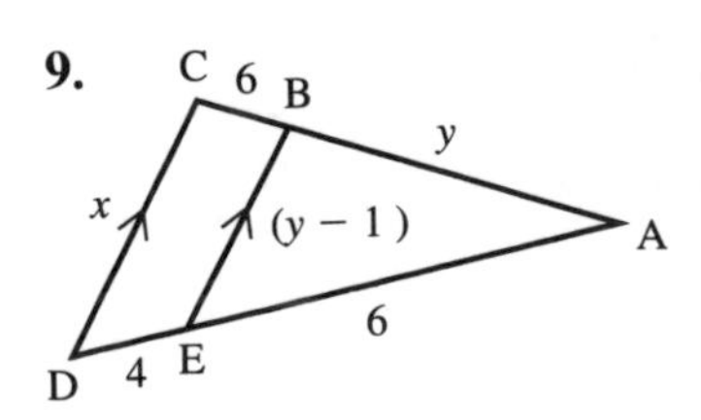

10. $B\widehat{A}C = D\widehat{B}C$

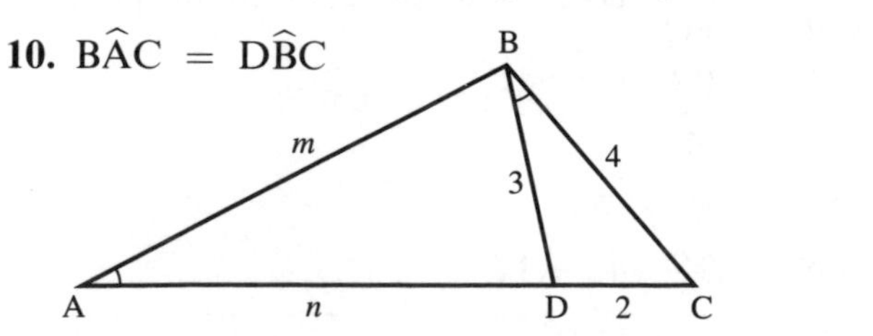

11.

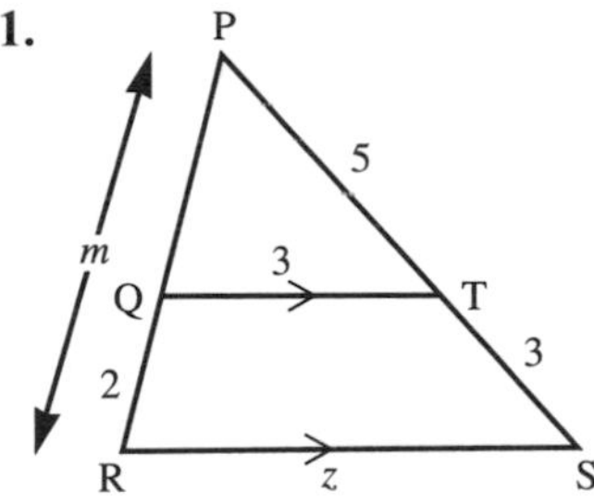

12.

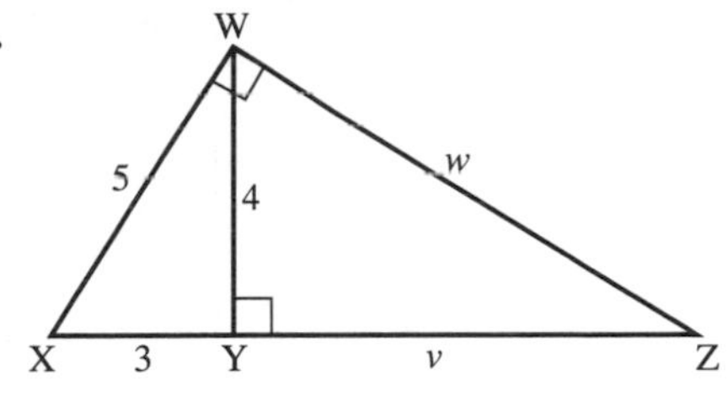

13. The drawing shows a rectangular picture 16 cm × 8 cm surrounded by a border of width 4 cm. Are the two rectangles similar?

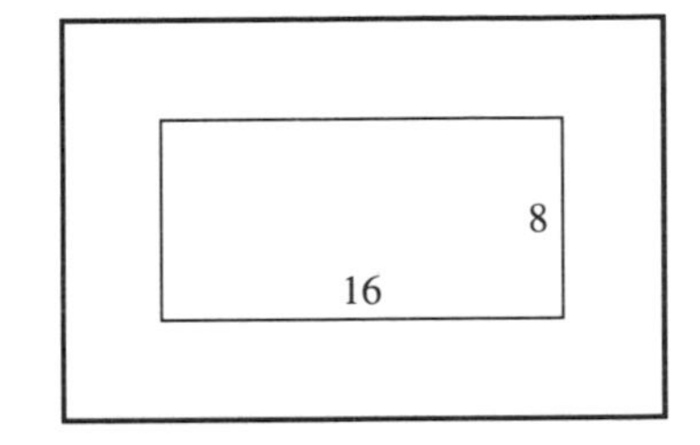

14. The diagonals of a trapezium ABCD intersect at O. AB is parallel to DC, AB = 3 cm and DC = 6 cm. If CO = 4 cm and OB = 3 cm, find AO and DO.

15. A tree of height 4 m casts a shadow of length 6·5 m. Find the height of a house casting a shadow 26 m long.

16. Triangles ABC and EBD are similar but DE is *not* parallel to AC. Work out the length x.

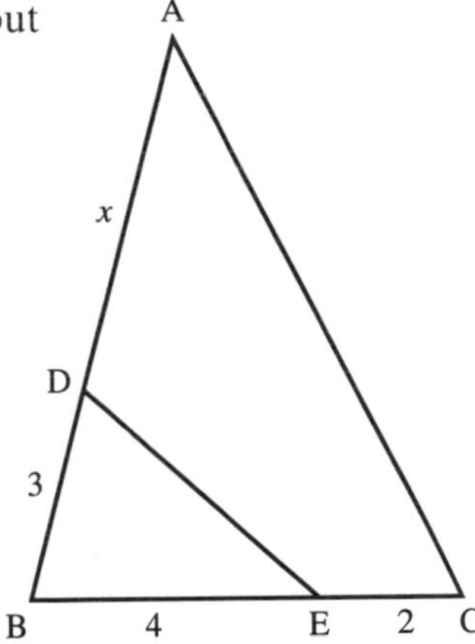

17. Which of the following *must* be similar to each other.
(a) Two equilateral triangles.
(b) Two rectangles.
(c) Two isosceles triangles.
(d) Two squares.
(e) Two regular pentagons.
(f) Two kites.
(g) Two rhombuses.
(h) Two circles.

18. In a triangle ABC, a line is drawn parallel to BC to meet AB at D and AC at E. DC and BE meet at X. Prove that
(a) the triangles ADE and ABC are similar
(b) the triangles DXE and BXC are similar
(c) $\dfrac{AD}{AB} = \dfrac{EX}{XB}$

19. In the diagram $A\hat{B}C = A\hat{D}B = 90°$, $AD = p$ and $DC = q$.

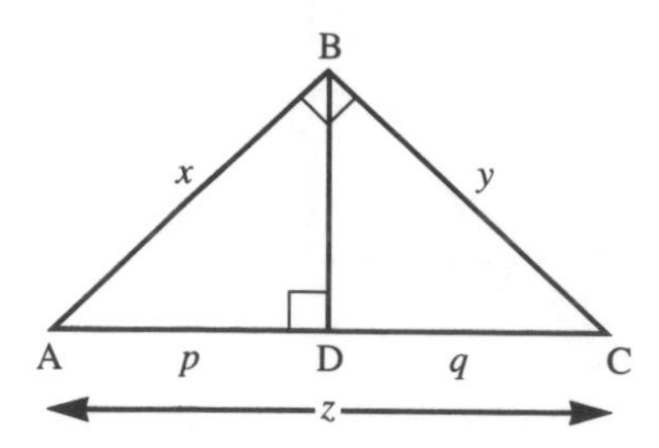

(a) Use similar triangles ABD and ABC to show that $x^2 = pz$.
(b) Find a similar expression for y^2
(c) Add the expressions for x^2 and y^2 and hence prove Pythagoras' theorem.

20. A rectangle 11 cm by 6 cm is similar to a rectangle 2 cm by x cm. Find the two possible values of x.

21. From the rectangle ABCD a square is cut off to leave rectangle BCEF.

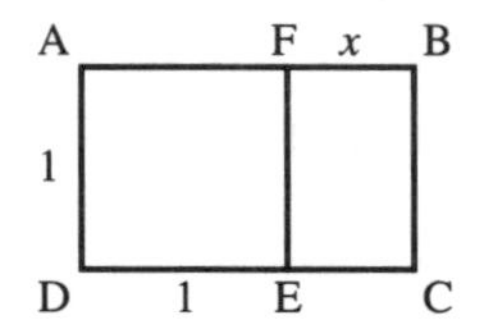

Rectangle BCEF is similar to ABCD. Find x and hence state the ratio of the sides of rectangle ABCD. ABCD is called the Golden Rectangle and is an important shape in architecture.

3.6 Trigonometry

Trigonometry is used to calculate sides and angles in triangles.

Right-angled triangles

The side opposite the right angle is called the *hypotenuse* (use hyp.). It is the longest side.
The side opposite the marked angle m is called the *opposite* (use opp.).
The other side is called the *adjacent* (use adj.).
Consider two triangles, one of which is an enlargement of the other.
It is clear that, for the angle 30°, the

$$\text{ratio} = \frac{\text{opposite}}{\text{hypotenuse}} = \frac{6}{12} = \frac{2}{4} = \frac{1}{2}$$

This is the same for both triangles.

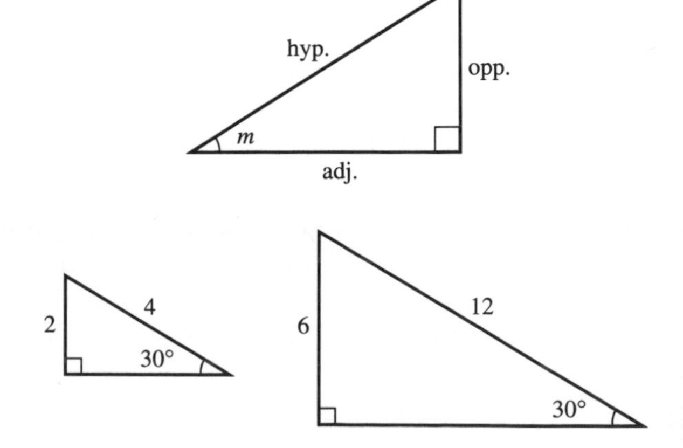

Sine, cosine, tangent

Three important ratios are defined for angle x.

$$\sin x = \frac{\text{opp.}}{\text{hyp.}} \quad \cos x = \frac{\text{adj.}}{\text{hyp.}} \quad \tan x = \frac{\text{opp.}}{\text{adj.}}$$

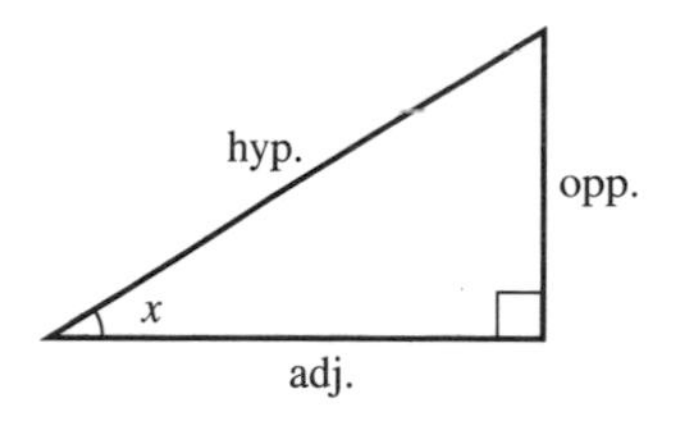

It is important to get the letters in the correct positions.

Some people find a simple sentence helpful where the first letters of each word describe sine, cosine or tangent, Hypotenuse, Opposite or Adjacent. An example is:

Silly Old Harry Caught A Herring Trawling Off Afghanistan

e.g. S O H $\sin = \frac{\text{opp.}}{\text{hyp.}}$

Finding the length of an unknown side

Example 1

Find the side marked x to 3 s.f.

Label the sides of the triangle hyp., opp., adj.

In this example, we only know opp. and adj., so use the tangent ratio.

$$\tan 25{\cdot}4° = \frac{\text{opp.}}{\text{adj.}} = \frac{x}{10}$$

Find tan 25·4° on a calculator: press [25·4] then [tan].

$$0{\cdot}4748 = \frac{x}{10}$$

Solve for x.

$$x = 10 \times 0{\cdot}4748 = 4{\cdot}748$$
$$x = 4{\cdot}75 \text{ cm (3 s.f.)}$$

Example 2

Find the side marked z.

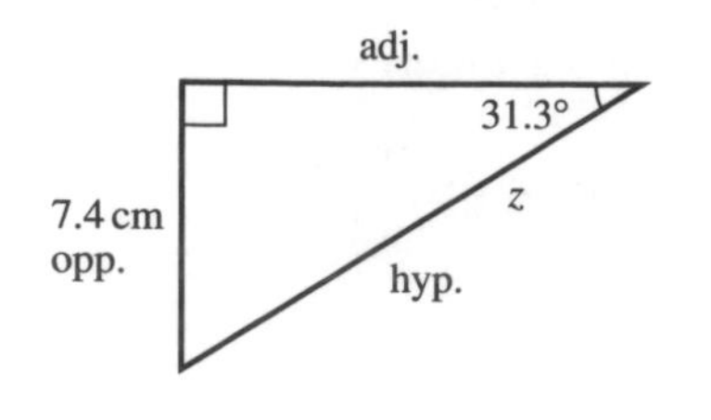

Label hyp., opp., adj.

$$\sin 31{\cdot}3° = \frac{\text{opp.}}{\text{hyp.}} = \frac{7{\cdot}4}{z}$$

Multiply by z. $\quad z \times (\sin 31{\cdot}3°) = 7{\cdot}4$

$$z = \frac{7{\cdot}4}{\sin 31{\cdot}3°}$$

On a calculator, press the keys as follows.

[7·4] [÷] [31·3] [sin] [=]

$$z = 14{\cdot}2\,\text{cm (3 s.f.)}$$

Exercise 12

In Questions **1** to **26**, find the length of the side marked with a letter. Give your answers to three significant figures.

1.
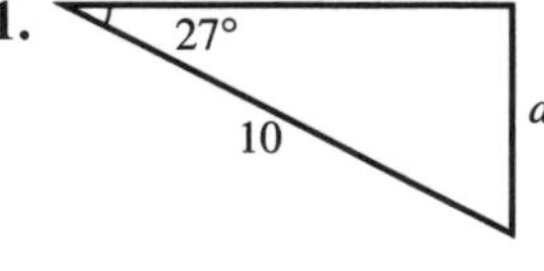

2.
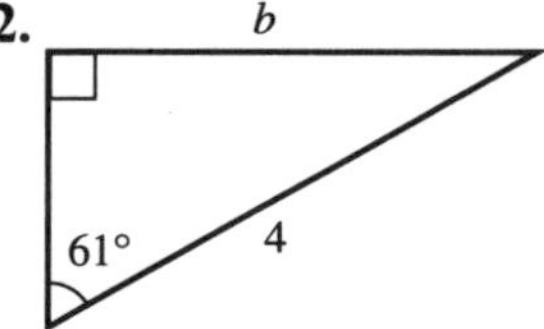

3.
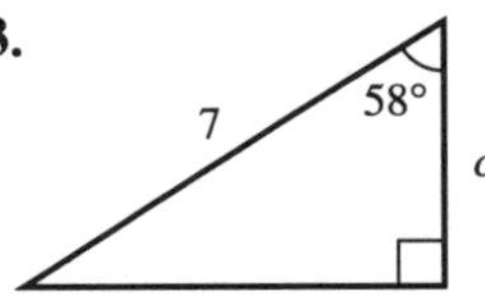

4.
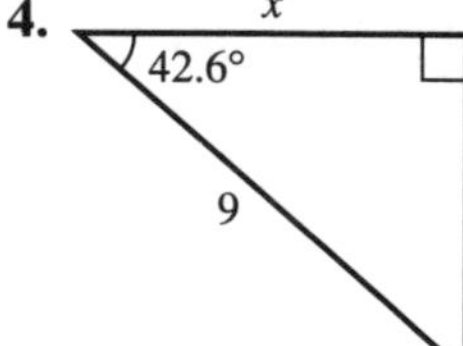

5.
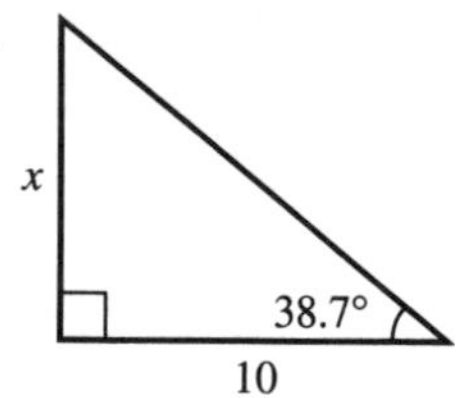

6.
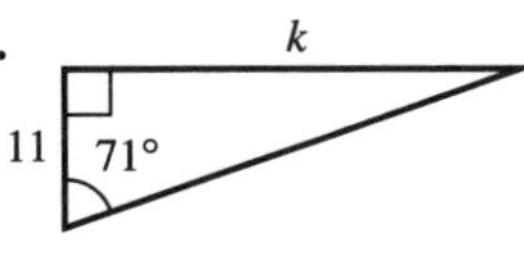

7.
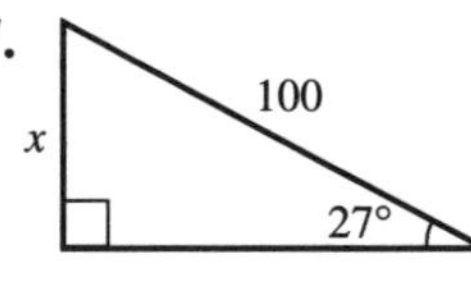

8.
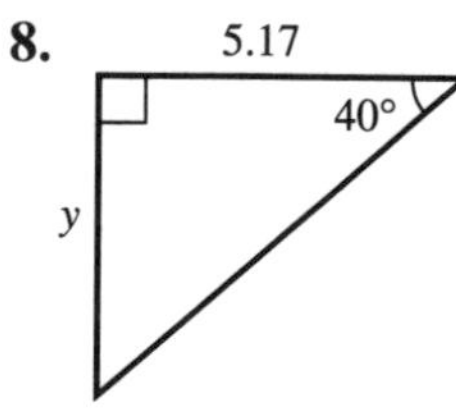

9.
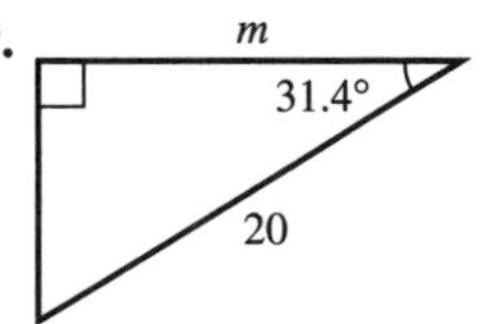

10.
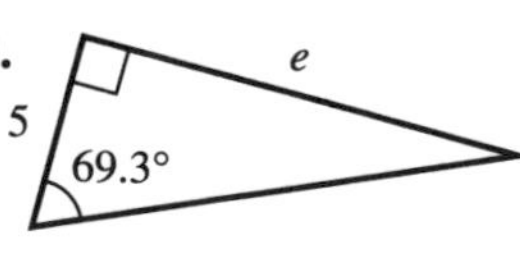

11.
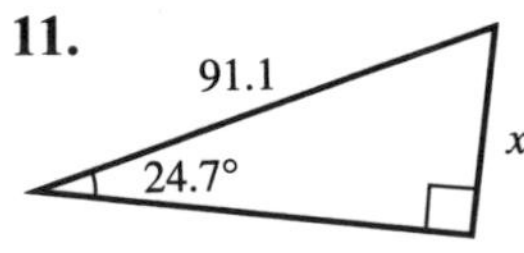

12.
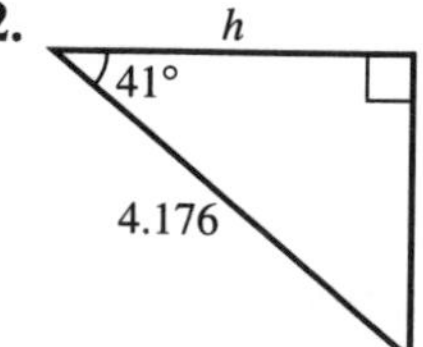

13.
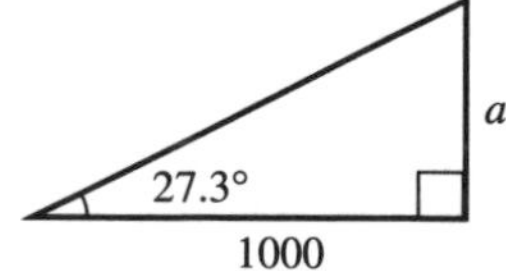

14.
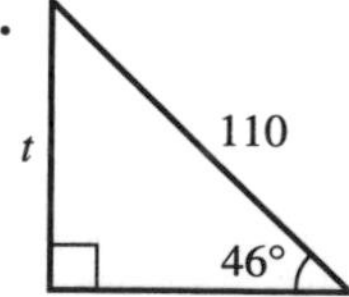

15.
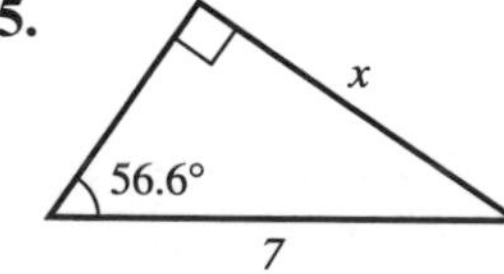

16.
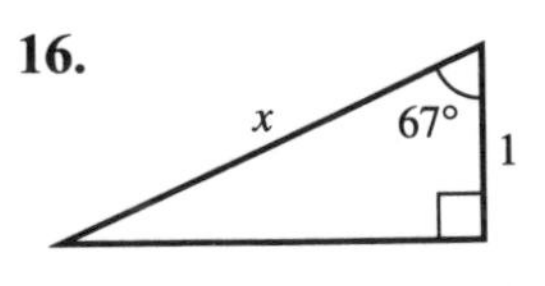

17.

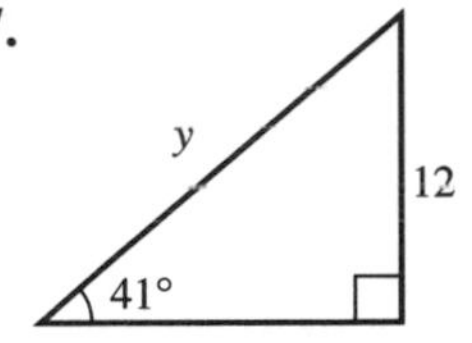

18.

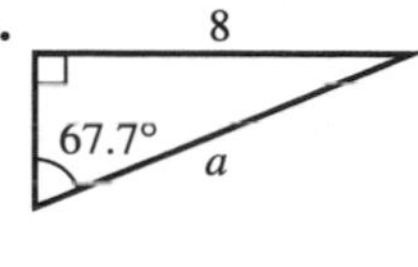

19.

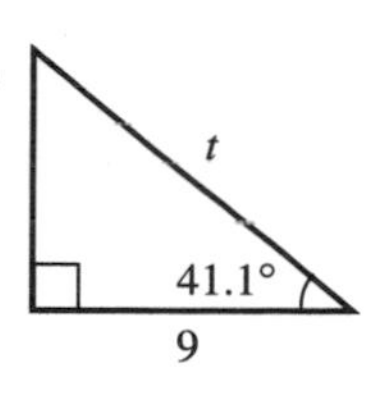

20.

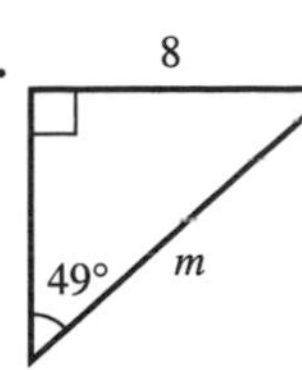

21.

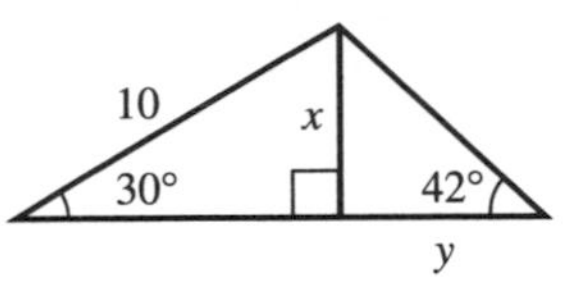

22.

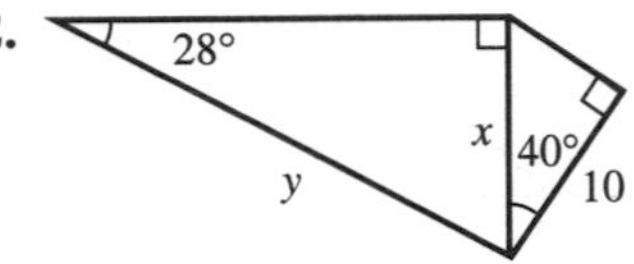

23.

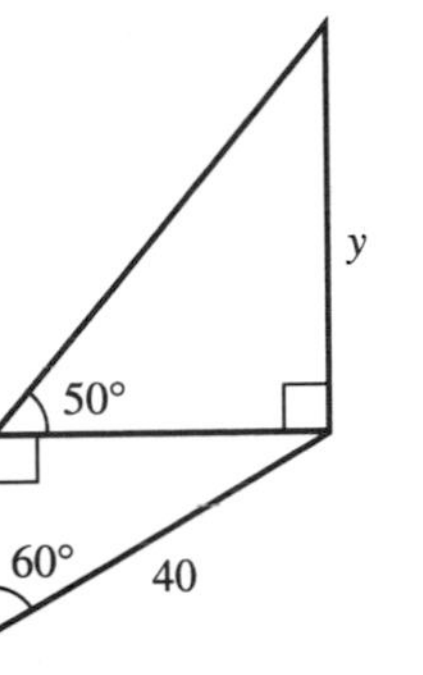

24.

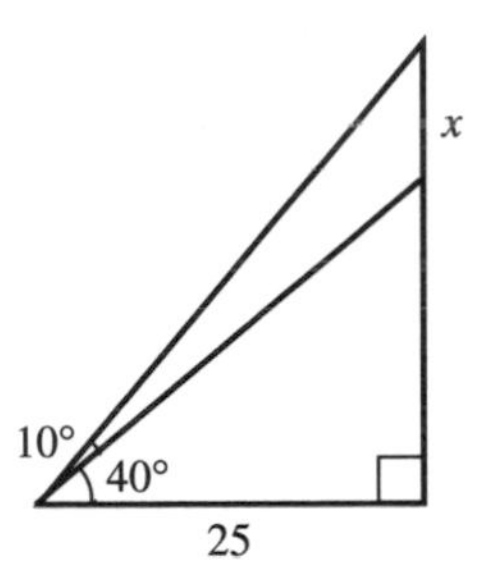

25.

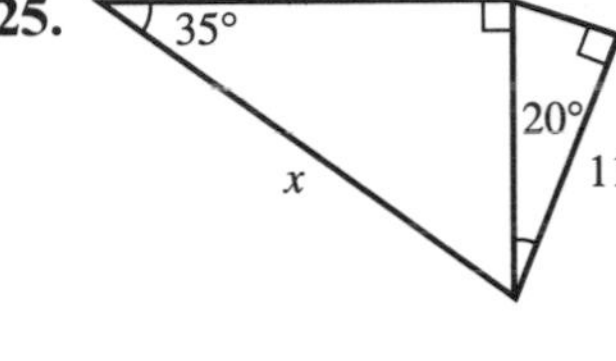

26.

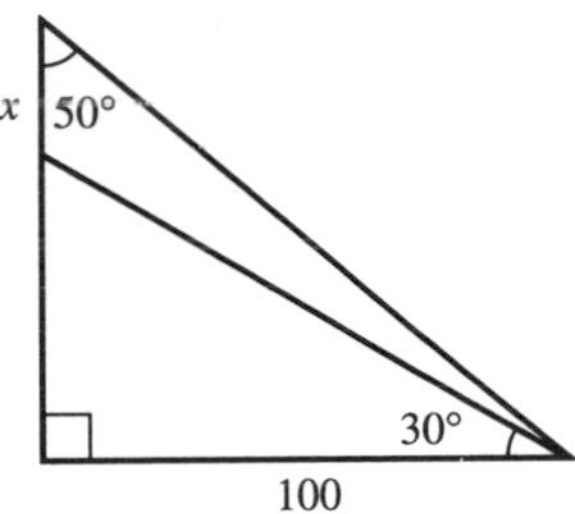

Finding an unknown angle

Example 3

Find the angle marked m to one decimal place.

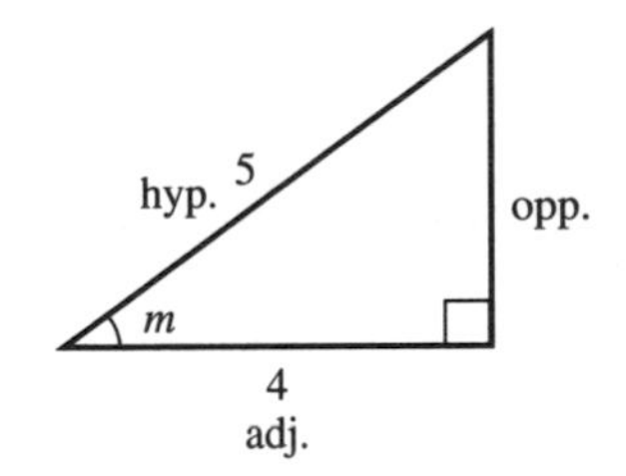

Label the sides of the triangle hyp., opp. and adj. in relation to angle m.

Here only adj. and hyp. are known, so use cosine.

$$\cos m = \frac{\text{adj.}}{\text{hyp.}} = \frac{4}{5}$$

Using a calculator, the angle m can be found as follows.

(a) Press [4] [÷] [5] [=]

(b) Press [INV] and then [cos]

This gives the angle as 36·869 898

$m = 36{\cdot}9°$ (1 d.p.)

Exercise 13

For Questions **1** to **15**, find the angle marked with a letter. All lengths are in cm.

1.

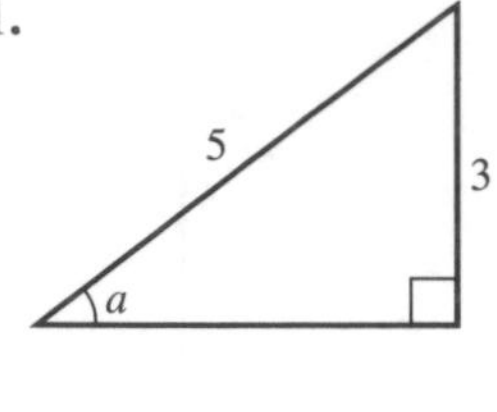

2.

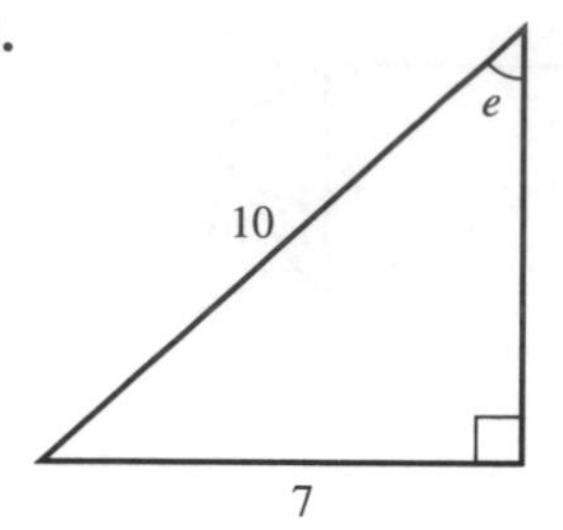

3.

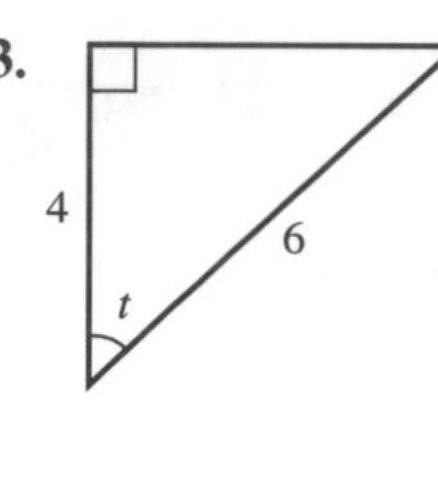

4.

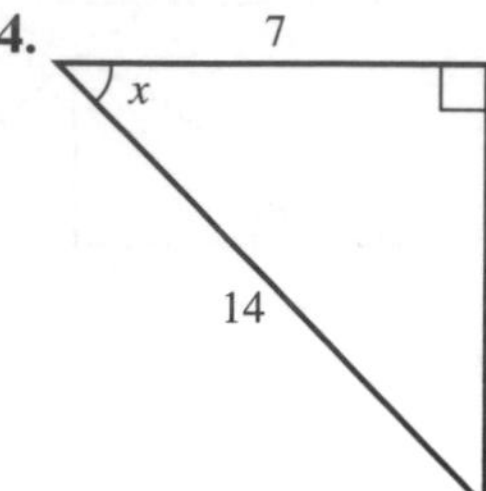

5.

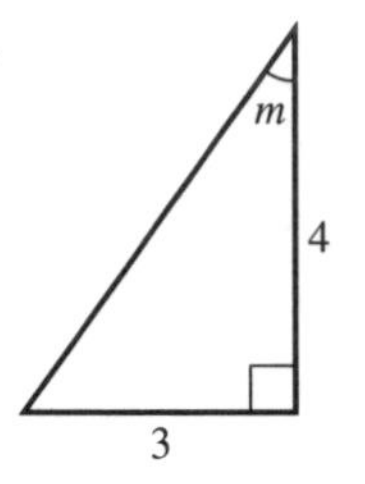

6.

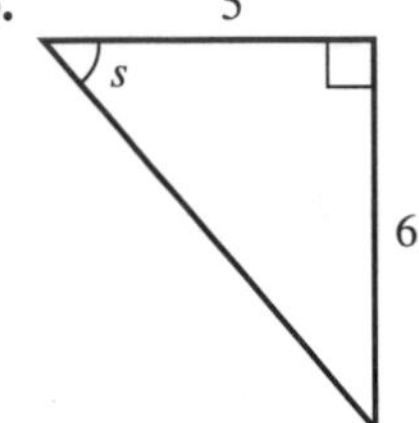

7.

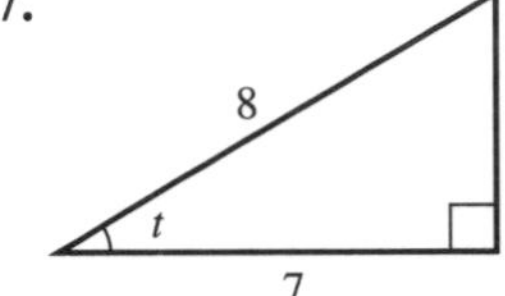

8.

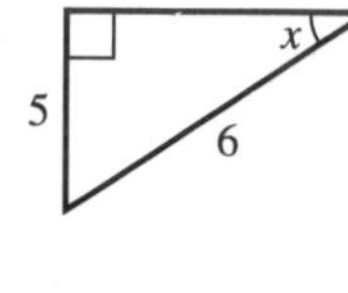

9.

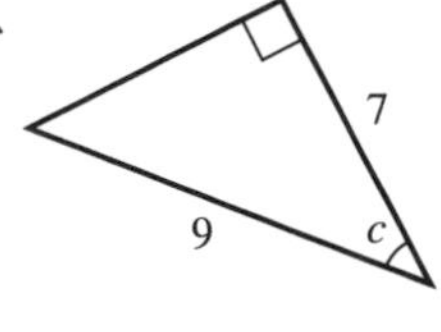

10.

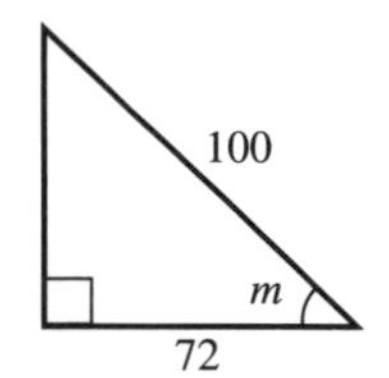

11.

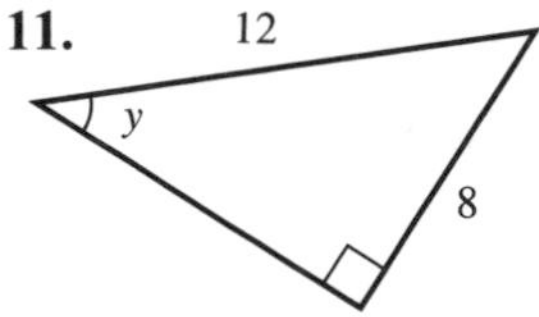

12.

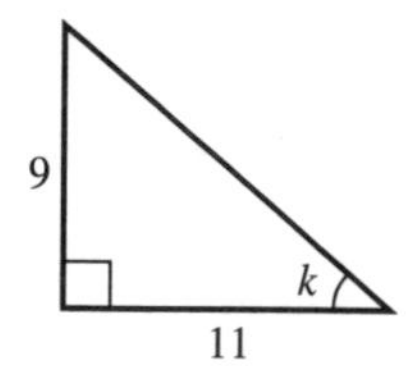

13.

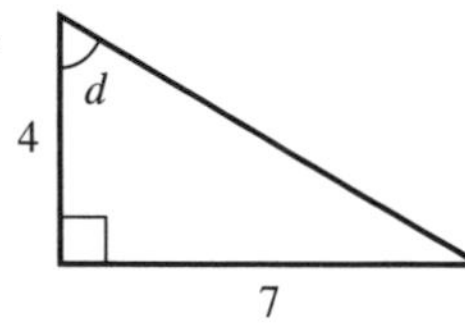

14.

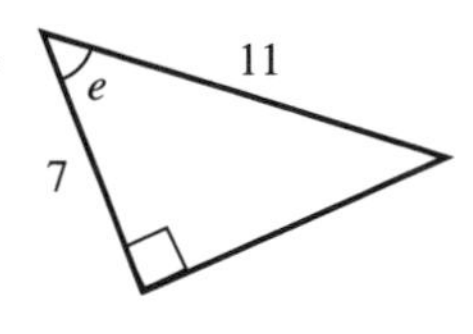

15.

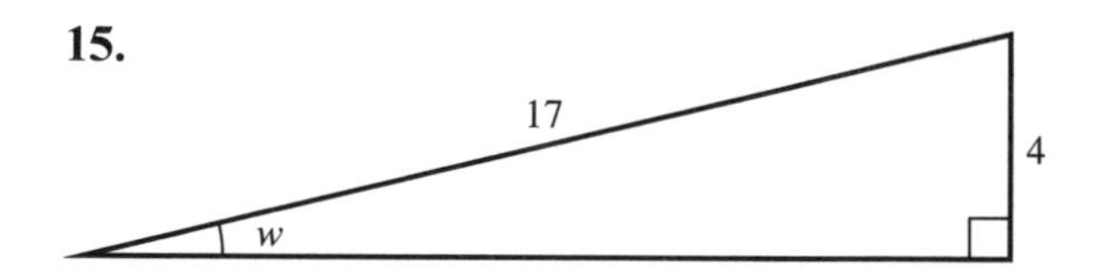

In Questions **16** to **20**, the triangle has a right angle at the middle letter.

16. In $\triangle$ABC, BC = 4, AC = 7. Find $\widehat{A}$.

17. In $\triangle$DEF, EF = 5, DF = 10. Find $\widehat{F}$.

18. In $\triangle$GHI, GH = 9, HI = 10. Find $\widehat{I}$.

19. In $\triangle$JKL, JL = 5, KL = 3. Find $\widehat{J}$.

20. In $\triangle$MNO, MN = 4, NO = 5. Find $\widehat{M}$.

In Questions **21** to **24**, find the angle x.

21.

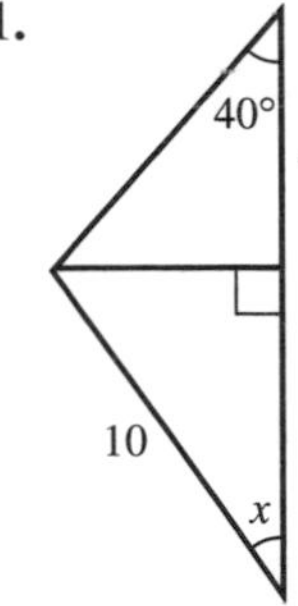

22.

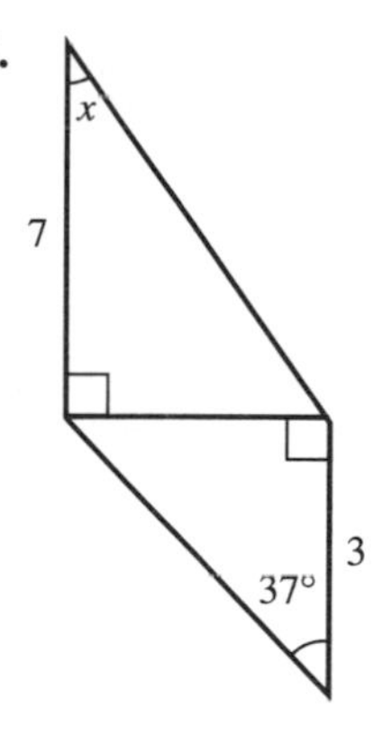

23.

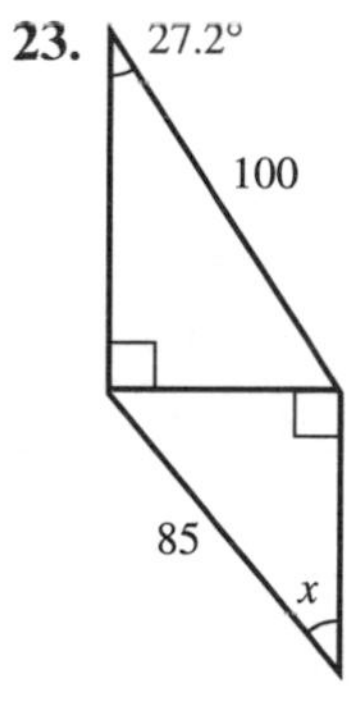

24.

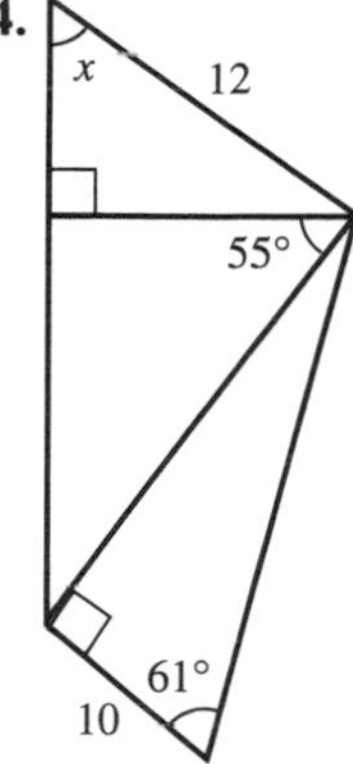

25. In △ABC
$B\hat{A}C = 42°$ and $AB = 12$ cm.

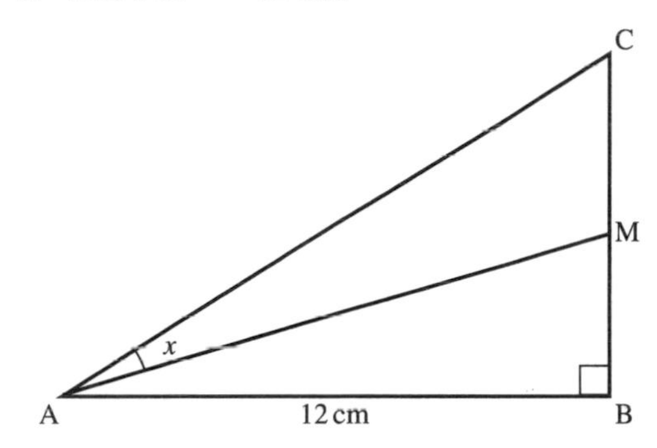

M is the mid-point of CB.
Find the angle x to 1 d.p.

Bearings

Bearings are measured *clockwise* from North.

(a)

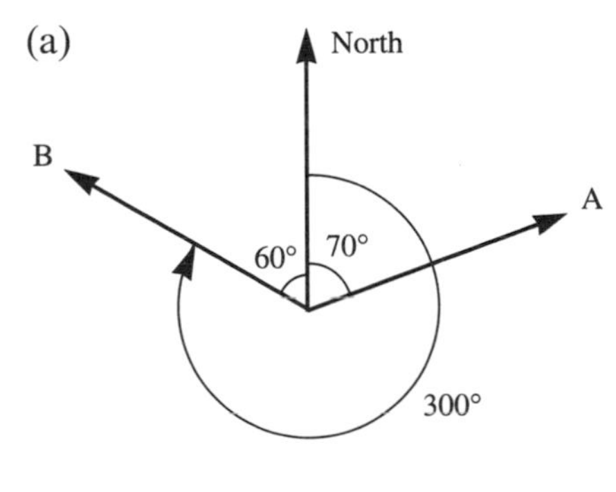

(b)

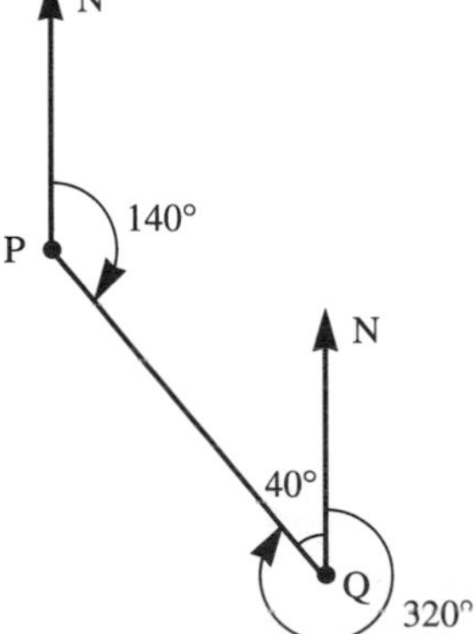

Ship A sails on a bearing 070°.
Ship B sails on a bearing 300°.

The bearing of Q from P is 140°.
The bearing of P from Q is 320°.

Angles of elevation and depression

(a)

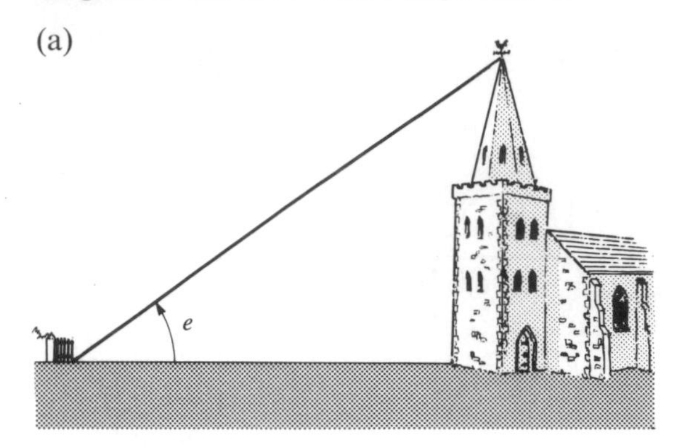

e is the angle of *elevation* of the Steeple from the Gate.

(b)

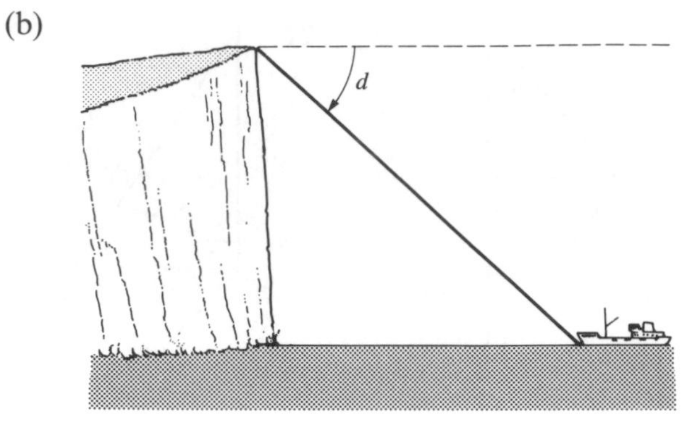

d is the angle of *depression* of the Boat from the Cliff top.

Exercise 14

In this exercise, start each question by drawing a clear diagram.

1. A ladder of length 6 m leans against a vertical wall so that the base of the ladder is 2 m from the wall. Calculate the angle between the ladder and the wall.

2. A ladder of length 8 m rests against a wall so that the angle between the ladder and the wall is 31°. How far is the base of the ladder from the wall?

3. Find TR if PR = 10 m and QT = 7 m.

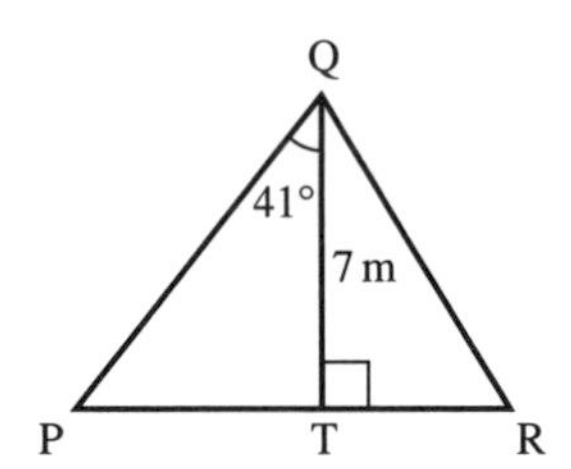

4. A ship sails 35 km on a bearing of 042°.
   (a) How far north has it travelled?
   (b) How far east has it travelled?

5. A ship sails 200 km on a bearing of 243·7°.
   (a) How far south has it travelled?
   (b) How far west has it travelled?

6. Find d.

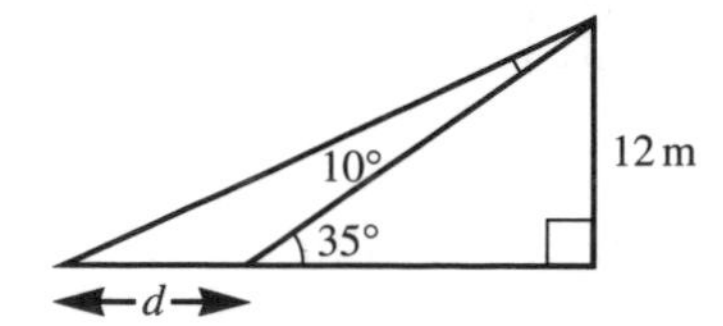

7. An aircraft flies 400 km from a point O on a bearing of 025° and then 700 km on a bearing of 080° to arrive at B.
(a) How far north of O is B?
(b) How far east of O is B?
(c) Find the distance and bearing of B from O.

8. An aircraft flies 500 km on a bearing of 100° and then 600 km on a bearing of 160°. Find the distance and bearing of the finishing point from the starting point.

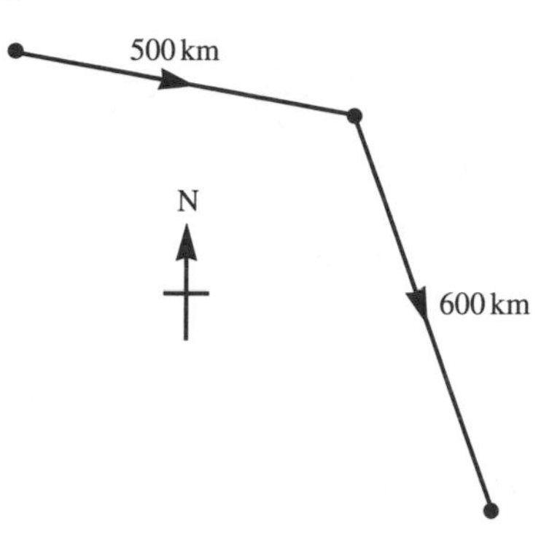

For Questions **9** to **12**, plot the points for each question on a sketch graph with x- and y-axes drawn to the same scale.

9. For the points A(5, 0) and B(7, 3), calculate the angle between AB and the x-axis.

10. For the points C(0, 2) and D(5, 9), calculate the angle between CD and the y-axis.

11. For the points A(3, 0), B(5, 2) and C(7, –2), calculate the angle BAC.

12. For the points P(2, 5), Q(5, 1) and R(0, –3), calculate the angle PQR.

13. From the top of a tower of height 75 m, a guard sees two prisoners, both due West of him.

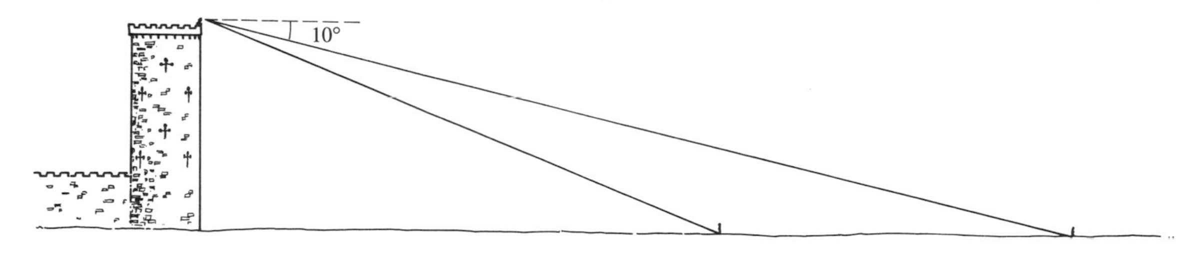

If the angles of depression of the two prisoners are 10° and 17°, calculate the distance between them.

14. An isosceles triangle has sides of length 8 cm, 8 cm and 5 cm. Find the angle between the two equal sides.

15. The angles of an isosceles triangle are 66°, 66° and 48°. If the shortest side of the triangle is 8·4 cm, find the length of one of the two equal sides.

16. A regular pentagon is inscribed in a circle of radius 7 cm.

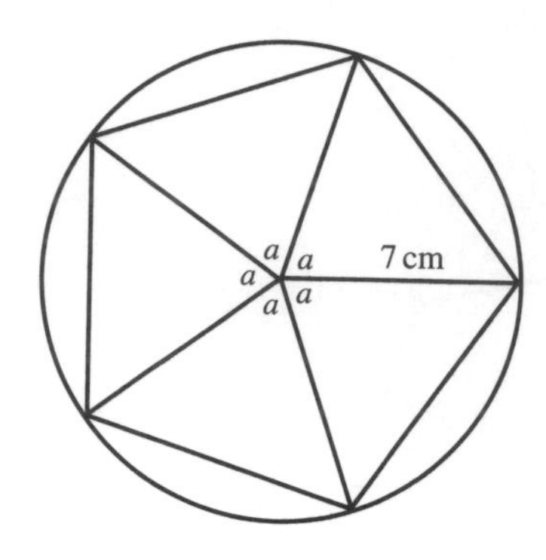

Find the angle a and then the length of a side of the pentagon.

17. Find the acute angle between the diagonals of a rectangle whose sides are 5 cm and 7 cm.

18. A kite flying at a height of 55 m is attached to a string which makes an angle of 55° with the horizontal. What is the length of the string?

19. A plane is flying at a constant height of 8000 m. It flies vertically above me and 30 seconds later the angle of elevation is 74°.

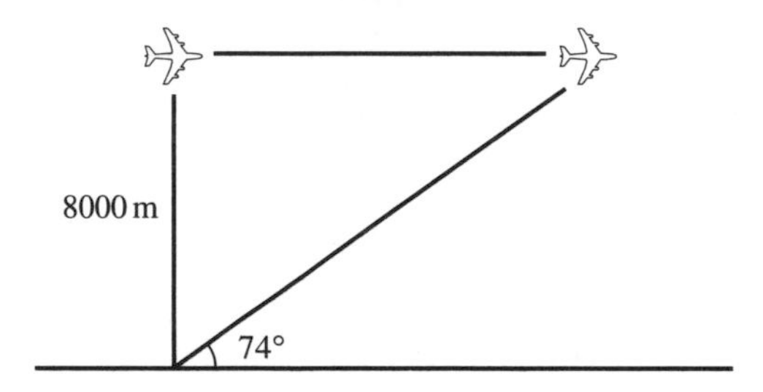

Find the speed of the plane in metres/second.

20. A rocket flies 10 km vertically, then 20 km at an angle of 15° to the vertical and finally 60 km at an angle of 26° to the vertical. Calculate the vertical height of the rocket at the end of the third stage.

21. Find x, given

AD = BC = 6 m.

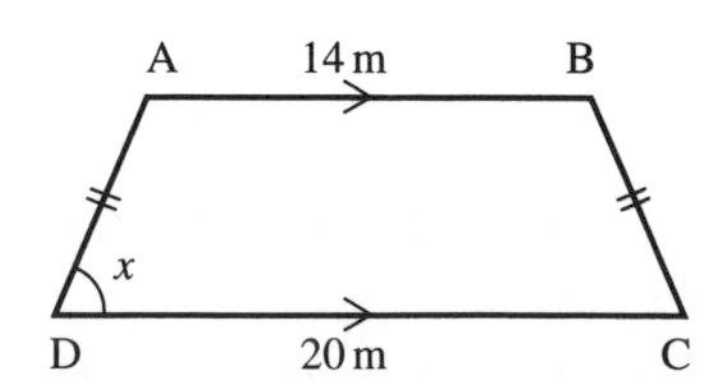

22. Ants can hear each other up to a range of 2 m. An ant A, 1 m from a wall sees her friend B about to be eaten by a spider. If the angle of elevation of B from A is 62°, will the spider have a meal or not? (Assume B escapes if he hears A calling.)

23. A hedgehog wishes to cross a road without being run over. He observes the angle of elevation of a lamp post on the other side of the road to be 27° from the edge of the road and 15° from a point 10 m back from the road. How wide is the road? If he can run at 1 m/s, how long will he take to cross?
If cars are travelling at 20 m/s, how far apart must they be if he is to survive?

24. A symmetrical drawbridge is shown below. When lowered, the roads AX and BY just meet in the middle. Calculate the length XY.

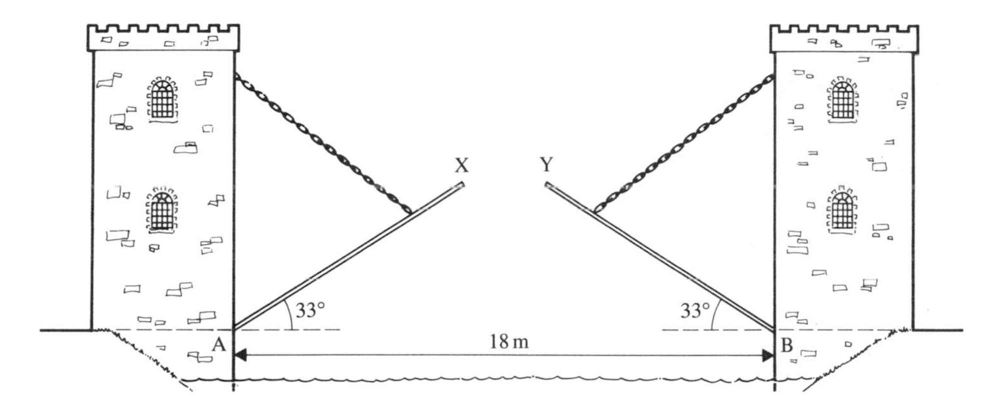

25. From a point 10 m from a vertical wall, the angles of elevation of the bottom and the top of a statue of Sir Isaac Newton, set in the wall, are 40° and 52°. Calculate the length of the statue.

26.† A rectangular paving stone 3 m by 1 m rests against a vertical wall as shown.

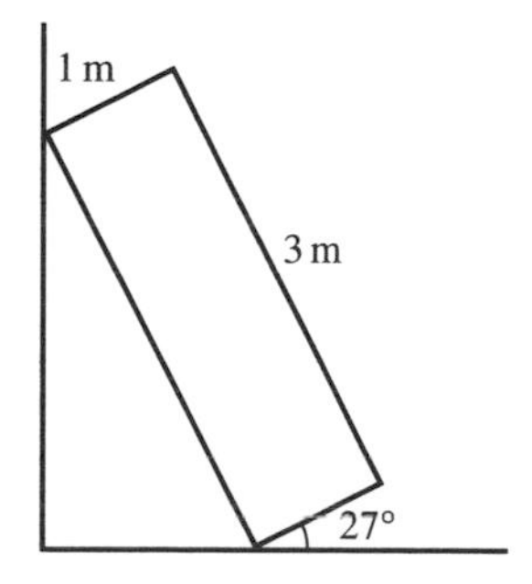

What is the height of the highest point of the stone above the ground?

27.† A rectangular piece of paper 30 cm by 21 cm is folded so that opposite corners coincide. How long is the crease?

28.† The diagram shows the cross section of a rectangular fish tank. When AB is inclined at 40°, the water just comes up to A.

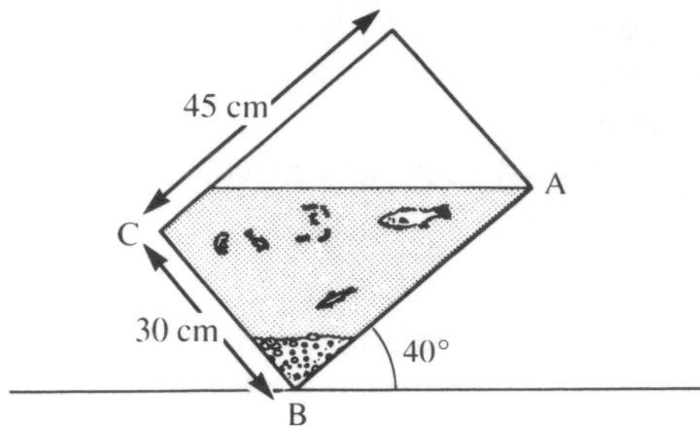

The tank is then lowered so that BC is horizontal. What is now the depth of water in the tank?

Three-dimensional problems

Always draw a large, clear diagram. It is often helpful to redraw the particular triangle which contains the length or angle to be found.

Example 4

A rectangular box with top WXYZ and base ABCD has AB = 6 cm, BC = 8 cm and WA = 3 cm.

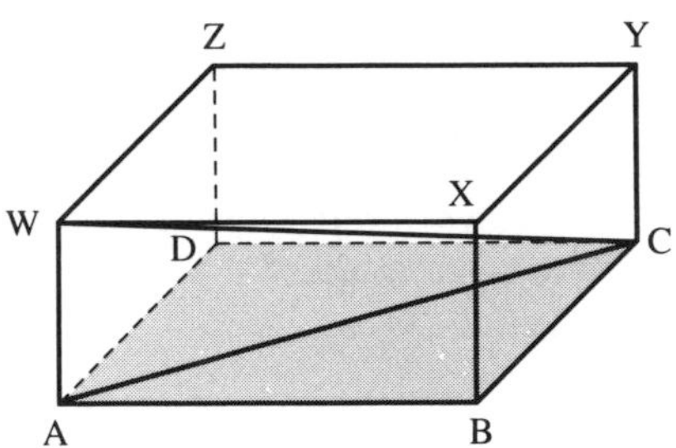

Calculate
(a) the length of AC
(b) the angle between WC and AC.

(a) Redraw triangle ABC

$$AC^2 = 6^2 + 8^2 = 100$$
$$AC = 10\,\text{cm}.$$

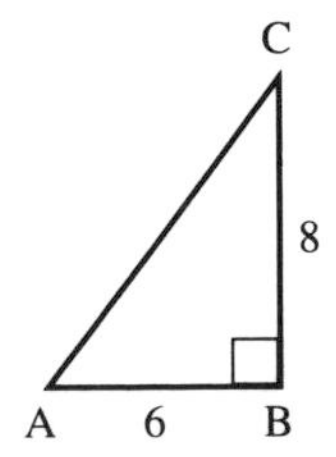

(b) Redraw triangle WAC

let $W\hat{C}A = \theta$

$$\tan\theta = \tfrac{3}{10}$$
$$\theta = 16{\cdot}7°.$$

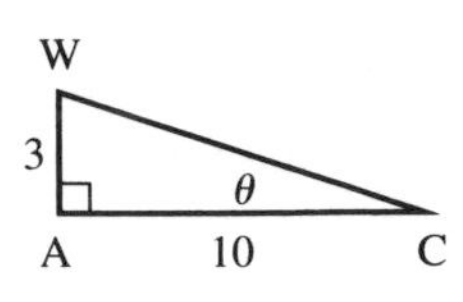

The angle between WC and AC is 16·7°.

Exercise 15

1. In the rectangular box shown, find
   (a) AC
   (b) AR
   (c) the angle between AC and AR.

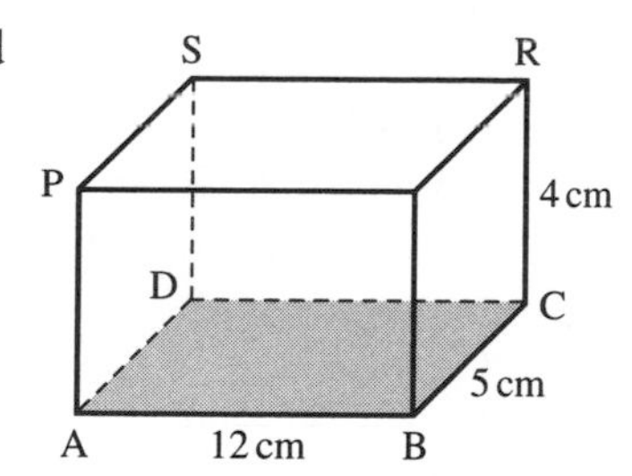

2. A vertical pole BP stands at one corner of a horizontal rectangular field as shown.

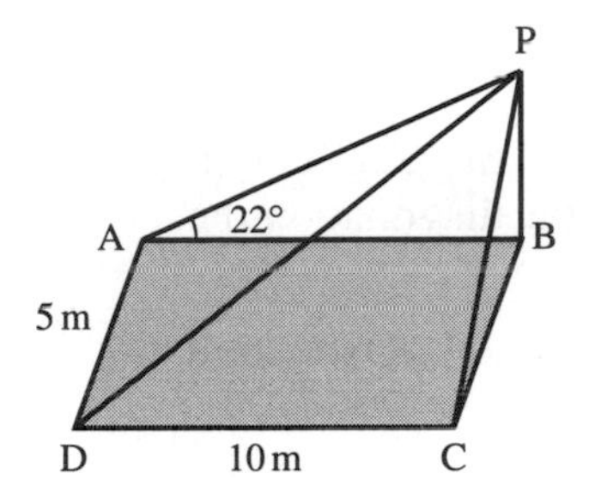

If AB = 10 m, AD = 5 m and the angle of elevation of P from A is 22°, calculate
(a) the height of the pole
(b) the angle of elevation of P from C
(c) the length of a diagonal of the rectangle ABCD
(d) the angle of elevation of P from D.

3. In the cube shown, find
   (a) BD
   (b) AS
   (c) BS
   (d) the angle SBD.

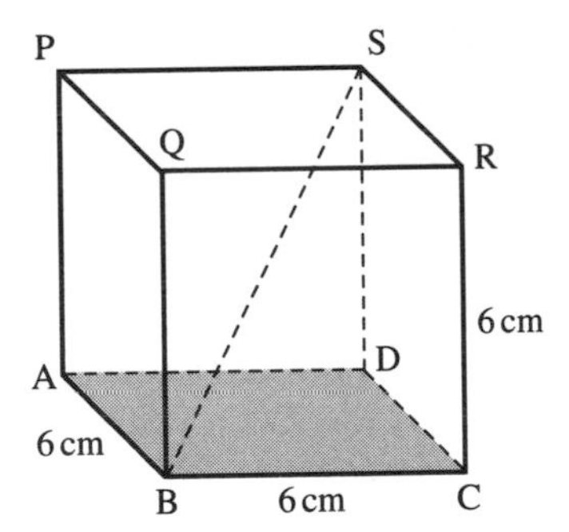

4. In the square-based pyramid, V is vertically above the middle of the base, AB = 10 cm and VC = 20 cm. Find
   (a) AC
   (b) the height of the pyramid
   (c) the angle between VC and the base ABCD.

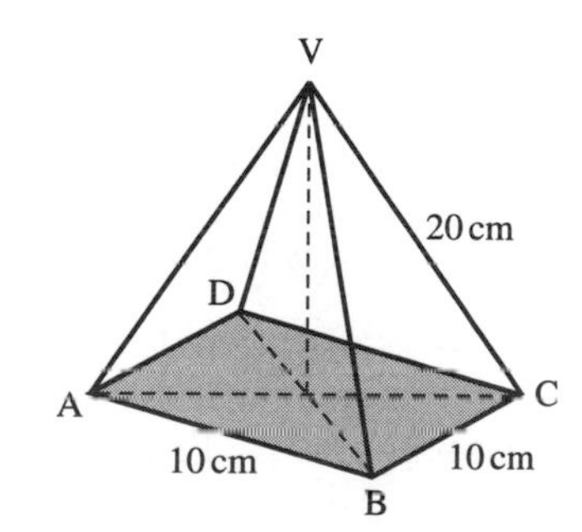

5. In the wedge shown, PQRS is perpendicular to ABRQ; PQRS and ABRQ are rectangles with AB = QR = 6 m, BR = 4 m, RS = 2 m. Find

(a) BS (b) AS

(c) angle BSR (d) angle ASR.

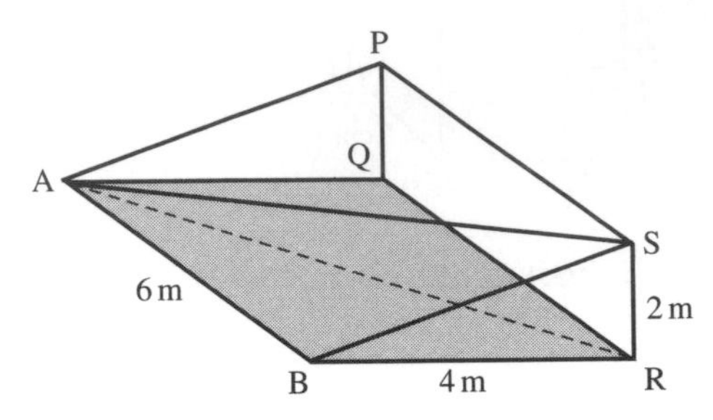

6. The edges of a box are 4 cm, 6 cm and 8 cm. Find the length of an internal diagonal and the angle it makes with the diagonal on the largest face.

7. In the diagram A, B and O are points in a horizontal plane and P is vertically above O, where OP = h m.

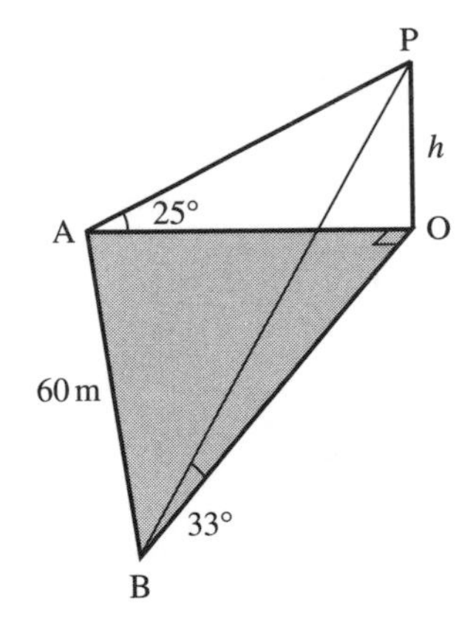

A is due West of O, B is due South of O and AB = 60 m. The angle of elevation of P from A is 25° and the angle of elevation of P from B is 33°.

(a) Find the length AO in terms of h

(b) Find the length BO in terms of h

(c) Find the value of h.

8.† The angle of elevation of the top of a tower is 38° from a point A due South of it. The angle of elevation of the top of the tower from another point B, due East of the tower is 29°. Find the height of the tower if the distance AB is 50 m.

9.† An observer at the top of a tower of height 15 m sees a man due West of him at an angle of depression 31°. He sees another man due South at an angle of depression 17°. Find the distance between the men.

Projections and planes

A projection is like a shadow on a surface or plane.

> The angle between a line and a plane is the angle between the line and its *projection* in the plane.

PA is a vertical pole standing at the vertex A of a horizontal rectangular field ABCD.

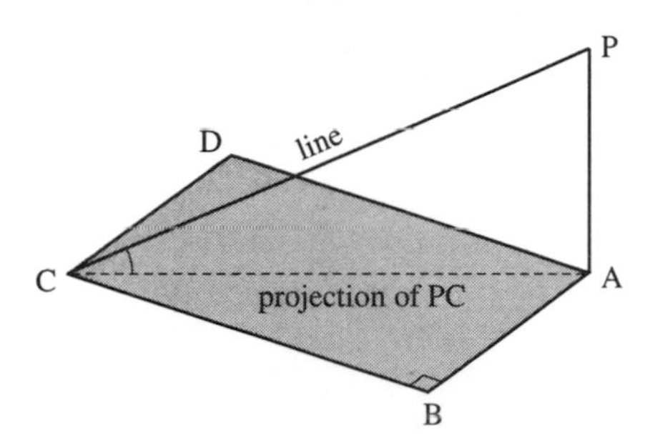

The angle between the line PC and the plane ABCD is found as follows:

(a) The projection of PC in the plane ABCD is AC

(b) The angle between the line PC and the plane ABCD is the angle PCA.

> ***The angle between two planes***
>
> (a) Find the common line where the two planes meet.
> (b) Find two lines at right angles to this common line, one in each plane.
> (c) The angle between these two lines is the angle between the planes.

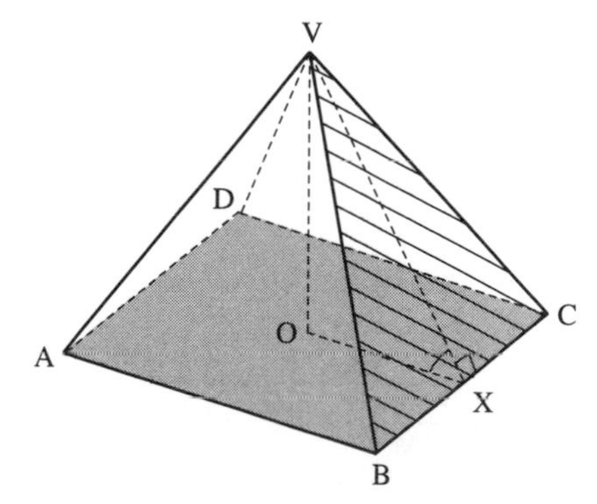

The angle between planes VBC and ABCD is found as follows:

(a) The line of intersection is BC.

(b) In plane VBC the line VX meets BC at 90°.
In plane ABCD the line OX meets BC at 90°.

(c) The angle between the planes is VXO.

Example 5

VABCD is a pyramid with a rectangular base ABCD in which AB = 12 cm and BC = 5 cm.

V is vertically above the centre of the rectangle and VA = VB = VC = VD = 10 cm.

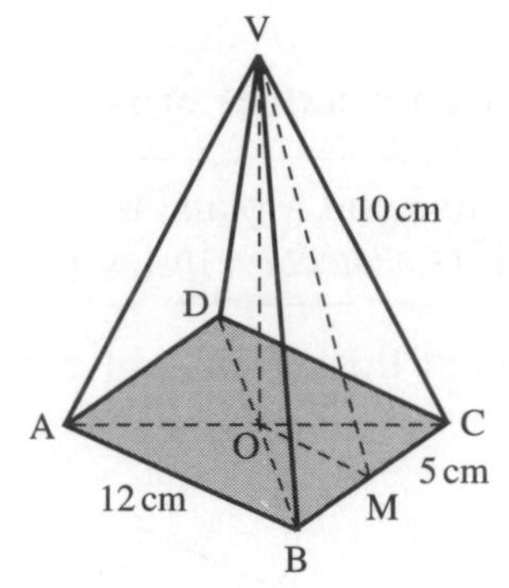

Find
(a) the angle between VA and the plane ABCD
(b) the angle between the planes VBC and ABCD.

(a) Find AC from △ABC

$$AC^2 = 12^2 + 5^2$$
$$AC = 13\,\text{cm}$$
$$\therefore \quad AO = 6{\cdot}5\,\text{cm}$$

Redraw △OVA

The projection of VA on the plane ABCD is AO.
The angle required is $V\hat{A}O$.

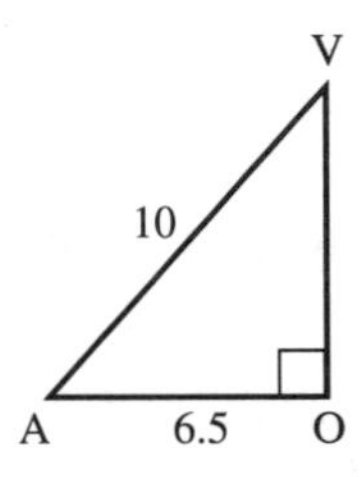

$$\cos V\hat{A}O = \frac{6{\cdot}5}{10} = 0{\cdot}65$$
$$\therefore \quad V\hat{A}O = 49{\cdot}5^\circ$$

The angle between VA and the plane ABCD is 49·5°.

(b) Introduce M at the mid-point of BC. The line of intersection of the planes in question is BC. OM and VM meet this line at right angles so the angle between planes ABCD and VBC is $V\hat{M}O$.

Redraw △VMO

$$OM = \frac{1}{2}AB = 6\,\text{cm}.$$

From △OVA,

$$VO^2 = 10^2 - 6{\cdot}5^2$$
$$VO = 7{\cdot}599\,\text{cm}.$$

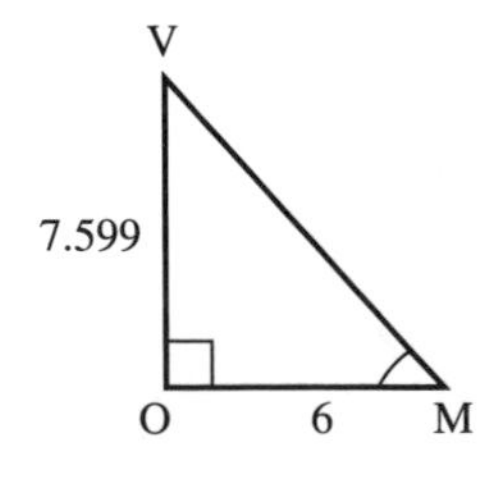

$$\tan V\hat{M}O = \frac{7{\cdot}599}{6}$$
$$\therefore \quad V\hat{M}O = 51{\cdot}7^\circ$$

The angle between planes VBC and ABCD is 51·7°.

Exercise 16

1. The figure shows a cuboid.
   Calculate
   (a) the lengths of AC and AY
   (b) the angle between AY and the plane ABCD
   (c) the angle between the planes WBCZ and ABCD.

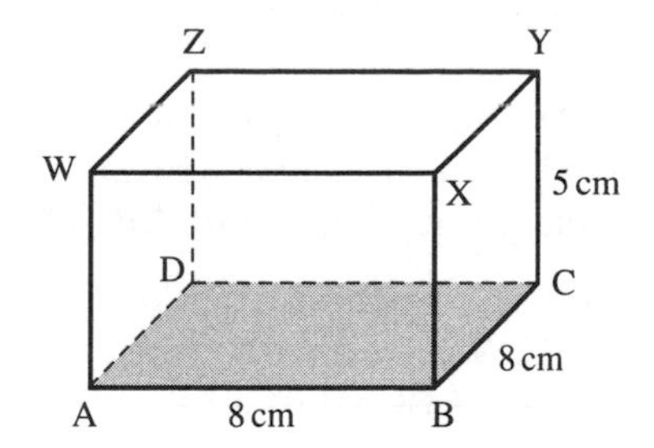

2. The figure shows a cuboid.
   Calculate
   (a) the lengths ZX and KX
   (b) the angle between NX and the plane WXYZ
   (c) the angle between KY and the plane KLWX.

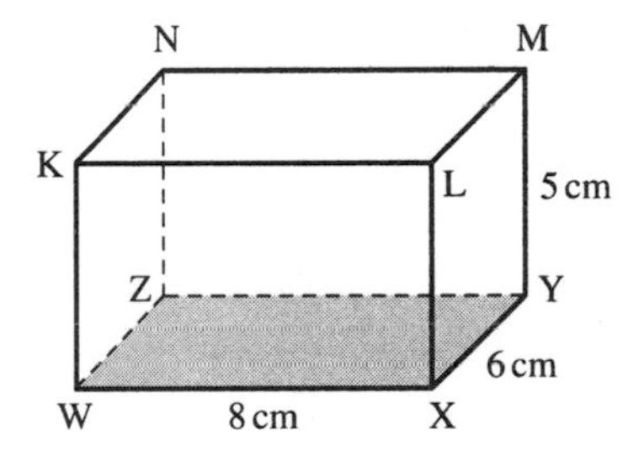

3. The figure shows a cuboid.
   Calculate
   (a) the angle between AG and the plane ABCD
   (b) the angle between the planes EBCH and ABCD.

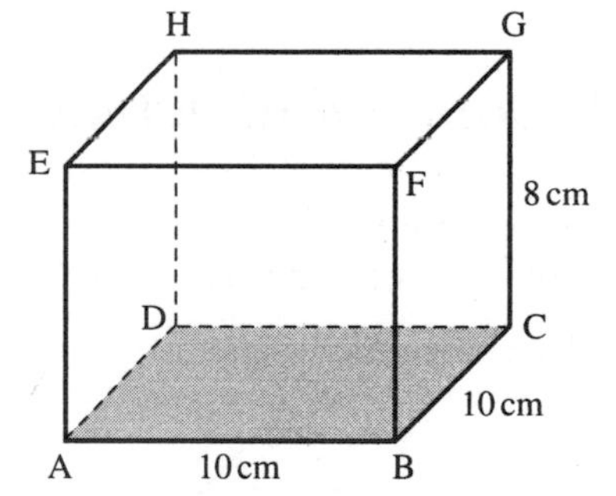

4. The pyramid VPQRS has a square base PQRS.
   VP = VQ = VR = VS = 12 cm and PQ = 9 cm.
   Calculate
   (a) the angle between VP and the plane PQRS
   (b) the angle between planes VPQ and PQRS.

5. The pyramid VABCD has a rectangular base ABCD.
   VA = VB = VC = VD = 15 cm, AB = 14 cm and BC = 8 cm.
   Calculate
   (a) the angle between VB and the plane ABCD
   (b) the angle between VX and the plane ABCD where X is the mid-point of BC
   (c) the angle between the planes VBC and ABCD.

6. The pyramid VABCD has a horizontal rectangular base ABCD in which AB = 20 m and BC = 14 m. V is vertically above B and VB = 7 m. Calculate
   (a) the angle between DV and the plane ABCD
   (b) the angle between AV and the plane ABCD
   (c) the angle between the planes VAD and ABCD.

7. The figure shows a triangular pyramid on a horizontal base ABC, V is vertically above B where VB = 10 cm, $A\hat{B}C = 90°$ and AB = BC = 15 cm.

   Calculate the angle between the planes AVC and ABC.

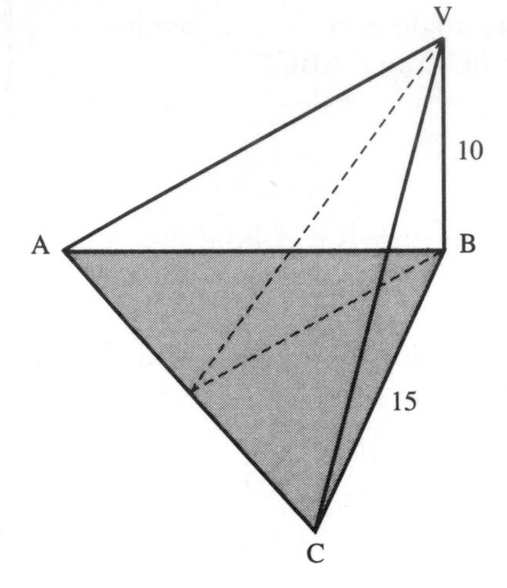

3.7 Congruent triangles

Two plane figures are congruent if one fits exactly on the other. The four types of congruence for triangles are as follows:

(a) Two sides and the included angle (S.A.S)

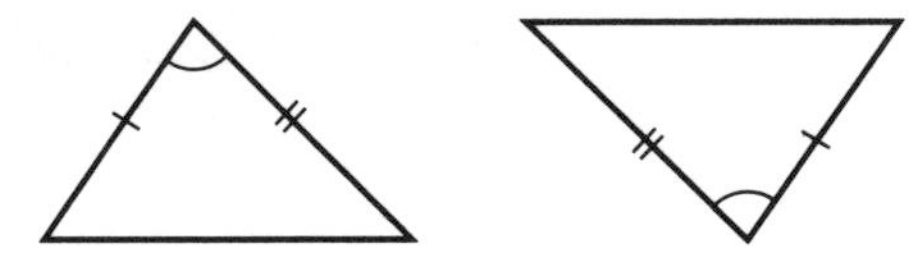

(b) Two angles and a corresponding side (A.A.S)

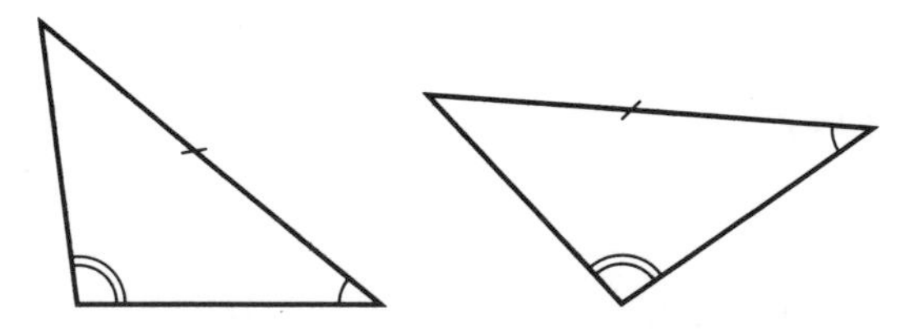

(c) Three sides (S.S.S)

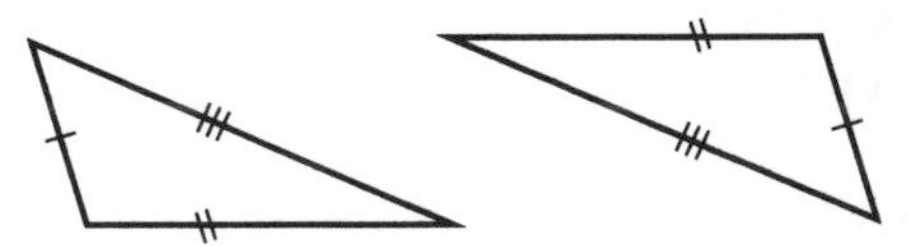

(d) Right angle, hypotenuse and one other side (R.H.S)

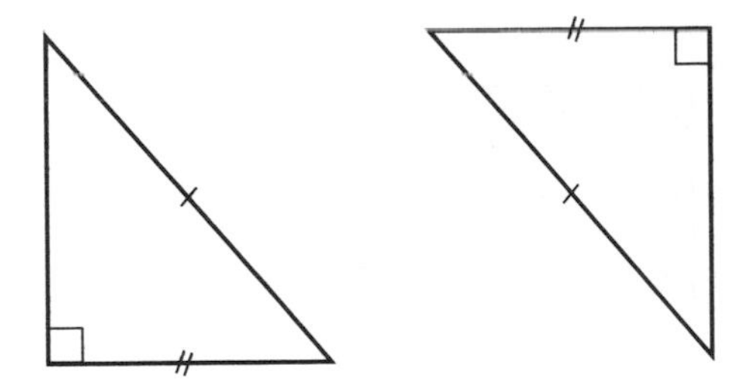

Exercise 17

For Questions **1** to **6**, decide whether the pair of triangles are congruent. If they are congruent, state which conditions for congruency are satisfied.

1.

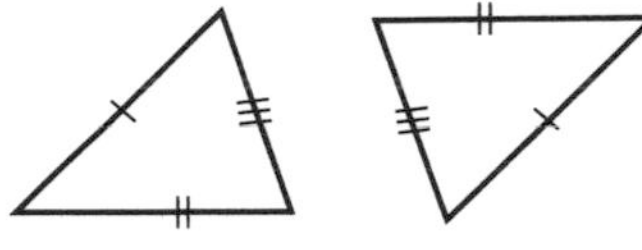

2.

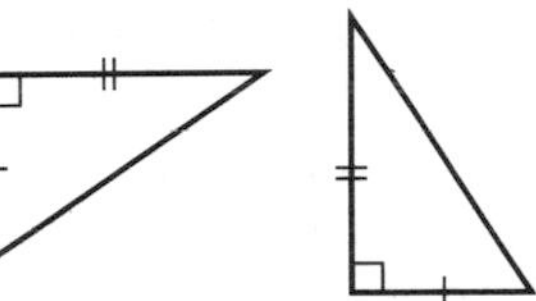

3.

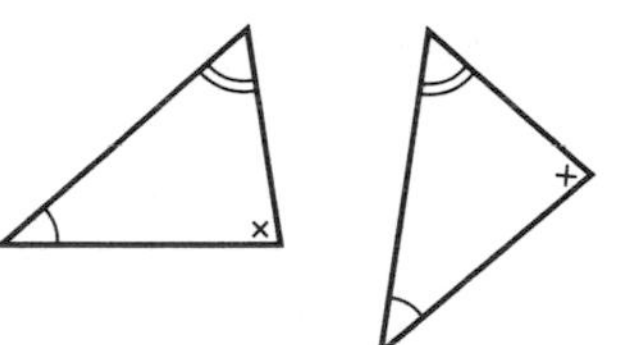

4.

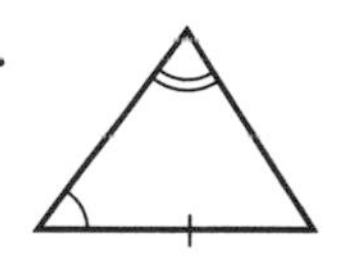

5.

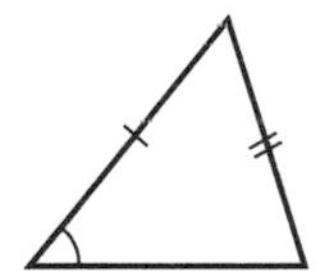

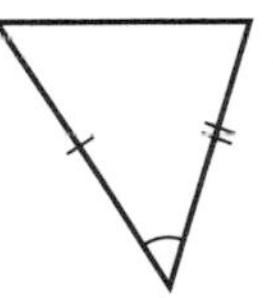

6.

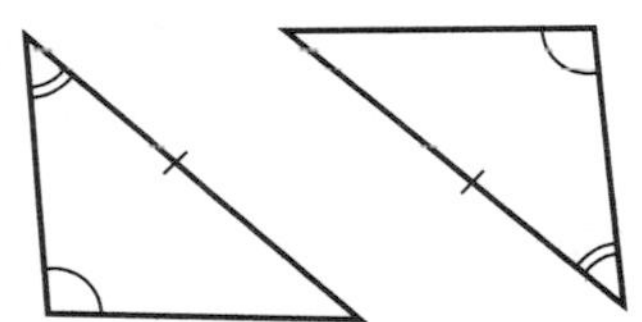

7. Triangle LMN is isosceles with LM = LN; X and Y are points on LM, LN respectively such that LX = LY. Prove that triangles LMY and LNX are congruent.

8. ABCD is a quadrilateral and a line through A parallel to BC meets DC at X. If $\widehat{D} = \widehat{C}$, prove that △ADX is isosceles.

9. XYZ is a triangle with XY = XZ. The bisectors of angles Y and Z meet the opposite sides in M and N respectively. Prove that YM = ZN.

10. In the diagram, DX = XC, DV = ZC and the lines AB and DC are parallel. Prove that
(a) AX = BX
(b) AC = BD
(c) triangles DBZ and CAV are congruent.

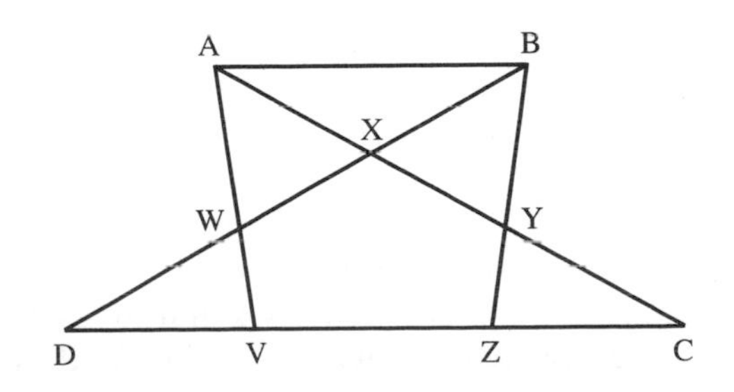

Examination exercise 3

1. The diagram shows a square garden of side 14 m. E, F, G and H are the mid-points of AB, BC, CD and DA respectively and EF, FG, GH and HE are arcs of a circle of radius 7 m.
The middle section of the garden is to be grassed as a lawn and the four side sections (shown shaded) are to be flower beds.

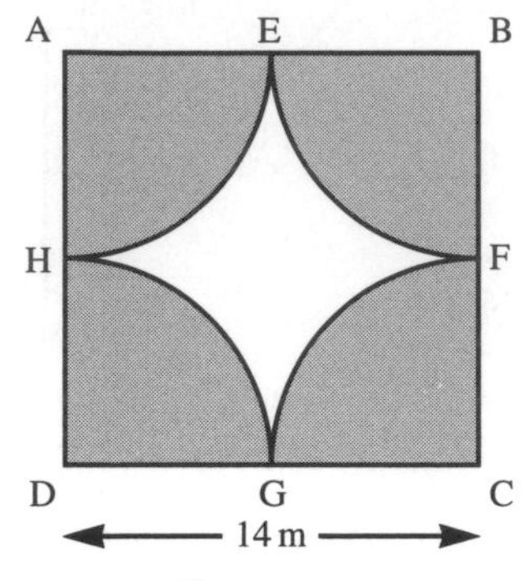

Calculate
   (a) the total area of the flower beds [take π as $\frac{22}{7}$ or 3·14],
   (b) the area of the lawn,
   (c) the amount of grass seed needed given that 80 grams of seed are needed per square metre. [W]

2. A metal ingot is in the form of a solid cylinder of length 7 cm and radius 3 cm.
   (a) Calculate the volume, in cm^3, of the ingot.
   The ingot is to be melted down and used to make cylindrical coins of thickness 3 mm and radius 12 mm.
   (b) Calculate the volume, in mm^3, of each coin.
   (c) Calculate the number of coins which can be made from the ingot, assuming that there is no wastage of metal. [M]

3.

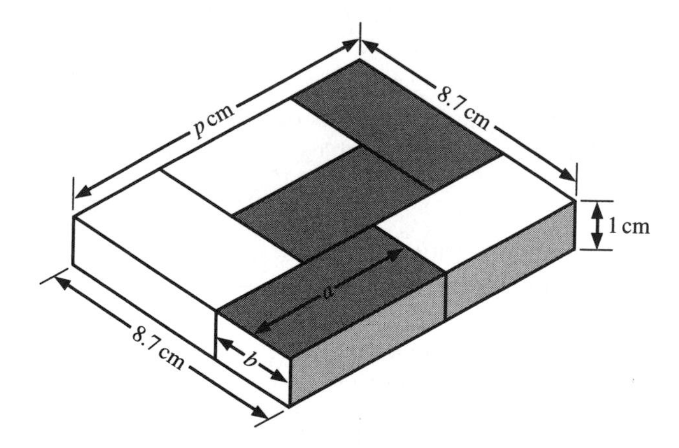

The drawing shows six identically shaped rectangular blocks of modelling clay — it is not drawn to scale. The length of each block is a and its width is b. Find the width b and hence the values of a and p. [N]

4. A rectangular match box measures 12 cm by 5 cm by 3 cm. Each match is a cuboid, 5 cm by 2 mm by 2 mm. What is the greatest number of matches which can be fitted into the box? [M]

5. The diagram shows a cuboid.
The coordinates of point A are (5, 0, 0).
The coordinates of point B are (5, 0, 4).
The coordinates of point C are (3, 3, 0).
What are the coordinates of point D? [N]

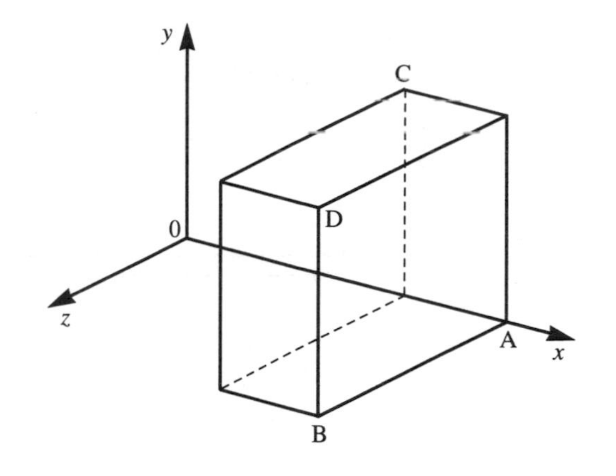

6. (a) Four of the vertices of a cuboid are at the points (1, 0, 0) (3, 0, 0) (1, 4, 0) and (1, 0, 3). Write down the coordinates of the other four vertices of this cuboid.
(b) Calculate the length of the line from the origin (0, 0, 0) to the point (3, 4, 12).

7. The diagram shows a box in the form of a cuboid, of internal dimensions 24 cm by 7 cm by 60 cm. Emma wishes to place the longest possible thin straight rod inside the box.
(a) Show a possible position of the rod in the diagram.
(b) Calculate the length of the rod. [M]

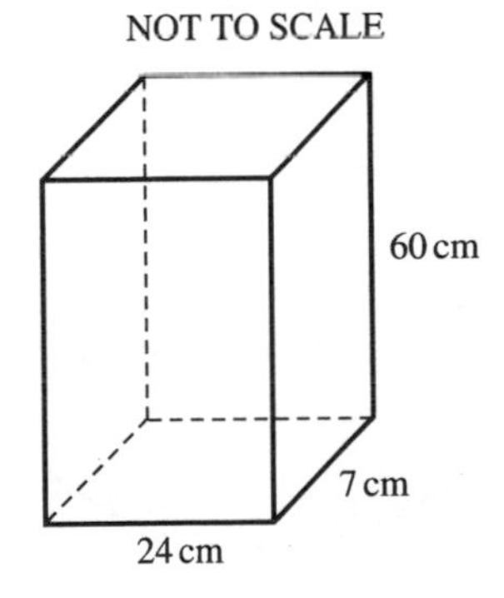

8. The diagram shows two sightings made from a point O of a helicopter flying at a height of 1600 metres. At the first sighting, the helicopter was due East of O and the angle of elevation was 58°. One minute later it was still due East of O but the angle of elevation was 45°. Calculate the speed of the helicopter in kilometres per hour. [N]

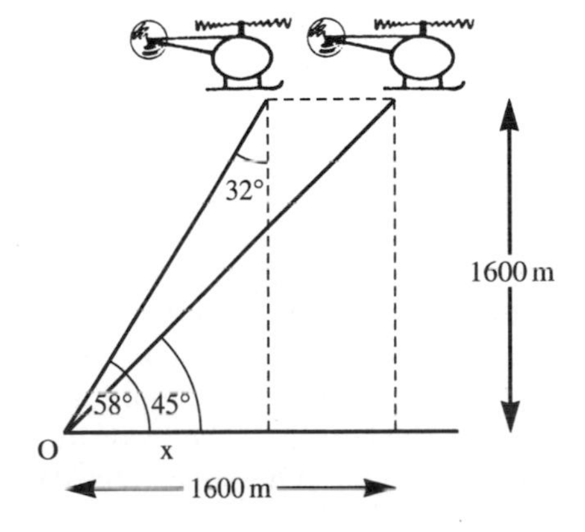

9.

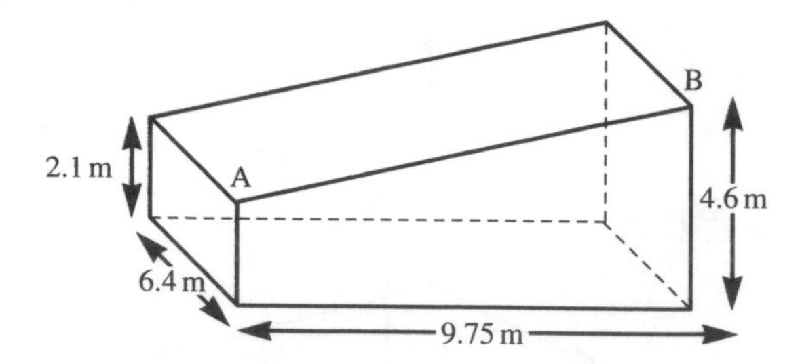

The diagram represents a squash court. The end walls and the floor are rectangular. The four walls are vertical and made of glass. The floor is horizontal and made of wood. The top is open.

(a) Find the total area, in m^2, of glass needed to build the court. Give your answer correct to three significant figures.

(b) Find the angle of elevation of B from A. Give your answer correct to the nearest degree. [M]

10.

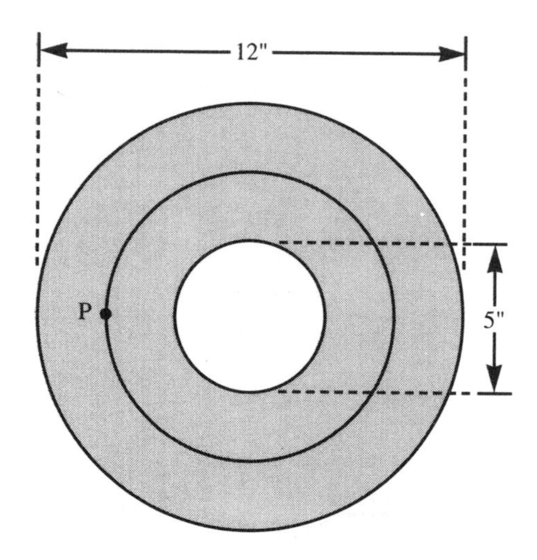

A 12-inch LP record has a circle of diameter 5 inches in the centre which contains the record label. The 12-inch record is played at $33\frac{1}{3}$ revolutions per minute (rpm). The playing time of one side is exactly 21 minutes.

(a) How many revolutions does the record make in this time?

The playing surface of the record is shaded in the diagram. The point P lies on a circle halfway between the inner and outer circumferences of the playing surface.

(b) Calculate the circumference of the circle through P.

The length of the recording track = circumference of the circle through $P \times$ the number of revolutions.

(c) Calculate the length of the recording track in miles correct to 2 decimal places. [One mile = 63 360 inches] [N]

11. Winston Jones makes a path across his rectangular front garden, as shown in the diagram.

AB = 10 m, AF = 2 m, FE = 3 m.

The edges of AC and FD of the path are parallel.

(a) Find the length of ED.
(b) Calculate the area of the path. [M]

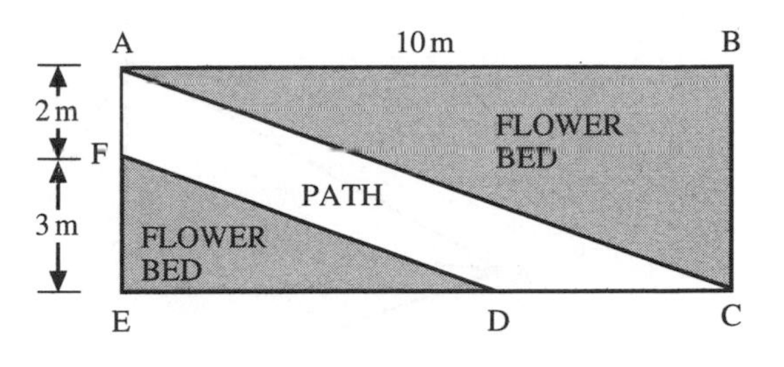

12. In Figure 1, OAB is a quadrant of a circle, radius r, and OPA is a semicircle drawn on OA as diameter.

(a) Calculate
  (i) the area of the semicircle OPA, in terms of r and π,
  (ii) the area of the shaded part of Figure 1, in terms of r and π.
(b) What is the size of angle OPA? Give your reason.

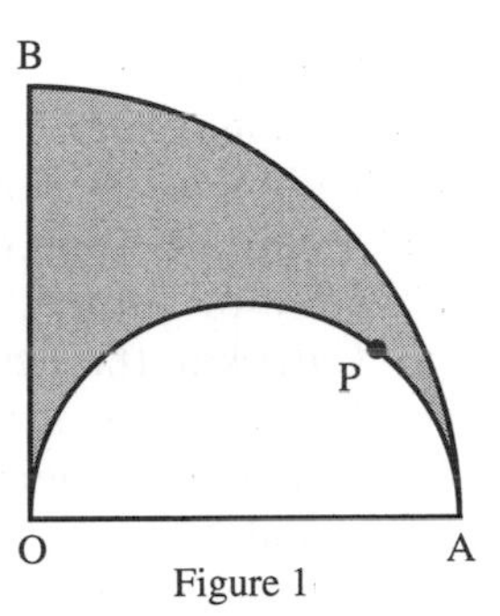

Figure 1

(c) In Figure 2, P has been moved to P′, so that BP′A is a straight line. Explain why a circle with OB as diameter will pass through P′. [N]

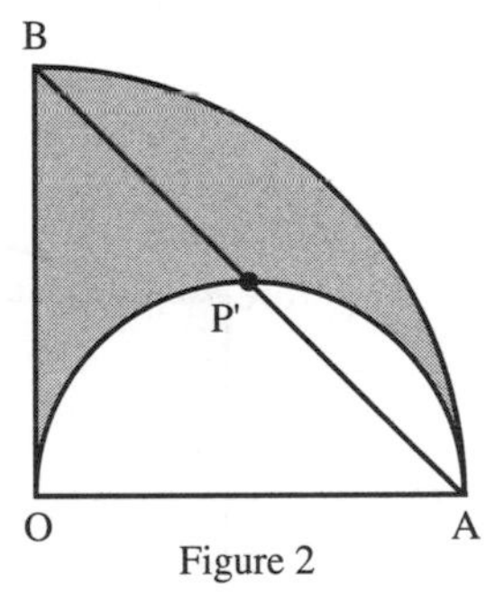

Figure 2

13. The horizontal base of the solid pyramid VABC is an isosceles triangle ABC. The edge VC is vertical. AC = BC = 13 cm, AB = 10 cm and VC = 8 cm.

(a) Calculate the size of the angle between the edge VB and the horizontal base.
(b) Calculate the size of angle ABC.
(c) Calculate the length of the line drawn from C to meet AB at 90°.
(d) Calculate the size of the angle between the face AVB and the horizontal base.

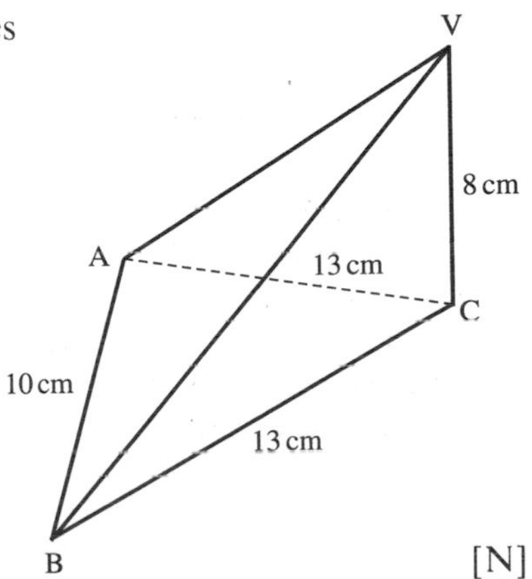

[N]

14.

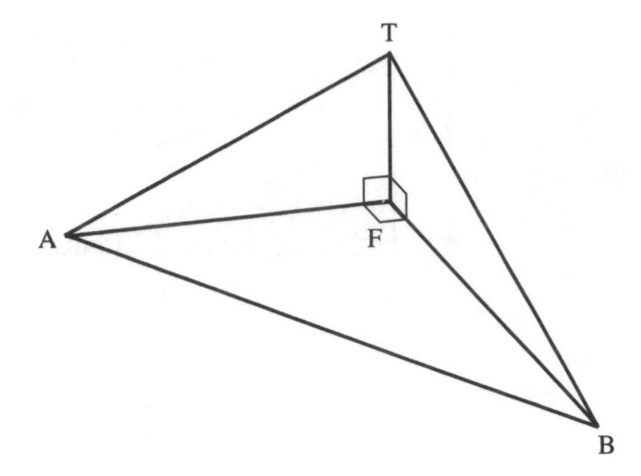

Alan is standing at the point A which is 600 m due West of the point F, the foot of a vertical television mast FT. His friend Brett is standing at the point B, 400 m due South of F.
The points A, B and F are in a horizontal plane. The angle of elevation of T from A is 13°.

(a) Calculate the height of the mast.
(b) Calculate the angle of elevation of T from B.

Given that C is the point on the line AB which is nearest to F,

(c) calculate
  (i) the length of CF
  (ii) the angle of elevation of T from C. [L]

15.

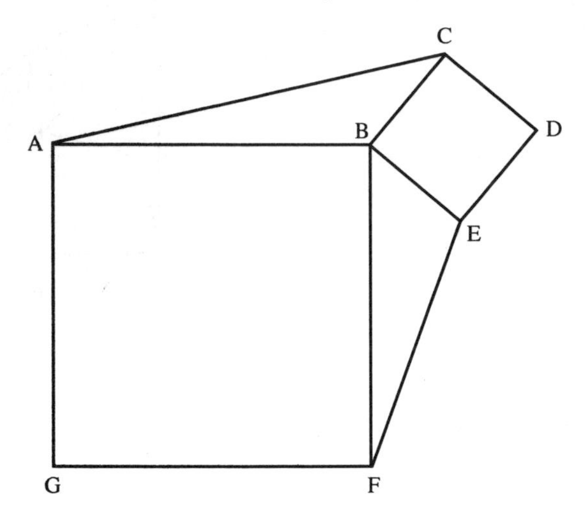

Squares ABFG and BCDE are drawn on the sides of an obtuse-angled triangle ABC. Are any of the following statements true? Give your reasons.

(i) Triangle ABC is congruent to triangle FBE.
(ii) Triangle ABC is similar to triangle FBE.
(iii) Triangle ABC is equal in area to triangle FBE. [N]

4 Algebra 2

4.1 Changing the subject of a formula

The operations involved in solving ordinary linear equations are exactly the same as the operations required in changing the subject of a formula.

Example 1

(a) Solve the equation $3x + 1 = 12$.

(b) Make x the subject of the formula $Mx + B = A$.

(c) Make y the subject of the formula $x(y - a) = e$.

(a)
$$\begin{aligned} 3x + 1 &= 12 \\ 3x &= 12 - 1 \\ x &= \frac{12 - 1}{3} = \frac{11}{3} \end{aligned}$$

(b)
$$\begin{aligned} Mx + B &= A \\ Mx &= A - B \\ x &= \frac{A - B}{M} \end{aligned}$$

(c)
$$\begin{aligned} x(y - a) &= e \\ xy - xa &= e \\ xy &= e + xa \\ y &= \frac{e + xa}{x} \end{aligned}$$

Exercise 1

Make x the subject of the following:

1. $Ax = B$
2. $Nx = T$
3. $Mx = K$
4. $xy = 4$
5. $9x = \mathrm{T} + \mathrm{N}$
6. $Ax = B - R$
7. $Cx = R + T$
8. $Lx = N - R^2$
9. $R - S^2 = Nx$
10. $x + 5 = 7$
11. $x + B = S$
12. $N = x + D$
13. $N^2 + x = T$
14. $L + x = N + M$
15. $x - R = A$
16. $x - A = E$
17. $F = x - B$
18. $F^2 = x - B^2$

Make y the subject of the following:

19. $L = y - B$
20. $N = y - T$
21. $3y + 1 = 7$
22. $2y - 4 = 5$
23. $Ay + C = N$
24. $By + D = L$
25. $Dy + E = F$
26. $Ny - F = H$
27. $Vy + m = Q$
28. $ty - m = n + a$
29. $V^2y + b = c$
30. $r = ny - 6$
31. $3(y - 1) = 5$
32. $A(y + B) = C$
33. $D(y + E) = F$
34. $h(y + n) = a$
35. $b(y - d) = q$
36. $n = r(y + t)$

Example 2

(a) Solve the equation $\frac{3a+1}{2} = 4$.

(b) Make a the subject of the formula $\frac{na+b}{m} = n$.

(c) Make a the subject of the formula $x - na = y$

(a)
$$\begin{aligned} \frac{3a+1}{2} &= 4 \\ 3a+1 &= 8 \\ 3a &= 7 \\ a &= \frac{7}{3} \end{aligned}$$

(b)
$$\begin{aligned} \frac{na+b}{m} &= n \\ na+b &= mn \\ na &= mn-b \\ a &= \frac{mn-b}{n} \end{aligned}$$

(c) $x - na = y$

Make the 'a' term positive

$$\begin{aligned} x &= y+na \\ x-y &= na \\ \frac{x-y}{n} &= a \end{aligned}$$

Exercise 2

Make a the subject.

1. $\frac{a}{4} = 3$
2. $\frac{a}{5} = 2$
3. $\frac{a}{D} = B$
4. $\frac{a}{B} = T$
5. $\frac{a}{N} = R$
6. $b = \frac{a}{m}$
7. $\frac{a-2}{4} = 6$
8. $\frac{a-A}{B} = T$
9. $\frac{a-D}{N} = A$
10. $\frac{a+Q}{N} = B^2$
11. $g = \frac{a-r}{e}$
12. $\frac{2a+1}{5} = 2$
13. $\frac{Aa+B}{C} = D$
14. $\frac{na+m}{p} = q$
15. $\frac{ra-t}{S} = v$
16. $\frac{za-m}{q} = t$
17. $\frac{m+Aa}{b} = c$
18. $A = \frac{Ba+D}{E}$
19. $n = \frac{ea-f}{h}$
20. $q = \frac{ga+b}{r}$
21. $6 - a = 2$
22. $7 - a = 9$
23. $5 = 7 - a$
24. $A - a = B$
25. $C - a = E$
26. $D - a = H$
27. $n - a = m$
28. $t = q - a$
29. $b = s - a$
30. $v = r - a$
31. $t = m - a$
32. $5 - 2a = 1$
33. $T - Xa = B$
34. $M - Na = Q$
35. $V - Ma = T$
36. $L = N - Ra$
37. $r = v^2 - ra$
38. $t^2 = w - na$
39. $n - qa = 2$
40. $\frac{3-4a}{2} = 1$
41. $\frac{5-7a}{3} = 2$
42. $\frac{B-Aa}{D} = E$
43. $\frac{D-Ea}{N} = B$
44. $\frac{h-fa}{b} = x$
45. $\frac{v^2-ha}{C} = d$
46. $\frac{M(a+B)}{N} = T$
47. $\frac{f(Na-e)}{m} = B$
48. $\frac{T(M-a)}{E} = F$
49. $\frac{y(x-a)}{z} = t$
50. $\frac{k^2(m-a)}{x} = x$

Formulas with fractions

Example 3

(a) Solve the equation $\dfrac{4}{z} = 7$.

(b) Make z the subject of the formula $\dfrac{n}{z} = k$.

(c) Make t the subject of the formula $\dfrac{x}{t} + m = a$.

(a) $$\frac{4}{z} = 7$$
$$4 = 7z$$
$$\frac{4}{7} = z$$

(b) $$\frac{n}{z} = k$$
$$n = kz$$
$$\frac{n}{k} = z$$

(c) $$\frac{x}{t} + m = a$$
$$\frac{x}{t} = a - m$$
$$x = (a - m)t$$
$$\frac{x}{(a - m)} = t$$

Exercise 3

Make a the subject.

1. $\dfrac{7}{a} = 14$ **2.** $\dfrac{5}{a} = 3$ **3.** $\dfrac{B}{a} = C$ **4.** $\dfrac{T}{a} = X$

5. $t = \dfrac{v}{a}$ **6.** $\dfrac{n}{a} = \sin 20°$ **7.** $\dfrac{7}{a} = \cos 30°$ **8.** $\dfrac{B}{a} = x$

9. $\dfrac{v}{a} = \dfrac{m}{s}$ **10.** $\dfrac{t}{b} = \dfrac{m}{a}$ **11.** $\dfrac{B}{a + D} = C$ **12.** $\dfrac{Q}{a - C} = T$

13. $\dfrac{V}{a - T} = D$ **14.** $\dfrac{L}{Ma} = B$ **15.** $\dfrac{N}{Ba} = C$ **16.** $\dfrac{m}{ca} = d$

17. $t = \dfrac{b}{c - a}$ **18.** $x = \dfrac{z}{y - a}$

Make x the subject

19. $\dfrac{2}{x} + 1 = 3$ **20.** $\dfrac{5}{x} - 2 = 4$ **21.** $\dfrac{A}{x} + B = C$ **22.** $\dfrac{V}{x} + G = H$

23. $\dfrac{r}{x} - t = n$ **24.** $q = \dfrac{b}{x} + d$ **25.** $t = \dfrac{m}{x} - n$ **26.** $h = d - \dfrac{b}{x}$

27. $C - \dfrac{d}{x} = e$ **28.** $r - \dfrac{m}{x} = e^2$ **29.** $t^2 = b - \dfrac{n}{x}$ **30.** $\dfrac{d}{x} + b = mn$

31. $3M = M + \dfrac{N}{P + x}$ **32.** $A = \dfrac{B}{c + x} - 5A$ **33.** $\dfrac{m^2}{x} - n = -p$ **34.** $t = w - \dfrac{q}{x}$

Squares and square roots

Example 4

Make x the subject of the formulas.

(a)
$$\begin{aligned} \sqrt{(x^2+A)} &= B \\ x^2+A &= B^2 \text{ (square both sides)} \\ x^2 &= B^2-A \\ x &= \pm\sqrt{(B^2-A)} \end{aligned}$$

(b)
$$\begin{aligned} (Ax-B)^2 &= M \\ Ax-B &= \pm\sqrt{M} \text{ (square root both sides)} \\ Ax &= B\pm\sqrt{M} \\ x &= \frac{B\pm\sqrt{M}}{A} \end{aligned}$$

Exercise 4

Make x the subject

1. $\sqrt{x}=2$
2. $\sqrt{(x-2)}=3$
3. $\sqrt{(x+C)}=D$
4. $\sqrt{(ax+b)}=c$
5. $b=\sqrt{(gx-t)}$
6. $\sqrt{(d-x)}=t$
7. $c=\sqrt{(n-x)}$
8. $g=\sqrt{(c-x)}$
9. $\sqrt{(Ax+B)}=\sqrt{D}$
10. $x^2=g$
11. $x^2=B$
12. $x^2-A=M$

Make k the subject

13. $C-k^2=m$
14. $mk^2=n$
15. $\frac{kz}{a}=t$
16. $n=a-k^2$
17. $\sqrt{(k^2-A)}=B$
18. $t=\sqrt{(m+k^2)}$
19. $A\sqrt{(k+B)}=M$
20. $\sqrt{\left(\frac{N}{k}\right)}=B$
21. $\sqrt{(a^2-k^2)}=t$
22. $2\pi\sqrt{(k+t)}=4$
23. $\sqrt{(ak^2-b)}=C$
24. $k^2+b=x^2$

Collecting from both sides

Example 5

Make x the subject of the formulas.

(a)
$$\begin{aligned} Ax-B &= Cx+D \\ Ax-Cx &= D+B \text{ (} x \text{ terms on one side)} \\ x(A-C) &= D+B \text{ (factorise)} \\ x &= \frac{D+B}{A-C} \end{aligned}$$

(b)
$$\begin{aligned} x+a &= \frac{x+b}{c} \\ c(x+a) &= x+b \\ cx+ca &= x+b \\ cx-x &= b-ca \text{ (} x \text{ terms on one side)} \\ x(c-1) &= b-ca \text{ (factorise)} \\ x &= \frac{b-ca}{c-1} \end{aligned}$$

Exercise 5

Make y the subject

1. $5(y-1)=2(y+3)$
2. $7(y-3)=4(3-y)$
3. $Ny+B=D-Ny$
4. $My-D=E-2My$
5. $ay+b=3b+by$
6. $my-c=e-ny$
7. $xy+4=7-ky$
8. $Ry+D=Ty+C$
9. $ay-x=z+by$

10. $m(y+a)=n(y+b)$

11. $x(y-b)=y+d$

12. $\dfrac{a-y}{a+y}=b$

13. $\dfrac{1-y}{1+y}=\dfrac{c}{d}$

14. $\dfrac{M-y}{M+y}=\dfrac{a}{b}$

15. $m(y+n)=(x-n)$

16. $y+m=\dfrac{2y-5}{m}$

17. $y-n=\dfrac{y+2}{n}$

18. $y+b=\dfrac{ay+e}{b}$

19. $\dfrac{ay+x}{x}=4-y$

20. $c-dy=e-ay$

21. $y(a-c)=by+d$

22. $y(m+n)=a(y+b)$

23. $t-ay=s-by$

24. $\dfrac{y+x}{y-x}=3$

25. $\dfrac{v-y}{v+y}=\dfrac{1}{2}$

26. $y(b-a)=a(y+b+c)$

27. $\sqrt{\left(\dfrac{y+x}{y-x}\right)}=2$

28. $\sqrt{\left(\dfrac{z+y}{z-y}\right)}=\dfrac{1}{3}$

29. $\sqrt{\left[\dfrac{m(y+n)}{y}\right]}=p$

30. $n-y=\dfrac{4y-n}{m}$

Exercise 6

Make the letter in square brackets the subject.

1. $ax+by+c=0$ [x]

2. $\sqrt{\{a(y^2-b)\}}=e$ [y]

3. $\dfrac{\sqrt{(k-m)}}{n}=\dfrac{1}{m}$ [k]

4. $a-bz=z+b$ [z]

5. $\dfrac{x+y}{x-y}=2$ [x]

6. $\sqrt{\left(\dfrac{a}{z}-c\right)}=e$ [z]

7. $lm+mn+a=0$ [n]

8. $t=2\pi\sqrt{\left(\dfrac{d}{g}\right)}$ [d]

9. $t=2\pi\sqrt{\left(\dfrac{d}{g}\right)}$ [g]

10. $\sqrt{(x^2+a)}=2x$ [x]

11. $\sqrt{\left\{\dfrac{b(m^2+a)}{e}\right\}}=t$ [m]

12. $\sqrt{\left(\dfrac{x+1}{x}\right)}=a$ [x]

13. $a+b-mx=0$ [m]

14. $\sqrt{(a^2+b^2)}=x^2$ [a]

15. $\dfrac{a}{k}+b=\dfrac{c}{k}$ [k]

16. $a-y=\dfrac{b+y}{a}$ [y]

17. $G=4\pi\sqrt{(x^2+T^2)}$ [x]

18. $M(ax+by+c)=0$ [y]

19. $x=\sqrt{\left(\dfrac{y-1}{y+1}\right)}$ [y]

20. $a\sqrt{\left(\dfrac{x^2-n}{m}\right)}=\dfrac{a^2}{b}$ [x]

21. $\dfrac{M}{N}+E=\dfrac{P}{N}$ [N

22. $\dfrac{Q}{P-x}=R$ [x]

23. $\sqrt{(z-ax)}=t$ [a]

24. $e+\sqrt{(x+f)}=g$ [x]

4.2 Direct and inverse proportion

Direct proportion

(a) When you buy petrol, the more you buy the more money you have to pay. So if 2·2 litres costs 121p, then 4·4 litres will cost 242p.
We say the cost of petrol is *directly proportional* to the quantity bought.
To show that quantities are proportional, we use the symbol '$\propto$'.
So in our example if the cost of petrol is c pence and the number of litres of petrol is l, we write

$$c \propto l$$

The '$\propto$' sign can always be replaced by '$= k$' where k is a constant.

So $c = kl$

From above, if $c = 121$ when $l = 2{\cdot}2$

then $121 = k \times 2{\cdot}2$

$$k = \frac{121}{2{\cdot}2} = 55$$

We can then write $c = 55l$, and this allows us to find the value of c for any value of l, and *vice versa*.

(b) If a quantity z is proportional to a quantity x, we have

$z \propto x$ or $z = kx$

Two other expressions are sometimes used when quantities are directly proportional. We could say

'z varies as x'

or 'z varies directly as x'.

The graph connecting z and x is a straight line which passes through the origin.

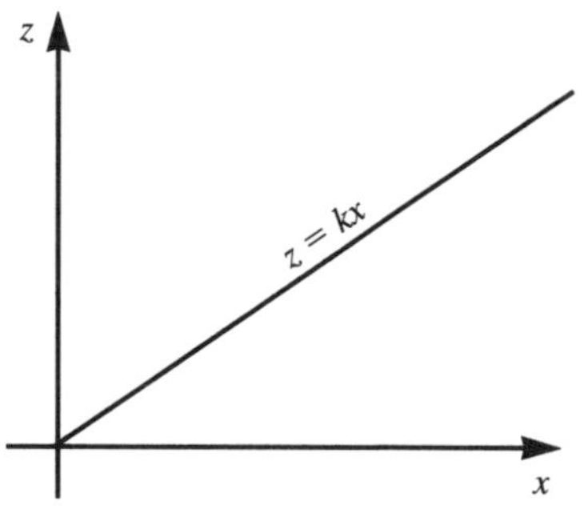

Example 1

y varies as z, and $y = 2$ when $z = 5$; find

(a) the value of y when $z = 6$
(b) the value of z when $y = 5$

Because $y \propto z$, then $y = kz$ where k is a constant.

$y = 2$ when $z = 5$

$\therefore \quad 2 = k \times 5$

$k = \frac{2}{5}$

So $\quad y = \frac{2}{5}z$

(a) When $z = 6$, $y = \frac{2}{5} \times 6 = 2\frac{2}{5}$.

(b) When $y = 5$, $5 = \frac{2}{5}z$; $z = \frac{25}{2} = 12\frac{1}{2}$

Example 2

The value V of a diamond is proportional to the square of its weight W. If a diamond weighing 10 grams is worth £200, find

(a) the value of a diamond weighing 30 grams
(b) the weight of a diamond worth £5000.

$V \propto W^2$

or $V = kW^2$ where k is a constant.

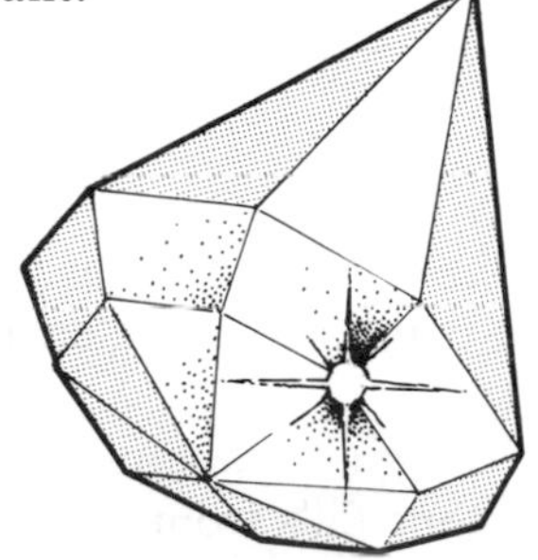

$V = 200$ when $W = 10$

$\therefore \quad 200 = k \times 10^2$

$k = 2$

So $V = 2W^2$

(a) When $W = 30$,

$V = 2 \times 30^2 = 2 \times 900$

$V = £1800$

So a diamond of weight 30 grams is worth £1800.

(b) When $V = 5000$,

$5000 = 2 \times W^2$

$$W^2 = \frac{5000}{2} = 2500$$

$W = \sqrt{2500} = 50$

So a diamond of value £5000 weighs 50 grams.

Exercise 7

1. Rewrite the statement connecting each pair of variables using a constant k instead of '$\propto$'.
(a) $S \propto e$ (b) $v \propto t$ (c) $x \propto z^2$
(d) $y \propto \sqrt{x}$ (e) $T \propto \sqrt{L}$

2. y varies as t. If $y = 6$ when $t = 4$, calculate
(a) the value of y, when $t = 6$
(b) the value of t, when $y = 4$.

3. z is proportional to m. If $z = 20$ when $m = 4$, calculate
(a) the value of z, when $m = 7$
(b) the value of m, when $z = 55$.

4. A varies directly as r^2. If $A = 12$, when $r = 2$, calculate
(a) the value of A, when $r = 5$
(b) the value of r, when $A = 48$.

5. Given that $z \propto x$, copy and complete the table.

x	1	3		$5\frac{1}{2}$
z	4		16	

6. Given that $V \propto r^3$, copy and complete the table.

r	1	2		$1\frac{1}{2}$
V	4		256	

7. The pressure of the water P at any point below the surface of the sea varies as the depth of the point below the surface d. If the pressure is 200 newtons/cm^2 at a depth of 3 m, calculate the pressure at a depth of 5 m.

8. The distance d through which a stone falls from rest is proportional to the square of the time taken t. If the stone falls 45 m in 3 seconds, how far will it fall in 6 seconds? How long will it take to fall 20 m?

9. The energy E stored in an elastic band varies as the square of the extension x. When the elastic is extended by 3 cm, the energy stored is 243 joules. What is the energy stored when the extension is 5 cm? What is the extension when the stored energy is 36 joules?

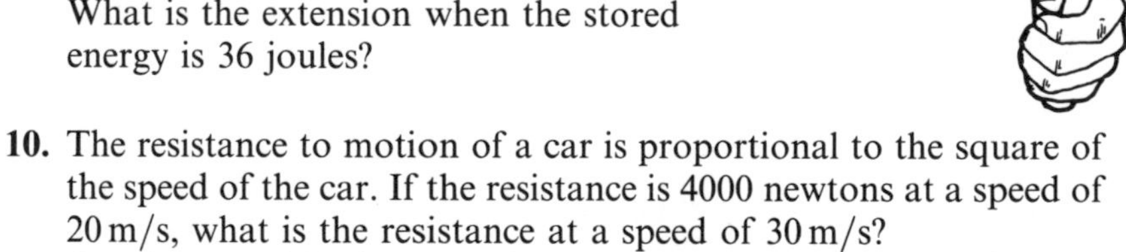

10. The resistance to motion of a car is proportional to the square of the speed of the car. If the resistance is 4000 newtons at a speed of 20 m/s, what is the resistance at a speed of 30 m/s? At what speed is the resistance 6250 newtons?

11. In an experiment, measurements of w and p were taken.

w	2	5	7
p	1·6	25	68·6

Which of these laws fits the results?

$p \propto w$ $\quad p \propto w^2$ $\quad p \propto w^3$

12. A road research organisation recently claimed that the damage to road surfaces was proportional to the fourth power of the axle load. The axle load of a 44-ton HGV is about 15 times that of a car. Calculate the ratio of the damage to road surfaces made by a 44-ton HGV and a car.

Inverse proportion

To travel a distance of 200 m at 10 m/s, the time taken is 20 s.
To travel the same distance at 20 m/s, the time taken is 10 s.
As you *double* the speed, you *halve* the time taken.
For a fixed journey distance, the time taken is *inversely proportional* to the speed of travel.
If t is inversely proportional to s we write

$$s \propto \frac{1}{t}$$

or $s = k \times \frac{1}{t}$

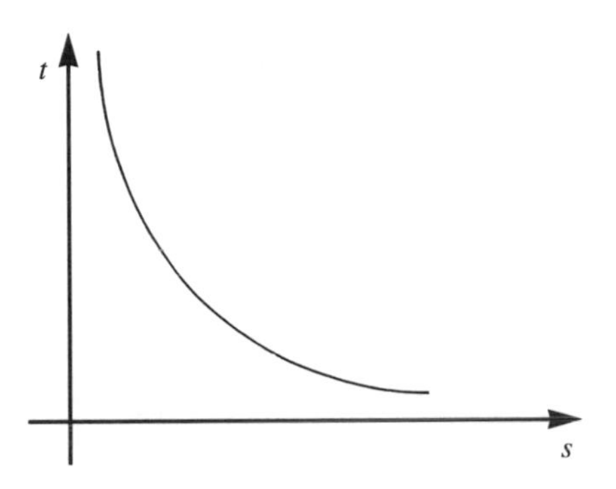

Notice that the product $s \times t$ is constant.
The graph connecting s and t is a curve.

The shape of the curve is the same as $y = \frac{1}{x}$.

Note: Sometimes we write 'x varies inversely as y'.
It means the same as 'x is inversely proportional to y'.

> *Example 3*
>
> z is inversely proportional to t^2 and $z = 4$ when $t = 1$.
> Calculate z when $t = 2$
>
> We have $z \propto \frac{1}{t^2}$ or $z = k \times \frac{1}{t^2}$ (k is a constant)
>
> $z = 4$ when $t = 1$,
>
> $\therefore\ 4 = k\left(\frac{1}{1^2}\right)$
>
> so $k = 4$
>
> $\therefore\ z = 4 \times \frac{1}{t^2}$
>
> When $t = 2$, $z = 4 \times \frac{1}{2^2} = 1$.

Exercise 8

1. Rewrite the statements connecting the variables using a constant of variation, k.

(a) $x \propto \frac{1}{y}$ (b) $s \propto \frac{1}{t^2}$ (c) $t \propto \frac{1}{\sqrt{q}}$

(d) m varies inversely as w

(e) z is inversely proportional to t^2.

2. b varies inversely as e. If $b = 6$ when $e = 2$, calculate
(a) the value of b when $e = 12$
(b) the value of e when $b = 3$.

3. x is inversely proportional to y^2. If $x = 4$ when $y = 3$, calculate
(a) the value of x when $y = 1$
(b) the value of y when $x = 2\frac{1}{4}$.

4. p is inversely proportional to $\sqrt{y}$. If $p = 1{\cdot}2$ when $y = 100$, calculate
(a) the value of p when $y = 4$
(b) the value of y when $p = 3$.

5. Given that $z \propto \frac{1}{y}$, copy and complete the table:

y	2	4		$\frac{1}{4}$
z	8		16	

6. Given that $v \propto \frac{1}{t^2}$, copy and complete the table:

t	2	5		10
v	25		$\frac{1}{4}$	

7. e varies inversely as $(y - 2)$. If $e = 12$ when $y = 4$, find
(a) e when $y = 6$ (b) y when $e = \frac{1}{2}$

8. The volume V of a given mass of gas varies inversely as the pressure P. When $V = 2\,\text{m}^3$, $P = 500\,\text{N/m}^2$. Find the volume when the pressure is $400\,\text{N/m}^2$. Find the pressure when the volume is $5\,\text{m}^3$.

9. The number of hours N required to dig a certain hole is inversely proportional to the number of men available x.

When 6 men are digging, the hole takes 4 hours. Find the time taken when 8 men are available. If it takes $\frac{1}{2}$ hour to dig the hole, how many men are there?

10. The force of attraction F between two magnets varies inversely as the square of the distance d between them. When the magnets are 2 cm apart, the force of attraction is 18 newtons. How far apart are they if the attractive force is 2 newtons?

11. The life expectancy L of a rat varies inversely as the square of the density d of poison distributed around his home.

When the density of poison is $1\,g/m^2$ the life expectancy is 50 days. How long will he survive if the density of poison is
(a) $5\,g/m^2$? (b) $\frac{1}{2}\,g/m^2$?

12. When cooking snacks in a microwave oven, a French chef assumes that the cooking time is inversely proportional to the power used. The five levels on his microwave have the powers shown in the table.

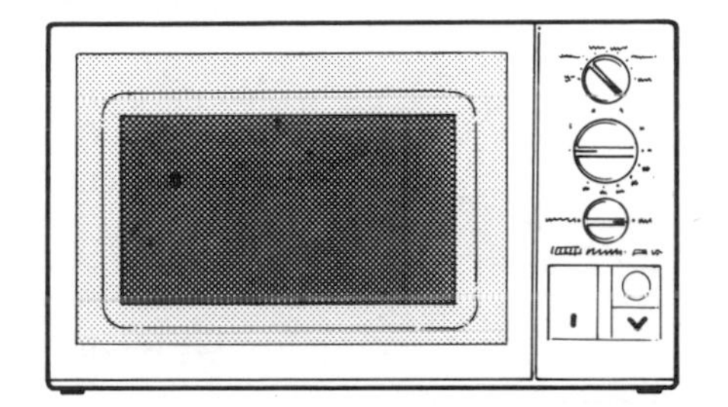

Level	Power used
Full	600 W
Roast	400 W
Simmer	200 W
Defrost	100 W
Warm	50 W

(a) Escargots de Bourgogne take 5 minutes on 'Simmer'. How long will they take on 'Warm'?
(b) Escargots à la Provençale are normally cooked on 'Roast' for 3 minutes. How long will they take on 'Full'?

13. Given $z = \dfrac{k}{x^n}$, find k and n, then copy and complete the table.

x	1	2	4	
z	100	$12\frac{1}{2}$		$\frac{1}{10}$

14. Given $y = \dfrac{k}{\sqrt[n]{v}}$, find k and n, then copy and complete the table.

v	1	4	36	
y	12	6		$\frac{3}{25}$

[$\sqrt[n]{v}$ means the nth root of v]

4.3 General straight line law $y = mx + c$

Gradient constant m

The gradient of a straight line is a measure of how steep it is.

Example 1

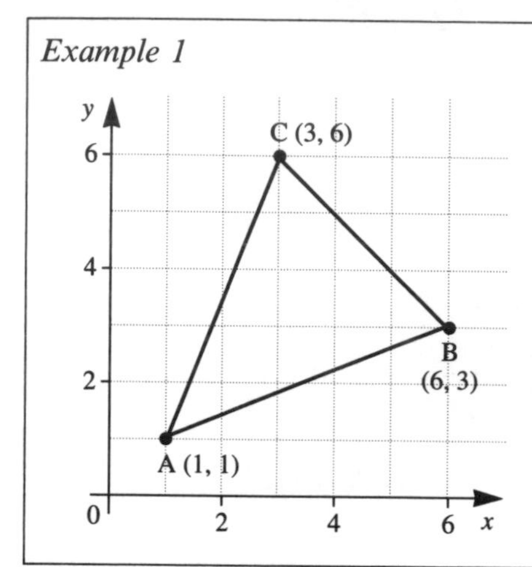

Gradient of line AB $= \dfrac{3-1}{6-1} = \dfrac{2}{5}$.

Gradient of line AC $= \dfrac{6-1}{3-1} = \dfrac{5}{2}$.

Gradient of line BC $= \dfrac{6-3}{3-6} = -1$.

A line which slopes upwards to the right has a *positive* gradient.
A line which slopes upwards to the left has a *negative* gradient.

$$\text{Gradient} = \frac{\text{difference in } y\text{-coordinates}}{\text{difference in } x\text{-coordinates}}$$

Exercise 9

1. Find the gradient of AB, BC, AC.

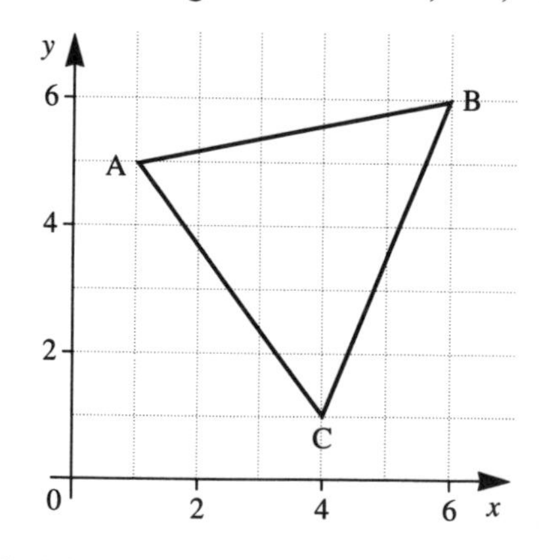

2. Find the gradient of PQ, PR, QR.

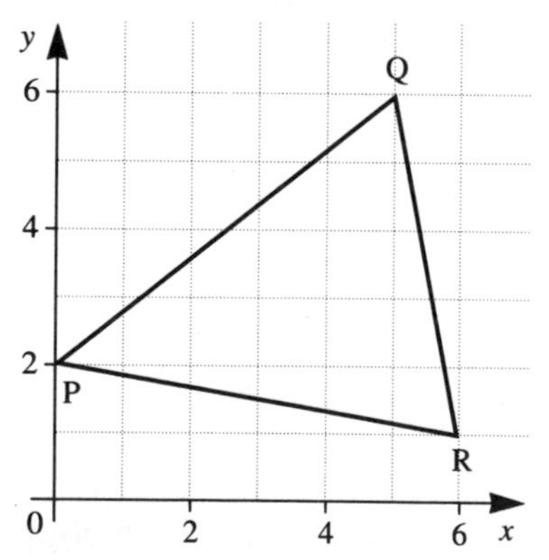

3. Find the gradient of the lines joining the following pairs of points:
(a) $(5, 2) \rightarrow (7, 8)$ (b) $(-1, 3) \rightarrow (1, 6)$
(c) $(\frac{1}{2}, 1) \rightarrow (\frac{3}{4}, 2)$ (d) $(3{\cdot}1, 2) \rightarrow (3{\cdot}2, 2{\cdot}5)$

4. Find the value of a if the line joining the points $(3a, 4)$ and $(a, -3)$ has a gradient of 1.

5. (a) Write down the gradient of the line joining the points $(2m, n)$ and $(3, -4)$,
(b) Find the value of n if the line is parallel to the x-axis,
(c) Find the value of m if the line is parallel to the y-axis.

Intercept constant c

Here are two straight lines.

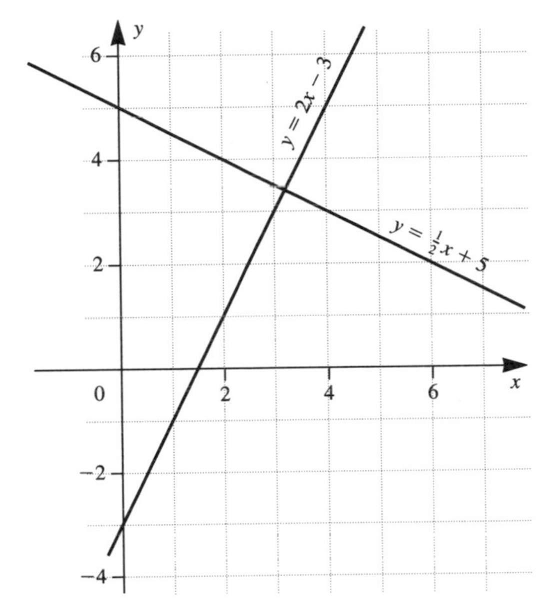

For $y = 2x - 3$, the gradient is 2 and the y-intercept is -3.

For $y = -\frac{1}{2}x + 5$ the gradient is $-\frac{1}{2}$ and the y-intercept is 5.

These two lines illustrate a general rule.
When the equation of a straight line is written in the form

$y = mx + c$

the gradient of the line is m and the intercept on the y-axis is c.

Example 2

Draw the line $y = 2x + 3$ on a *sketch* graph.

The word 'sketch' implies that we do not plot a series of points but simply show the position and slope of the line.

The line $y = 2x + 3$ has a gradient of 2 and cuts the y-axis at (0, 3)

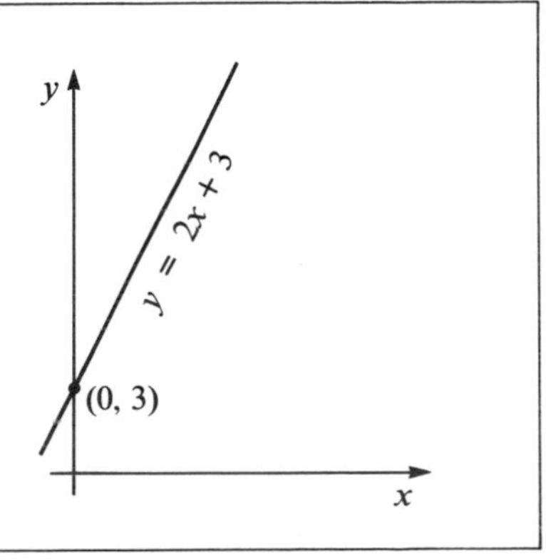

Example 3

Draw the line $x + 2y - 6 = 0$ on a sketch graph.

(a) Rearrange the equation to make y the subject.

$$x + 2y - 6 = 0$$
$$2y = -x + 6$$
$$y = -\tfrac{1}{2}x + 3.$$

(b) The line has a gradient of $-\frac{1}{2}$ and cuts the y-axis at (0, 3).

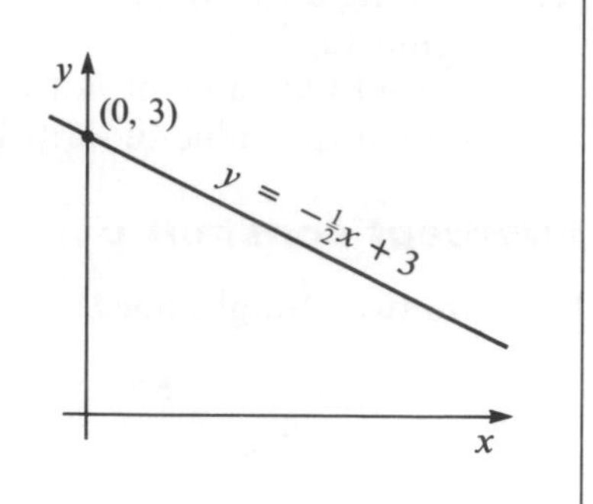

Exercise 10

In Questions **1** to **20**, find the gradient of the line and the intercept on the y-axis. Hence draw a small sketch graph of each line.

1. $y = x + 3$ **2.** $y = x - 2$ **3.** $y = 2x + 1$

4. $y = 2x - 5$ **5.** $y = 3x + 4$ **6.** $y = \frac{1}{2}x + 6$

7. $y = 3x - 2$ **8.** $y = 2x$ **9.** $y = \frac{1}{4}x - 4$

10. $y = -x + 3$ **11.** $y = 6 - 2x$ **12.** $y = 2 - x$

13. $y + 2x = 3$ **14.** $3x + y + 4 = 0$ **15.** $2y - x = 6$

16. $3y + x - 9 = 0$ **17.** $4x - y = 5$ **18.** $3x - 2y = 8$

19. $10x - y = 0$ **20.** $y - 4 = 0$

21. Find the equations of the lines A and B.

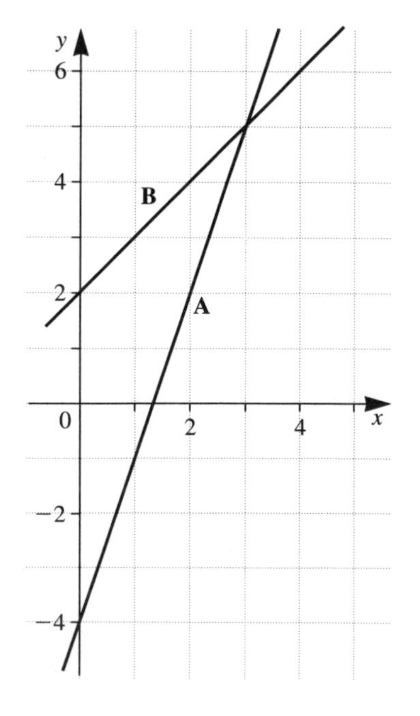

22. Find the equations of the lines C and D.

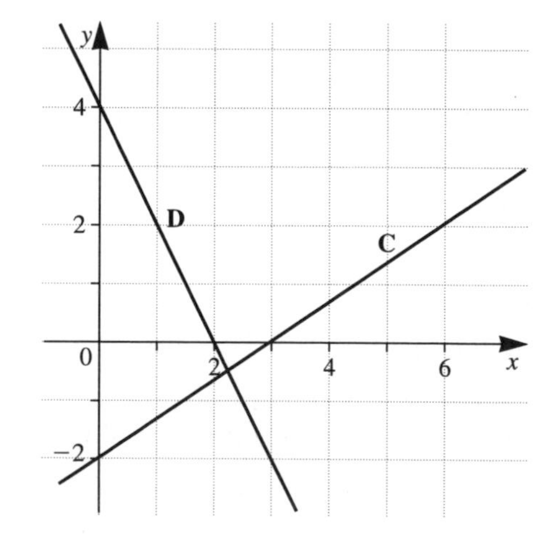

23. The sketch represents a section of the curve $y = x^2 - 2x - 8$.
Calculate
(a) the coordinates of A and of B,
(b) the gradient of the line AB,
(c) the equation of the straight line AB.

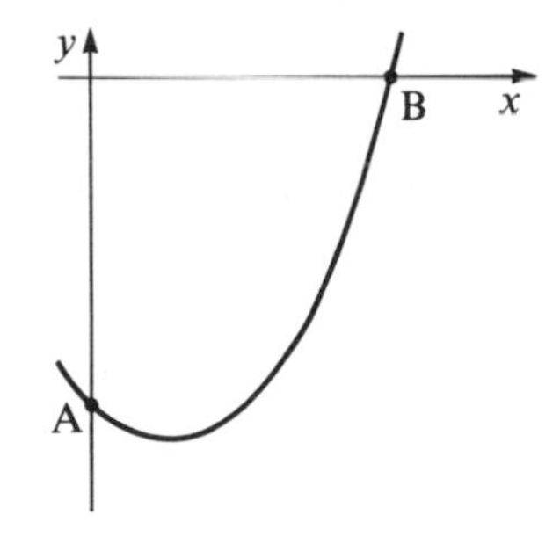

Finding the equation of a line

Example 4

Find the equation of the straight line which passes through (1, 3) and (3, 7).

(a) Let the equation of the line take the form $y = mx + c$.

The gradient, $m = \dfrac{7-3}{3-1} = 2$

so we may write the equation as

$y = 2x + c$ ...[1]

(b) Since the line passes through (1, 3), substitute 3 for y and 1 for x in [1].

$\therefore \quad 3 = 2 \times 1 + c$
$\quad\quad 1 = c$

The equation of the line is $y = 2x + 1$

Exercise 11

In Questions **1** to **10** find the equation of the line which

1. Passes through (0, 7) at a gradient of 3

2. Passes through (0, −9) at a gradient of 2

3. Passes through (0, 5) at a gradient of −1

4. Passes through (2, 3) at a gradient of 2

5. Passes through (2, 11) at a gradient of 3

6. Passes through (4, 3) at a gradient of −1

7. Passes through (6, 0) at a gradient of $\frac{1}{2}$

8. Passes through (2, 1) and (4, 5)

9. Passes through (5, 4) and (6, 7)

10. Passes through (0, 5) and (3, 2)

Finding a law

A ball is released from the top of a building. Its position is recorded every second using an electronic photoflash camera.
Here are the results, giving distance fallen d and time taken t.

t	0	1	2	3
d	0	4·9	19·6	44·1

By plotting d against t, a curve is obtained.

It is not easy to find the equation (the law) connecting d and t from a curve.

Try plotting d against t^2.

t	0	1	2	3
t^2	0	1	4	9
d	0	4·9	19·6	44·1

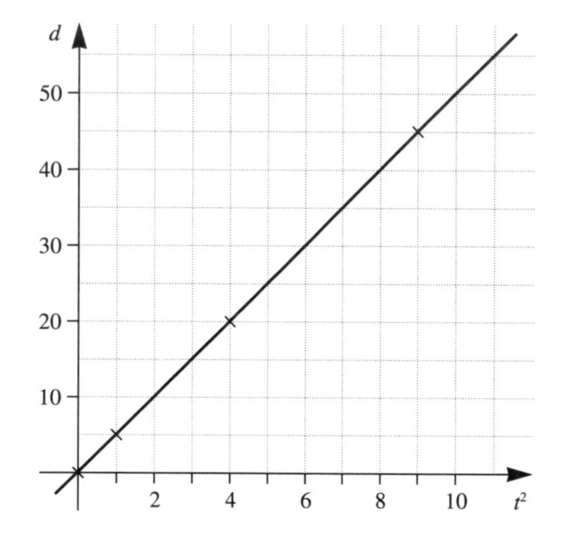

This is a straight line with gradient 4·9 and intercept 0 on the d-axis.

$\therefore$ the law connecting d and t is
$d = 4{\cdot}9t^2 + 0$
or $d = 4{\cdot}9t^2$

Note: We can only find the law from a straight line graph. It is not easy to discover an equation from a curve.

Exercise 12

1. Two variables Z and X are thought to be connected by an equation of the form $Z = aX + c$. Here are some values of Z and X.

X	1	2	3·6	4·2	6·4
Z	4	6·5	10·5	12	17·5

Draw a graph with X on the horizontal axis using a scale of 2 cm to 1 unit and Z on the vertical axis with a scale of 1 cm to 1 unit. Use your graphs to find a and c and hence write down the equation relating Z and X.

2. In an experiment, the following measurements of the variables q and t were taken.

q	0·5	1·0	1·5	2·0	2·5	3·0
t	3·85	5·0	6·1	7·0	7·75	9·1

A scientist suspects that q and t are related by an equation of the form $t = mq + c$, (m and c constants). Plot the values obtained from the experiment and draw the line of best fit through the points. Plot q on the horizontal axis with a scale of 4 cm to 1 unit, and t on the vertical axis with a scale of 2 cm to 1 unit. Find the gradient and intercept on the t-axis and hence estimate the values of m and c.

3. In an experiment, the following measurements of p and z were taken:

z	1·2	2·0	2·4	3·2	3·8	4·6
p	11·5	10·2	8·8	7·0	6·0	3·5

Plot the points on a graph with z on the horizontal axis and draw the line of best fit through the points. Hence estimate the values of n and k if the equation relating p and z is of the form $p = nz + k$.

4. In an experiment the following measurements of t and z were taken:

t	1·41	2·12	2·55	3·0	3·39	3·74
z	3·4	3·85	4·35	4·8	5·3	5·75

Draw a graph, plotting t^2 on the horizontal axis and z on the vertical axis, and hence confirm that the equation connecting t and z is of the form $z = mt^2 + c$. Find approximate values for m and c.

5. Two variables T and V are thought to be related by an equation which is either

$V = a\sqrt{T} + b$ or $V = \dfrac{a}{T} + b$

Here are some values of T and V

T	1	2	4	10
V	15·5	9·5	6·5	4·7

Draw a suitable graph (or graphs) to determine which equation is the correct one.
What are the values of a and b?

6. Two engineers measured the power P (kilowatts) required for a car to travel at various speeds V (km/h). Here are their results.

V	40	60	80	100	120	140
P	13·5	30·5	54	84·5	122	166

John thinks the formula connecting P and V is of the form $P = mV + c$.
Steve thinks the formula is $\sqrt{P} = kV$.
Steve says that John must be wrong, just by looking at the results. Explain why John's formula cannot be correct.
Draw a graph with V on the horizontal axis (1 cm = 20 units) and $\sqrt{P}$ on the vertical axis (1 cm = 1 unit).
Draw a line of best fit to confirm that Steve's formula is correct and estimate the value of the constant k.

4.4 Real life graphs

Travel graphs

Exercise 13

1. The graph shows the journeys made by a van and a car starting at York, travelling to Durham and returning to York.
   (a) For how long was the van stationary during the journey?
   (b) At what time did the car first overtake the van?

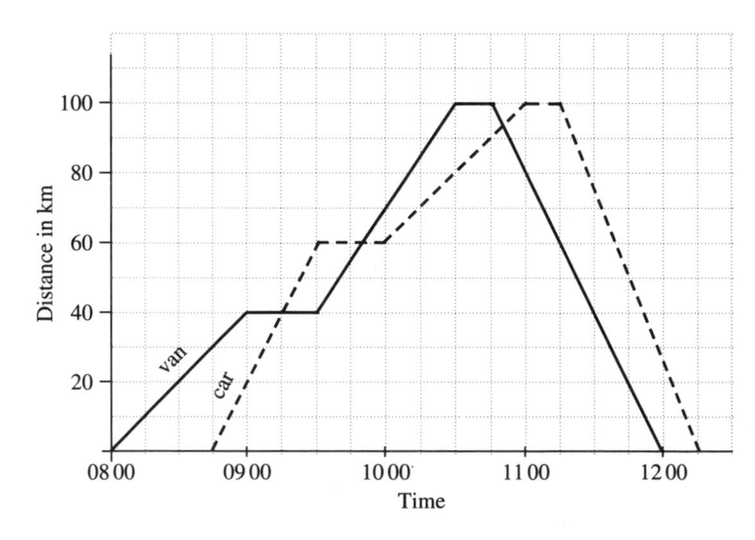

   (c) At what speed was the van travelling between 09 30 and 10 00?
   (d) What was the greatest speed attained by the car during the entire journey?
   (e) What was the average speed of the car over its entire journey?

2. The graph shows the journeys of a bus and a car along the same road. The bus goes from Leeds to Darlington and back to Leeds. The car goes from Darlington to Leeds and back to Darlington.

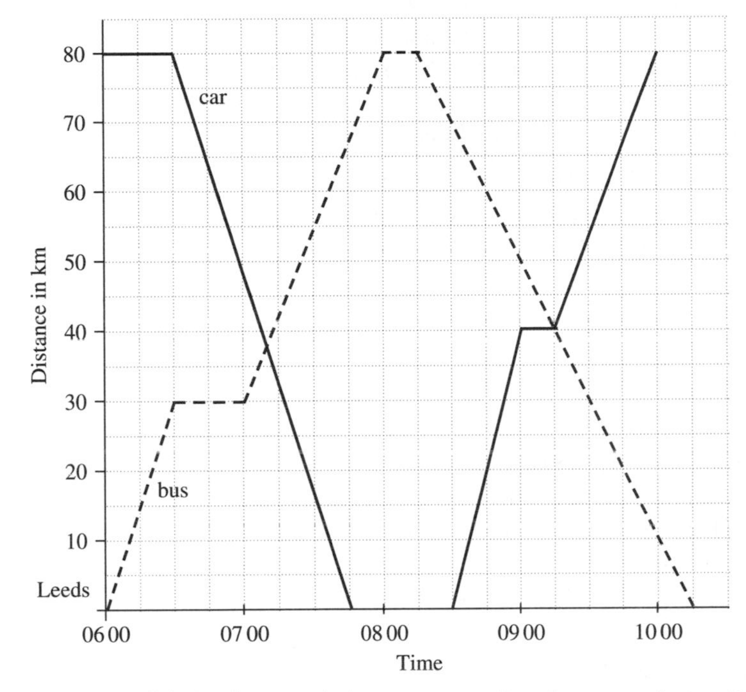

(a) When did the bus and the car meet for the second time?
(b) At what speed did the car travel from Darlington to Leeds?
(c) What was the average speed of the bus over its entire journey?
(d) Approximately how far apart were the bus and the car at 09 45?
(e) What was the greatest speed attained by the car during its entire journey?

In Questions **3, 4, 5,** draw a travel graph to illustrate the journey described. Draw axes with the same scales as in Question **2**.

3. (a) Mrs Chuong leaves home at 08 00 and drives at a speed of 50 km/h. After $\frac{1}{2}$ hour she reduces her speed to 40 km/h and continues at this speed until 09 30. She stops from 09 30 until 10 00 and then returns home at a speed of 60 km/h.
(b) Use a graph to find the approximate time at which she arrives home.

4. (a) Mr Coe leaves home at 09 00 and drives at a speed of 20 km/h. After $\frac{3}{4}$ hour he increases his speed to 45 km/h and continues at this speed until 10 45. He stops from 10 45 until 11 30 and then returns home at a speed of 50 km/h.
(b) Use the graph to find the approximate time at which he arrives home.

5. (a) At 10 00 Akram leaves home and cycles to his grandparents' house which is 70 km away. He cycles at a speed of 20 km/h until 11 15, at which time he stops for $\frac{1}{2}$ hour. He then completes the journey at a speed of 30 km/h. At 11 45 Akram's sister, Hameeda, leaves home and drives her car at 60 km/h. Hameeda also goes to her grandparents' house and uses the same road as Akram.
   (b) At approximately what time does Hameeda overtake Akram?

6. A boat can travel at a speed of 20 km/h in still water. The current in a river flows at 5 km/h so that downstream the boat travels at 25 km/h and upstream it travels at only 15 km/h.

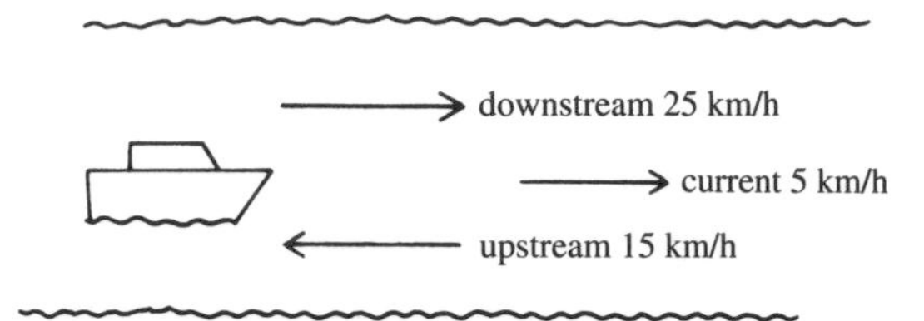

The boat has only enough fuel to last 3 hours. The boat leaves its base and travels downstream. Draw a distance-time graph and draw lines to indicate the outward and return journeys. After what time must the boat turn round so that it can get back to base without running out of fuel?

7. The boat in Question **6** sails in a river where the current is 10 km/h and it has fuel for four hours. At what time must the boat turn round this time if it is not to run out of fuel?

Interpreting graphs

Exercise 14

1. The graph shows the number of pupils on the premises of a school one day.
   The graph tells you some interesting things. Referring to the points A, B, C, D, E, describe briefly what happened during the day. Give an explanation of what you think might have happened.

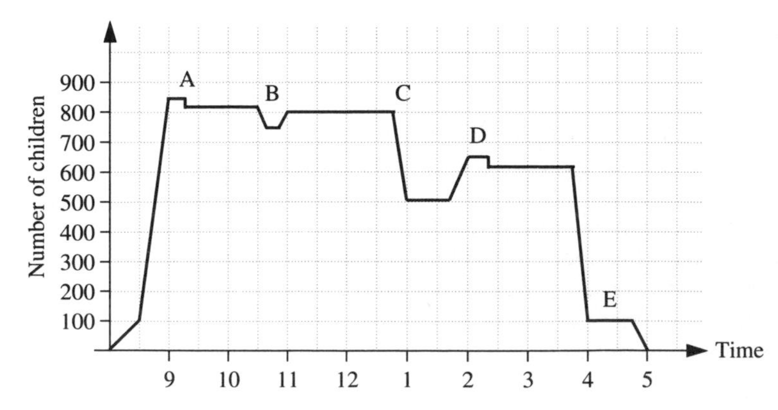

2. The graph shows the amount of income tax paid on various yearly incomes.

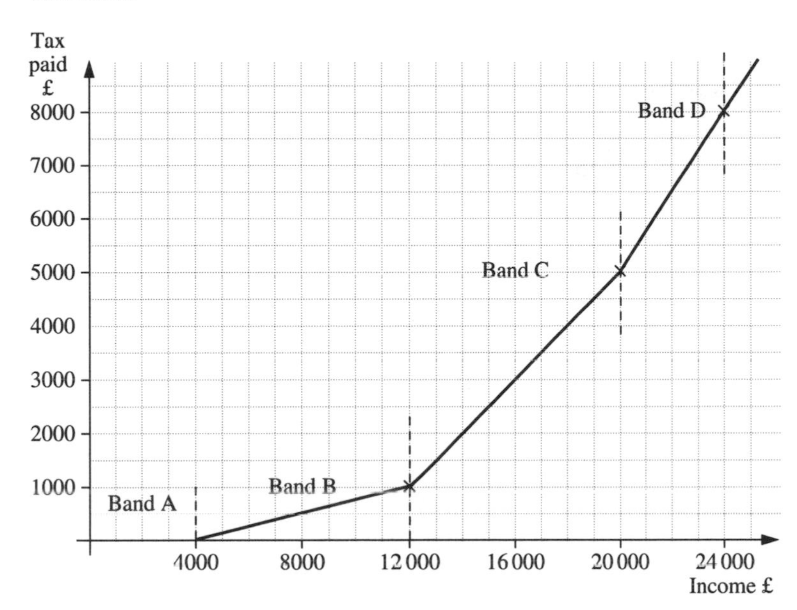

(a) How much tax is paid by a woman earning £14 000?

(b) What is the income of a man who pays £6500 in tax?

(c) Jason earns £20 000 and Neil earns £24 000. How much more tax does Neil pay than Jason?

(d) At what percentage rate is tax paid on incomes over £20 000?

3. A car travels along a motorway and the amount of petrol in its tank is monitored as shown on the graph.

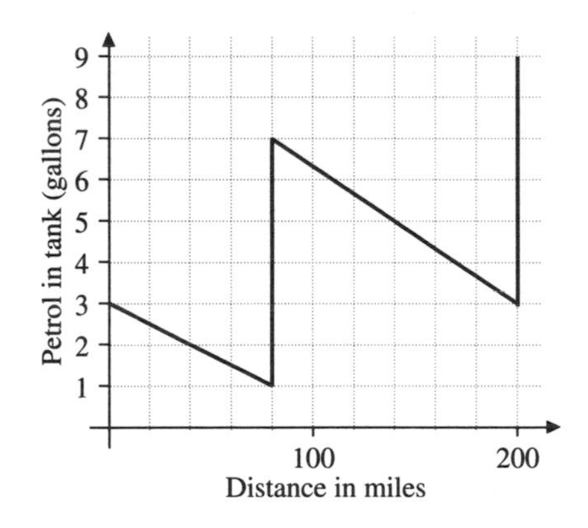

(a) How much petrol was bought at the first stop?

(b) What was the petrol consumption in miles per gallon:
   (i) before the first stop,
   (ii) between the two stops?

(c) What was the average petrol consumption over the 200 miles?

After it leaves the second service station the car encounters road works and slow traffic for the next 20 miles. Its petrol consumption is reduced to 20 m.p.g. After that, the road clears and the car travels a further 75 miles during which time the consumption is 30 m.p.g. Draw the graph above and extend it to show the next 95 miles. How much petrol is in the tank at the end of the journey?

4. The graph shows how the share price of the chemical firm ICI varied over a period of weeks. The share price is the price in pence paid for one share in the company.

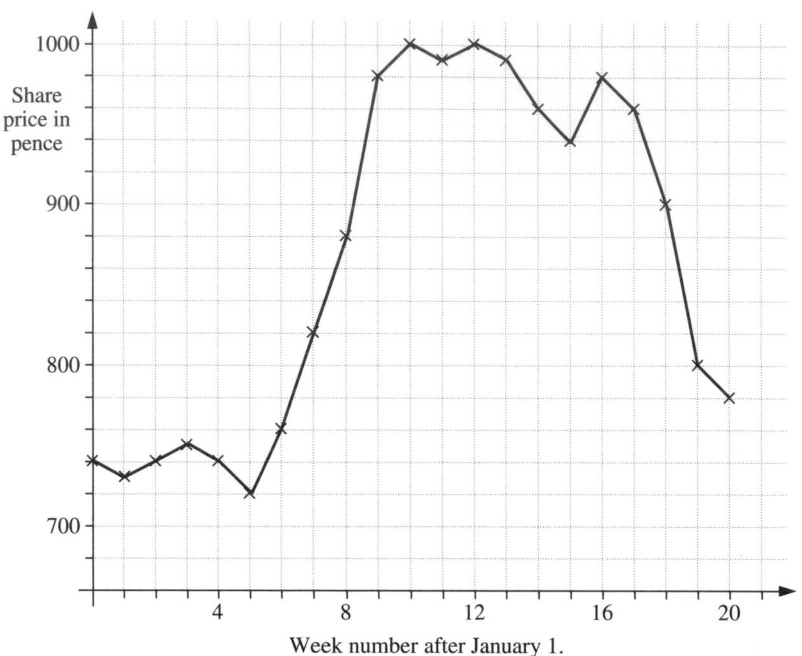

(a) What was the share price in Week 4?

(b) Naomi bought 200 shares in Week 6 and sold them all in Week 18. How much profit did she make?

(c) In Week 2 Mr Gibson consults a very accurate fortune teller who can predict the share price over coming weeks. What is the maximum profit he could make on 5000 shares?

(d) When there is a full moon, the fortune teller's predictions can be fairly disastrous. What is the maximum *loss* Mr Gibson could make?

5. A firm makes a profit of P thousand pounds from producing x thousand tiles.
Corresponding values of P and x are given below

x	0	0·5	1·0	1·5	2·0	2·5	3·0
P	−1·0	0·75	2·0	2·75	3·0	2·75	2·0

Using a scale of 4 cm to one unit on each axis, draw the graph of P against x. [Plot x on the horizontal axis]. Use your graph to find

(a) the number of tiles the firm should produce in order to make the maximum profit.

(b) the minimum number of ties that should be produced to cover the cost of production.

(c) the range of values of x for which the profit is more than £2850.

6. A small firm increases its monthly expenditure on advertising and records its monthy income from sales.

Month	Expenditure (£)	Income (£)
1	100	280
2	200	450
3	300	560
4	400	630
5	500	680
6	600	720
7	700	740

Draw a graph to display this information.
(a) Is it wise to spend £100 per month on advertising?
(b) Is it wise to spend £700 per month?
(c) What is the most sensible level of expenditure on advertising?

Sketch graphs

Exercise 15

1. Which of the graphs A to D below best fits the following statement: 'Unemployment is still rising but by less each month.'

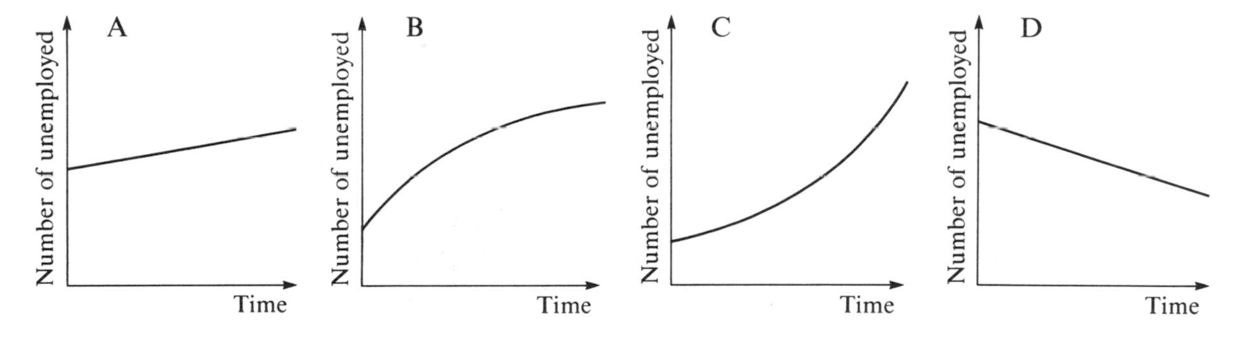

2. Which of the graphs A to D below best fits each of the following statements:
   (a) The birthrate was falling but is now steady.
   (b) Unemployment, which rose slowly until 1980, is now rising rapidly.
   (c) Inflation, which has been rising steadily, is now beginning to fall.
   (d) The price of gold has fallen steadily over the last year.

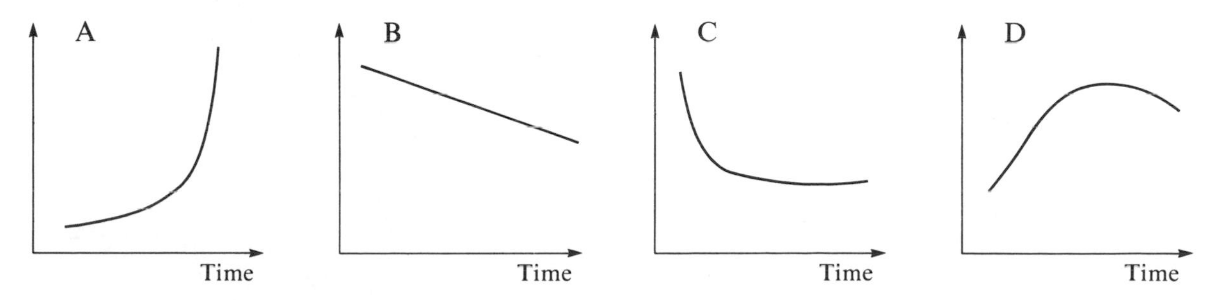

3. Which of the graphs A to D best fits the following statement: 'The price of oil was rising more rapidly in 1993 than at any time in the previous ten years.'

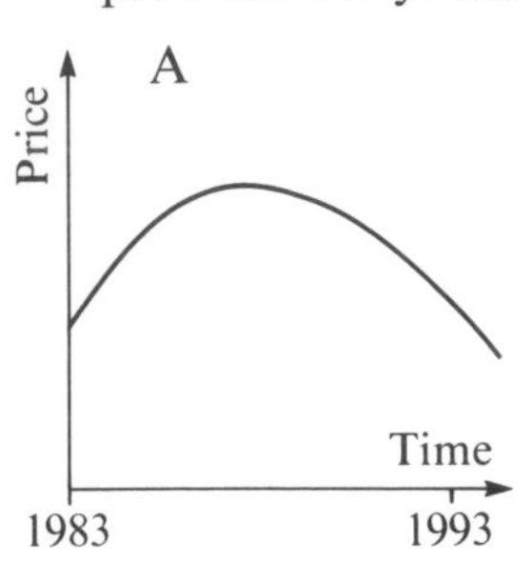

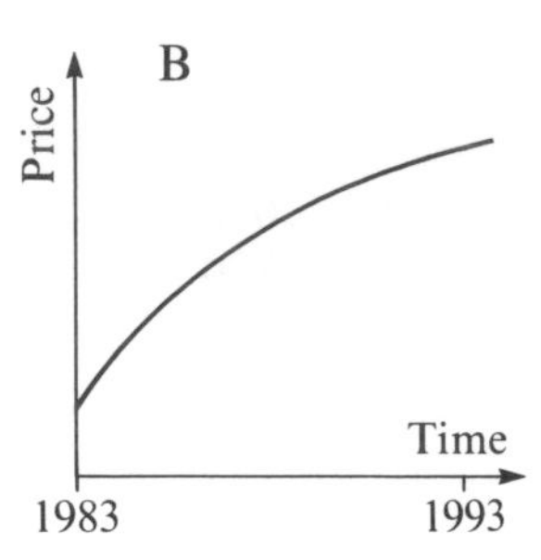

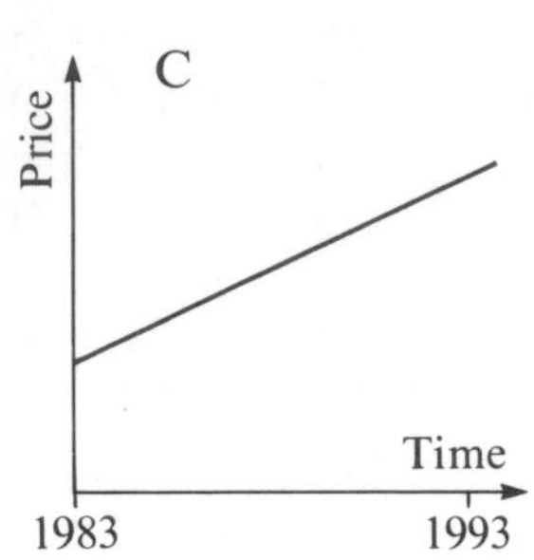

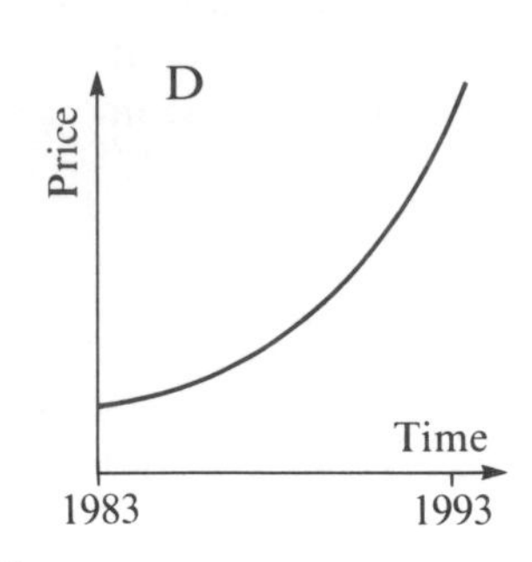

4. A car hire firm makes the charges shown. Draw a sketch graph showing 'miles travelled' across the page and 'total cost' up the page. Assume the car is hired for just one day.

5. A car in a Grand Prix goes around the circuit shown. Draw a sketch graph showing speed against time for one lap of the circuit. [Not the first lap. Why not?]

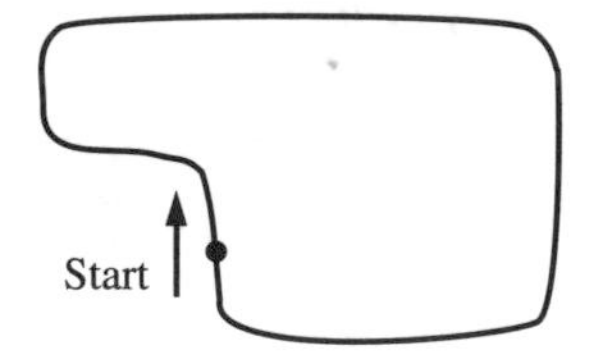

6. Mr Gibson organises a trip for pupils from his school. He hires a coach for £100 and decides to divide the cost equally between the pupils on the trip. Sketch a graph showing the 'number of pupils' across the page and the 'cost per pupil' up the page. Is it a linear graph?

7. The graph shows the motion of three cars A, B and C along the same road.
Answer the following questions giving estimates where necessary.
(a) Which car is in front after
 (i) 10 s, (ii) 20 s?
(b) When is B in the front?
(c) When are B and C going at the same speed?
(d) When are A and C going at the same speed?
(e) Which car is going fastest after 5 s?
(f) Which car starts slowly and then goes faster and faster?

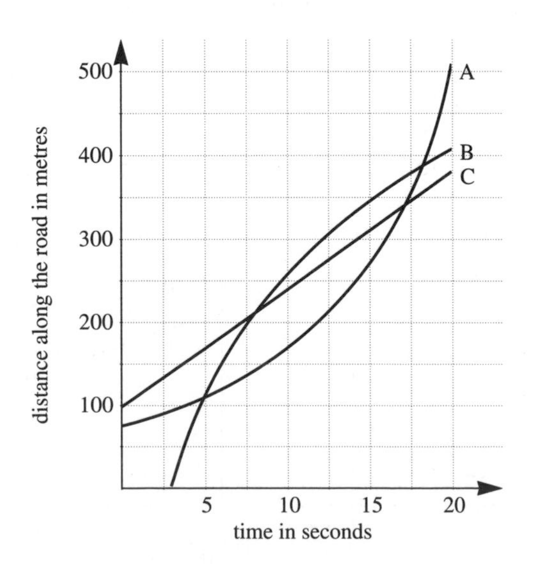

8. Three girls Hanna, Fateema and Carine took part in an egg and spoon race. Describe what happened, giving as many details as possible.

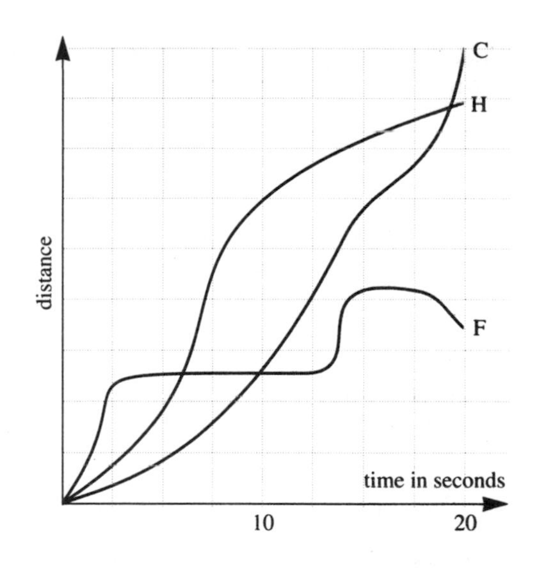

4.5 Curved graphs

Common curves

It is helpful to know the general shape of some of the more common curves.

Quadratic curves have an x^2 term as the highest power of x.

e.g. $y = 2x^2 - 3x + 7$ and $y = 5 + 2x - x^2$

(a) When the x^2 term is positive, the curve is ∪-shaped.

(b) When the x^2 term is negative the curve is an inverted ∩.

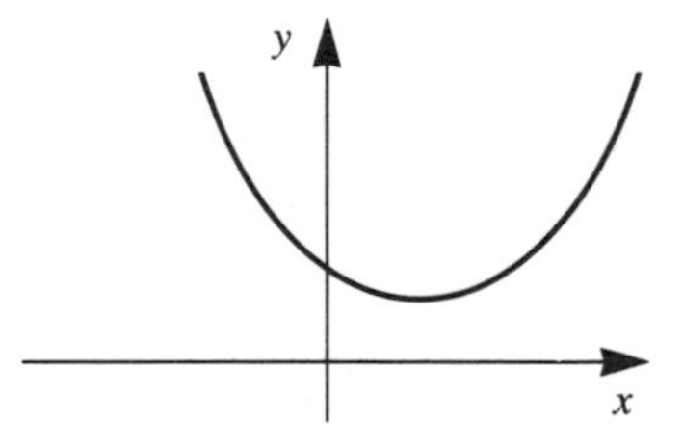

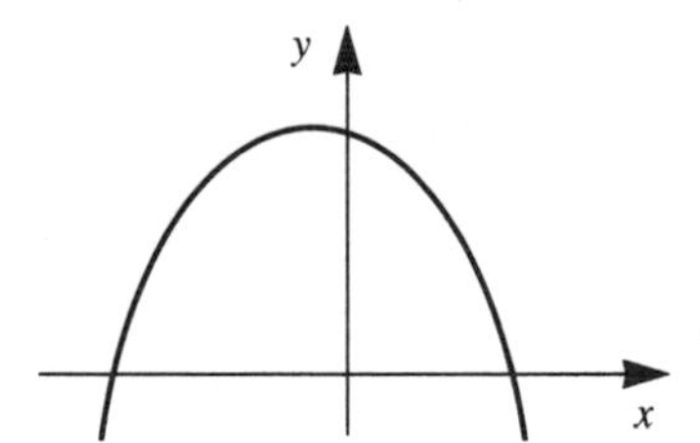

Cubic curves have an x^3 term as the highest power of x.

e.g. $y = x^3 + 7x - 4$ and $y = 8x - 4x^3$

(a) When the x^3 term is positive, the curve can be like one of the two shown below. Notice that as x gets larger, so does y.

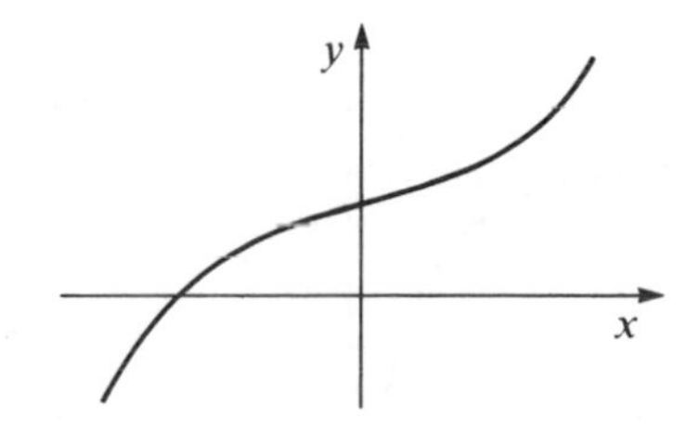

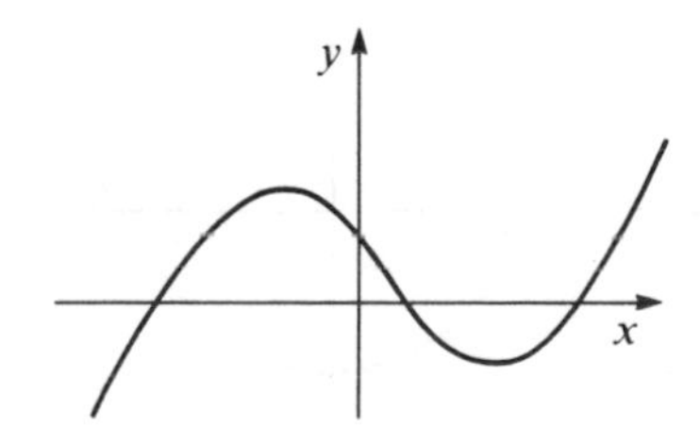

(b) When the x^3 term is negative, the curve can be like one of the two shown below. Notice that as x gets large, y is large but negative.

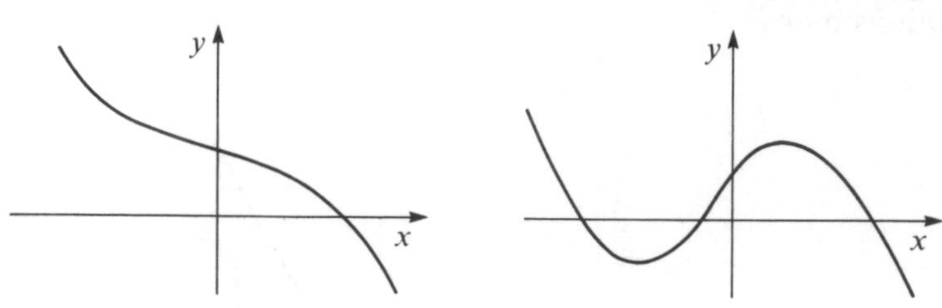

Reciprocal curves have a $\frac{1}{x}$ term.

e.g. $y = \frac{12}{x}$ and

$y = \frac{6}{x} + 5$

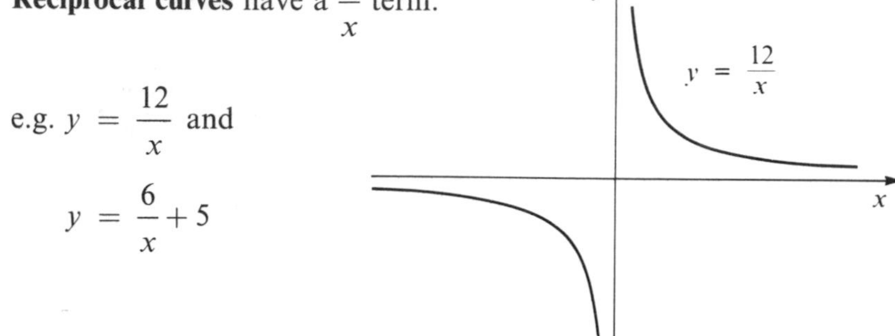

The curve has a break at $x = 0$. The x-axis and the y-axis are called *asymptotes* to the curve. The curve gets very near but never actually touches the asymptotes.

Exercise 16

1. What sort of curves are these? State as much information as you can.

(a)

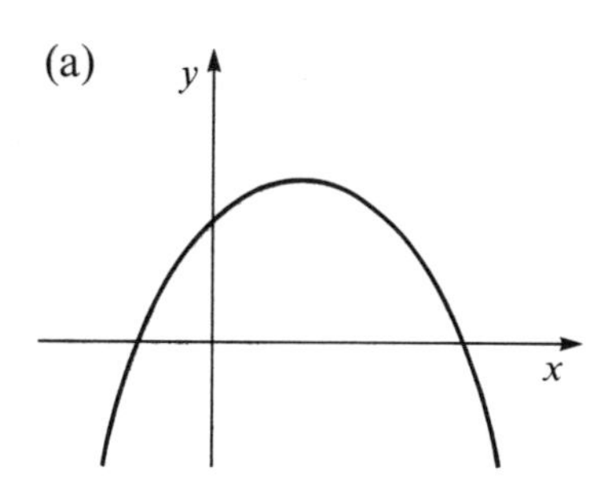

(b)

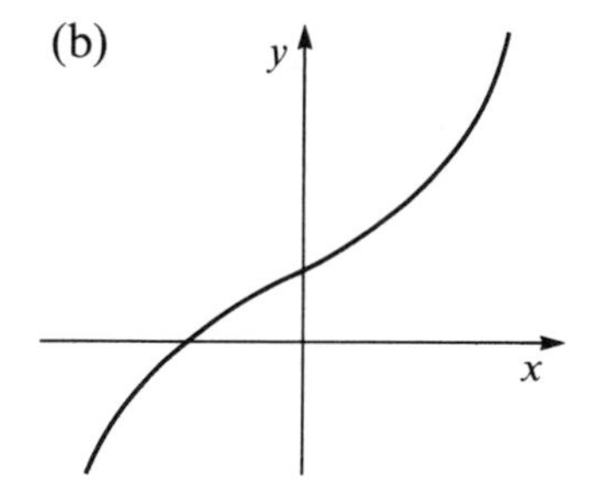

(c)

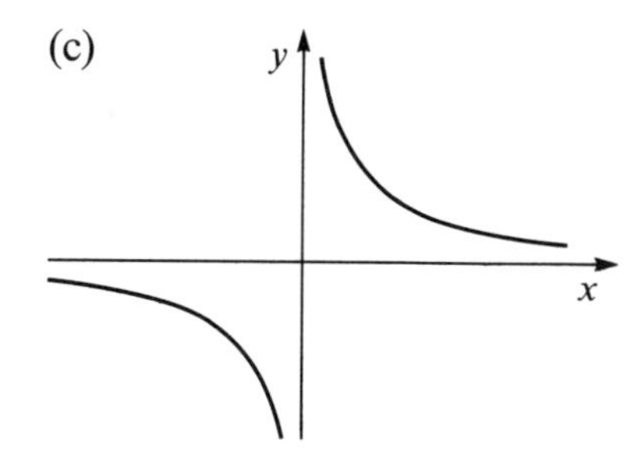

(d)

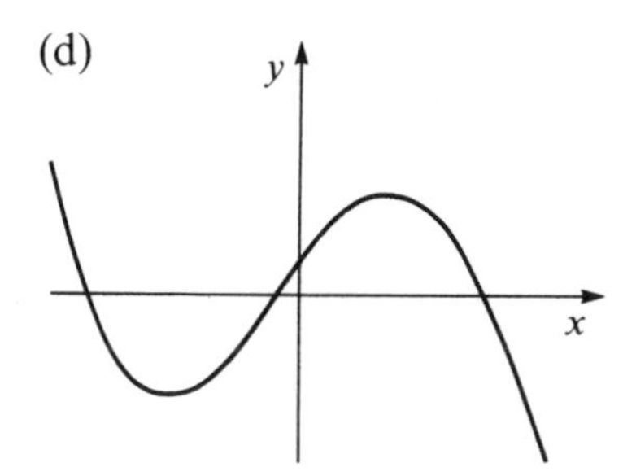

(e)

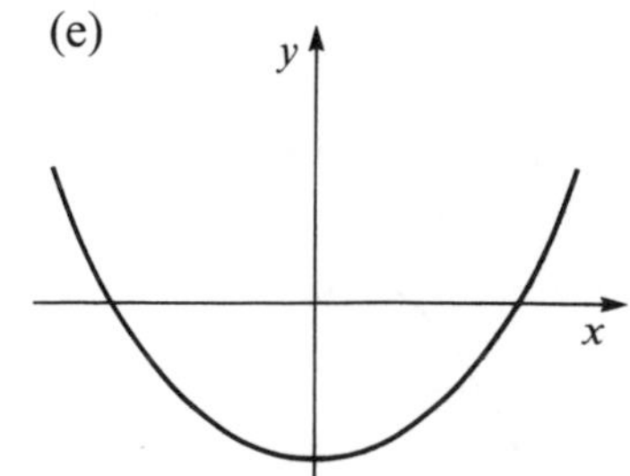

(f)

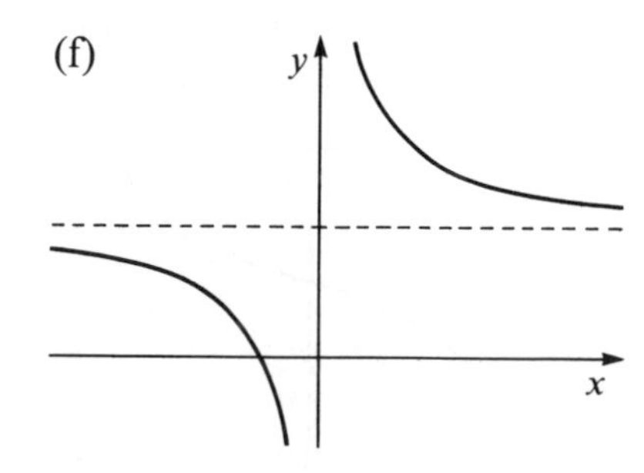

2. Draw the general shape of the following curves.

(a) $y = 3x^2 - 7x + 11$ (b) $y = 5 - x^3$

(c) $y = \dfrac{100}{x}$ (d) $y = 8x - x^2$

(e) $y = 10x^3 + 7x - 2$ (f)† $y = \dfrac{1}{x^2}$

3. Here are the equations of the six curves in Question 1, but not in the correct order.

(i) $y = \dfrac{8}{x}$ (ii) $y = 2x^3 + x + 2$

(iii) $y = 5 + 3x - x^2$ (iv) $y = x^2 - 6$

(v) $y = \dfrac{10}{x} + 4$ (vi) $y = 12 + 11x - 2x^2 - x^3$

Decide which equations fit the curves (a) to (f).

Plotting curved graphs

Example 1

Draw the graph of the function

$y = 2x^2 + x - 6$, for $-3 \leqslant x \leqslant 3$.

(a)

x	−3	−2	−1	0	1	2	3
$2x^2$	18	8	2	0	2	8	18
x	−3	−2	−1	0	1	2	3
−6	−6	−6	−6	−6	−6	−6	−6
y	9	0	−5	−6	−3	4	15

(b) Draw and label axes using suitable scales.
(c) Plot the points and draw a smooth curve through them with a pencil.
(d) Check any points which interrupt the smoothness of the curve.
(e) Label the curve with its equation.

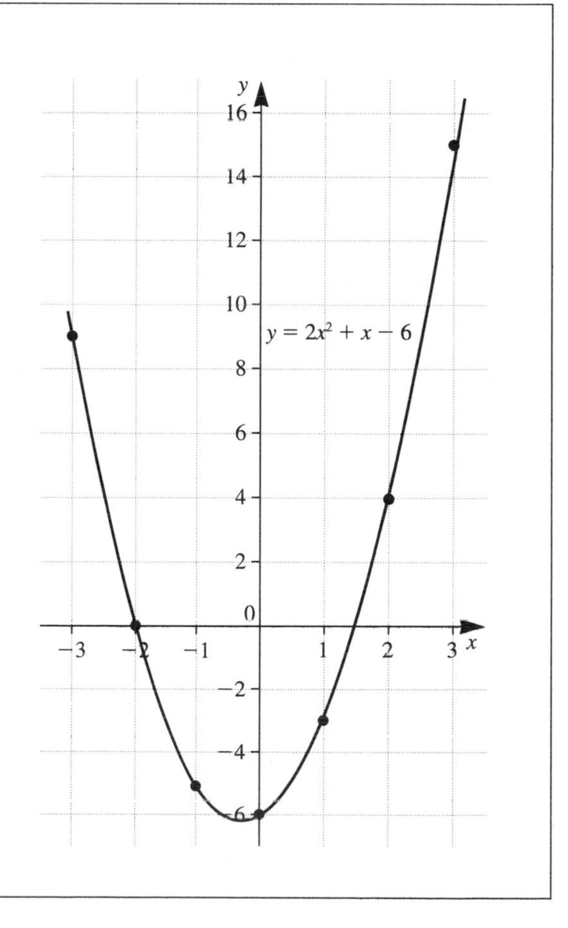

Exercise 17

Draw the graphs of the following functions using a scale of 2 cm for 1 unit on the x-axis and 1 cm for 1 unit on the y-axis.

1. $y = x^2 + 2x$, for $-3 \leqslant x \leqslant 3$

2. $y = x^2 + 4x$, for $-3 \leqslant x \leqslant 3$

3. $y = x^2 - 3x$, for $-3 \leqslant x \leqslant 3$

4. $y = x^2 + 2$, for $-3 \leqslant x \leqslant 3$

5. $y = x^2 - 7$, for $-3 \leqslant x \leqslant 3$

6. $y = x^2 + x - 2$, for $-3 \leqslant x \leqslant 3$

7. $y = x^2 + 3x - 9$, for $-4 \leqslant x \leqslant 3$

8. $y = x^2 - 3x - 4$, for $-2 \leqslant x \leqslant 4$

9. $y = x^2 - 5x + 7$, for $0 \leqslant x \leqslant 6$

10. $y = 2x^2 - 6x$, for $-1 \leqslant x \leqslant 5$

11. $y = 2x^2 + 3x - 6$, for $-4 \leqslant x \leqslant 2$

12. $y = 3x^2 - 6x + 5$, for $-1 \leqslant x \leqslant 3$

13. $y = 2 + x - x^2$, for $-3 \leqslant x \leqslant 3$

14. $f(x) = 1 - 3x - x^2$, for $-5 \leqslant x \leqslant 2$

15. $f(x) = 3 + 3x - x^2$, for $-2 \leqslant x \leqslant 5$

16. $f(x) = 7 - 3x - 2x^2$, for $-3 \leqslant x \leqslant 3$

17. $f(x) = 6 + x - 2x^2$, for $-3 \leqslant x \leqslant 3$

18. $f: x \rightarrow 8 + 2x - 3x^2$, for $-2 \leqslant x \leqslant 3$

19. $f: x \rightarrow x(x - 4)$, for $-1 \leqslant x \leqslant 6$

20. $f: x \rightarrow (x + 1)(2x - 5)$, for $-3 \leqslant x \leqslant 3$.

Example 2

Draw the graph of $y = \frac{12}{x} + x - 6$, for $1 \leqslant x \leqslant 8$.

Use the graph to find approximate values for

(a) the minimum value of $\frac{12}{x} + x - 6$

(b) the value of $\frac{12}{x} + x - 6$, when $x = 2{\cdot}25$.

Here is the table of values.

x	1	2	3	4	5	6	8
$\frac{12}{x}$	12	6	4	3	2·4	2	1·5
x	1	2	3	4	5	6	8
-6	-6	-6	-6	-6	-6	-6	-6
y	7	2	1	1	1·4	2	3·5

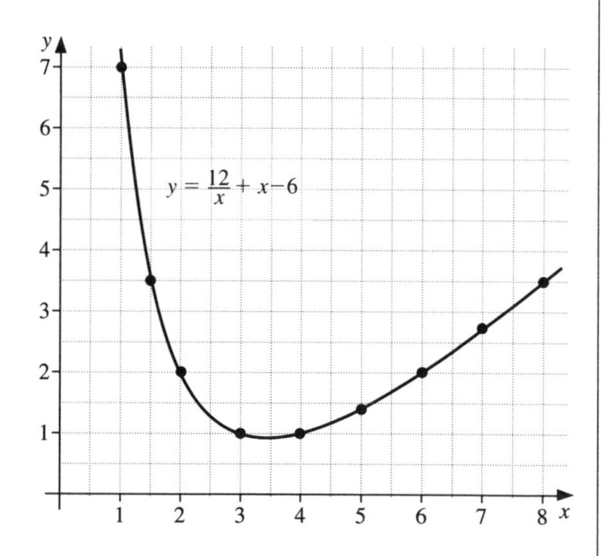

(a) From the graph, the minimum value of $\frac{12}{x} + x - 6$ (i.e. y) is approximately 0·9.

(b) At $x = 2{\cdot}25$, y is approximately 1·6.

Exercise 18

Draw the following curves. The scales given are for one unit of x and y.

1. $y = \dfrac{12}{x}$, for $1 \leqslant x \leqslant 10$. (Scales: 1 cm for x and y)

2. $y = \dfrac{9}{x}$, for $1 \leqslant x \leqslant 10$. (Scales: 1 cm for x and y)

3. $y = \dfrac{12}{x+1}$, for $0 \leqslant x \leqslant 8$. (Scales: 2 cm for x, 1 cm for y)

4. $y = \dfrac{8}{x-4}$, for $-4 \leqslant x \leqslant 3{\cdot}5$. (Scales: 2 cm for x, 1 cm for y)

5. $y = \dfrac{15}{3-x}$, for $-4 \leqslant x \leqslant 2$. (Scales: 2 cm for x, 1 cm for y)

6. $y = \dfrac{x}{x+4}$, for $-3{\cdot}5 \leqslant x \leqslant 4$. (Scales: 2 cm for x and y)

7. $y = \dfrac{3x}{5-x}$, for $-3 \leqslant x \leqslant 4$. (Scales: 2 cm for x, 1 cm for y)

8. $y = \dfrac{x+8}{x+1}$, for $0 \leqslant x \leqslant 8$. (Scales: 2 cm for x and y)

9. $y = \dfrac{x-3}{x+2}$, for $-1 \leqslant x \leqslant 6$. (Scales: 2 cm for x and y)

10. $y = \dfrac{10}{x} + x$, for $1 \leqslant x \leqslant 7$. (Scales: 2 cm for x, 1 cm for y)

11. $y = \dfrac{12}{x} - x$, for $1 \leqslant x \leqslant 7$. (Scales: 2 cm for x, 1 cm for y)

12. $y = 5 + 3x - x^2$, for $-2 \leqslant x \leqslant 5$. (Scales: 2 cm for x, 1 cm for y)
    Find
    (a) the maximum value of the function $5 + 3x - x^2$,
    (b) the two values of x for which $y = 2$.

13. $y = \dfrac{15}{x} + x - 7$, for $1 \leqslant x \leqslant 7$. (Scales: 2 cm for x and y)
    From your graph find
    (a) the minimum value of y,
    (b) the y-value when $x = 5{\cdot}5$.

14. $y = x^3 - 2x^2$, for $0 \leqslant x \leqslant 4$. (Scales: 2 cm for x, $\frac{1}{2}$ cm for y)
    From your graph find
    (a) the y-value at $x = 2{\cdot}5$
    (b) the x-value at $y = 15$.

15. $y = \frac{1}{10}(x^3 + 2x + 20)$, for $-3 \leqslant x \leqslant 3$. (Scales: 2 cm for x and y)
    From your graph find
    (a) the x-value where $x^3 + 2x + 20 = 0$,
    (b) the x-value where $y = 3$.

16.† Draw the graph of

$$y = \frac{x}{x^2 + 1}, \text{ for } -6 \leqslant x \leqslant 6. \text{ (Scales: 1 cm for } x, \text{ 10 cm for } y)$$

17.† Draw the graph of

$$E = \frac{5000}{x} + 3x \text{ for } 10 \leqslant x \leqslant 80.$$

(Scales: 1 cm to 5 units for x and 1 cm to 25 units for E)
From the graph find,
(a) the minimum value of E,
(b) the value of x corresponding to this minimum value.
(c) the range of values of x for which E is less than 275.

Mixed questions

Exercise 19

1. A rectangle has a perimeter of 14 cm and length x cm. Show that the width of the rectangle is $(7 - x)$ cm and hence that the area A of the rectangle is given by the formula $A = x(7 - x)$. Draw the graph, plotting x on the horizontal axis with a scale of 2 cm to 1 unit, and A on the vertical axis with a scale of 1 cm to 1 unit. Take x from 0 to 7. From the graph find,
(a) the area of the rectangle when $x = 2{\cdot}25$ cm,
(b) the dimensions of the rectangle when its area is 9 cm^2,
(c) the maximum area of the rectangle,
(d) the length and width of the rectangle corresponding to the maximum area,
(e) what shape of rectangle has the largest area.

2. A farmer has 60 m of wire fencing which he uses to make a rectangular pen for his sheep. He uses a stone wall as one side of the pen so the wire is used for only 3 sides of the pen.

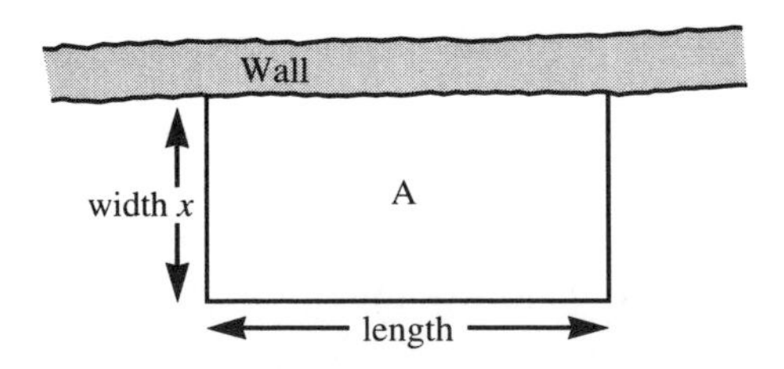

If the width of the pen is x m, what is the length (in terms of x)?
What is the area A of the pen?
Draw a graph with area A on the vertical axis and the width x on the horizontal axis. Take values of x from 0 to 30.
What dimensions should the pen have if the farmer wants to enclose the largest possible area?

3. A ball is thrown in the air so that t seconds after it is thrown, its height h metres above its starting point is given by the function $h = 25t - 5t^2$. Draw the graph of the function of $0 \leqslant t \leqslant 6$, plotting t on the horizontal axis with a scale of 2 cm to 1 second, and h on the vertical axis with a scale of 2 cm for 10 metres. Use the graph to find,
   (a) the time when the ball is at its greatest height,
   (b) the greatest height reached by the ball,
   (c) the interval of time during which the ball is at a height of more than 30 m.

4. The velocity v m/s of a missile t seconds after launching is given by the equation $v = 54t - 2t^3$.
   Draw a graph, plotting t on the horizontal axis with a scale of 2 cm to 1 second, and v on the vertical axis with a scale of 1 cm for 10 m/s. Take values of t from 0 to 5.
   Use the graph to find,
   (a) the maximum velocity reached,
   (b) the time taken to accelerate to a velocity of 70 m/s,
   (c) the interval of time during which the missile is travelling at more than 100 m/s.

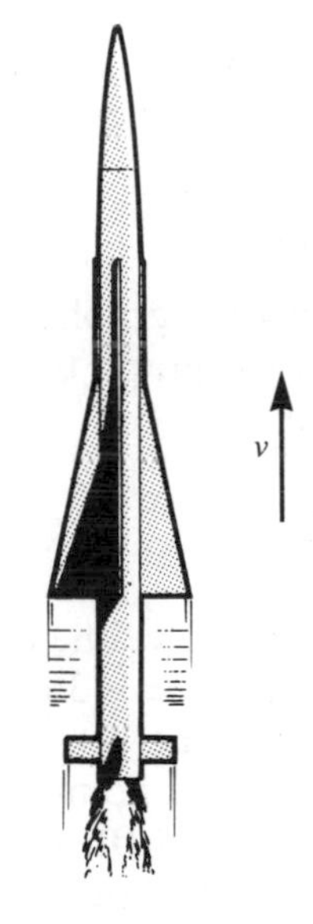

5. Draw the graph of $y = 2^x$, for $-4 \leqslant x \leqslant 4$.
   (Scales: 2 cm for x, 1 cm for y)

6. Draw the graph of $y = 3^x$, for $-3 \leqslant x \leqslant 3$.
   (Scales: 2 cm for x, $\frac{1}{2}$ cm for y)

7. Consider the equation $y = \dfrac{1}{x}$.

   When $x = \frac{1}{2}$, $y = \frac{1}{\frac{1}{2}} = 2$. When $x = \frac{1}{100}$, $y = \frac{1}{\frac{1}{100}} = 100$.

   As the denominator of $\dfrac{1}{x}$ gets smaller, the answer gets larger.
   An 'infinitely small' denominator gives an 'infinitely large' answer.

   We write $\dfrac{1}{0} \rightarrow \infty$ '$\dfrac{1}{0}$ tends to an infinitely large number.'

   Draw the graph of $y = \dfrac{1}{x}$ for
   $x = -4, -3, -2, -1, -0{\cdot}5, -0{\cdot}25, 0{\cdot}25, 0{\cdot}5, 1, 2, 3, 4$
   (Scales: 2 cm for x and y)

8. Draw the graph of $y = x + \dfrac{1}{x}$ for
   $x = -4, -3, -2, -1, -0{\cdot}5, -0{\cdot}25, 0{\cdot}25, 0{\cdot}5, 1, 2, 3, 4$
   (Scales: 2 cm for x and y)

9.† Draw the graph of $y = \dfrac{2^x}{x}$, for $-4 \leqslant x \leqslant 7$, including $x = -0{\cdot}5$, $x = 0{\cdot}5$.
(Scales: 1 cm to 1 unit for x and y)

4.6 Graphical solution of equations

Accurately drawn graphs enable approximate solutions to be found for a wide range of equations, many of which are impossible to solve exactly by other methods.

Example 1

Draw the graph of the function $y = 2x^2 - x - 3$ for $-2 \leqslant x \leqslant 3$.
Use the graph to find approximate solutions to the following equations

(a) $2x^2 - x - 3 = 6$ (b) $2x^2 - x = x + 5$

(a) To solve the equation
$2x^2 - x - 3 = 6$,
the line $y = 6$ is drawn.
At the points of intersection (A and B), y simultaneously equals both 6 and $(2x^2 - x - 3)$.
So we may write
$2x^2 - x - 3 = 6$
The solutions are the x-values of the points A and B.
i.e. $x = -1{\cdot}9$ and $x = 2{\cdot}4$ approx.

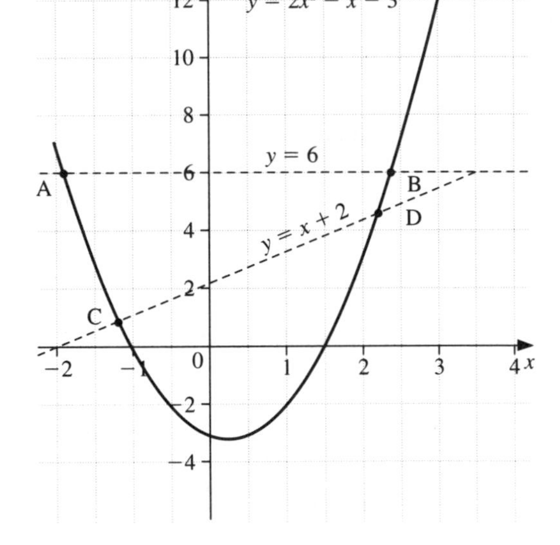

(b) To solve the equation $2x^2 - x = x + 5$, we rearrange the equation to obtain the function $(2x^2 - x - 3)$ on the left-hand side. In this case, subtract 3 from both sides.
$2x^2 - x - 3 = x + 5 - 3$
$2x^2 - x - 3 = x + 2$

If we now draw the line $y = x + 2$, the solutions of the equation are given by the x-values of C and D, the points of intersection.

i.e. $x = -1{\cdot}2$ and $x = 2{\cdot}2$ approx.

It is important to rearrange the equation to be solved so that the function already plotted is on one side.

Example 2

Assuming that the graph of $y = x^2 - 3x + 1$ has been drawn, find the equation of the line which should be drawn to solve the equation $x^2 - 4x + 3 = 0$

Rearrange $x^2 - 4x + 3 = 0$ in order to obtain $(x^2 - 3x + 1)$ on the left-hand side.

$$x^2 - 4x + 3 = 0$$

add x $\quad x^2 - 3x + 3 = x$

subtract 2 $\quad x^2 - 3x + 1 = x - 2$

Therefore draw the line $y = x - 2$ to solve the equation.

Exercise 20

1. In the diagram shown, the graphs of $y = x^2 - 2x - 3$, $y = -2$ and $y = x$ have been drawn.

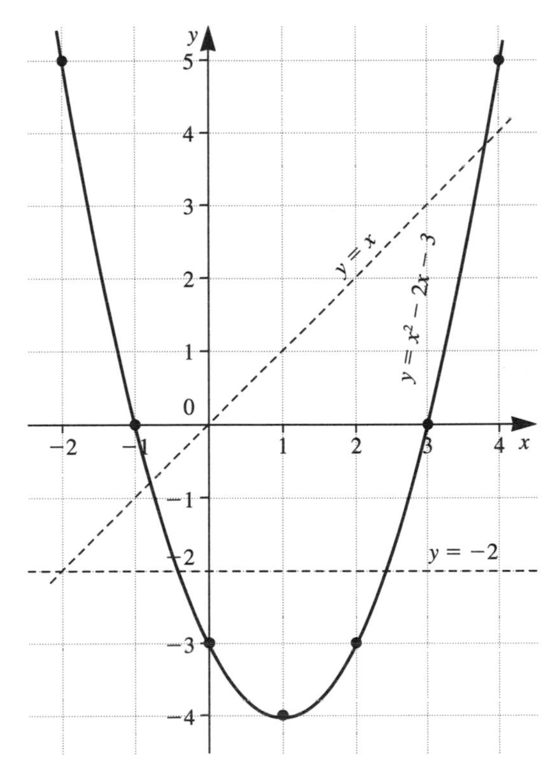

Use the graphs to find approximate solutions to the following equations.

(a) $x^2 - 2x - 3 = -2$ (b) $x^2 - 2x - 3 = x$

(c) $x^2 - 2x - 3 = 0$ (d) $x^2 - 2x - 1 = 0$

In Questions **2** to **4**, use a scale of 2 cm to 1 unit for x and 1 cm to 1 unit for y.

2. Draw the graphs of the functions $y = x^2 - 2x$ and $y = x + 1$ for $-1 \leqslant x \leqslant 4$. Hence find approximate solutions of the equation $x^2 - 2x = x + 1$.

3. Draw the graphs of the functions $y = x^2 - 3x + 5$ and $y = x + 3$ for $-1 \leqslant x \leqslant 5$. Hence find approximate solutions of the equation $x^2 - 3x + 5 = x + 3$.

4. Draw the graphs of the functions $y = 6x - x^2$ and $y = 2x + 1$ for $0 \leqslant x \leqslant 5$. Hence find approximate solutions of the equation $6x - x^2 = 2x + 1$.

In Questions **5** to **7**, do *not* draw any graphs.

5. Assuming the graph of $y = x^2 - 5x$ has been drawn, find the equation of the line which should be drawn to solve the equations
(a) $x^2 - 5x = 3$ (b) $x^2 - 5x = -2$
(c) $x^2 - 5x = x + 4$ (d) $x^2 - 6x = 0$
(e) $x^2 - 5x - 6 = 0$

6. Assuming the graph of $y = x^2 + x + 1$ has been drawn, find the equation of the line which should be drawn to solve the equations
(a) $x^2 + x + 1 = 6$ (b) $x^2 + x + 1 = 0$
(c) $x^2 + x - 3 = 0$ (d) $x^2 - x + 1 = 0$
(e) $x^2 - x - 3 = 0$

7. Assuming the graph of $y = 6x - x^2$ has been drawn, find the equation of the line which should be drawn to solve the equations
(a) $4 + 6x - x^2 = 0$ (b) $4x - x^2 = 0$
(c) $2 + 5x - x^2 = 0$ (d) $x^2 - 6x = 3$
(e) $x^2 - 6x = -2$

For Questions **8** to **10**, use scales of 2 cm to 1 unit for x and 1 cm to 1 unit for y.

8. Draw the graph of $y = x^2 - 2x + 2$ for $-2 \leqslant x \leqslant 4$. By drawing other graphs, solve the equations
(a) $x^2 - 2x + 2 = 8$ (b) $x^2 - 2x + 2 = 5 - x$
(c) $x^2 - 2x - 5 = 0$

9. Draw the graph of $y = x^2 - 7x$ for $0 \leqslant x \leqslant 7$. Draw suitable straight lines to solve the equations
(a) $x^2 - 7x + 9 = 0$ (b) $x^2 - 5x + 1 = 0$

10. Draw the graph of $y = 2x^2 + 3x - 9$ for $-3 \leqslant x \leqslant 2$. Draw suitable straight lines to find approximate solutions of the equations
(a) $2x^2 + 3x - 4 = 0$ (b) $2x^2 + 2x - 9 = 1$

11.† Draw the graph of $y = \dfrac{18}{x}$ for $1 \leqslant x \leqslant 10$, using scales of 1 cm to one unit on both axes. Use the graph to solve approximately

(a) $\dfrac{18}{x} = x + 2$ (b) $\dfrac{18}{x} + x = 10$ (c) $x^2 = 18$

12.† Draw the graph of $y = \frac{1}{2}x^2 - 6$ for $-4 \leqslant x \leqslant 4$, taking 2 cm to 1 unit on each axis.

(a) Use your graph to solve approximately the equation $\frac{1}{2}x^2 - 6 = 1$.

(b) Using tables or a calculator confirm that your solutions are approximately $\pm\sqrt{14}$ and explain why this is so.

(c) Use your graph to find the square roots of 8.

13.† Draw the graph of $y = 6 - 2x - \frac{1}{2}x^3$ for $x = \pm 2, \pm 1\frac{1}{2}, \pm 1, \pm\frac{1}{2}, 0$. Take 4 cm to 1 unit for x and 1 cm to 1 unit for y.

Use your graph to find approximate solutions of the equations

(a) $\frac{1}{2}x^3 + 2x - 6 = 0$ (b) $x - \frac{1}{2}x^3 = 0$

Confirm that two of the solutions to the equation in part (b) are $\pm\sqrt{2}$ and explain why this is so.

14.† Draw the graph of $y = 2^x$ for $-4 \leqslant x \leqslant 4$, taking 2 cm to one unit for x and 1 cm to one unit for y. Find approximate solutions to the equations

(a) $2^x = 6$ (b) $2^x = 3x$ (c) $x2^x = 1$

Find also the approximate value of $2^{2 \cdot 5}$.

15.† Draw the graph of $\dfrac{x^4}{4^x}$, for $x = -1, -\frac{3}{4}, -\frac{1}{2}, -\frac{1}{4}, 0, \frac{1}{4}, \frac{1}{2}, \frac{3}{4}$, 1, 1·5, 2, 2·5, 3, 4, 5, 6, 7.

(Scales: 2 cm to 1 unit for x, 5 cm to 1 unit for y)

(a) For what values of x is the gradient of the function zero?

(b) For what values of x is $y = 0 \cdot 5$?

4.7 Indices

Six basic rules

1. $a^n \times a^m = a^{n+m}$

$7^2 \times 7^4 = 7.7 \times 7.7.7.7 = 7^6$

so $7^2 \times 7^4 = 7^{2+4} = 7^6$

2. $a^n \div a^m = a^{n-m}$

$3^5 \div 3^2 = \dfrac{3.3.3.\not{3}.\not{3}}{\not{3}.\not{3}} = 3^3$

so $3^5 \div 3^2 = 3^{5-2} = 3^3$

3. $(a^n)^m = a^{nm}$

$(2^2)^3 = (2.2)(2.2)(2.2) = 2^6$

so $(2^2)^3 = 2^{2\times 3} = 2^6$

4. $a^{-n} = \dfrac{1}{a^n}$

Consider this sequence

2^3	2^2	2^1	2^0	2^{-1}	2^{-2}
↓	↓	↓	↓	↓	↓
8	4	2	1	$\frac{1}{2}$	$\frac{1}{4}$

Check: $5^3 \div 5^5 = \dfrac{\not{5}.\not{5}.\not{5}}{\not{5}.\not{5}.\not{5}.5.5} = \dfrac{1}{5^2} = 5^{-2}$

5. $a^{\frac{1}{n}}$ means 'the nth root of a'

$3^{\frac{1}{2}}.3^{\frac{1}{2}} = \sqrt{3}.\sqrt{3}$
Using Rule 1:
$3^{\frac{1}{2}}.3^{\frac{1}{2}} = 3^1$

6. $a^{\frac{m}{n}}$ means 'the nth root of a raised to the power m'.

$4^{\frac{3}{2}} = (\sqrt{4})^3 = 8$

Example 1

Simplify
(a) $x^7.x^{13}$
(b) $x^3 \div x^7$
(c) $(x^4)^3$
(d) $(3x^2)^3$
(e) $3y^2 \times 4y^3$

(a) $x^7.x^{13} = x^{7+13} = x^{20}$
(b) $x^3 \div x^7 = x^{3-7} = x^{-4} = \dfrac{1}{x^4}$
(c) $(x^4)^3 = x^{12}$
(d) $(3x^2)^3 = 3^3.(x^2)^3 = 27x^6$
(e) $3y^2 \times 4y^3 = 12y^5$

Shown opposite is the graph of $y = 2^x$

A calculator with a button marked [x^y] can be used to work out difficult powers.

e.g. for $2^{1\cdot7}$ press [2] [x^y] [1·7] [=]

$2^{1\cdot7} = 3\cdot249$ (4 s.f.)

Some calculators have a button marked [$x^{\frac{1}{y}}$] to work out roots of numbers.

e.g. the fourth root of 100 is $100^{\frac{1}{4}}$

Press [100] [$x^{\frac{1}{y}}$] [4] [=]

$100^{\frac{1}{4}} = 3\cdot162$ (4 s.f.)

You could press [100] [x^y] [0·25] [=] instead.

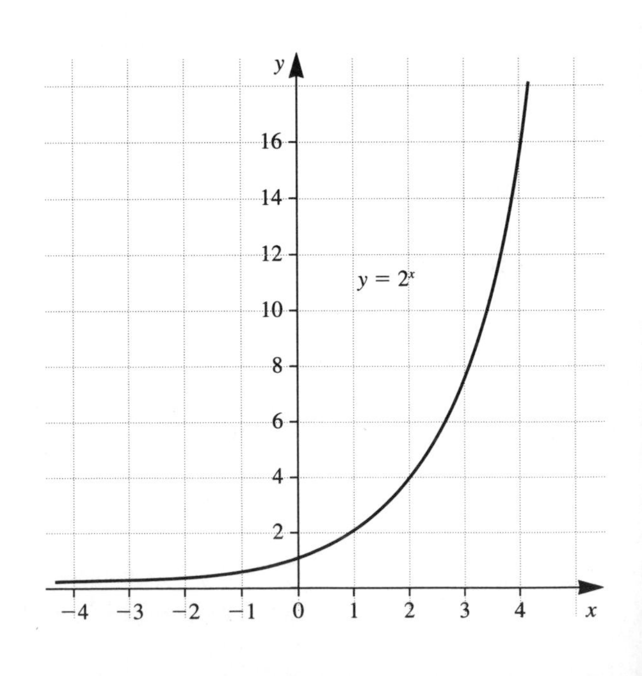

Exercise 21

Express in index form

1. $3 \times 3 \times 3 \times 3$
2. $4 \times 4 \times 5 \times 5 \times 5$
3. $3 \times 7 \times 7 \times 7$
4. $2 \times 2 \times 2 \times 7$
5. $\dfrac{1}{10 \times 10 \times 10}$
6. $\dfrac{1}{2 \times 2 \times 3 \times 3 \times 3}$
7. $\sqrt{15}$
8. $\sqrt[3]{3}$
9. $\sqrt[5]{10}$
10. $(\sqrt{5})^3$

Simplify

11. $x^3 \times x^4$
12. $y^6 \times y^7$
13. $z^7 \div z^3$
14. $z^{50} \times z^{50}$
15. $m^3 \div m^2$
16. $e^{-3} \times e^{-2}$
17. $y^{-2} \times y^4$
18. $w^4 \div w^{-2}$
19. $y^{\frac{1}{2}} \times y^{\frac{1}{2}}$
20. $(x^2)^5$
21. $x^{-2} \div x^{-2}$
22. $w^{-3} \times w^{-2}$
23. $w^{-7} \times w^2$
24. $x^3 \div x^{-4}$
25. $(a^2)^4$
26. $(k^{\frac{1}{2}})^6$
27. $e^{-4} \times e^4$
28. $x^{-1} \times x^{30}$
29. $(y^4)^{\frac{1}{2}}$
30. $(x^{-3})^{-2}$
31. $z^2 \div z^{-2}$
32. $t^{-3} \div t$
33. $(2x^3)^2$
34. $(4y^5)^2$
35. $2x^2 \times 3x^2$
36. $5y^3 \times 2y^2$
37. $5a^3 \times 3a$
38. $(2a)^3$
39. $3x^3 \div x^3$
40. $8y^3 \div 2y$
41. $10y^2 \div 4y$
42. $8a \times 4a^3$
43. $(2x)^2 \times (3x)^3$
44. $4z^4 \times z^{-7}$
45. $6x^{-2} \div 3x^2$
46. $5y^3 \div 2y^{-2}$
47. $(x^2)^{\frac{1}{2}} \div (x^{\frac{1}{3}})^3$
48. $7w^{-2} \times 3w^{-1}$
49. $(2n)^4 \div 8n^0$
50. $4x^{\frac{3}{2}} \div 2x^{\frac{1}{2}}$

Example 2

Evaluate

(a) $9^{\frac{1}{2}}$	(a) $9^{\frac{1}{2}} = \sqrt{9} = 3$
(b) 5^{-1}	(b) $5^{-1} = \frac{1}{5}$
(c) $4^{-\frac{1}{2}}$	(c) $4^{-\frac{1}{2}} = \dfrac{1}{4^{\frac{1}{2}}} = \dfrac{1}{\sqrt{4}} = \dfrac{1}{2}$
(d) $25^{\frac{3}{2}}$	(d) $25^{\frac{3}{2}} = (\sqrt{25})^3 = 5^3 = 125$
(e) $(5^{\frac{1}{2}})^3.5^{\frac{1}{2}}$	(e) $(5^{\frac{1}{2}})^3.5^{\frac{1}{2}} = 5^{\frac{3}{2}}.5^{\frac{1}{2}} = 5^2 = 25$
(f) 7^0	(f) $7^0 = 1 \left[\text{consider } \dfrac{7^3}{7^3} = 7^{3-3} = 7^0 = 1\right]$

Note $\quad a^0 = 1$ for any non-zero value of a.

Exercise 22

Evaluate the following without a calculator.

1. $3^2 \times 3$ **2.** 100^0 **3.** 3^{-2}
4. $(5^{-1})^{-2}$ **5.** $4^{\frac{1}{2}}$ **6.** $16^{\frac{1}{2}}$
7. $81^{\frac{1}{2}}$ **8.** $8^{\frac{1}{3}}$ **9.** $9^{\frac{3}{2}}$
10. $27^{\frac{1}{3}}$ **11.** $9^{-\frac{1}{2}}$ **12.** $8^{-\frac{1}{3}}$
13. $1^{\frac{5}{2}}$ **14.** $25^{-\frac{1}{2}}$ **15.** $1000^{\frac{1}{3}}$
16. $2^{-2} \times 2^5$ **17.** $2^4 \div 2^{-1}$ **18.** $8^{\frac{2}{3}}$
19. $27^{-\frac{2}{3}}$ **20.** $4^{-\frac{3}{2}}$ **21.** $36^{\frac{1}{2}} \times 27^{\frac{1}{3}}$
22. $10\,000^{\frac{1}{4}}$ **23.** $100^{\frac{3}{2}}$ **24.** $(100^{\frac{1}{2}})^{-3}$
25. $(9^{\frac{1}{2}})^{-2}$ **26.** $(-16{\cdot}371)^0$ **27.** $81^{\frac{1}{4}} \div 16^{\frac{1}{4}}$
28. $(5^{-4})^{\frac{1}{2}}$ **29.** $1000^{-\frac{1}{3}}$ **30.** $(4^{-\frac{1}{2}})^2$

31. $8^{-\frac{2}{3}}$ **32.** $100^{\frac{5}{2}}$ **33.** $1^{\frac{4}{3}}$
34. 2^{-5} **35.** $(0{\cdot}01)^{\frac{1}{2}}$ **36.** $(0{\cdot}04)^{\frac{1}{2}}$
37. $(2{\cdot}25)^{\frac{1}{2}}$ **38.** $(7{\cdot}63)^0$ **39.** $3^5 \times 3^{-3}$
40. $(3\frac{3}{8})^{\frac{1}{3}}$ **41.** $(11\frac{1}{9})^{-\frac{1}{2}}$ **42.** $(\frac{1}{8})^{-2}$
43. $(\frac{1}{1000})^{\frac{2}{3}}$ **44.** $(\frac{9}{25})^{-\frac{1}{2}}$ **45.** $(10^{-6})^{\frac{1}{3}}$

Example 3

Solve the equations (a) $2^x = 16$

(b) $5^y = \frac{1}{25}$

We use the principle that if

$$a^x = a^y \text{ then } x = y$$

(a) $2^x = 16 = 2^4$

$\therefore\ x = 4$

(b) $5^y = \frac{1}{25} = 5^{-2}$

$\therefore\ y = -2$

Exercise 23

Solve the equations for x

1. $2^x = 8$ **2.** $3^x = 81$ **3.** $5^x = \frac{1}{5}$
4. $10^x = \frac{1}{100}$ **5.** $3^{-x} = \frac{1}{27}$ **6.** $4^x = 64$
7. $6^{-x} = \frac{1}{6}$ **8.** $100\,000^x = 10$ **9.** $12^x = 1$

10. $10^x = 0{\cdot}0001$ **11.** $2^x + 3^x = 13$ **12.** $(\frac{1}{2})^x = 32$
13. $5^{2x} = 25$ **14.** $1\,000\,000^{3x} = 10$

15. Find two solutions of the equation $x^2 = 2^x$

16.† Use a calculator to find solutions correct to three significant figures.
(a) $x^x = 100$ (b) $x^x = 10\,000$

17.† The last digit of 7^3 is 3 [$7^3 = 343$].
(a) Copy and complete the table below, which gives the last digit of 7^n.

n	1	2	3	4	5	6	7	8	9	10
Last digit of 7^n	7	9	3		7			1		

(b) Write down the last digit of:
(i) 7^{48} (ii) 7^{101} (iii) 49^{35}

18.† It is given that $10^x = 3$ and $10^y = 7$. What is the value of 10^{x+y}?

19.† In a laboratory we start with 2 cells in a dish. The number of cells in the dish doubles every 30 minutes.
(a) How many cells are in the dish after four hours?
(b) After what time are there 2^{13} cells in the dish?
(c) After $10\frac{1}{2}$ hours there are 2^{22} cells in the dish and an experimental fluid is added which eliminates half of the cells. How many cells are left?

20.† It is given that $x + \dfrac{1}{x} = 1$, where x is not zero. Show that $x^2 = x - 1$ and $x^3 = x^2 - x$. Use these two expressions to show that $x^3 = -1$.
Using this value for x^3, find the value of x^6 and show that $x^7 = x$. Hence show that $x^6 + \dfrac{1}{x^6} = 2$ and $x^7 + \dfrac{1}{x^7}$.
Deduce the value of $x^{60} + \dfrac{1}{x^{60}}$ and of $x^{61} + \dfrac{1}{x^{61}}$.

21.† Consider the function $\mathrm{f}(x) = \left(1 + \dfrac{1}{x}\right)^x$

So
$$\mathrm{f}(1) = \left(1 + \frac{1}{1}\right)^1 = 2$$
$$\mathrm{f}(2) = \left(1 + \frac{1}{2}\right)^2 = 2{\cdot}25$$

(a) Find f(3), f(4)
(b) Find f(10), f(1000), f(10000)
(c) Find the $\boxed{e^x}$ button on your calculator and press $\boxed{1}$ $\boxed{e^x}$
(d) Complete the sentence:

'As $x \to \infty, \mathrm{f}(x) \to$ '

Examination exercise 4

1. The cost, c pounds, of making x cars is made up of three charges:

A charge of a pounds per car for labour.
A charge of b pounds per car for materials.
A fixed charge of d pounds independent of the number of cars made.

(a) Write down the formula for the cost c of making x cars.
(b) Make x the subject of the formula.
(c) Find the number of cars that can be made for a cost of £71 200 when $a = 2600$, $b = 3000$, $d = 4000$. [S]

2. A machinist in a clothing factory is paid a weekly wage consisting of a fixed sum of money plus a bonus for each skirt made.
The Table shows the weekly wage earned for different numbers of skirts made.

Number of skirts made	10	20	30	40	50	60
Wage earned in £	40	55	70	85	100	115

(a) Using a scale of 2 cm for 10 units on each axis, draw clearly labelled axes. Plot the points for the values given in the table. Draw a straight line through these points.
(b) Use your graph to find
  (i) the weekly wage earned when the machinist made 34 skirts,
  (ii) the smallest number of skirts to be made in one week to earn more than £95,
  (iii) the fixed sum of money paid each week,
  (iv) the bonus paid for each skirt.

In the first week at work, a machinist made 30 skirts. During the next week the machinists' output increased by 20%.
(c) Find
  (i) the number of skirts made in the second week,
  (ii) how much more the machinist earned in the second week than in the first week. [L]

3. (a) Solve the simultaneous equations

$$3x - 4y = 10$$
$$5x + 7y = 3.$$

(b) In a Physics experiment three variables, u, v and f are connected by the formula

$$f = \frac{uv}{u + v}.$$

Make v the subject of the formula. [L]

4. The voltage, V volts, available from a 12 volt battery of internal resistance r ohms when connected to apparatus of resistance R ohms, is given by

$$V = \frac{12R}{(r + R)}$$

(a) Find V when $r = 1{\cdot}5$ and $R = 6$.
(b) Express R in terms of V and r. [M]

5. The current, I amps, in a circuit containing a resistor of resistance R ohms, is given by

$$I = \frac{12}{R + 2}$$

Express R in terms of I. [M]

6. The formula for the surface area, A, of a closed circular cylinder of radius r and height h is

$$A = 2\pi r(r + h).$$

(a) Make h the subject of the formula.
(b) Find the radius, in cm to one decimal place, of a cylinder whose surface area is $120\pi\,\text{cm}^2$ and whose height is 12 cm.
[L]

7. A stone is dropped down a water well.
The time taken in seconds, for the stone to drop from ground level to the water level is measured. The depth of a well is proportional to the square of the time taken for a stone to drop down it. One day, when the water level in the well was 100 m below ground level, the time taken was 4 seconds. After heavy rain another stone took 3 seconds to drop from ground level to the water level. What was the depth from ground level to water level after the heavy rain? [L]

8. (a) To what power must x^2 be raised to give

(i) x^6? (ii) $\frac{1}{x^2}$? (iii) x^{100}?

(b) Write down the square root of $121x^{16}$. [N]

9. (a) $216^{\frac{1}{3}} = 6^x$ find x.

(b) $\frac{5^{-2}}{5^{-4}} = 5^y$ find y.

(c) Evaluate $\sqrt{\frac{17{\cdot}85 - 12{\cdot}52}{27{\cdot}84 - 1{\cdot}84^5}}$. [S]

10. This table is reproduced from the Highway Code.

Speed	*Braking distance*	*Thinking distance*	*Overall stopping distance*
30 mph	45 feet	30 feet	75 feet
50 mph	125 feet	50 feet	175 feet
70 mph	245 feet	70 feet	315 feet

It is known that the overall stopping distance D (feet) and the speed v (miles per hour), are related by the formula

$$D = av + bv^2$$

where a and b are constants.

(a) Use some of the information given above to form two equations connecting D and v.

(b) Solve these two equations simultaneously, and hence show that

$$D = v + \frac{v^2}{20}.$$

(c) Complete the table below which shows values of D corresponding to the given values of v.

v	0	20	30	40	50	60	70	80
D	0		75		175		315	

(d) Draw a graph of D against v for values of v from 0 to 80.

(e) The driver of a car involved in a motor car accident told the police, 'As I passed through the traffic lights I was going at about 40 mph. It was then that I saw a man step off the pavement. I put the brakes on straightaway but could not avoid hitting him.'
The distance from the traffic lights to the place where the man was knocked down is 210 feet.
From your graph, estimate the minimum speed of the car as it passed through the traffic lights. [N]

11. The rectangle ABCD has a perimeter of 12 cm and side AD = x cm.

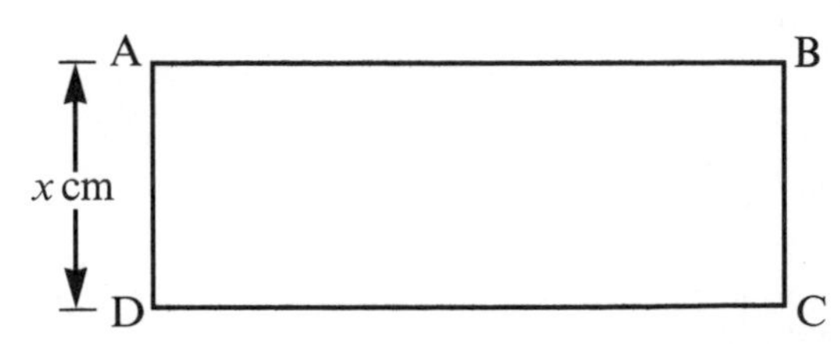

(a) Write down, in terms of x, an expression for the area of the rectangle.

(b) The rectangle ABCD has an area of $6{\cdot}44\,\text{cm}^2$. Show that $x^2 - 6x + 6{\cdot}44 = 0$.

(c) Complete the table of values for $y = x^2 - 6x + 6{\cdot}44$.
Draw the graph of y against x. Take x from 0 to 6 and y from -4 to 8.

x	0	1	2	3	4	5	6
y		1·44	−1·56	−2·56	−1·56		

(d) Read off from your graph an approximate value for the width of the rectangle. [N]

12. Graph paper must be used for this question.

(a) (i) Either by copying and completing the following table, or otherwise, find the values of y, for the given values of x from 1 to 6, when

$$y = 4x^2 - 28x + 41.$$

x	1	2	3	4	5	6
$4x^2$	4			64		144
$-28x$	−28			−112		−168
$+41$	+41			+41		41
y	17			−7		17

(ii) Using a scale of 2 centimetres to represent 1 unit on the x-axis and a scale of 2 centimetres to represent 5 units on the y-axis, draw the graph of $y = 4x^2 - 28x + 41$ for values of x from 1 to 6.

(iii) Draw in the line of symmetry on your graph and calculate the minimum value of y.

(b) (i) Show how the equation $4x^3 - 28x^2 + 41x - 12 = 0$ may be transformed into the form

$$4x^2 - 28x + 41 = \frac{12}{x} \quad (x \neq 0).$$

(ii) By drawing on the same axes an appropriate graph, find the two solutions of $4x^3 - 28x^2 + 41x - 12 = 0$ which lie between 1 and 6. [M]

13. In an experiment, this table of values was obtained for x and y.

x	2	3	4·5	5·5	6
y	1	2·4	4·25	5·6	6·4

Plot the points and find the equation which you think best fits these points. [N]

5 Shape and space 2

5.1 Similar figures: surface area and volume

Areas of similar shapes

The two rectangles shown are similar.
The ratio of their corresponding sides is k.

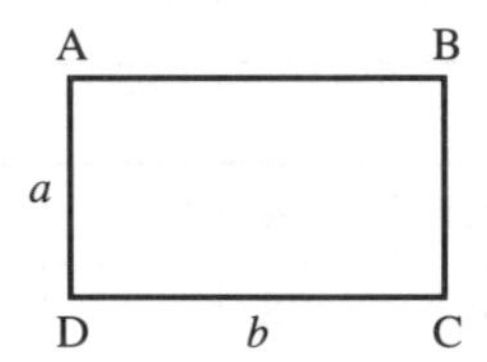

area of ABCD $= ab$
area of WXYZ $= ka \,.\, kb = k^2ab$.

$$\therefore \quad \frac{\text{area of WXYZ}}{\text{area ABCD}} = \frac{k^2ab}{ab} = k^2$$

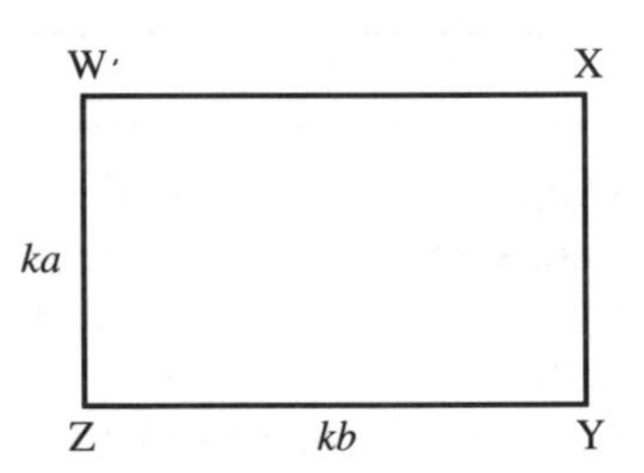

This illustrates an important general rule for all similar shapes:

If two figures are similar and the ratio of corresponding sides is k, then the ratio of their areas is k^2.

Note. k may be called the *linear scale factor*.

This result also applies for the surface areas of similar three-dimensional objects.

Example 1

In triangle ABC, XY is parallel to BC and $\frac{\text{AB}}{\text{AX}} = \frac{3}{2}$. If the area of triangle AXY is $4\,\text{cm}^2$, find the area of triangle ABC.

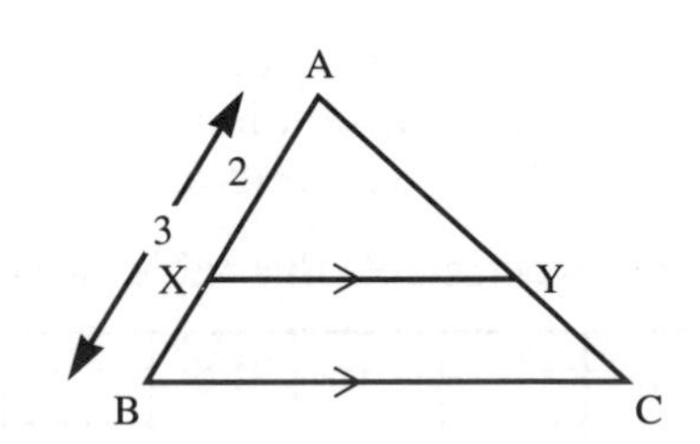

The triangles ABC and AXY are similar.

Ratio of corresponding sides $(k) = \frac{3}{2}$

$\therefore$ Ratio of areas $(k^2) = \frac{9}{4}$

$\therefore$ Area of $\triangle$ABC $= \frac{9}{4} \times$ (area of $\triangle$AXY)

$= \frac{9}{4} \times (4) = 9\,\text{cm}^2$.

Example 2

Two similar triangles have areas of $18\,\text{cm}^2$ and $32\,\text{cm}^2$ respectively. If the base of the smaller triangle is 6 cm, find the base of the larger triangle.

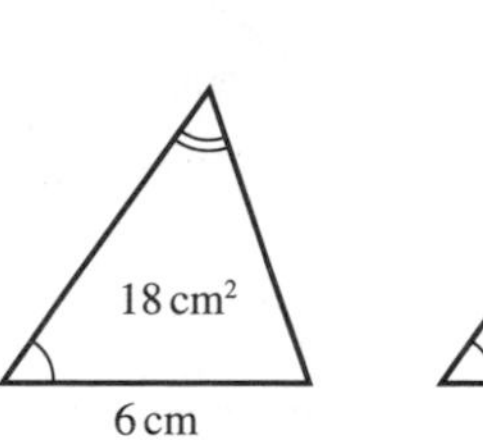

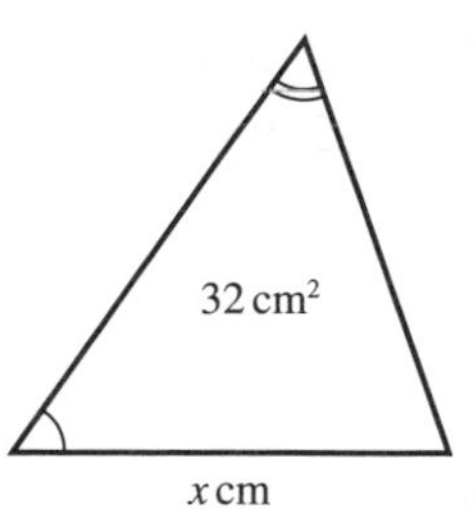

$$\text{Ratio of areas } (k^2) = \frac{32}{18} = \frac{16}{9}$$

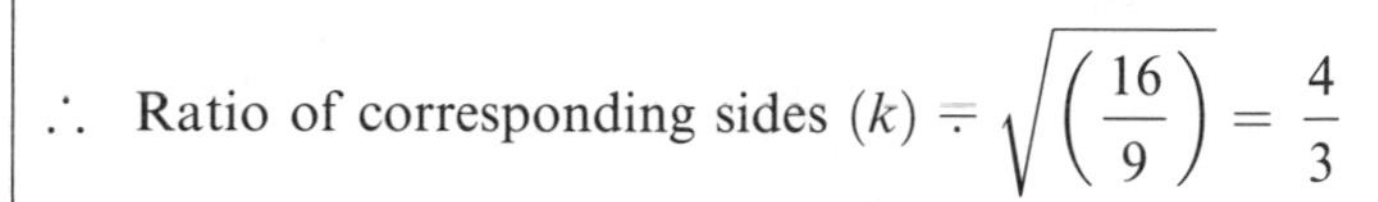

$$\therefore \quad \text{Ratio of corresponding sides } (k) = \sqrt{\left(\frac{16}{9}\right)} = \frac{4}{3}$$

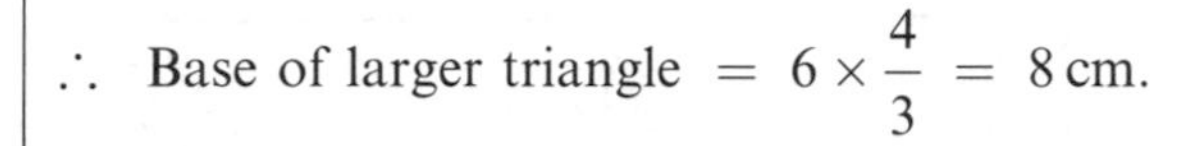

$$\therefore \quad \text{Base of larger triangle} = 6 \times \frac{4}{3} = 8\,\text{cm}.$$

Exercise 1

In this exercise, a number written inside a figure represents the area of the shape in cm^2. Numbers on the outside give linear dimensions in cm. In each case the shapes are similar. In Questions **1** to **6**, find the unknown area A.

1.

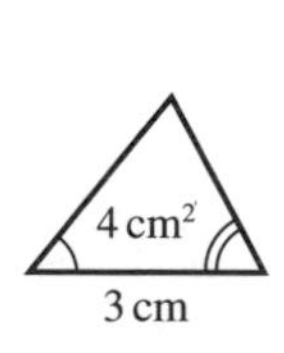

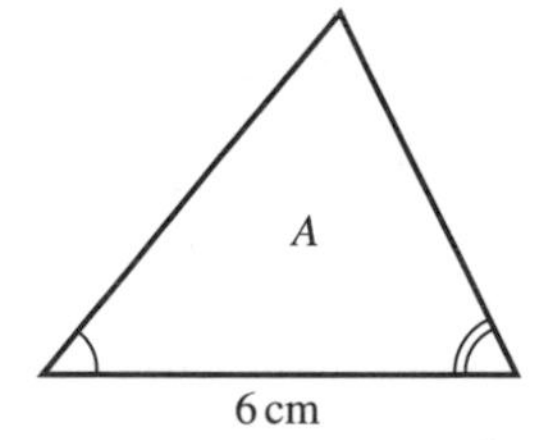

2.

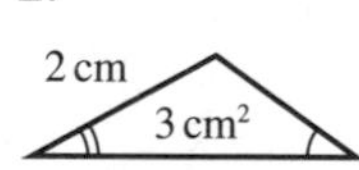

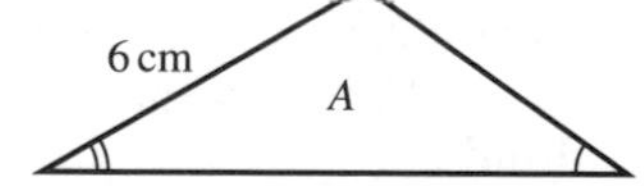

3.

2 cm, 6 cm, 5 cm²; A, 9 cm, 3 cm

4.

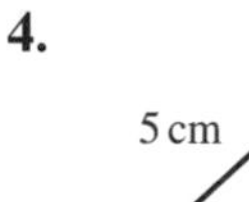

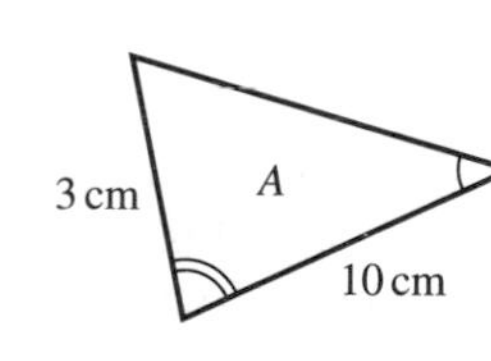

5.

A, 8 cm, 16 cm; 18 cm², 6 cm, 3 cm

6.

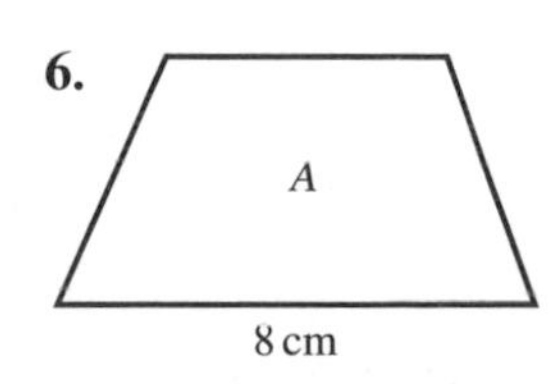

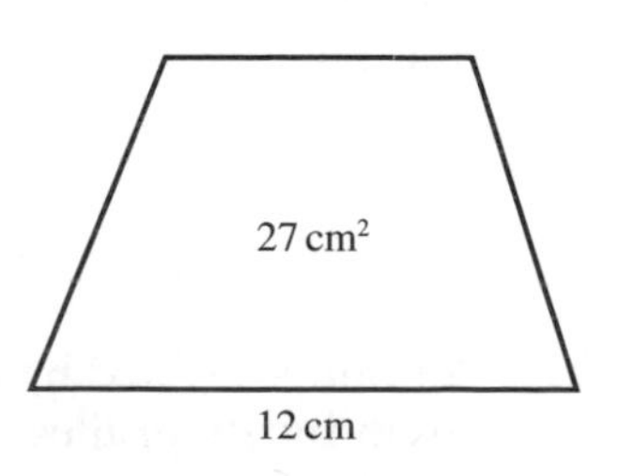

In Questions **7** to **10**, find the lengths marked for each pair of similar shapes.

7.

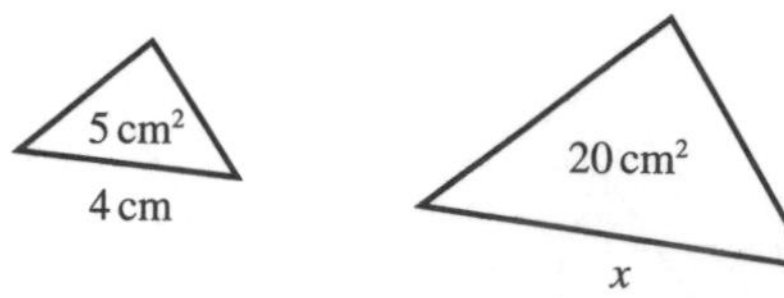

8.

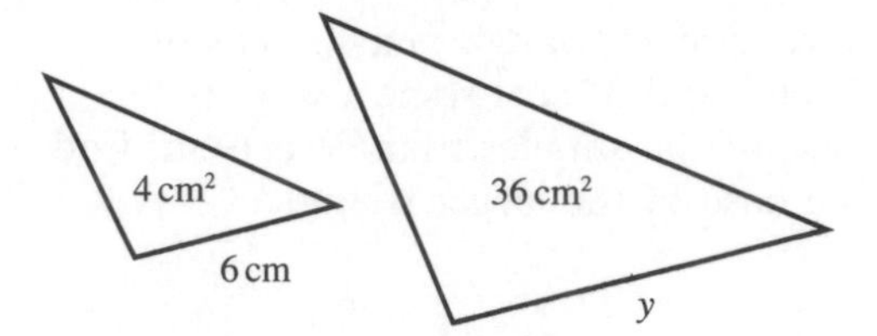

9.

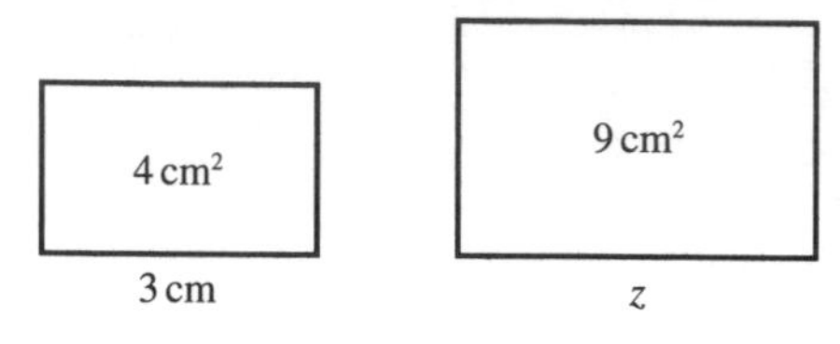

10.

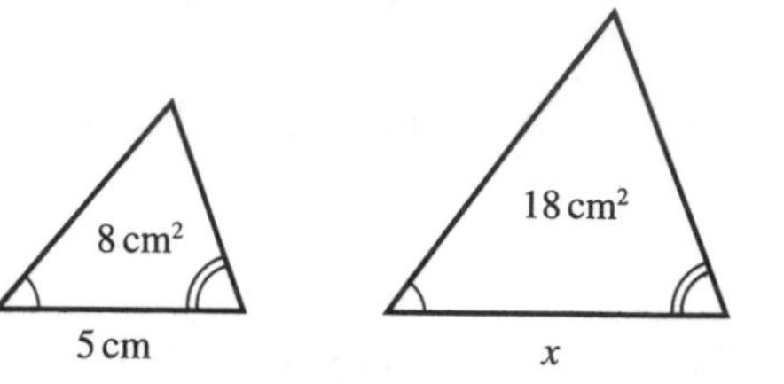

11. Given AD = 3 cm, AB = 5 cm and area of $\triangle$ADE = 6 cm^2.
Find:
(a) area of $\triangle$ABC
(b) area of DECB.

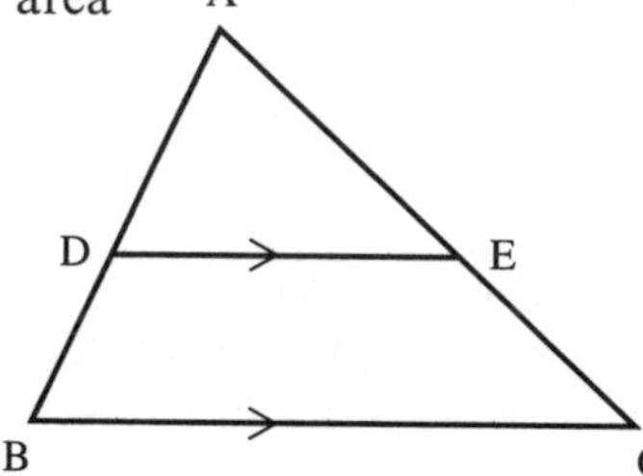

12. Given XY = 5 cm, MY = 2 cm and area of $\triangle$MYN = 4 cm^2.
Find:
(a) area of $\triangle$XYZ
(b) area of MNZX.

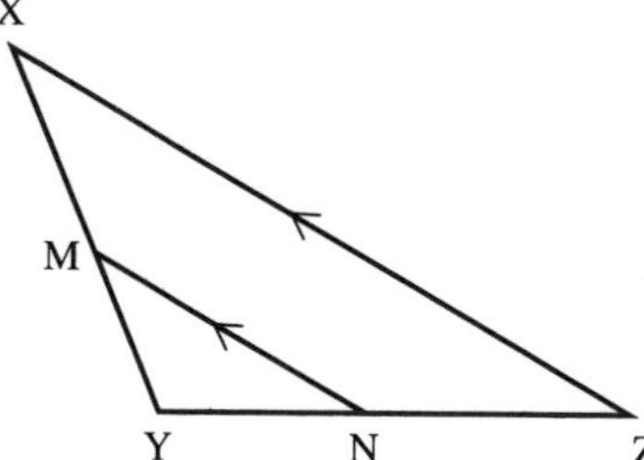

13. The triangles ABC and EBD are similar (AC and DE are *not* parallel).
If AB = 8 cm, BE = 4 cm and the area of $\triangle$DBE = 6 cm^2,
find the area of $\triangle$ABC.

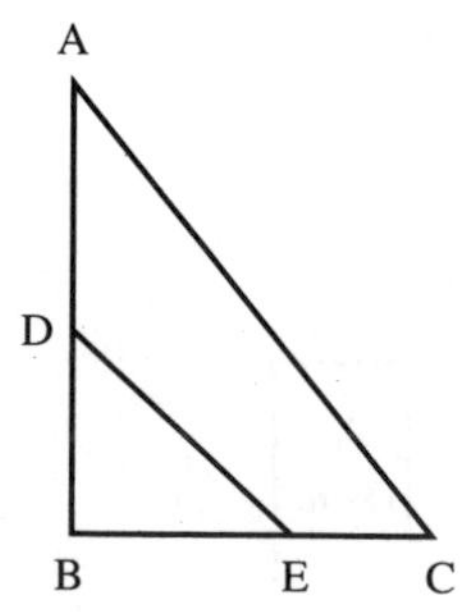

14. A floor is covered by 600 tiles which are 10 cm by 10 cm. How many 20 cm by 20 cm tiles are needed to cover the same floor?

15. A wall is covered by 160 tiles which are 15 cm by 15 cm. How many 10 cm by 10 cm tiles are needed to cover the same wall?

16. When potatoes are peeled do you lose more peel or less when big potatoes are used as opposed to small ones?

17. A supermarket offers cartons of cheese with 10% extra for the same price.

The old carton has a radius of 6 cm and a thickness of 1·2 cm. The new carton has the same thickness. Calculate its radius.

18.† Rectangle ABCD is inscribed in $\triangle PQR$. The length of AD is one third the length of the perpendicular from Q to PR.

Calculate the ratio $\left(\dfrac{\text{area ABCD}}{\text{area PQR}}\right)$

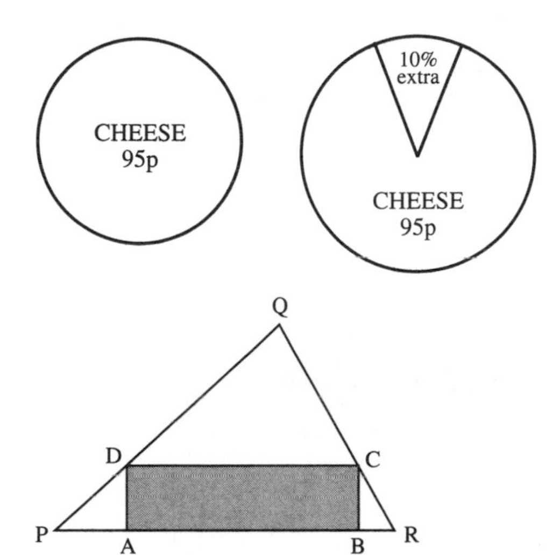

Volumes of similar objects

When solid objects are similar, one is an accurate enlargement of the other. If two objects are similar and the ratio of corresponding sides is k, then the ratio of their volumes is k^3.
A line has one dimension, and the scale factor is used once.
An area has two dimensions, and the scale factor is used twice.
A volume has three dimensions, and the scale factor is used three times.

Example 3

Two similar cylinders have heights of 3 cm and 6 cm respectively. If the volume of the smaller cylinder is 30 cm³, find the volume of the larger cylinder.

ratio of heights (k) $= \frac{6}{3}$ (linear scale factor)
$= 2$

$\therefore$ ratio of volumes (k^3) $= 2^3$
$= 8$

and volume of larger cylinder $= 8 \times 30$
$= 240\ \text{cm}^3$.

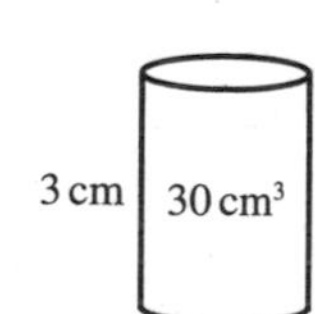

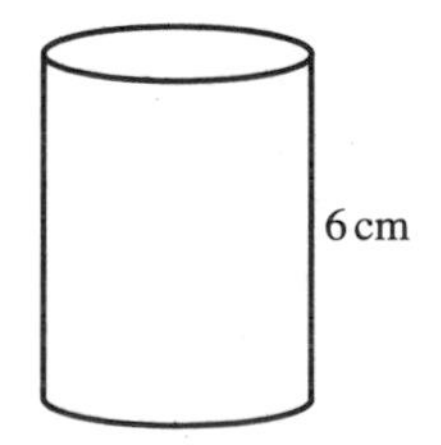

Example 4

Two similar spheres made of the same material have weights of 32 kg and 108 kg respectively. If the radius of the larger sphere is 9 cm, find the radius of the smaller sphere.

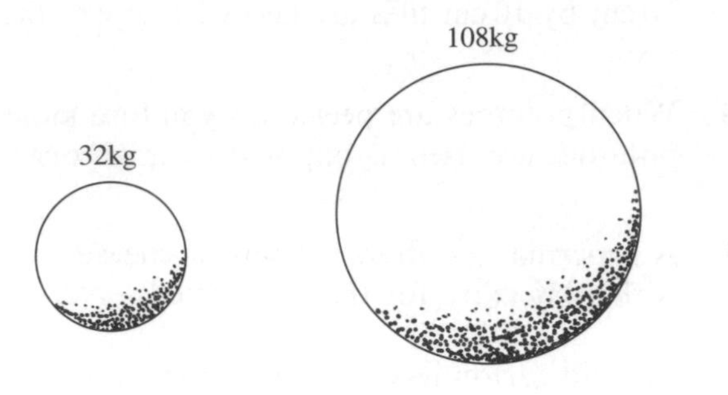

We may take the ratio of weights to be the same as the ratio of volumes.

$$\text{ratio of volumes } (k^3) = \frac{32}{108} = \frac{8}{27}$$

$$\text{ratio or corresponding lengths } (k) = \sqrt[3]{\left(\frac{8}{27}\right)} = \frac{2}{3}$$

$$\therefore \quad \text{radius of smaller sphere} = \frac{2}{3} \times 9 = 6\,\text{cm}.$$

Exercise 2

In this exercise, the objects are similar and a number written inside a figure represents the volume of the object in cm^3.
Numbers on the outside give linear dimensions in cm. In Questions **1** to **8**, find the unknown volume V.

1.

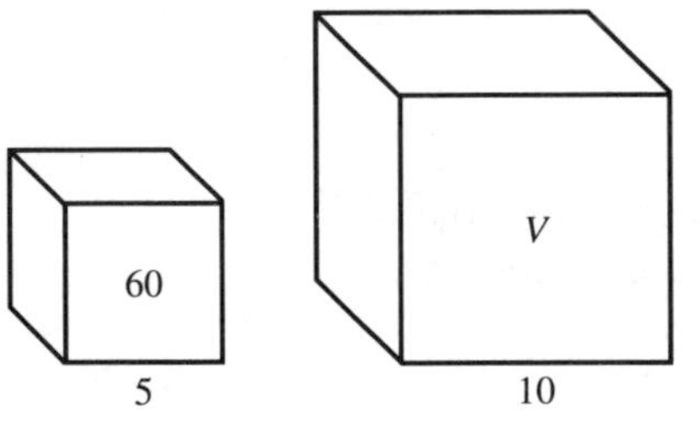

2.

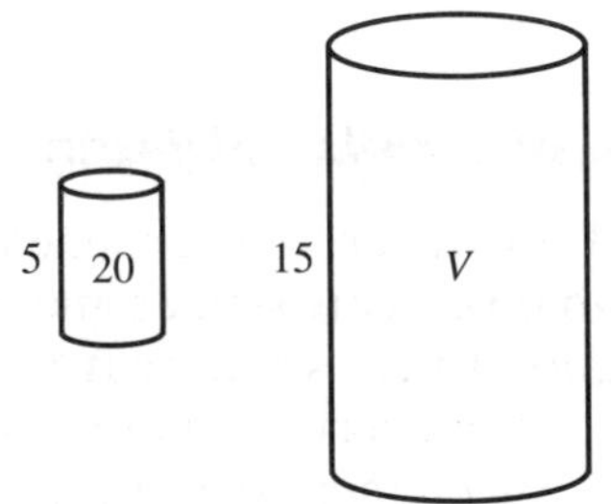

3.

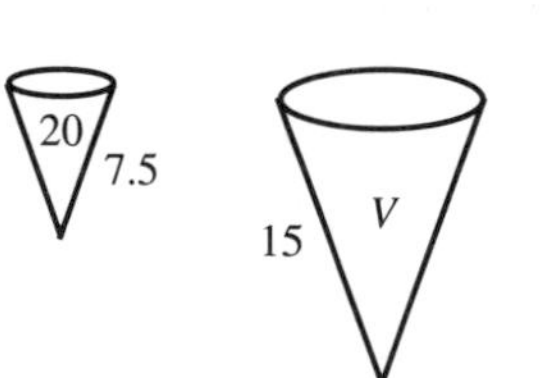

4.

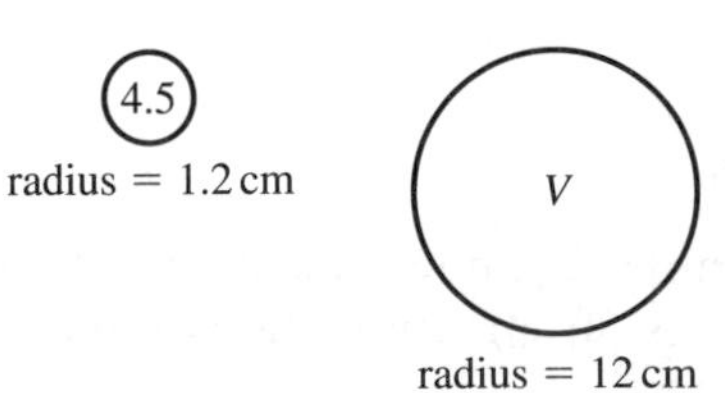

5.

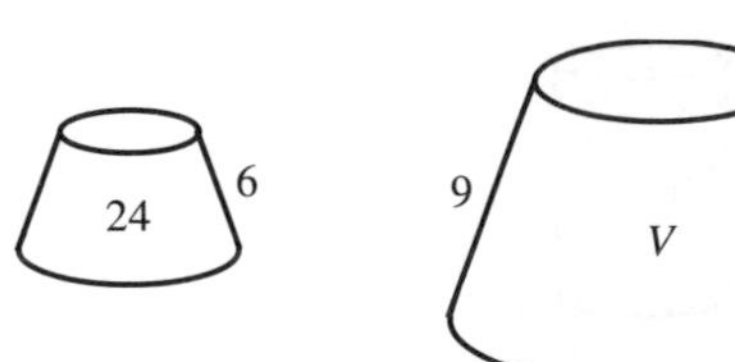

6.

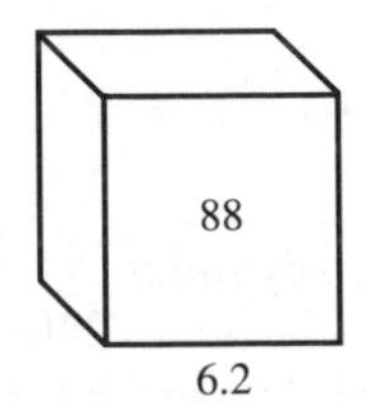

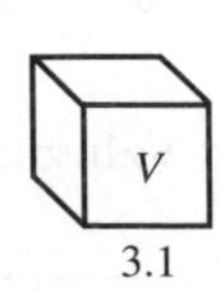

7.

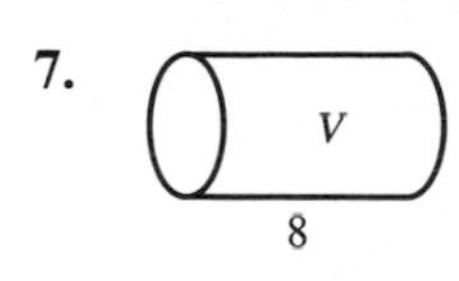

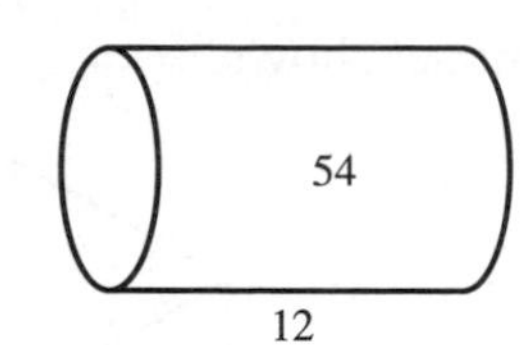

8.

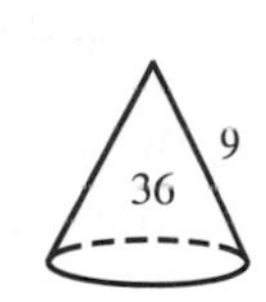

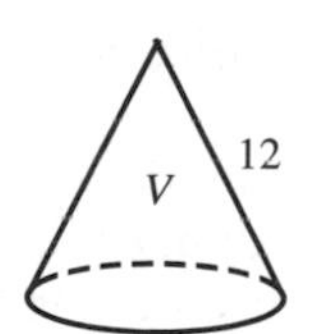

In Questions **9** to **14**, find the lengths marked by a letter.

9.

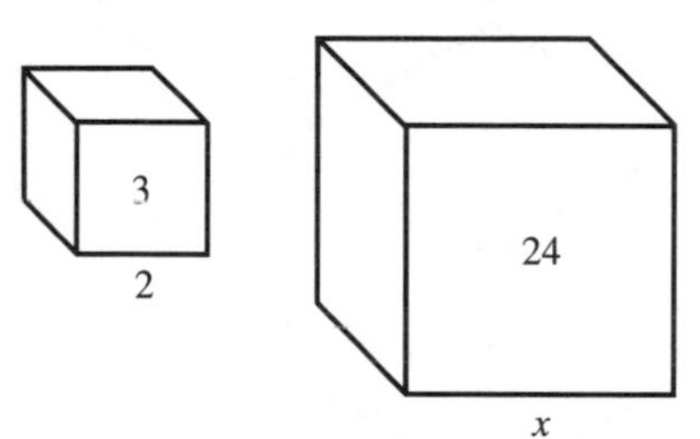

10.

y 270

11.

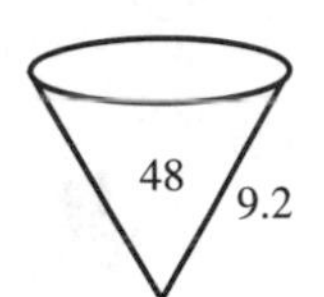

12.

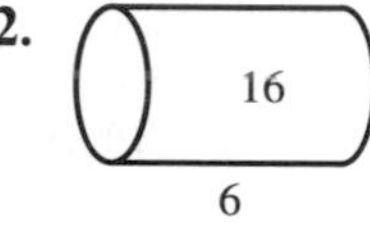

54
m

13.

81 h

14.

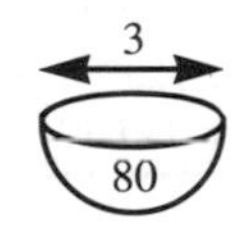

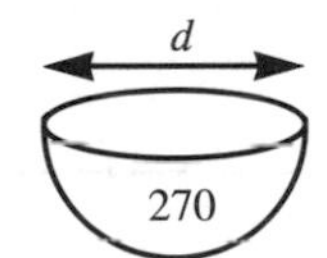

15. Two similar jugs have heights of 4 cm and 6 cm respectively. If the capacity of the smaller jug is 50 cm^3, find the capacity of the larger jug

16. Two similar cylindrical tins have base radii of 6 cm and 8 cm respectively. If the capacity of the larger tin is 252 cm^3, find the capacity of the small tin.

17. Two solid metal spheres have masses of 5 kg and 135 kg respectively. If the radius of the smaller one is 4 cm, find the radius of the larger one.

18. Two similar cones have surface areas in the ratio 4 : 9. Find the ratio of:
(a) their lengths, (b) their volumes.

19. The area of the bases of two similar glasses are in the ratio 4 : 25. Find the ratio of their volumes.

20. Four unifix cubes are joined together to make the model shown.

A new model is made from the same size cubes but all the dimensions are three times as big. How many cubes are needed?

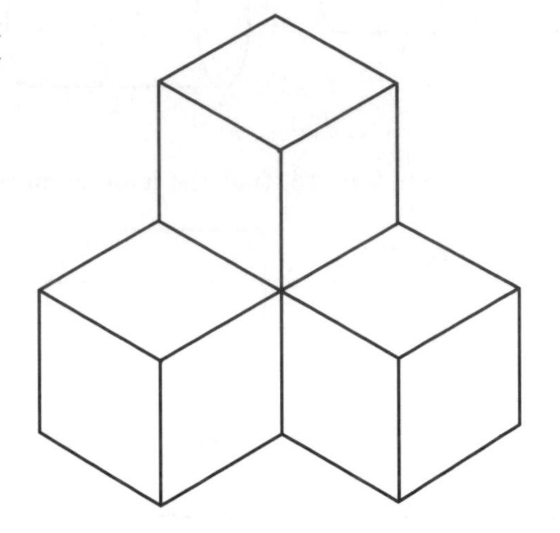

21. Sam has a model of a pirate ship which is made to a scale of 1 : 50.

Copy and complete the table using the given units.

	On model	On actual ship
Length	42 cm	m
Capacity of hold	500 cm^3	m^3
Area of sails	cm^2	175 m^2
Number of cannon	12	
Deck area	370 cm^2	m^2

22. Two solid spheres have surface areas of 5 cm^2 and 45 cm^2 respectively and the mass of the smaller sphere is 2 kg. Find the mass of the larger sphere.

23. The masses of two similar objects are 24 kg and 81 kg respectively. If the surface area of the larger object is 540 cm^2, find the surface area of the smaller object.

24. A cylindrical can has a circumference of 40 cm and a capacity of 4·8 litres. Find the capacity of a similar cylinder of circumference 50 cm.

25. A container has a surface area of 5000 cm^2 and a capacity of 12·8 litres. Find the surface area of a similar container which has a capacity of 5·4 litres.

5.2 Arcs and sectors

Arc length

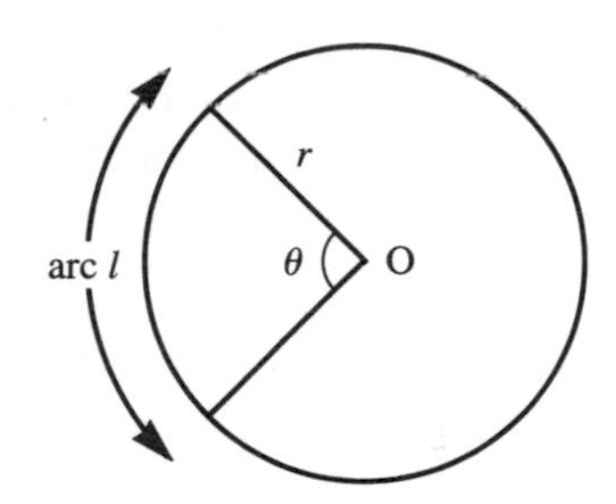

$$\text{Arc length, } l = \frac{\theta}{360} \times 2\pi r$$

We take a fraction of the whole circumference depending on the angle at the centre of the circle.

Sector area

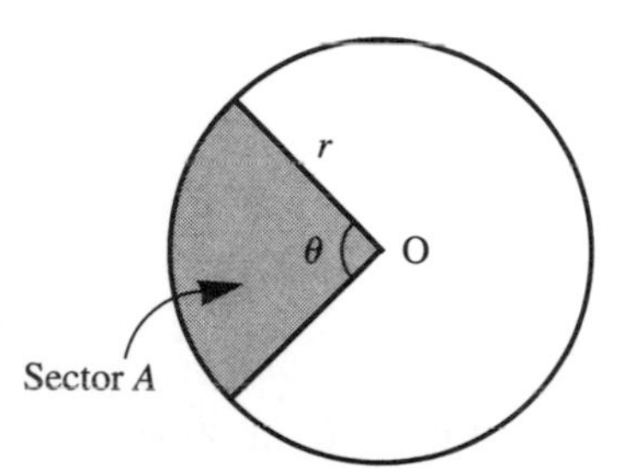

$$\text{Sector area, } A = \frac{\theta}{360} \times \pi r^2$$

We take a fraction of the whole area depending on the angle at the centre of the circle.

Example 1

Find the length of an arc which subtends an angle of 140° at the centre of a circle of radius 12 cm. (Take $\pi = \frac{22}{7}$.)

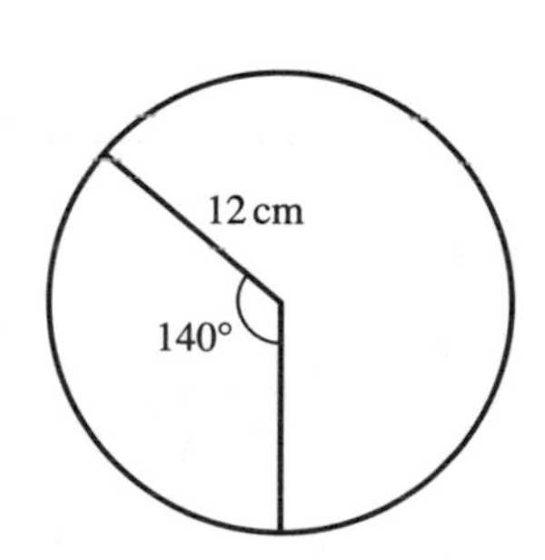

$$\begin{aligned}\text{Arc length} &= \frac{140}{360} \times 2 \times \frac{22}{7} \times 12 \\ &= \frac{88}{3} \\ &= 29\tfrac{1}{3}\text{ cm.}\end{aligned}$$

Example 2

A sector of a circle of radius 10 cm has an area of 25 cm^2.

Find the angle at the centre of the circle.

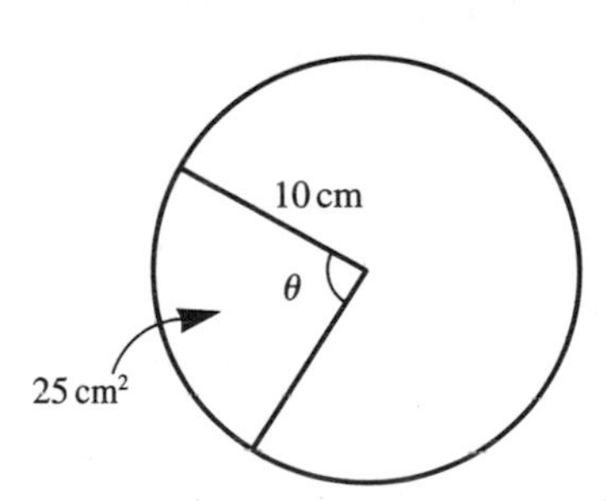

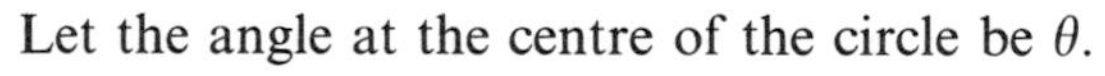

Let the angle at the centre of the circle be θ.

$$\frac{\theta}{360} \times \pi \times 10^2 = 25$$

$$\therefore \quad \theta = \frac{25 \times 360}{\pi \times 100}$$

$$\theta = 28{\cdot}6° \text{ (1 d.p.)}$$

The angle at the centre of the circle is 28·6°

Exercise 3

[Use the π button on a calculator unless told otherwise]

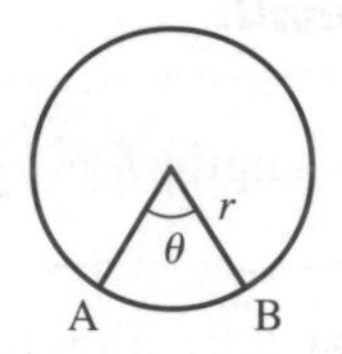

1. Arc AB subtends an angle θ at the centre of circle radius r. Find the arc length and sector area when
   (a) $r = 4$ cm, $\theta = 30°$ (b) $r = 10$ cm, $\theta = 45°$
   (c) $r = 2$ cm, $\theta = 235°$

In Questions **2** and **3** find the total area of the shape.

2. OA = 2 cm, OB = 3 cm, OC = 5 cm, OD = 3 cm.

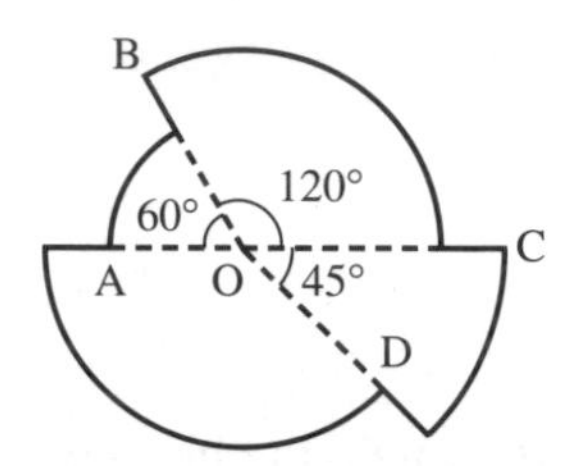

3. ON = 6 cm, OM = 3 cm, OL = 2 cm, OK = 6 cm.

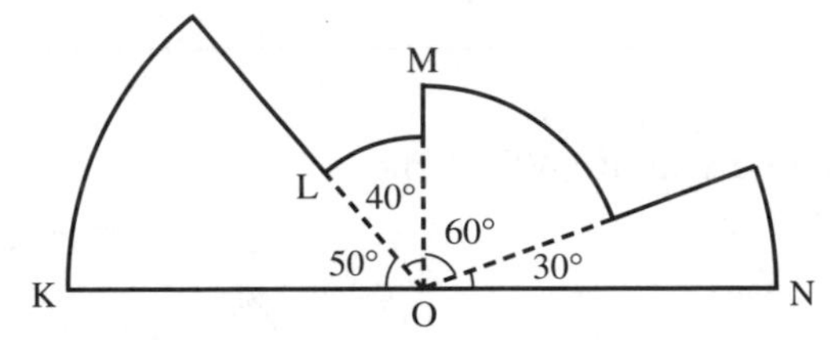

4. Find the shaded area.

(a)

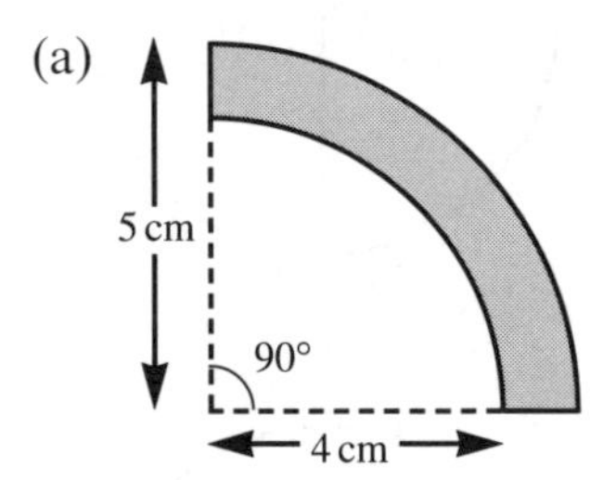

(b)

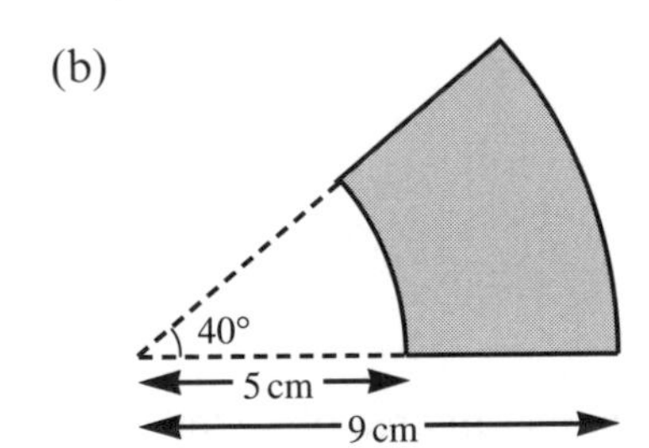

5. The lengths of the minor and major arcs of a circle are 5·2 cm and 19·8 cm respectively. Find
   (a) the radius of the circle
   (b) the angle subtended at the centre by the minor arc.

6. The length of the minor arc AB of a circle, centre O, is 2π cm and the length of the major arc is 22π cm. Find
   (a) the radius of the circle,
   (b) the acute angle AOB.

7. A wheel of radius 10 cm is turning at a rate of 5 revolutions per minute. Calculate
   (a) the angle through which the wheel turns in 1 second
   (b) the distance moved by a point on the rim in 2 seconds.

8. In the diagram the arc length is l and the sector area is A.

(a) Find θ, when $r = 5\,\text{cm}$ and $l = 7{\cdot}5\,\text{cm}$

(b) Find θ, when $r = 2\,\text{m}$ and $A = 2\,\text{m}^2$

(c) Find r, when $\theta = 55°$ and $l = 6\,\text{cm}$.

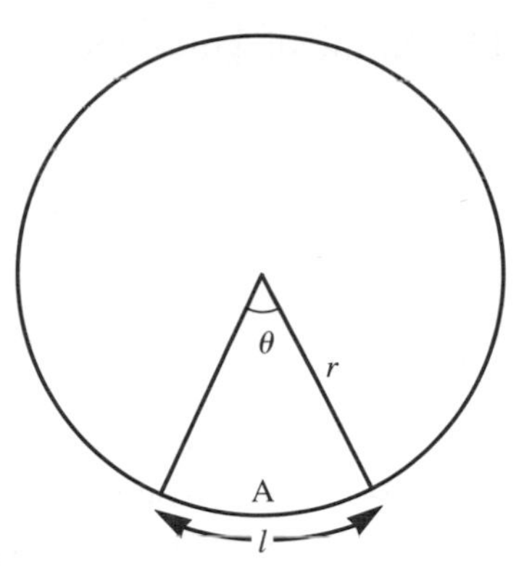

9. Two parallel lines are drawn 2 cm from the centre of a circle of radius 4 cm. Calculate the area shaded.

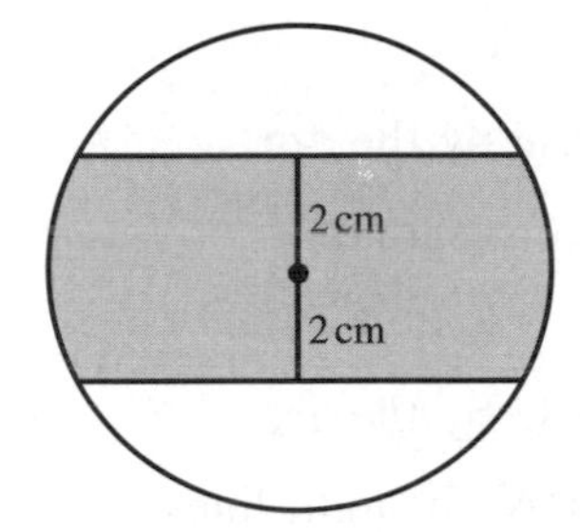

10. A long time ago Mr Gibson found an island shaped as a triangle with three straight shores of length 3 km, 4 km and 5 km. He declared an 'exclusion zone' around his island and forbade anyone to come within 1 km of his shore. What was the area of his exclusion zone?

11. ABCDE is a regular pentagon. Calculate the size of angle x.

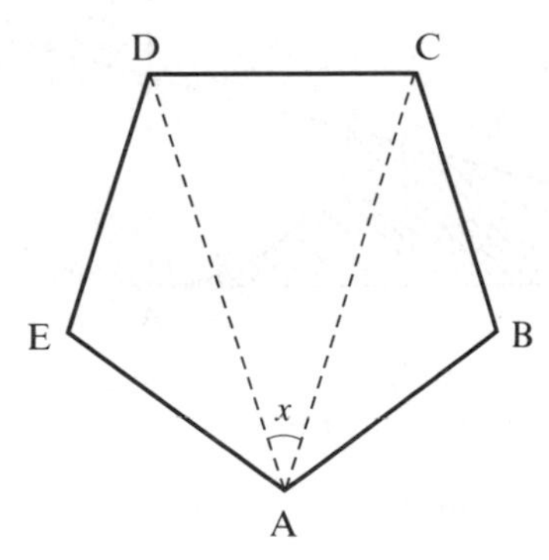

In this diagram, arcs are drawn with centres at the opposite corners of the pentagon, e.g. arc CD has centre at A. The radius of each arc is 5 cm. Show that the perimeter of the shape is 5π cm.

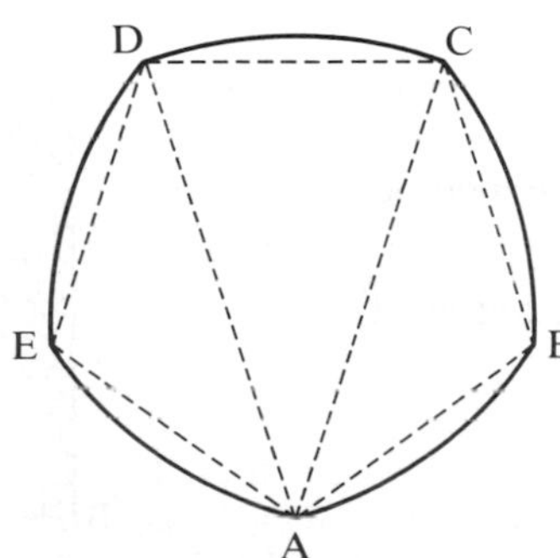

12. In the diagram, the arc length is l and the sector area is A.

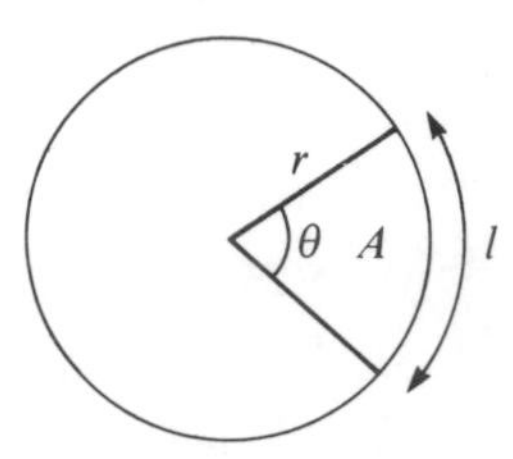

(a) Find l, when $\theta = 72°$ and $A = 15\,\text{cm}^2$
(b) Find l, when $\theta = 135°$ and $A = 162\,\text{m}^2$
(c) Find A, when $l = 11\,\text{cm}$ and $r = 5{\cdot}2\,\text{cm}$

13.† The length of an arc of a circle is 12 cm. The corresponding sector area is $108\,\text{cm}^2$. Find
(a) the radius of the circle
(b) the angle subtended at the centre of the circle by the arc.

14.† The length of an arc of a circle is 7·5 cm. The corresponding sector area is $37{\cdot}5\,\text{cm}^2$. Find
(a) the radius of the circle
(b) the angle subtended at the centre of the circle by the arc.

15.† In the diagram, AB is a tangent to the circle at A. Straight lines BCD and ACE pass through the centre of the circle C. Angle ACB = $x°$ and the radius of the circle is 1 unit.

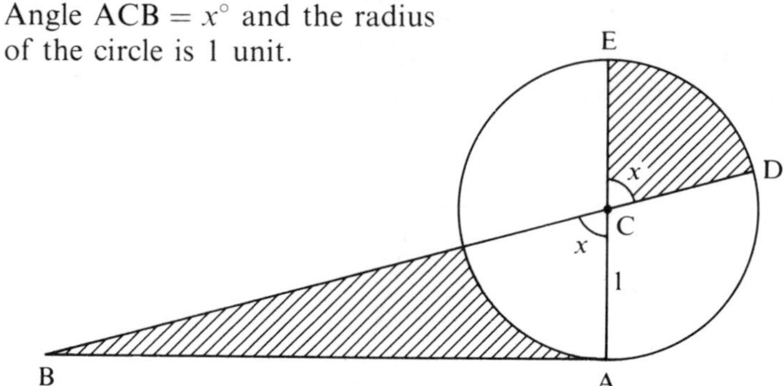

The two shaded areas are equal.
(a) Show that x satisfies the equation $x = \left(\dfrac{90}{\pi}\right) \tan x$
(b) Use trial and improvement to find a solution for x correct to 1 decimal place.

16.† (Hard, unless you can find a neat solution!)
In the diagram:
BC is a diameter of the semicircle
$A\hat{B}C = 90°$
Shaded area ① = shaded area ②
Angle ACB = x
Show that $\tan x = \dfrac{\pi}{4}$.
Hence find angle x.

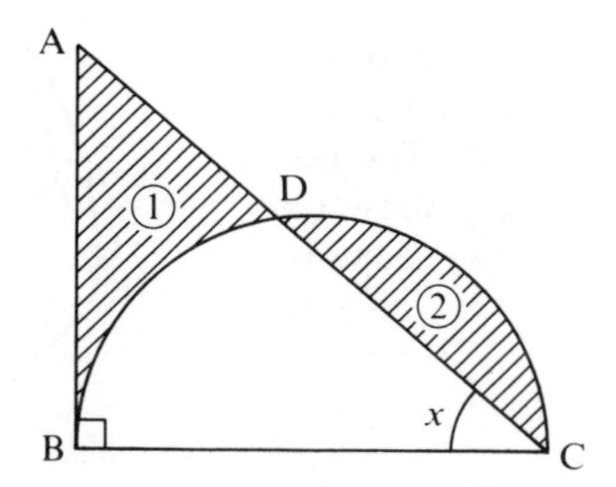

Segments

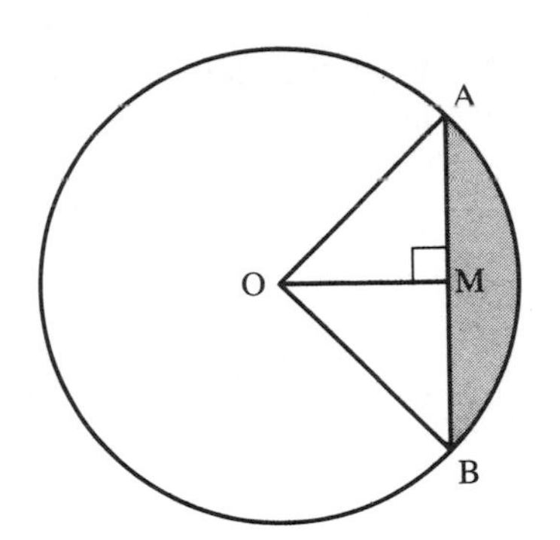

The line AB is a chord. The area of a circle cut off by a chord is called a *segment*. In the diagram the *minor* segment is shaded and the *major* segment is unshaded.

(a) The line from the centre of a circle to the mid-point M of a chord *bisects* the chord at *right angles*.
(b) The line from the centre of a circle to the mid-point of a chord bisects the angle subtended by the chord at the centre of the circle.

Example 3

XY is a chord of length 12 cm of a circle of radius 10 cm, centre O. Calculate
(a) the angle XOY
(b) the area of the minor segment cut off by the chord XY.

Let the mid-point of XY be M

$$\therefore \quad MY = 6\text{ cm}$$
$$\sin M\hat{O}Y = \frac{6}{10}$$
$$\therefore \quad M\hat{O}Y = 36{\cdot}87^\circ$$
$$\therefore \quad X\hat{O}Y = 2 \times 36{\cdot}87 = 73{\cdot}74^\circ$$

area of minor segment = area of sector XOY − area of △XOY

$$\text{area of sector XOY} = \frac{73{\cdot}74}{360} \times \pi \times 10^2 = 64{\cdot}32\text{ cm}^2.$$
$$\text{area of }\triangle XOY = \tfrac{1}{2} \times 10 \times 10 \times \sin 73{\cdot}74^\circ = 48{\cdot}00\text{ cm}^2$$
$$\therefore \quad \text{Area of minor segment} = 64{\cdot}32 - 48{\cdot}00 = 16{\cdot}3\text{ cm}^2 \text{ (3 S.F.)}$$

Exercise 4

Use the 'π' button on a calculator

1. The chord AB subtends an angle of 130° at the centre O. The radius of the circle is 8 cm. Find
(a) the length of AB,
(b) the area of sector OAB,
(c) the area of triangle OAB,
(d) the area of the minor segment (shown shaded).

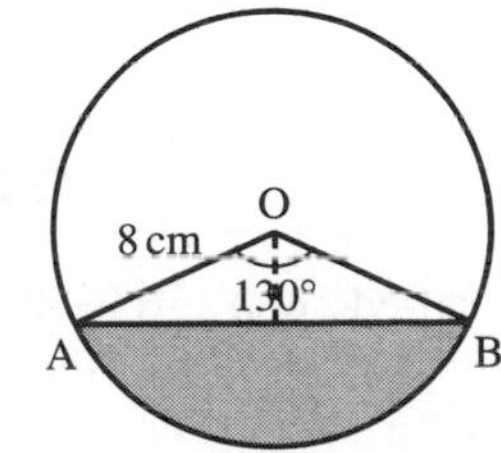

2. Find the shaded area when,
(a) $r = 6\,\text{cm}$, $\theta = 70°$
(b) $r = 14\,\text{cm}$, $\theta = 104°$
(c) $r = 5\,\text{cm}$, $\theta = 80°$

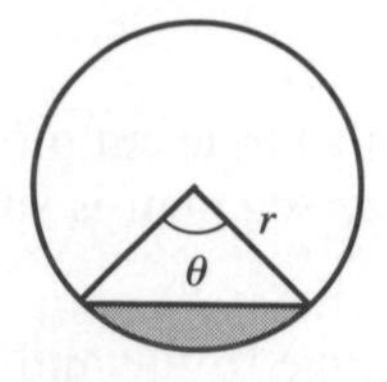

3. Find θ and hence the shaded area when,
(a) $\text{AB} = 10\,\text{cm}$, $r = 10\,\text{cm}$
(b) $\text{AB} = 8\,\text{cm}$, $r = 5\,\text{cm}$

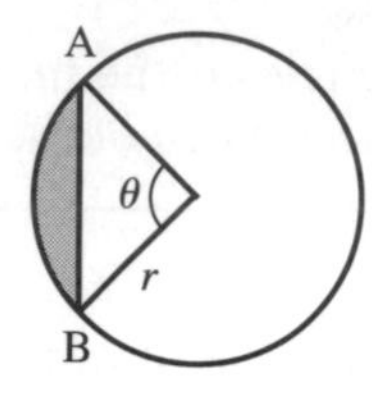

4. How far is a chord of length 8 cm from the centre of a circle of radius 5 cm?

5. How far is a chord of length 9 cm from the centre of a circle of radius 6 cm?

6. The diagram shows the cross section of a cylindrical pipe with water lying in the bottom.
(a) If the maximum depth of the water is 2 cm and the radius of the pipe is 7 cm, find the area shaded.
(b) What is the *volume* of water in a length of 30 cm?

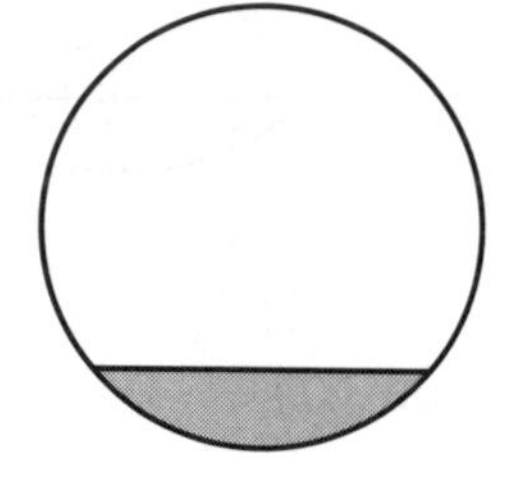

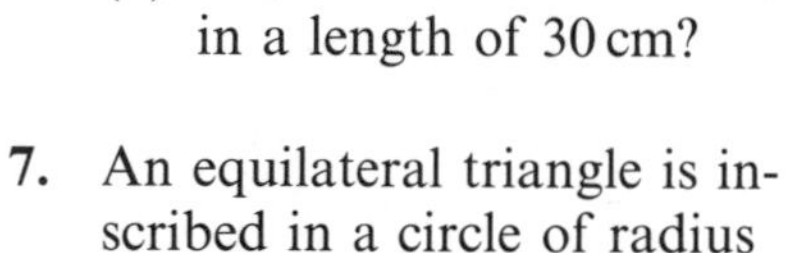

7. An equilateral triangle is inscribed in a circle of radius 10 cm. Find
(a) the area of the triangle,
(b) the area shaded.

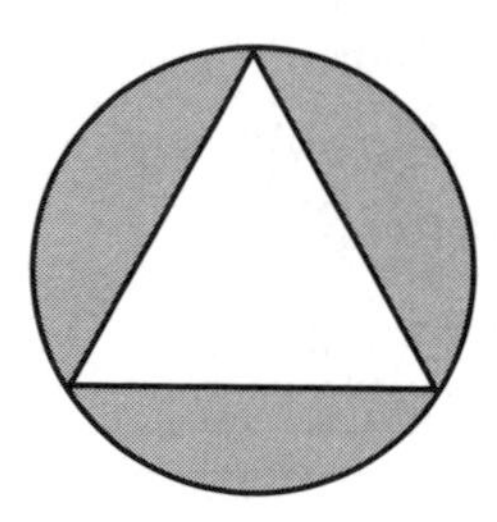

8. A regular hexagon is circumscribed by a circle of radius 6 cm. Find the area shaded.

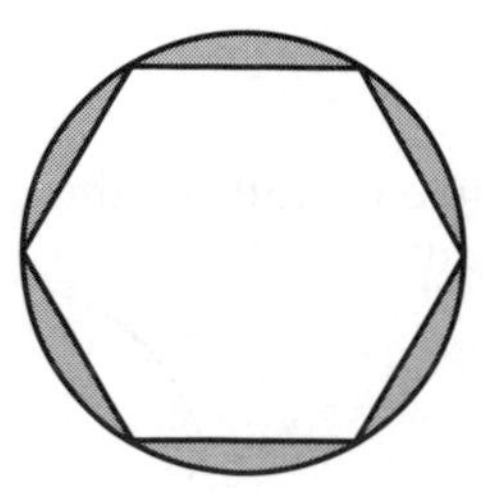

9. A regular octagon is circumscribed by a circle of radius r cm. Find the area enclosed between the circle and the octagon. (Give the answer in terms of r.)

10.† Find the radius of the circle
(a) when $\theta = 90°$, $A = 20\,\text{cm}^2$
(b) when $\theta = 30°$, $A = 35\,\text{cm}^2$
(c) when $\theta = 150°$, $A = 114\,\text{cm}^2$

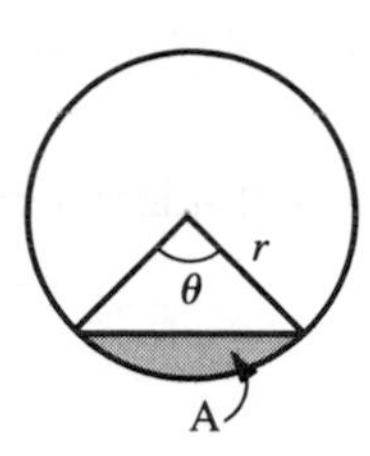

11.† The diagram shows a regular pentagon of side 10 cm with a star inside. Calculate the area of the star.

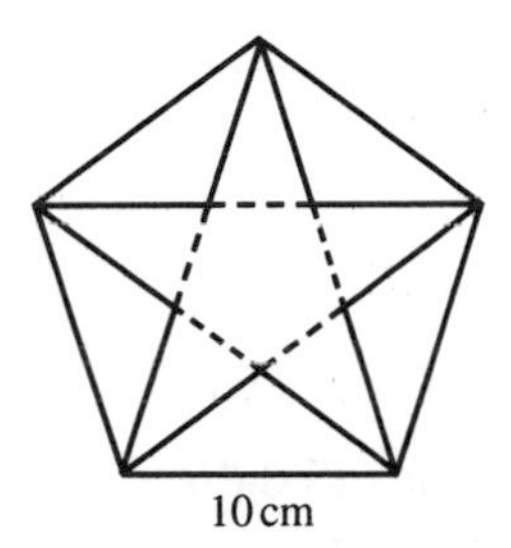

5.3 Sphere, cone, cylinder

Sphere

Volume $= \frac{4}{3}\pi r^3$

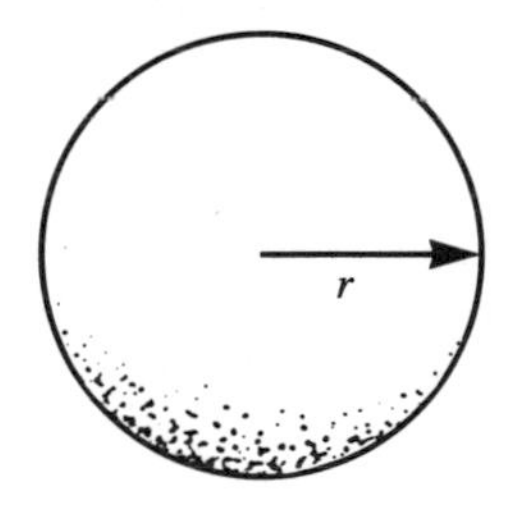

Cone

Volume $= \frac{1}{3}\pi r^2 h$

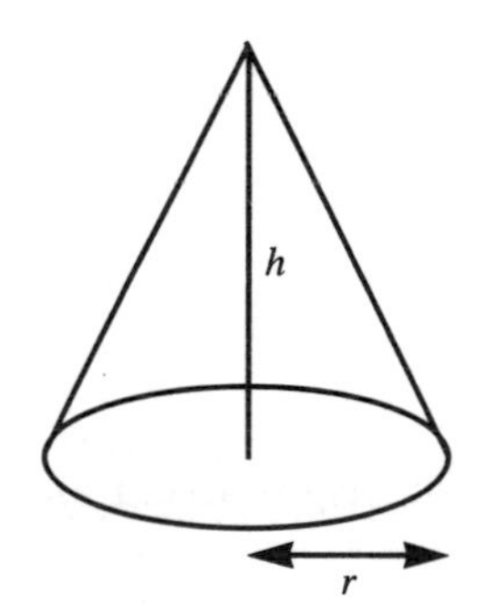

Cylinder

(a) Volume $= \pi r^2 h$
(b) Curved surface area $= 2\pi rh$

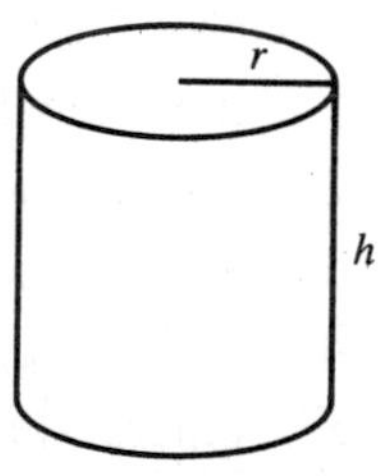

Example 1

Calculate the radius of a sphere of volume $500\,\text{cm}^3$.

Let the radius of the sphere be r cm

$$\frac{4}{3}\pi r^3 = 500$$

$$r^3 = \frac{3 \times 500}{4\pi}$$

$$r = \sqrt[3]{\left(\frac{3 \times 500}{4\pi}\right)} = 4{\cdot}92 \text{ (3 S.F.)}$$

The radius of the sphere is 4·92 cm.

Example 2

Calculate the *total* surface area of a solid cylinder of radius 3 cm and height 8 cm.

$$\begin{aligned}
\text{Curved surface area} &= 2\pi rh \\
&= 2 \times \pi \times 3 \times 8 \\
&= 48\pi \text{ cm}^2 \\
\text{Area of two ends} &= 2 \times \pi r^2 \\
&= 2 \times \pi \times 3^2 \\
&= 18\pi \text{ cm}^2 \\
\text{Total surface area} &= (48\pi + 18\pi) \text{ cm}^2 \\
&= 207 \text{ cm}^2 \ \ (3 \text{ S.F.})
\end{aligned}$$

Exercise 5

In Questions **1** to **6**, find the volumes of the following objects.

1. cone: height = 5 cm, radius = 2 cm

2. sphere: radius = 5 cm

3. sphere: radius = 10 cm

4. cone: height = 6 cm, radius = 4 cm

5. sphere: diameter = 8 cm

6. cylinder: height = 7 cm, radius = 5 cm

7. Find the curved surface area of a cylinder of height 13 cm and radius 9·5 cm.

8. Find the volume of a hemisphere of radius 5 cm.

9. A cone is attached to a hemisphere of radius 4 cm. If the total height of the object is 10 cm, find its volume.

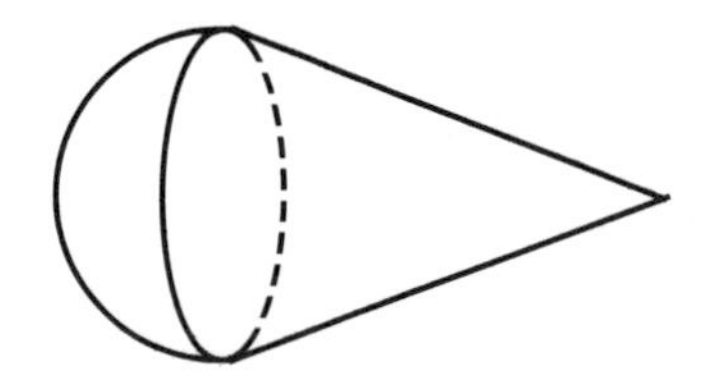

10. A toy consists of a cylinder of diameter 6 cm 'sandwiched' between a hemisphere and a cone of the same diameter. If the cone is of height 8 cm and the cylinder is of height 10 cm, find the total volume of the toy.

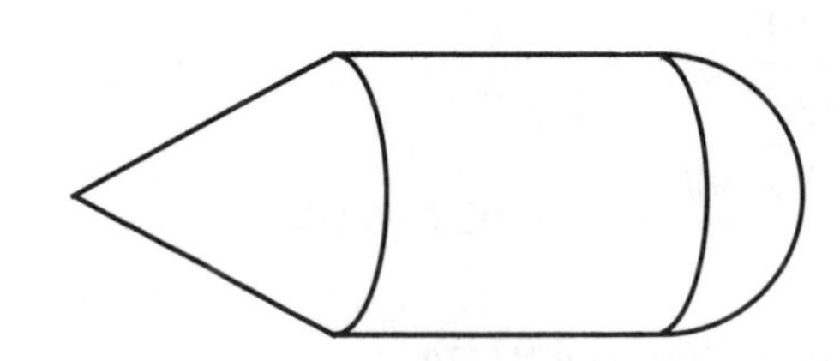

11. A solid cylinder of height 10 cm and radius 4 cm is to be plated with material costing £11 per cm^2. Find the cost of the plating.

12. A tin of paint covers a surface area of $60\,m^2$ and costs £4·50. Find the cost of painting the outside surface of a cylindrical gas holder of height 30 m and radius 18 m. The top of the gas holder is a flat circle.

13. A solid wooden cylinder of height 8 cm and radius 3 cm is cut in two along a vertical axis of symmetry. Calculate the total surface area of the two pieces.

14. Find the radius of a sphere of volume $60\,cm^3$.

15. Find the height of a cone of volume 2·5 litre and radius 10 cm.

16. Find the height of a solid cylinder of radius 1 cm and *total* surface area $28\,cm^2$.

17. A solid metal cube of side 6 cm is recast into a solid sphere. Find the radius of the sphere.

18. A hollow spherical vessel has internal and external radii of 6 cm and 6·4 cm respectively. Calculate the weight of the vessel if it is made of metal of density $10\,g/cm^3$.

19. Water is flowing into an inverted cone, of diameter and height 30 cm, at a rate of 4 litres per minute. How long, in seconds, will it take to fill the cone?

20. Builders' sand is tipped into the shape of a cone where the height and radius are both 1·4 m.
(a) Calculate the volume of the sand, correct to 1 d.p.
(b) A further $2\,m^3$ is added to the pile. This now forms a larger cone, but the height still equals the radius. Calculate the height of this larger pile correct to 1 d.p.

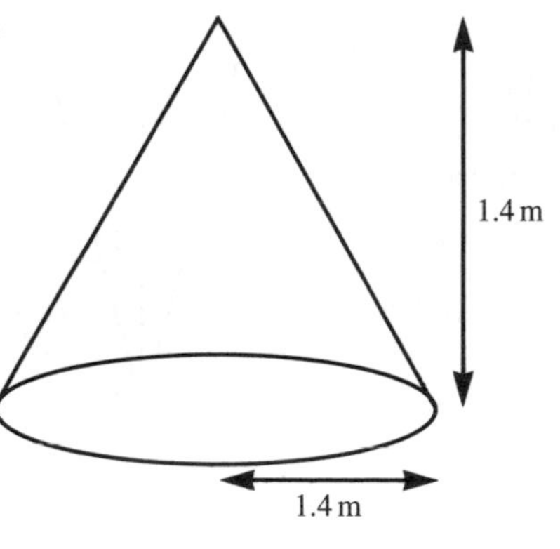

21. Metal spheres of radius 2 cm are packed into a rectangular box of internal dimensions 16 cm × 8 cm × 8 cm. When 16 spheres are packed the box is filled with a preservative liquid. Find the volume of this liquid.

22. A solid metal sphere is recast into many smaller spheres. Calculate the number of the smaller spheres if the initial and final radii are as follows:
(a) initial radius = 10 cm, final radius = 2 cm
(b) initial radius = 1 m, final radius = $\frac{1}{3}$ cm.

23. A spherical ball is immersed in water contained in a vertical cylinder.

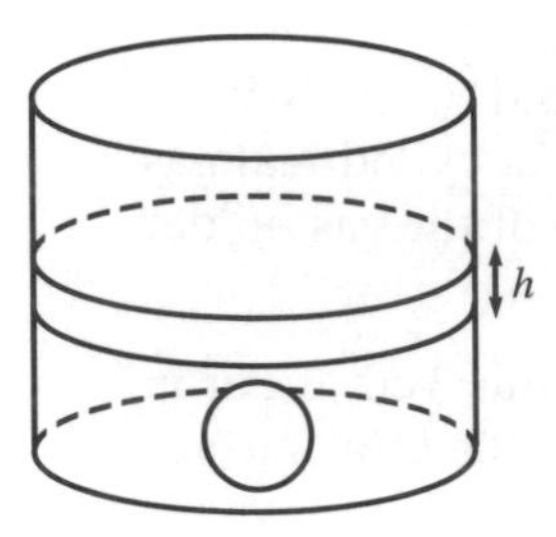

Assuming the water covers the ball, calculate the rise in the water level if
(a) sphere radius = 3 cm, cylinder radius = 10 cm
(b) sphere radius = 2 cm, cylinder radius = 5 cm.

24. A spherical ball is immersed in water contained in a vertical cylinder. The rise in water level is measured in order to calculate the radius of the spherical ball. Calculate the radius of the ball in the following cases:
(a) cylinder of radius 10 cm, water level rises 4 cm.
(b) cylinder of radius 100 cm, water level rises 8 cm.

25. The diagram shows the cross section of an inverted cone of height MC = 12 cm. If AB = 6 cm and XY = 2 cm, use similar triangles to find the length NC. Hence find the volume of the cone of height NC.

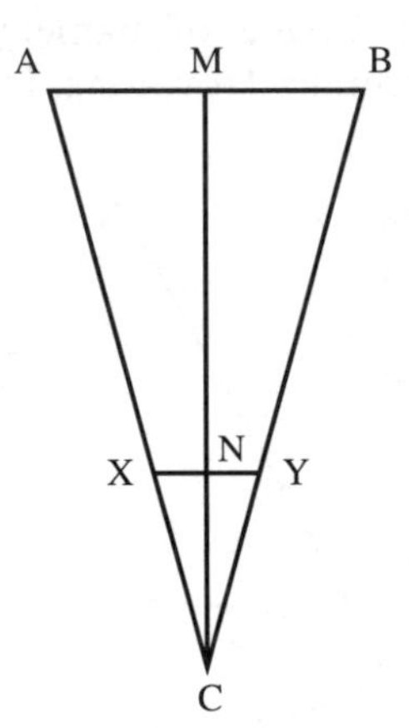

26. The diagram shows a sector of a circle of radius 10 cm.
(a) Find, as a multiple of π, the arc length of the sector.
The straight edges are brought together to make a cone.
Calculate
(b) the radius of the base of the cone.
(c) the vertical height of the cone.

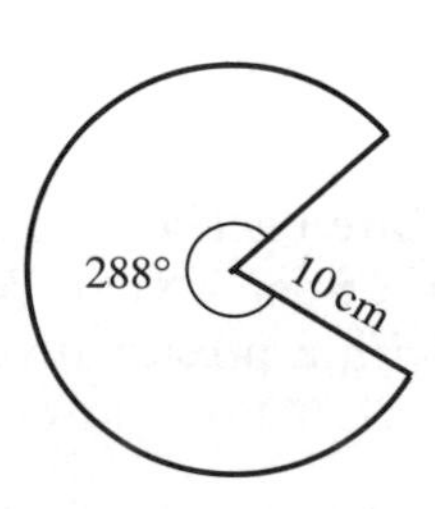

27.† A sphere passes through the eight corners of a cube of side 10 cm. Find the volume of the sphere.

5.4 Dimensions of formulas

Here are some formulas for finding volumes, areas and lengths met earlier in this book.

sphere: volume $= \frac{4}{3}\pi \boldsymbol{r}^3$

surface area $= 4\pi \boldsymbol{r}^2$

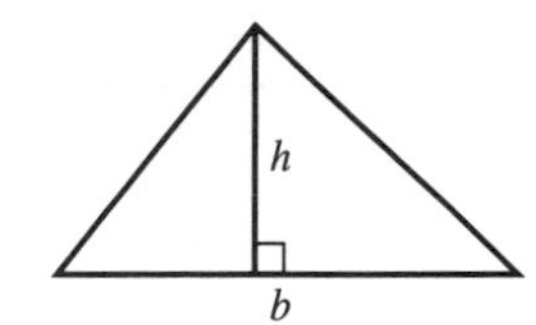

triangle: area $= \frac{1}{2}\boldsymbol{bh}$

cylinder: volume $= \pi \boldsymbol{r}^2\boldsymbol{h}$

curved surface area $= 2\pi \boldsymbol{rh}$

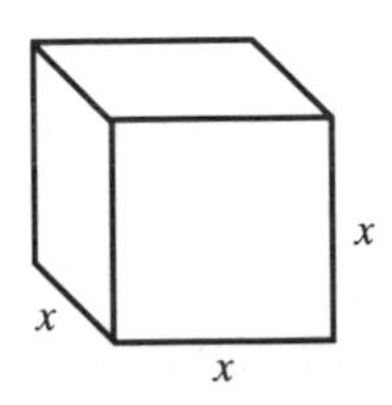

cube: volume $= \boldsymbol{x}^3$

surface area $= 6\boldsymbol{x}^2$

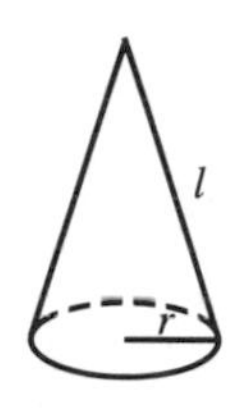

cone: volume $= \frac{1}{3}\pi \boldsymbol{r}^2\boldsymbol{h}$

curved surface area $= \pi \boldsymbol{rl}$

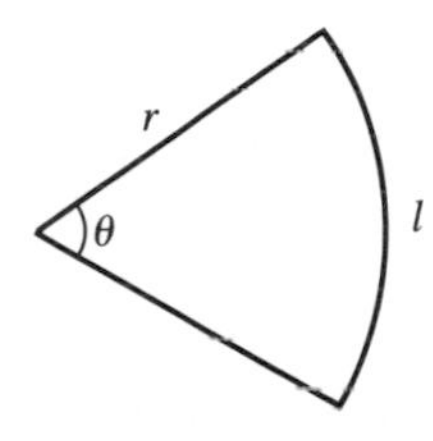

sector area $= \frac{\theta}{360} \times \pi \boldsymbol{r}^2$

arc length $= \frac{\theta}{360} \times 2\pi \boldsymbol{r}$

All the symbols in bold type are lengths. They have the *dimension* of length and are measured in cm, metres, km, etc. The other symbols are numbers (including π) or angles and have no dimensions.

(a) It is not hard to see that all the formulas for volume have *three* lengths multiplied together. They have three dimensions.
(b) All the formulas for area have *two* lengths multiplied together. They have two dimensions.
(c) Any formula for the length of an object will involve just *one* length (or one dimension).

It is quite possible that a formula can have more than one term. The formula $A = \pi r^2 + 3rd$ has two terms and each term has two dimensions.
It is *not* possible to have a mixture of terms some with, say, two dimensions and some with three dimensions.
So the formula $A = \pi r^2 + 3r^2 d$ could not possibly represent area. Nor could it represent volume.

The formula $z = \frac{2\pi r^2 h}{L}$ has three dimensions on the top line and one

dimension on the bottom. The dimensions can be 'cancelled' so the expression for z has only two dimensions and can only represent an area.
We can use these facts to check that any formula we may be using has the correct number of dimensions.

Example 1

Here are four formulas where the letters c, d, r represent lengths:

(a) $t = 3c^2$ (b) $k = \frac{\pi}{3}r^3 + 4r^2d$

(c) $m = \pi(c + d)$ (d) $f = 4c + 3cd$

State whether the formula gives:
(i) a length (ii) an area
(iii) a volume (iv) an impossible expression.

(a) $t = 3c^2 = 3c \times c$
This has *two* dimensions so t is an *area*.

(b) $k = \frac{\pi}{3}r^3 + 4r^2d.$
Both $\frac{\pi}{3}r^3$ and $4r^2d$ have *three* dimensions so k is a *volume*.

(c) $m = \pi(c + d) = \pi c + \pi d$
πc is a length and πd is a length.
So m is a length plus a length.
$\therefore$ m is a length.

(d) $f = 4c + 3cd$
$4c$ is a length.
$3cd$ is a length multiplied by a length and is an area.
So f is a length plus an area which is an *impossible* expression.

Exercise 6

The symbols a, b, d, h, l, r represent lengths while x and θ are angles.

1. State the number of dimensions for each of the following:

(a) πl^2 (b) $3\pi lr$ (c) $\frac{\pi}{2}b^2h$

(d) $\pi(a + b)$ (e) $\frac{ab + h^2}{6}$ (f) $abd \sin\theta$

2. Give the number of dimensions that a formula for each of the following should have.

(a) Total area of windows in a room.
(b) Volume of sand in a lorry.
(c) Area of a sports field.
(d) The diagonal of a rectangle.
(e) The capacity of an oil can.
(f) The perimeter of a trapezium.
(g) The number of people in a cinema.
(h) The surface area of the roof of a house.

3. From this list of expressions, choose the two that represent volume, the four that represent area and the one that represents a length. The other expressions are faulty.

(a) $\pi rh + \pi r^2$ (b) $5a + 6c$
(c) $3{\cdot}5\,abd$ (d) $4hl + \pi rh$
(e) $3r^2hl$ (f) $2\pi r(r + h)$
(g) $2(rb^2 + h^3)$ (h) $\frac{\pi}{2}(l + d)^2$

4. In Sam's notes, the formula for the volume of a container was written with the index over the r missing. The formula was

$V = \frac{\pi}{3} r^{\square} h$. What was the missing index?

5. Work out the missing index numbers in these formulas:

(a) Area $= 3(a^{\square} + bd)$ (b) Volume $= \frac{\pi L^{\square}}{3}$

(c) Length $= \frac{\pi r^{\square}}{3}$ (d) Area $= 3(a^{\square} + b^{\square} d)$

(e) Volume $= 2\pi(r^{\square} + b^{\square})$ (f) Area $= \frac{3\pi}{4}(a + b)^{\square}$

6. A physicist worked out a formula for the surface area of a complicated object and got

$S = 3\pi(a + b)^2 \sin x + \frac{\pi}{2} a$

Explain why the formula could not be correct.

5.5 Vectors

A vector quantity has both magnitude and direction; it is used to represent physical quantities such as velocity, force, acceleration, etc. The symbol for a vector is a bold letter, e.g. **a, x**. On a coordinate grid, the magnitude and direction of the vector can be shown by a column vector, e.g. $\begin{pmatrix} 2 \\ 1 \end{pmatrix}$, $\begin{pmatrix} 1 \\ -3 \end{pmatrix}$ where the upper number shows the distance across the page and the lower number shows the distance up the page.

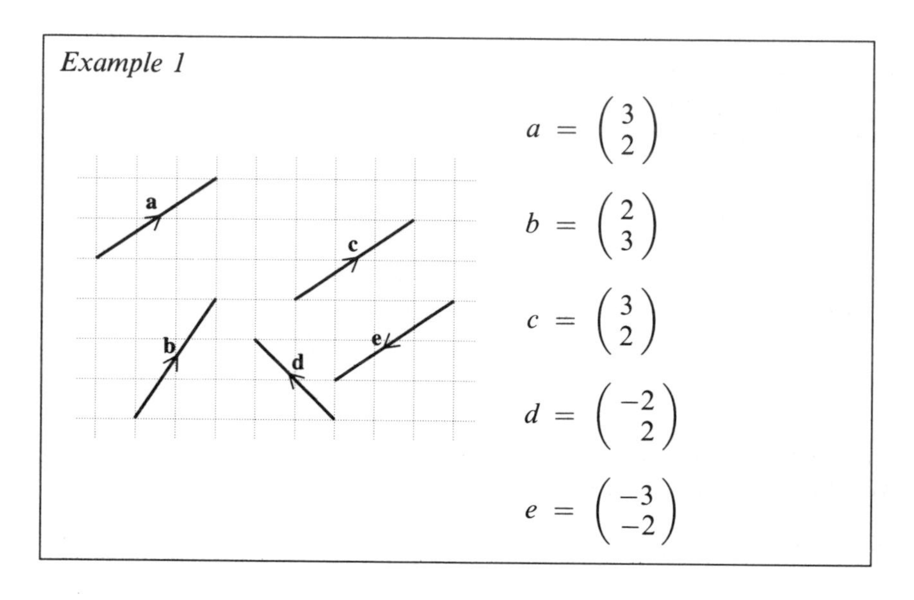

Equal vectors

Two vectors are equal if they have the same length *and* the same direction. The actual position of the vector on the diagram or in space is of no consequence.
Thus in Example 1, vectors **a** and **c** are equal because they have the same magnitude and direction. Even though vector **b** also has the same length (magnitude) as **a** and **c**, it is not equal to **a** or **c** because it acts in a different direction. Likewise, vector **e** has the same length as vector **c** but acts in the reverse direction and so cannot equal **c**.
Equal vectors have identical column vectors.

Addition of vectors

Vectors **a** and **b** are represented by the line segments shown below, they can be added by using the 'nose-to-tail' method to give a single equivalent vector.

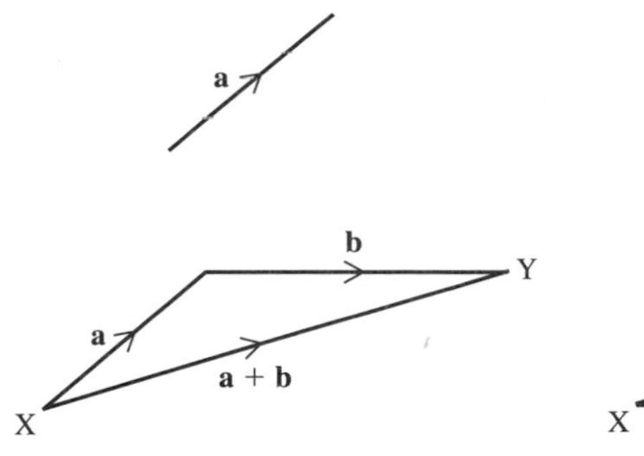

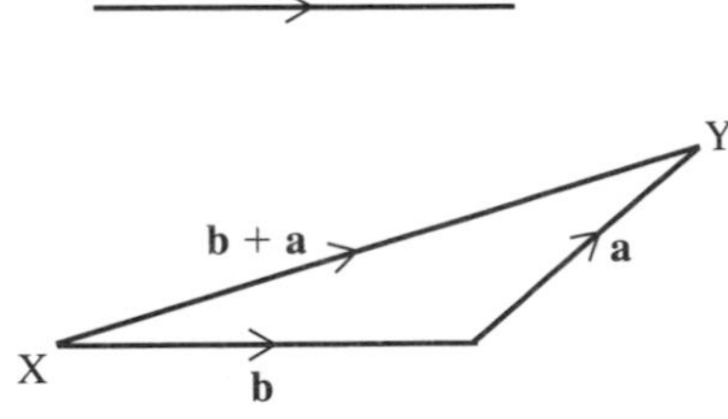

The 'tail' of vector **b** is joined to the 'nose' of vector **a**.

Alternatively the tail of **a** can be joined to the 'nose' of vector **b**.

In both cases the vector $\overrightarrow{XY}$ has the same length and direction and therefore $\mathbf{a} + \mathbf{b} = \mathbf{b} + \mathbf{a}$.

Multiplication by a scalar

A scalar quantity has magnitude but no direction (e.g. mass, volume, temperature). Ordinary numbers are scalars.

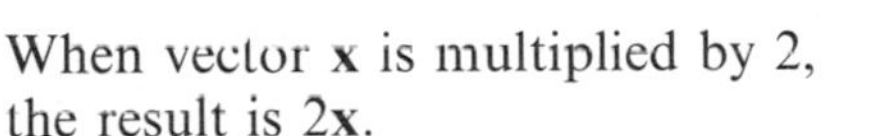

When vector **x** is multiplied by 2, the result is $2\mathbf{x}$.

When **x** is multiplied by -3 the result is $-3\mathbf{x}$.

Note
(1) The negative sign reverses the direction of the vector.
(2) The result $\mathbf{a} - \mathbf{b}$ is $\mathbf{a} + -\mathbf{b}$.
So, subtracting **b** is equivalent to adding the negative of **b**.

Example 2

The diagram on the right shows vectors **a** and **b**. Draw a diagram to show $\overrightarrow{OP}$ and $\overrightarrow{OQ}$ such that

$\overrightarrow{OP} = 3\mathbf{a} + \mathbf{b} \quad \overrightarrow{OQ} = -2\mathbf{a} - 3\mathbf{b}$

a
b

P
b
3a
3a + b
O
−2a − 3b
−2a
Q
−3b

Exercise 7

In Questions **1** to **15**, use the diagram below to describe the vectors given in terms of **c** and **d** where $\mathbf{c} = \overrightarrow{QN}$ and $\mathbf{d} = \overrightarrow{QR}$, e.g. $\overrightarrow{QS} = 2\mathbf{d}$, $\overrightarrow{TD} = \mathbf{c} + \mathbf{d}$.

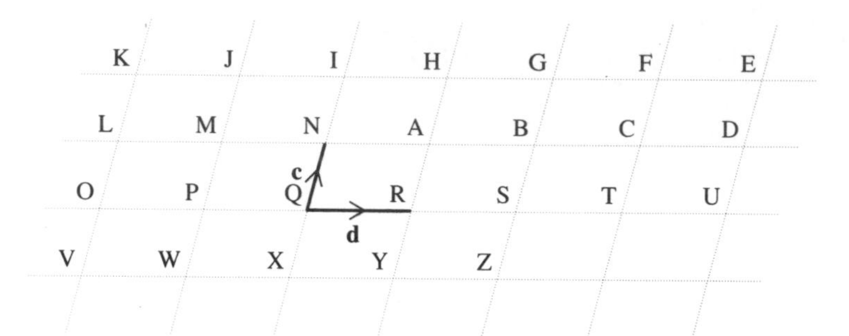

1. $\overrightarrow{AB}$	**2.** $\overrightarrow{SG}$	**3.** $\overrightarrow{VK}$	**4.** $\overrightarrow{KH}$	**5.** $\overrightarrow{OT}$
6. $\overrightarrow{WJ}$	**7.** $\overrightarrow{FH}$	**8.** $\overrightarrow{FT}$	**9.** $\overrightarrow{KV}$	**10.** $\overrightarrow{NQ}$
11. $\overrightarrow{OM}$	**12.** $\overrightarrow{SD}$	**13.** $\overrightarrow{PI}$	**14.** $\overrightarrow{YG}$	**15.** $\overrightarrow{OI}$

In Questions **16** to **21**, use the same diagram above to find vectors for the following in terms of the capital letters, starting from Q each time.

e.g. $3\mathbf{d} = \overrightarrow{QT}$, $\mathbf{c} + \mathbf{d} = \overrightarrow{QA}$.

16. $2\mathbf{c}$ **17.** $4\mathbf{d}$ **18.** $2\mathbf{c} + \mathbf{d}$ **19.** $2\mathbf{d} + \mathbf{c}$ **20.** $3\mathbf{d} + 2\mathbf{c}$ **21.** $2\mathbf{c} - \mathbf{d}$

In Questions **22** to **25**, write each vector in terms of **a**, **b**, or **a** and **b**.

22. (a) $\overrightarrow{BA}$
(b) $\overrightarrow{AC}$
(c) $\overrightarrow{DB}$
(d) $\overrightarrow{AD}$

A **a** B
b
D **2a** C

23. (a) $\overrightarrow{ZX}$
(b) $\overrightarrow{YW}$
(c) $\overrightarrow{XY}$
(d) $\overrightarrow{XZ}$

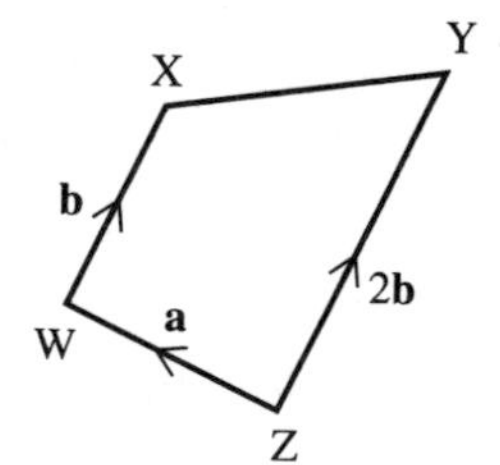

24. (a) $\overrightarrow{MK}$
(b) $\overrightarrow{NL}$
(c) $\overrightarrow{NK}$
(d) $\overrightarrow{KN}$

K **a** L
b
N **3a** M

25. (a) $\overrightarrow{FE}$
(b) $\overrightarrow{BC}$
(c) $\overrightarrow{FC}$
(d) $\overrightarrow{DA}$

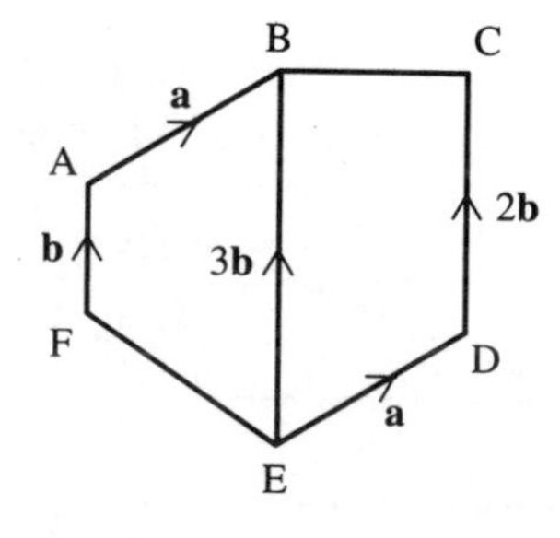

Example 3

In the diagram, OA = AP and BQ = 3OB.
N is the mid-point of PQ;
$\overrightarrow{OA} = \mathbf{a}$ and $\overrightarrow{OB} = \mathbf{b}$.

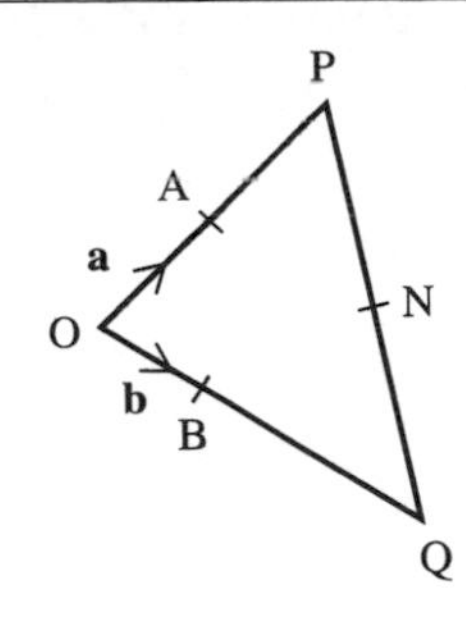

Express each of the following vectors in terms of **a**, **b**, or **a** and **b**.

(a) $\overrightarrow{AP}$ (b) $\overrightarrow{AB}$ (c) $\overrightarrow{OQ}$ (d) $\overrightarrow{PO}$ (e) $\overrightarrow{PQ}$ (f) $\overrightarrow{PN}$ (g) $\overrightarrow{ON}$ (h) $\overrightarrow{AN}$

(a) $\overrightarrow{AP} = \mathbf{a}$ (b) $\overrightarrow{AB} = -\mathbf{a} + \mathbf{b}$ (c) $\overrightarrow{OQ} = 4\mathbf{b}$ (d) $\overrightarrow{PO} = -2\mathbf{a}$

(e) $\overrightarrow{PQ} = \overrightarrow{PO} + \overrightarrow{OQ}$
$= 2\mathbf{a} + 4\mathbf{b}$

(f) $\overrightarrow{PN} = \frac{1}{2}\overrightarrow{PQ}$
$= -\mathbf{a} + 2\mathbf{b}$

(g) $\overrightarrow{ON} = \overrightarrow{OP} + \overrightarrow{PN}$
$= 2\mathbf{a} + (-\mathbf{a} + 2\mathbf{b})$
$= \mathbf{a} + 2\mathbf{b}$

(h) $\overrightarrow{AN} = \overrightarrow{AP} + \overrightarrow{PN}$
$= \mathbf{a} + (-\mathbf{a} + 2\mathbf{b})$
$= 2\mathbf{b}$

Exercise 8

In Questions **1** to **4**, $\overrightarrow{OA} = \mathbf{a}$ and $\overrightarrow{OB} = \mathbf{b}$. Copy each diagram and use the information given to express the following vectors in terms of **a**, **b** or **a** and **b**.

(a) $\overrightarrow{AP}$ (b) $\overrightarrow{AB}$ (c) $\overrightarrow{OQ}$ (d) $\overrightarrow{PO}$ (e) $\overrightarrow{PQ}$
(f) $\overrightarrow{PN}$ (g) $\overrightarrow{ON}$ (h) $\overrightarrow{AN}$ (i) $\overrightarrow{BP}$ (j) $\overrightarrow{QA}$

1. A, B and N are mid-points of OP, OB and PQ respectively.

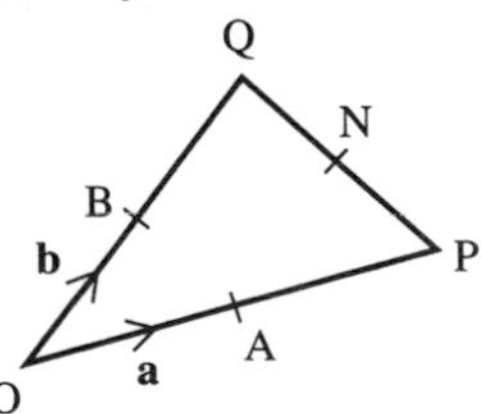

2. A and N are mid-points of OP and PQ; BQ = 2OB.

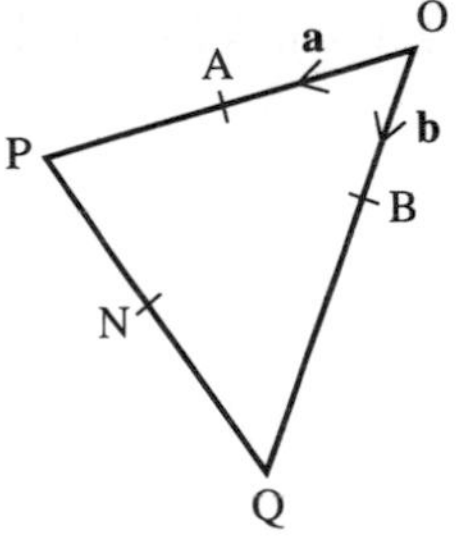

3. AP = 2OA, BQ = OB, PN = NQ.

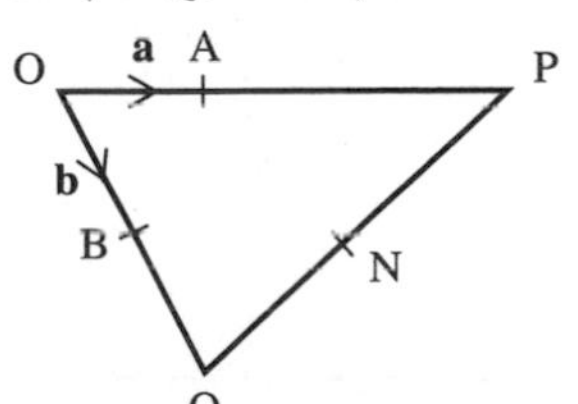

4. OA = 2AP, BQ = 3OB, PN = 2NQ.

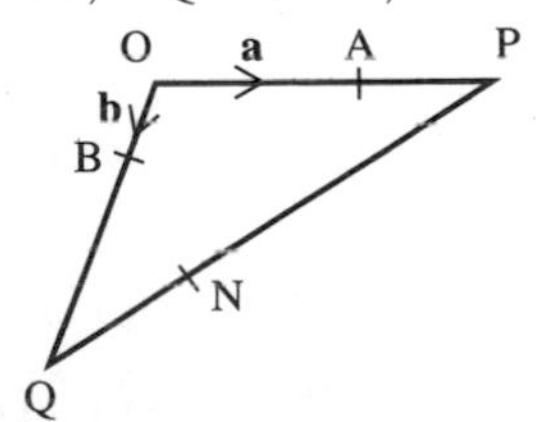

5. In $\triangle XYZ$, the mid-point of YZ is M. If $\overrightarrow{XY} = \mathbf{s}$ and $\overrightarrow{ZX} = \mathbf{t}$, find $\overrightarrow{XM}$ in terms of **s** and **t**.

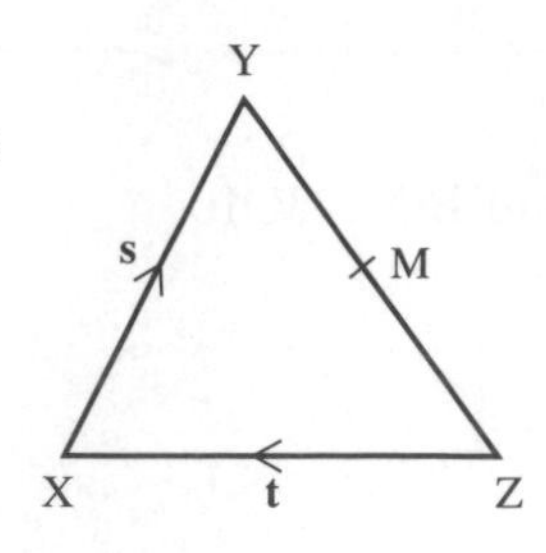

6. In $\triangle AOB$, AM : MB = 2 : 1. If $\overrightarrow{OA} = \mathbf{a}$, and $\overrightarrow{OB} = \mathbf{b}$, find $\overrightarrow{OM}$ in terms of **a** and **b**.

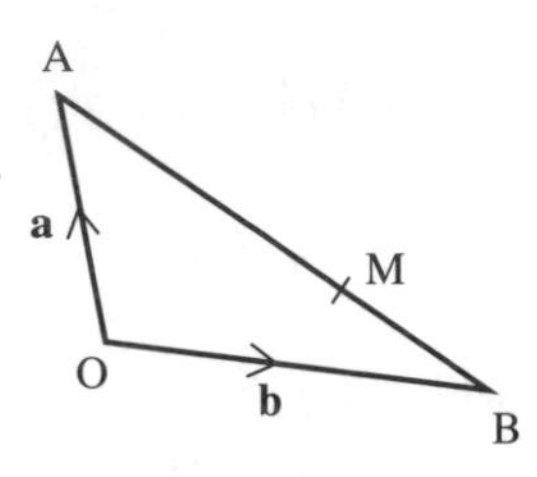

7. O is any point in the plane of the square ABCD. The vectors $\overrightarrow{OA}$, $\overrightarrow{OB}$, and $\overrightarrow{OC}$, are **a, b** and **c** respectively. Find the vector $\overrightarrow{OD}$, in terms of **a, b** and **c**.

8. ABCDEF is a regular hexagon with $\overrightarrow{AB}$, representing the vector **m** and $\overrightarrow{AF}$, representing the vector **n**. Find the vector representing $\overrightarrow{AD}$.

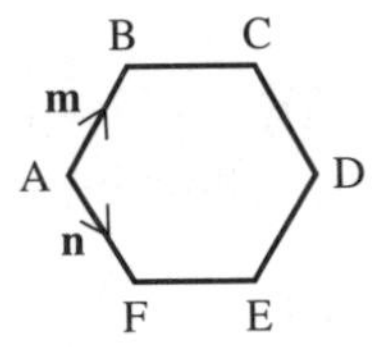

9. ABCDEF is a regular hexagon with centre O. $\overrightarrow{FA} = \mathbf{a}$ and $\overrightarrow{FB} = \mathbf{b}$.

Express the following vectors in terms of **a** and/or **b**.

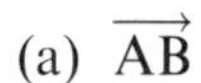

(a) $\overrightarrow{AB}$ (b) $\overrightarrow{FO}$ (c) $\overrightarrow{FC}$

(d) $\overrightarrow{BC}$ (e) $\overrightarrow{AO}$ (f) $\overrightarrow{FD}$

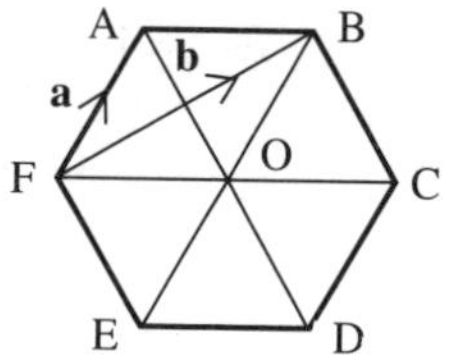

10. In the diagram, M is the mid-point of CD, BP : PM = 2 : 1, $\overrightarrow{AB} = \mathbf{x}$, and $\overrightarrow{AC} = \mathbf{y}$ and $\overrightarrow{AD} = \mathbf{z}$.

Express the following vectors in terms of **x**, **y** and **z**.

(a) $\overrightarrow{DC}$ (b) $\overrightarrow{DM}$ (c) $\overrightarrow{AM}$

(d) $\overrightarrow{BM}$ (e) $\overrightarrow{BP}$ (f) $\overrightarrow{AP}$

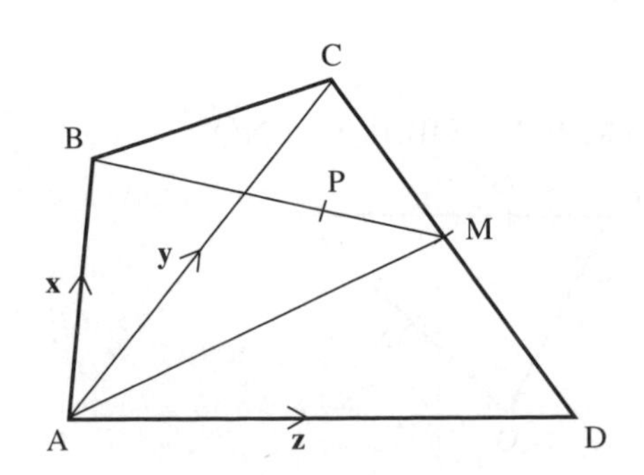

Vector applications

Resultant of forces

Forces are vectors since they have magnitude and direction. So forces can be added just like vectors.

Example 4

Suppose two forces of magnitude 8 N and 13 N act on a mass as shown.
What is the single force which would have the same effect?

First, add the two forces as vectors, joining them 'nose to tail'.
Then calculate the length and direction using Pythagoras' Theorem and a suitable trigonometric ratio.

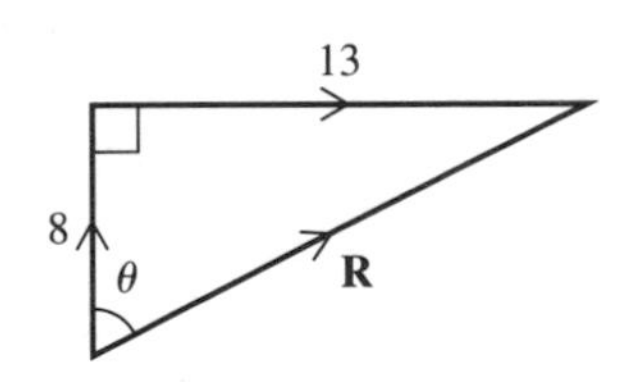

$$\mathbf{R}^2 = 8^2 + 13^2$$
$$\mathbf{R} = 15{\cdot}3\text{ N (3 s.f.)}$$
$$\tan\theta = \frac{13}{8}$$
$$\theta = 58{\cdot}4^\circ \text{ (1 d.p.)}$$

The resultant force is 15·3 N acting at 58·4° to the direction of the 8 N force.

Note. Because force is a vector, both its magnitude and its direction have to be found.

Example 5

Find the resultant of the two forces shown in the diagram.

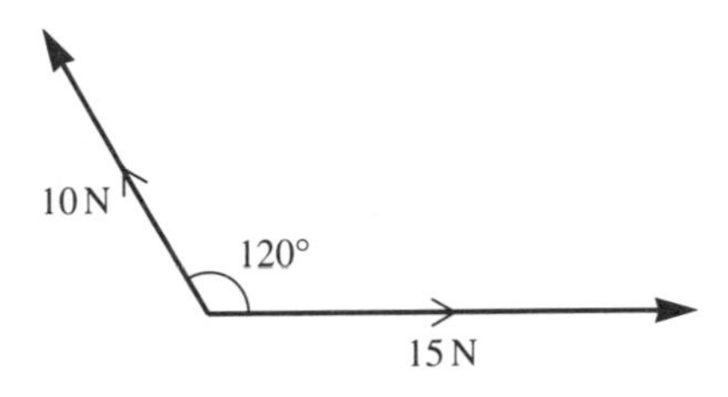

Add the two forces as vectors and consider the triangle formed.

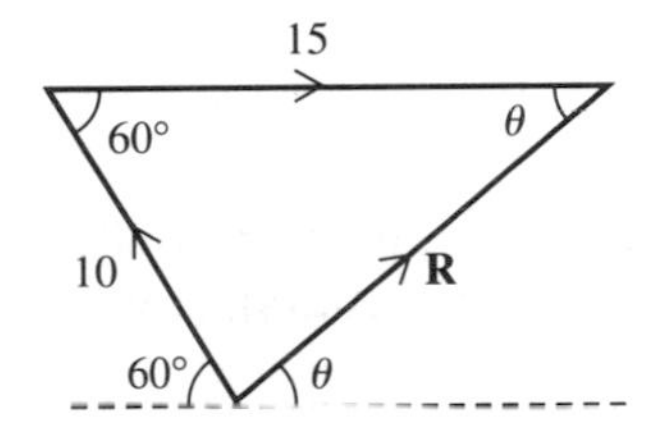

The magnitude and direction of the resultant **R** can be found:

either (a) from a scale drawing,
or (b) by the cosine rule and sine rule (see page 220).

Often the scale drawing method is quicker and quite accurate enough for many problems.

Method (b)

$$\mathbf{R}^2 = 10^2 + 15^2 - 2.10.15\cos 60^\circ$$
$$\Rightarrow \mathbf{R} = 13{\cdot}229$$
$$\frac{\sin\theta}{10} = \frac{\sin 60^\circ}{13{\cdot}229}$$
$$\Rightarrow \theta = 40{\cdot}9^\circ$$

The resultant force is 13·2 N and acts at 40·9° to the direction of the 15 N force.

Exercise 9

1. Find the resultant force. Give the direction of the resultant as a bearing. Up the page is North.

(a)

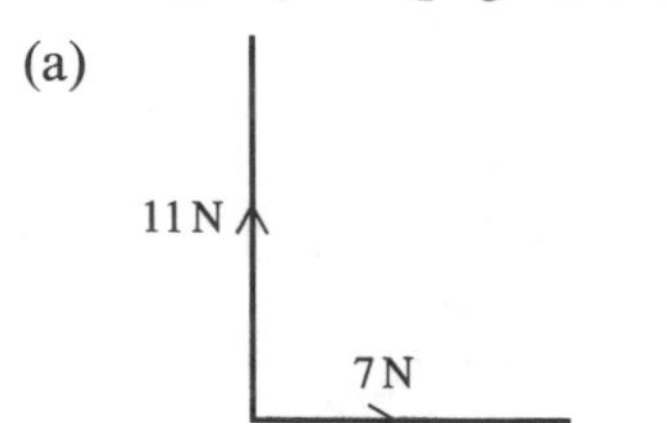

(b)

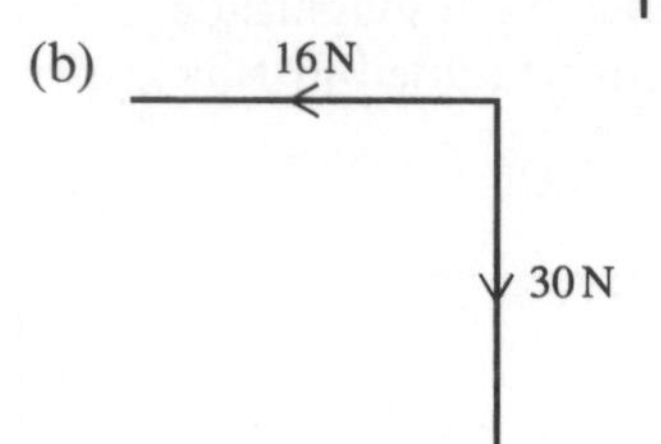

(c)

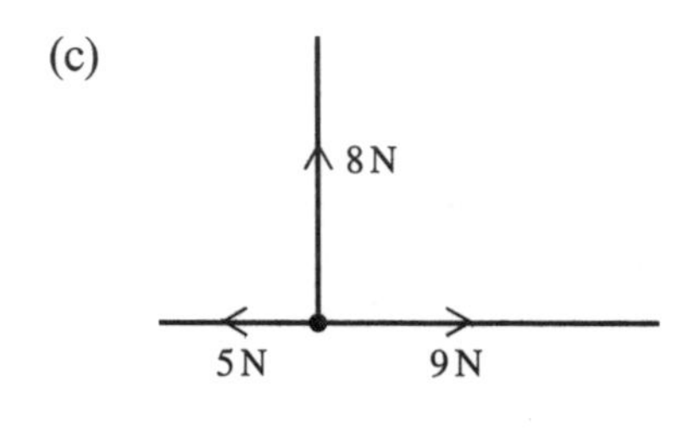

(d)

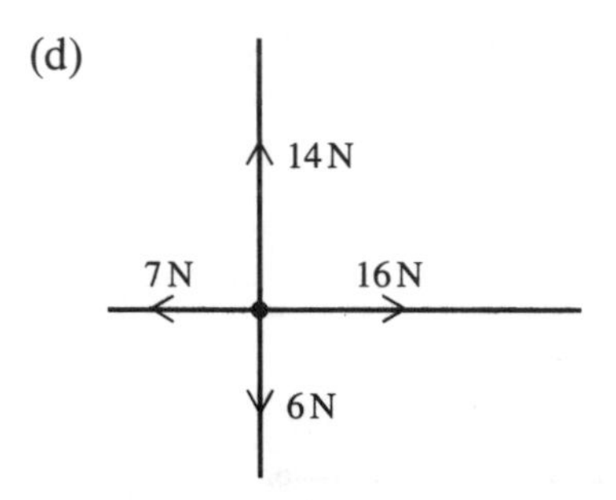

2. Find the resultant by scale drawing (let 1 N = 1 cm). Give the direction of the resultant relative to the vertical which is up the page.

(a) (b)

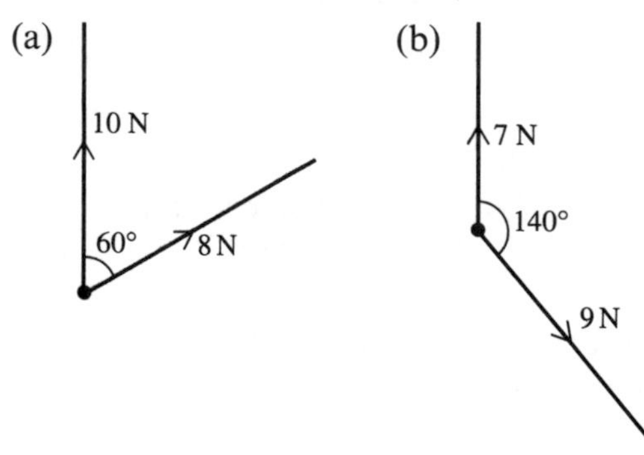

3. Find the resultant by calculation.

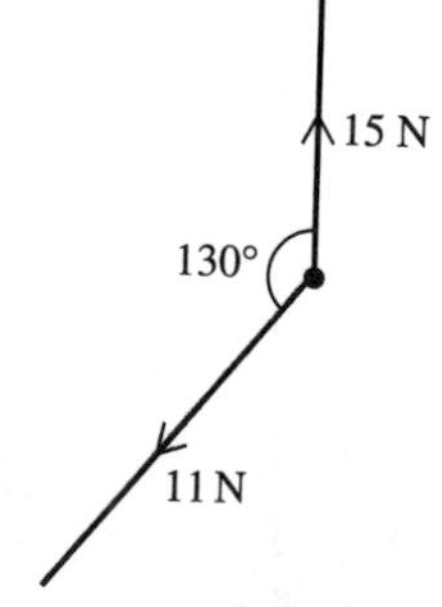

4. Two forces P and Q act on a mass. The resultant of P and Q is a force of magnitude 10 N and acts on a bearing 055°. Force P has magnitude 12 N and acts due North. Find the magnitude and bearing of force Q.

Velocity as a vector

Speed has only magnitude. Velocity has magnitude and direction, so it is a vector quantity.

Example 6

A man wants to row North across a river which flows due east at 3 m/s. He can row in still water at 2 m/s.

If he heads North straight across the river, in which direction will the boat actually move?

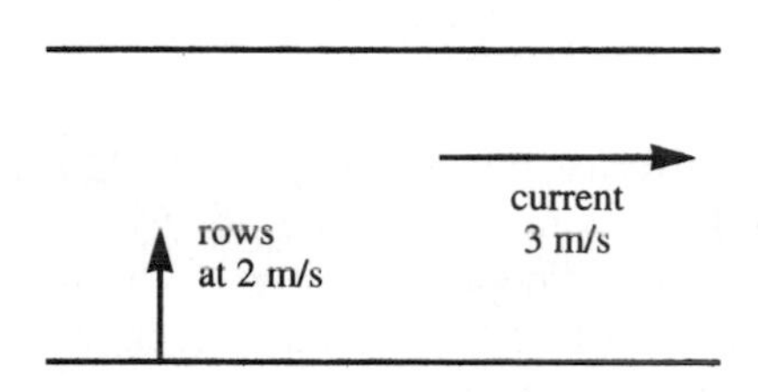

The man in the boat is rowing due North at 2 m/s, so the front of the boat will point due North throughout. We need to find the resultant of two velocities:

(actual boat velocity)
= (rowing velocity) + (water velocity)

We can add the velocities as vectors and then calculate the resultant velocity for the boat's actual route across.

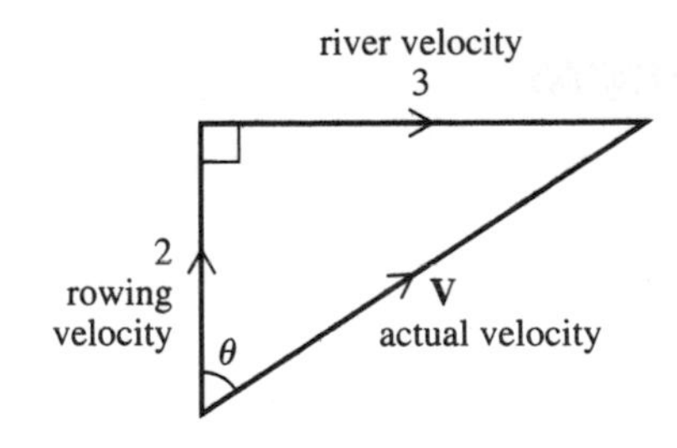

$$\mathbf{V}^2 = 2^2 + 3^2$$
$$\Longrightarrow \mathbf{V} = 3{\cdot}6\,\text{m/s (2 s.f.)}$$
$$\tan\theta = \frac{3}{2}$$
$$\theta = 56^\circ \text{ (nearest degree)}$$

The boat actually moves at 3·6 m/s on a bearing of 056°.

Example 7

An aircraft can fly at 300 mph in still air. The wind is blowing at 40 mph towards the South East. If the aircraft heads due North, what is its actual velocity relative to the ground?

(actual velocity of aircraft)
= (velocity of aircraft relative to air)
+ (velocity of air)

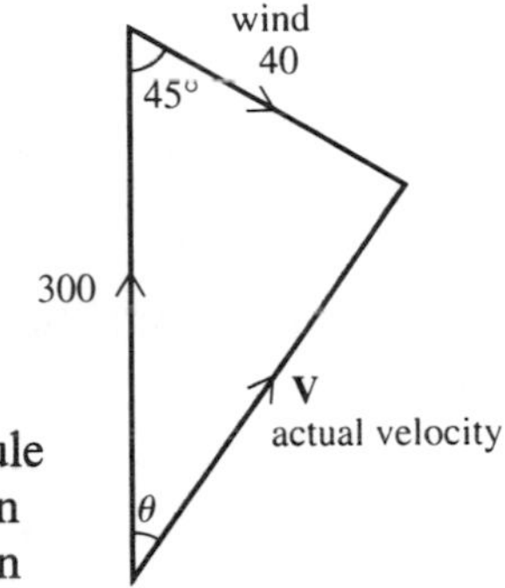

Using the cosine rule and the sine rule on page 220, we obtain

$$\mathbf{V} = 273\,\text{mph (3 s.f.)}$$
$$\theta = 6^\circ \text{ (to the nearest degree)}$$

The actual velocity of the aircraft is 273 mph on a bearing 006°.

Example 8

An aircraft can fly at 200 mph in still air (i.e. its airspeed is 200 mph) and the wind is blowing due west at 35 mph. In what direction should the aircraft head so that its actual velocity is due North?

Here is the velocity diagram:

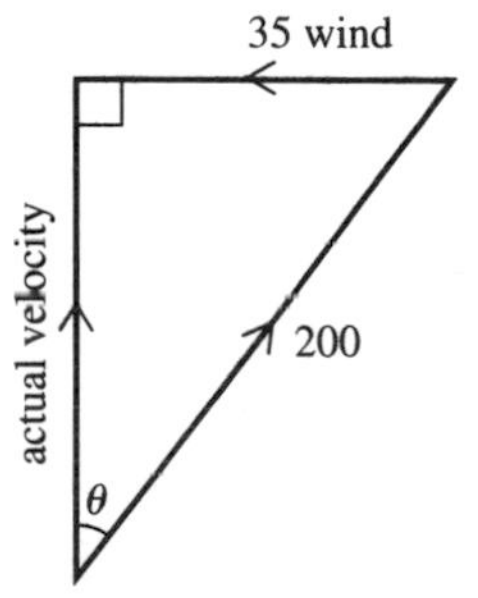

$$\sin\theta = \frac{35}{200}$$
$$\theta = 10^\circ$$
(nearest degree)

The aircraft should head on a bearing of 010° in order to fly due North.

Exercise 10

1. The speed of a boat in still water is 4 m/s. A river flows due East at 2 m/s. Find the actual velocity of the boat if it heads due North. (Give the magnitude and bearing).

2. The speed of an aircraft in still air is 280 km/h. The wind blows at 60 km/h due South. Find the actual velocity of the aircraft if it heads due East (give magnitude and bearing).

3. A boat can sail at 10 knots in still water and is heading South. The current is flowing at 3 knots towards the North East. In what direction and at what speed does the boat actually travel?

4. A river is flowing due East at 2·5 m/s. Sue can swim in still water at 1·5 m/s.

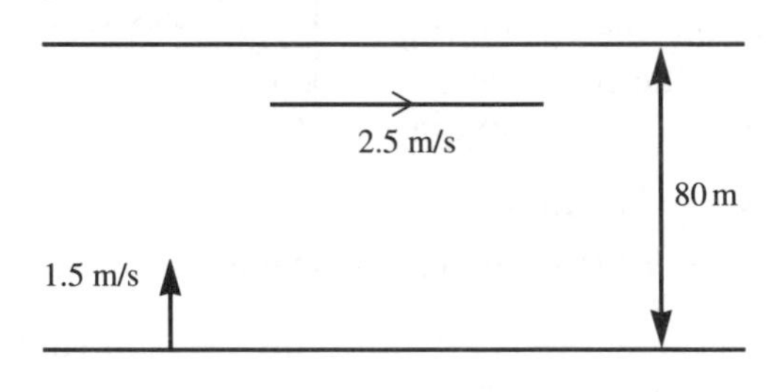

She heads due North across the river which is 80 m wide. How far downstream will she be when she reaches the opposite bank?

5. A ship can sail at 15 knots in still water and there is a current of 6 knots flowing due North. In what direction should the boat head so that it actually sails due East?

6. An aircraft can fly at 260 km/h in still air and the wind is blowing at 70 km/h towards the West. In what direction should the aircraft head so that its actual velocity is on a bearing 030°?

7. An aircraft has an airspeed of 220 km/h and the wind is blowing at 75 km/h from the East. The aircraft has to fly on a bearing of 350°. In what direction should the aircraft head?

8. An aircraft has an airspeed of 300 km/h and heads due North. The actual velocity of the aircraft relative to the ground is 280 km/h on a bearing of 030°. What is the speed of the wind?

5.6 Sine, cosine, tangent for any angle

So far we have used sine, cosine and tangent only in right-angled triangles. For angles greater than 90°, we will see that there is a close connection between trigonometric ratios and circles.

The circle on the right is of radius 1 unit with centre (0, 0). A point P with coordinates (x, y) moves round the circumference of the circle. The angle that OP makes with the positive x-axis as it turns in an anticlockwise direction is θ.

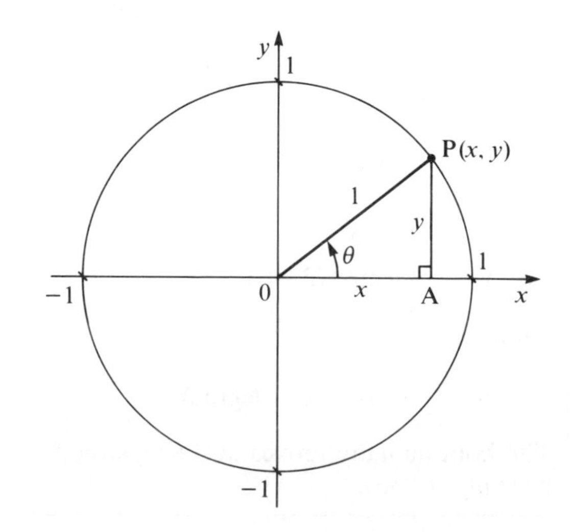

In triangle OAP, $\cos\theta = \frac{x}{1}$ and $\sin\theta = \frac{y}{1}$

The x-coordinate of P is $\cos\theta$
The y-coordinate of P is $\sin\theta$

This idea is used to define the cosine and the sine of any angle, including angles greater than 90°.
Here are two angles that are greater than 90°.

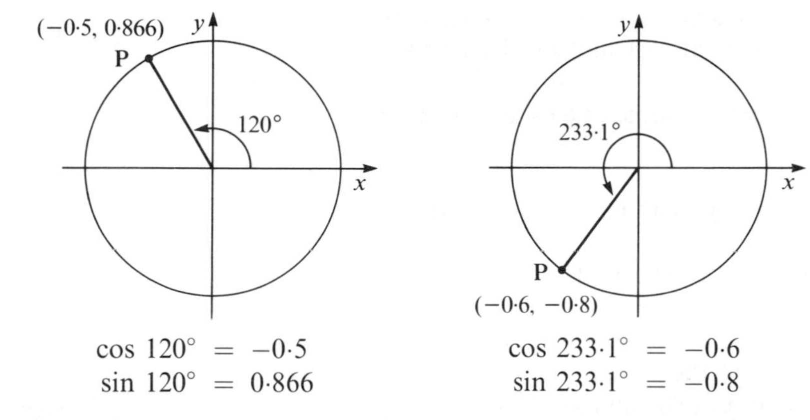

$\cos 120° = -0{\cdot}5$
$\sin 120° = 0{\cdot}866$

$\cos 233{\cdot}1° = -0{\cdot}6$
$\sin 233{\cdot}1° = -0{\cdot}8$

A graphics calculator can be used to show the graph of $y = \sin x$ for any range of angles. The graph below shows $y = \sin x$ for x from 0° to 360°. The curve above the x-axis has symmetry about $x = 90°$ and that below the x-axis has symmetry about $x = 270°$.

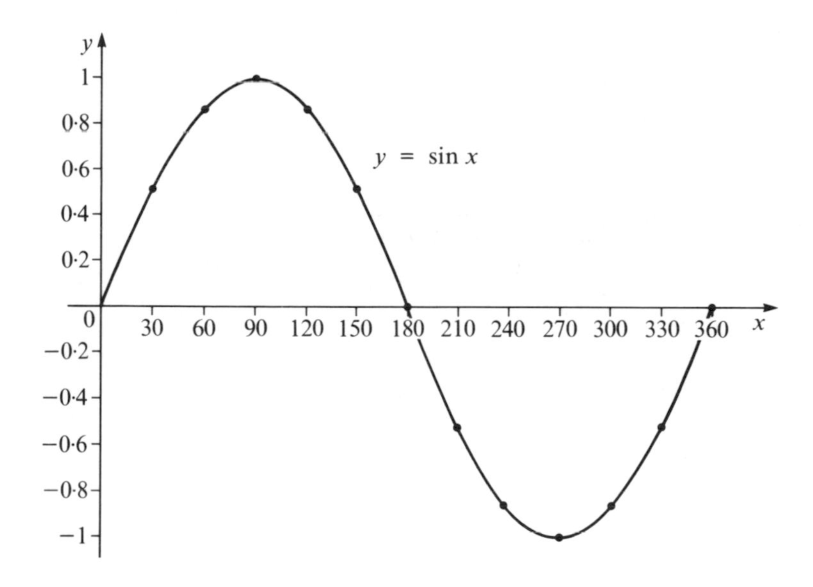

Note
$\sin 150° = \sin 30°$ and $\cos 150° = -\cos 30°$
$\sin 110° = \sin 70°$ $\quad \cos 110° = -\cos 70°$
$\sin 163° = \sin 17°$ $\quad \cos 163° = -\cos 17°$

or $\sin x = \sin (180° - x)$
or $\cos x = -\cos (180° - x)$

These two results are particularly important for use with obtuse angles ($90° < x < 180°$) in Section 5.7 when applying the sine formula or the cosine formula.

Exercise 11

1. (a) Use a calculator to find the cosine of all the angles 0°, 30°, 60°, 90°, 120°, ... 330°, 360°.
(b) Draw a graph of $y = \cos x$ for $0 \leqslant x \leqslant 360°$. Use a scale of 1 cm to 30° on the x-axis and 5 cm to 1 unit on the y-axis.

2. Draw the graph of $y = \sin x$, using the same angles and scales as in Question **1**.

3. Find the tangent of the angles 0°, 20°, 40°, 60°, ... 320°, 340°, 360°.
(a) Notice that we have deliberately omitted 90° and 270°. Why has this been done?
(b) Draw a graph of $y = \tan x$. Use a scale of 1 cm to 20° on the x-axis and 1 cm to 1 unit on the y-axis.
(c) Draw a vertical dotted line at $x = 90°$ and $x = 270°$. These lines are *asymptotes* to the curve. As the value of x approaches 90° from either side, the curve gets nearer and nearer to the asymptote but it *never* quite reaches it.

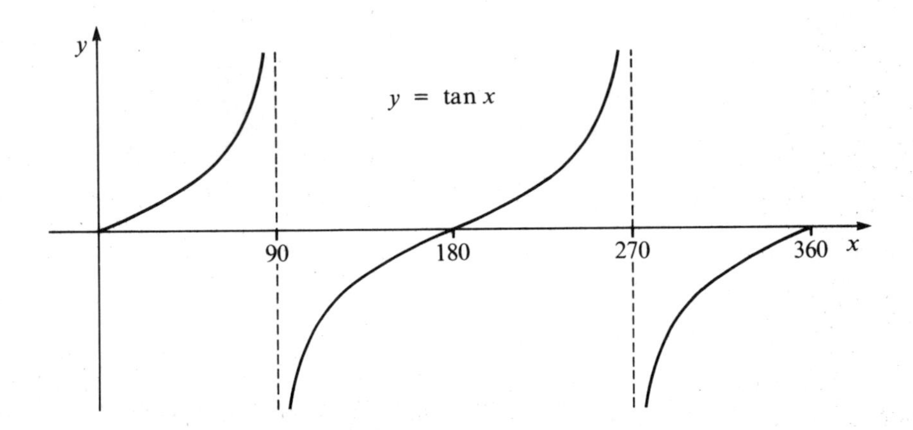

In Questions **4** to **15** do not use a calculator. Use the symmetry of the graphs $y = \sin x$, $y = \cos x$, and $y = \tan x$.
Angles are given to the nearest degree.

4. If sin 18° = 0·309, give another angle whose sine is 0·309.

5. If sin 27° = 0·454, give another angle whose sine is 0·454.

6. Give another angle which has the same sine as
(a) 40° (b) 70° (c) 130°

7. If cos 70° = 0·342, give another angle whose cosine is 0·342.

8. If cos 45° = 0·707, give another angle whose cosine is 0·707.

9. Give another angle which has the same cosine as
(a) $10°$ (b) $56°$ (c) $300°$

10. If $\tan 40° = 0{\cdot}839$, give another angle whose tangent is $0{\cdot}839$.

11. If $\sin 20° = 0{\cdot}342$, what other angle has a sine of $0{\cdot}342$?

12. If $\sin 98° = 0{\cdot}990$, give another angle whose sine is $0{\cdot}990$.

13. If $\tan 135° = -1$, give another angle whose tangent is -1.

14. If $\cos 120° = -0{\cdot}5$, give another angle whose cosine is $-0{\cdot}5$.

15. If $\tan 70° = 2{\cdot}75$, give another angle whose tangent is $2{\cdot}75$.

16. Find two values for x, between $0°$ and $360°$, if $\sin x = 0{\cdot}848$. Give each angle to the nearest degree.

17. If $\sin x = 0{\cdot}35$, find two solutions for x between $0°$ and $360°$.

18. If $\cos x = 0{\cdot}6$, find two solutions for x between $0°$ and $360°$.

19. Find two solutions between $0°$ and $360°$:
(a) $\sin x = 0{\cdot}72$ (b) $\cos x = 0{\cdot}3$
(c) $\tan x = 5$ (d) $\sin x = -0{\cdot}65$

20. Find *four* solutions of the equation $(\sin x)^2 = \frac{1}{4}$ for x between $0°$ and $360°$.

21. Draw the graph of $y = 2\sin x + 1$ for $0 \leqslant x \leqslant 180°$, taking 1 cm to $10°$ for x and 5 cm to 1 unit for y. Find approximate solutions to the equations
(a) $2\sin x + 1 = 2{\cdot}3$
(b) $\dfrac{1}{(2\sin x + 1)} = 0{\cdot}5$

22. Draw the graph of $y = 2\sin x + \cos x$ for $0 \leqslant x \leqslant 180°$, taking 1 cm to $10°$ for x and 5 cm to 1 unit for y.
(a) Solve approximately the equations
  (i) $2\sin x + \cos x = 1{\cdot}5$
  (ii) $2\sin x + \cos x = 0$
(b) Estimate the maximum value of y
(c) Find the value of x at which the maximum occurs.

23. Draw the graph of $y = 3\cos x - 4\sin x$ for $0° \leqslant x \leqslant 220°$, taking 1 cm to $10°$ for x and 2 cm to 1 unit for y.
Solve approximately the equations
(a) $3\cos x - 4\sin x + 1 = 0$
(b) $3\cos x = 4\sin x$

5.7 Sine and cosine rules

The sine rule enables us to calculate sides and angles in some triangles where there is not a right angle.

In triangle ABC, we use the convention that

a is the side opposite $\widehat{A}$
b is the side opposite $\widehat{B}$

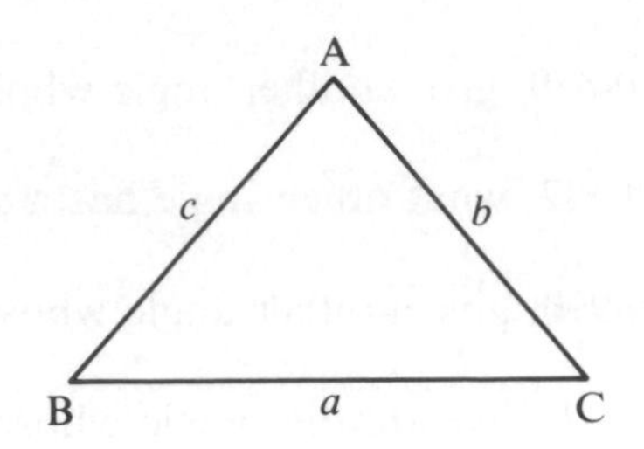

Sine rule

either $$\frac{a}{\sin A} = \frac{b}{\sin B} = \frac{c}{\sin C} \qquad \ldots [1]$$

or $$\frac{\sin A}{a} = \frac{\sin B}{b} = \frac{\sin C}{c} \qquad \ldots [2]$$

Use [1] when finding a *side*, and [2] when finding an *angle*.

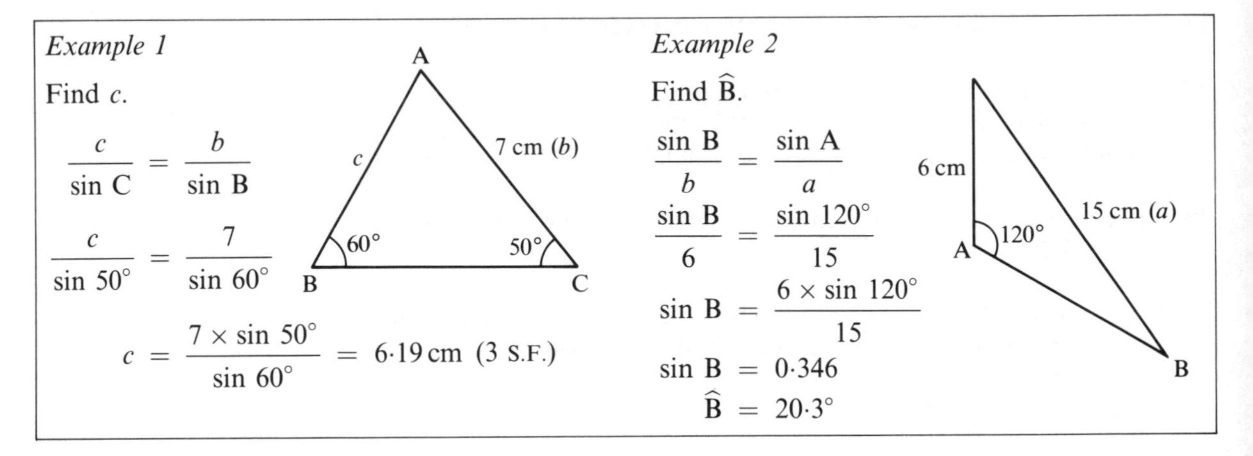

Example 1

Find c.

$$\frac{c}{\sin C} = \frac{b}{\sin B}$$

$$\frac{c}{\sin 50°} = \frac{7}{\sin 60°}$$

$$c = \frac{7 \times \sin 50°}{\sin 60°} = 6{\cdot}19\,\text{cm (3 S.F.)}$$

Example 2

Find $\widehat{B}$.

$$\frac{\sin B}{b} = \frac{\sin A}{a}$$

$$\frac{\sin B}{6} = \frac{\sin 120°}{15}$$

$$\sin B = \frac{6 \times \sin 120°}{15}$$

$$\sin B = 0{\cdot}346$$

$$\widehat{B} = 20{\cdot}3°$$

Reminder
For an obtuse angle x, we have $\sin x = \sin(180° - x)$

Exercise 12

For Questions **1** to **6**, find each side marked with a letter.

1.

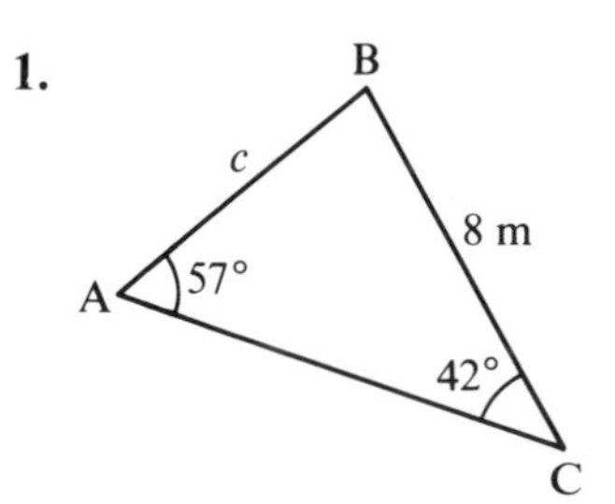

2.

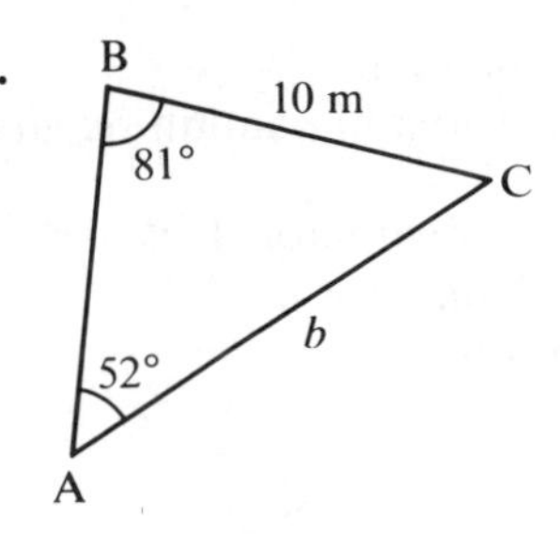

3.

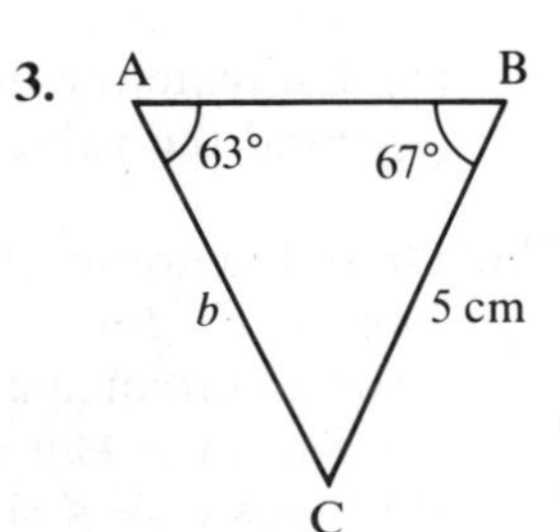

4.

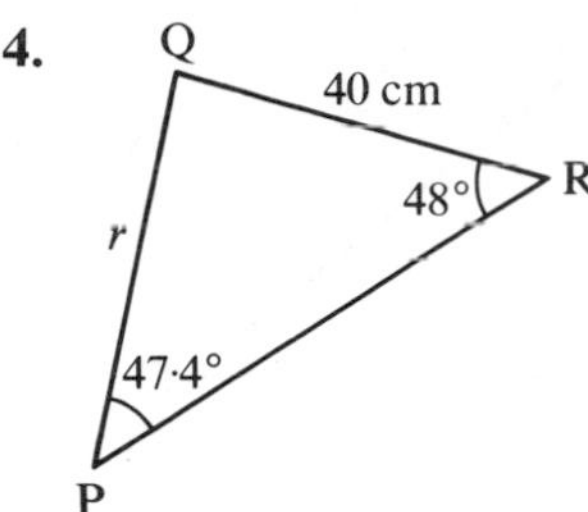

5.

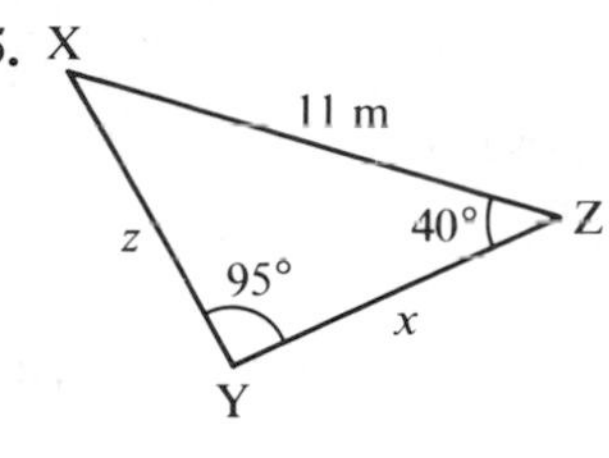

6.

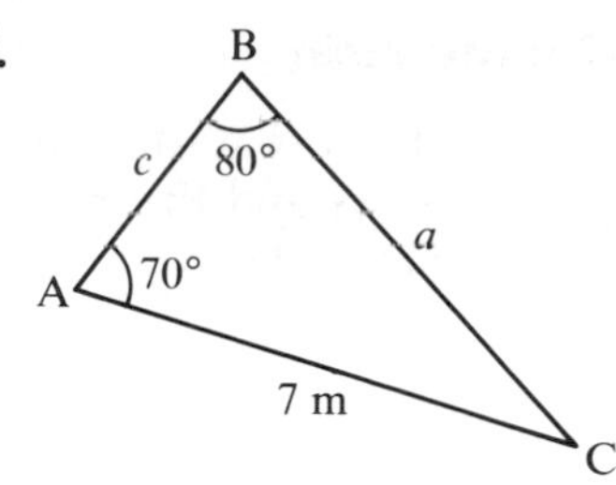

7. In $\triangle$ABC, $\widehat{A} = 61°$, $\widehat{B} = 47°$, AC = 7·2 cm. Find BC.

8. In $\triangle$XYZ, $\widehat{Z} = 32°$, $\widehat{Y} = 78°$, XY = 5·4 cm. Find XZ.

9. In $\triangle$PQR, $\widehat{Q} = 100°$, $\widehat{R} = 21°$, PQ = 3·1 cm. Find PR.

10. In $\triangle$LMN, $\widehat{L} = 21°$, $\widehat{N} = 30°$, MN = 7 cm. Find LN.

In Questions **11** to **18**, find each angle marked *. All lengths are in centimetres.

11.

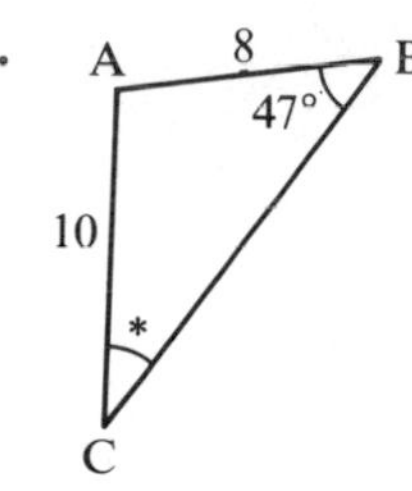

12.

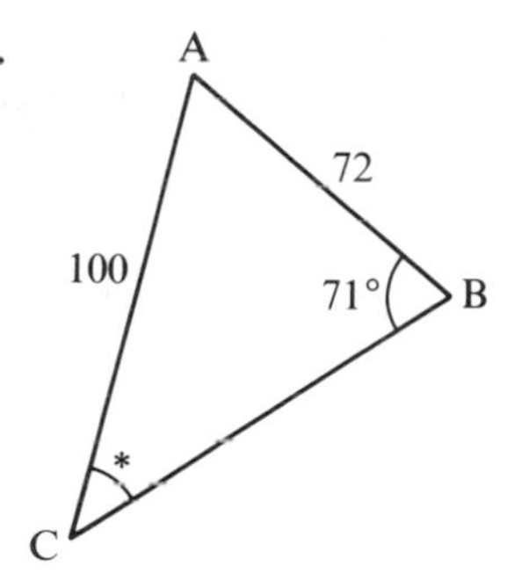

13.

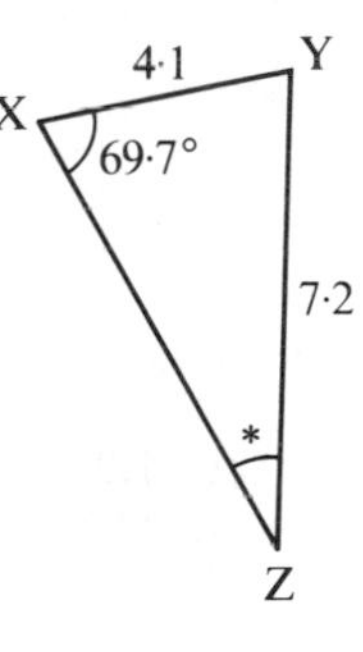

14.

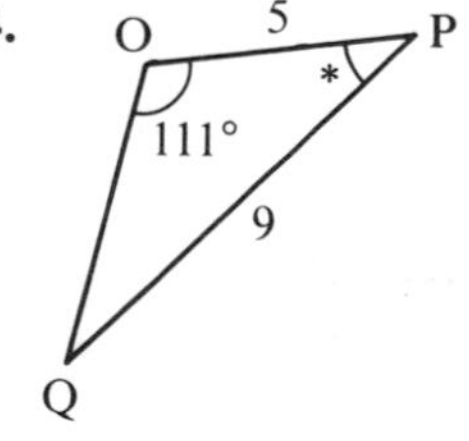

15.

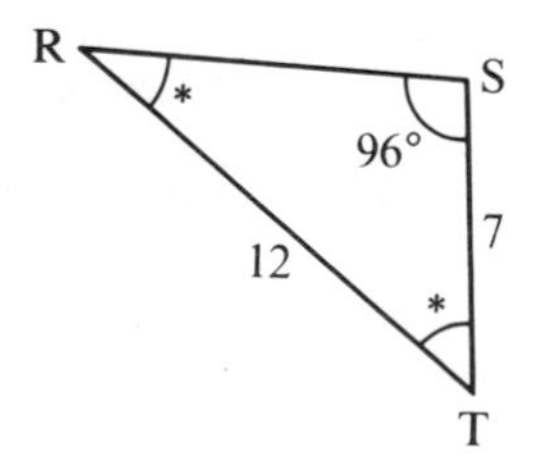

16.

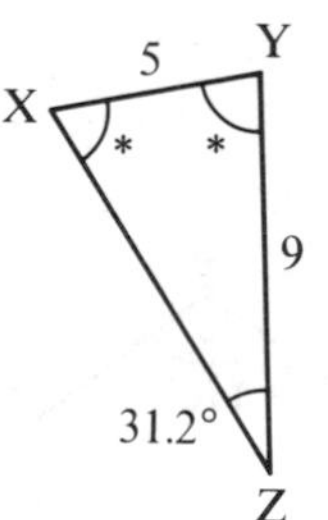

17.

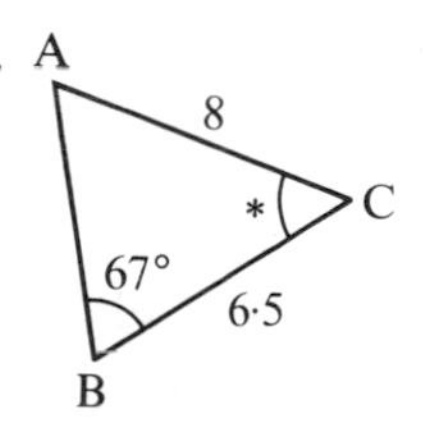

18.

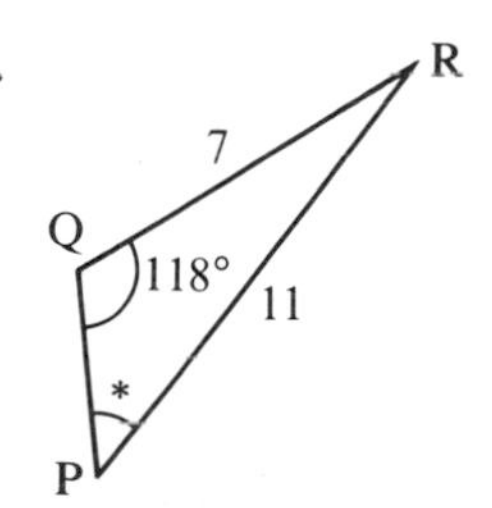

19. In $\triangle$ABC, $\widehat{A} = 62°$, BC = 8, AB = 7. Find $\widehat{C}$.

20. In $\triangle$XYZ, $\widehat{Y} = 97·3°$, XZ = 22, XY = 14. Find $\widehat{Z}$.

21. In $\triangle$DEF, $\widehat{D} = 58°$, EF = 7·2, DE = 5·4. Find $\widehat{F}$.

22. In $\triangle$LMN, $\widehat{M} = 127·1°$, LN = 11·2, LM = 7·3. Find $\widehat{L}$.

Cosine rule

We use the cosine rule when we know either
(a) two sides and the included angle or
(b) all three sides.

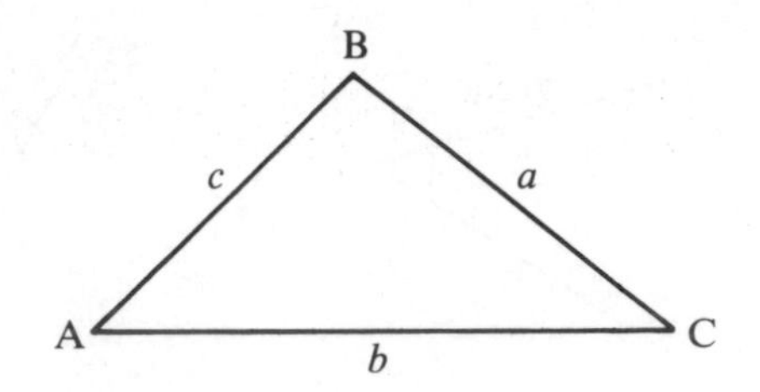

There are two forms.

1. To find the length of a side.

$$a^2 = b^2 + c^2 - (2bc \cos A)$$

or $$b^2 = c^2 + a^2 - (2ac \cos B)$$

or $$c^2 = a^2 + b^2 - (2ab \cos C)$$

2. To find an angle when given all three sides.

$$\cos A = \frac{b^2 + c^2 - a^2}{2bc}$$

or $$\cos B = \frac{a^2 + c^2 - b^2}{2ac}$$

or $$\cos C = \frac{a^2 + b^2 - c^2}{2ab}$$

For an obtuse angle x we have $\cos x = -\cos(180° - x)$

Examples $\cos 120° = -\cos 60°$
$\cos 142° = -\cos 38°$

Example 3

Find b.

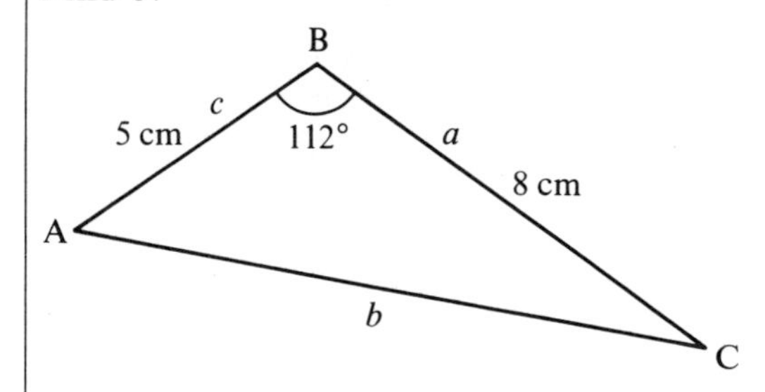

$$b^2 = a^2 + c^2 - (2ac \cos B)$$
$$b^2 = 8^2 + 5^2 - (2 \times 8 \times 5 \times \cos 112°)$$
$$b^2 = 64 + 25 - [80 \times (-0{\cdot}3746)]$$
$$b^2 = 64 + 25 + 29{\cdot}968$$

(Notice the change of sign for the obtuse angle)

$b = \sqrt{(118{\cdot}968)} = 10{\cdot}9$ cm (to 3 S.F.)

Example 4

Find angle C.

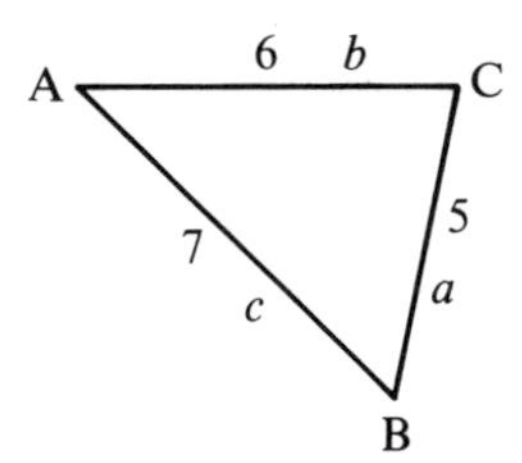

$$\cos C = \frac{a^2 + b^2 - c^2}{2ab}$$

$$\cos C = \frac{5^2 + 6^2 - 7^2}{2 \times 5 \times 6} = \frac{12}{60} = 0{\cdot}200$$

$\hat{C} = 78{\cdot}5°$ (to 1 d.p.)

Exercise 13

Find the sides marked $*$.

1.

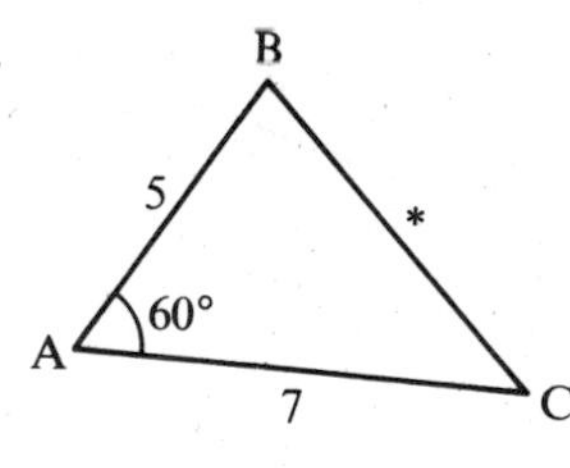

2.

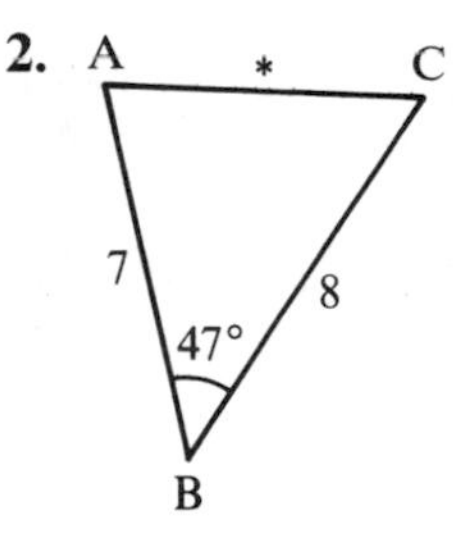

3.

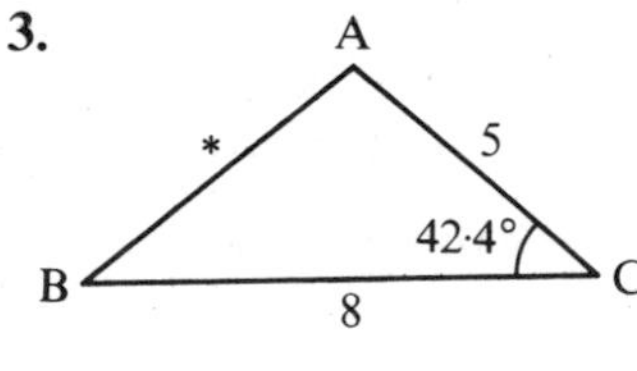

4.

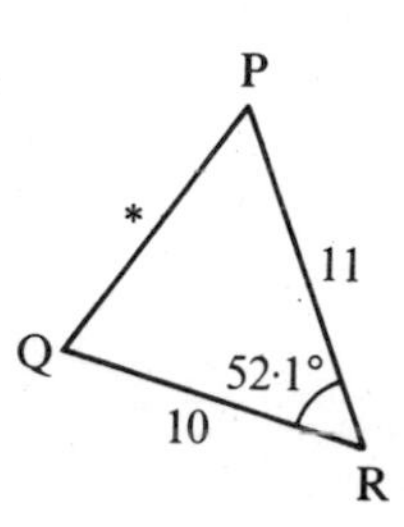

5.

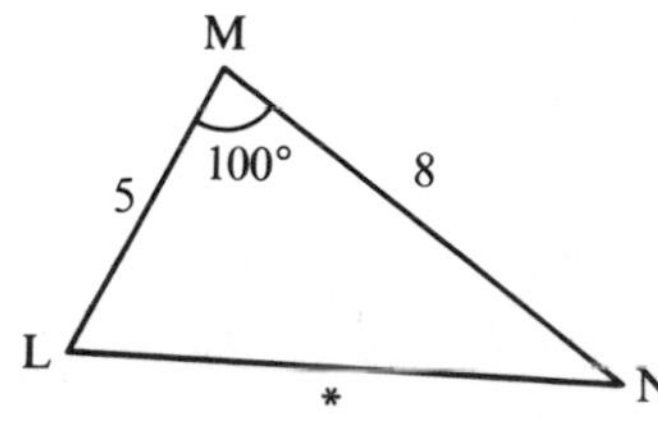

6.

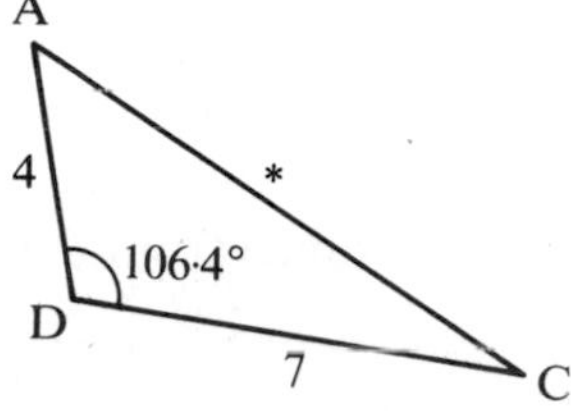

7. In $\triangle$ABC, AB $= 4$ cm, AC $= 7$ cm, $\widehat{\text{A}} = 57°$. Find BC.

8. In $\triangle$XYZ, XY $= 3$ cm, YZ $= 3$ cm, $\widehat{\text{Y}} = 90°$. Find XZ.

9. In $\triangle$LMN, LM $= 5{\cdot}3$ cm, MN $= 7{\cdot}9$ cm, $\widehat{\text{M}} = 127°$. Find LN.

10. In $\triangle$PQR, $\widehat{\text{Q}} = 117°$, PQ $= 80$ cm, QR $= 100$ cm. Find PR.

In Questions **11** to **16**, find the angles marked $*$.

11.

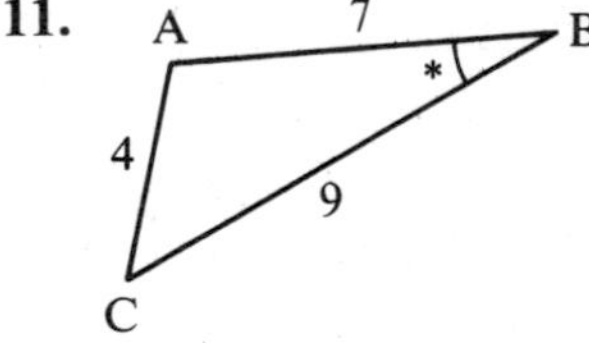

12.

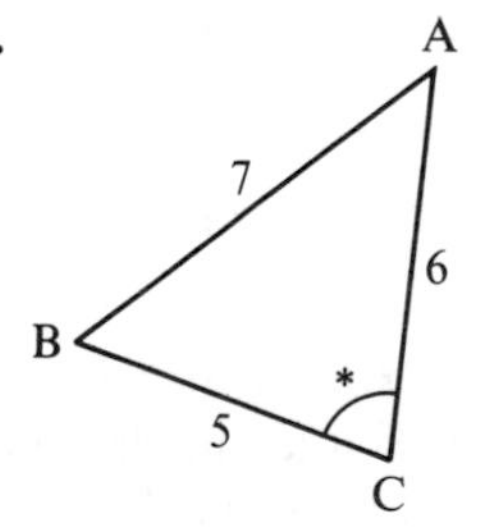

13.

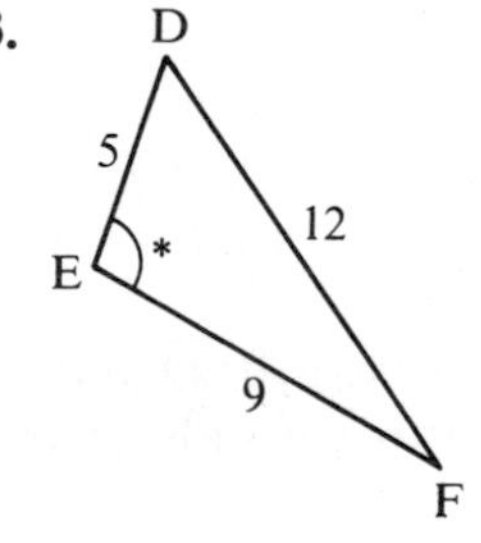

14.

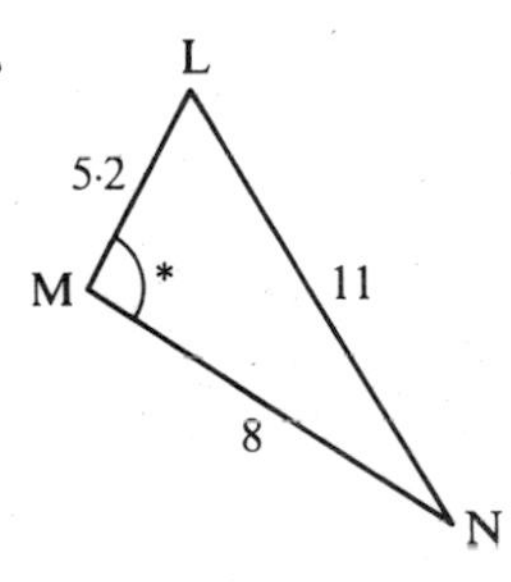

15.

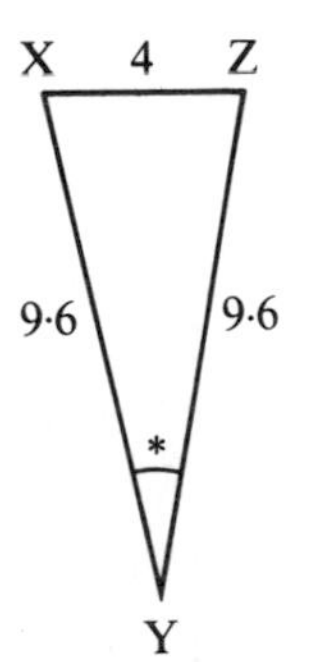

16.

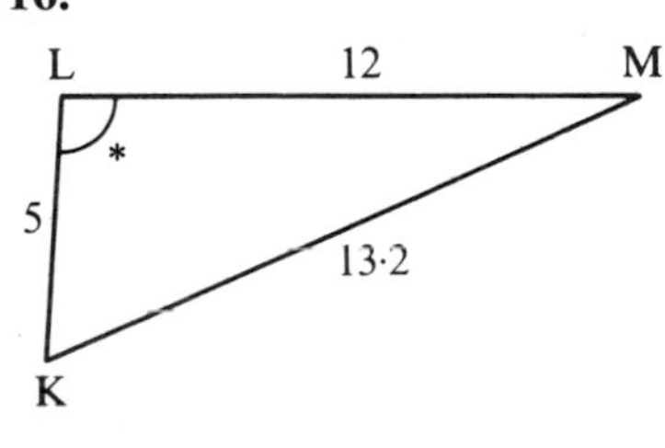

17. In $\triangle$ABC, $a = 4{\cdot}3$, $b = 7{\cdot}2$, $c = 9$. Find $\widehat{\text{C}}$.

18. In $\triangle$DEF, $d = 30$, $e = 50$, $f = 70$. Find $\widehat{\text{E}}$.

19. In $\triangle$PQR, $p = 8$, $q = 14$, $r = 7$. Find $\widehat{\text{Q}}$.

20. In $\triangle$LMN, $l = 7$, $m = 5$, $n = 4$. Find $\widehat{\text{N}}$.

Example 5

A ship sails from a port P a distance of 7 km on a bearing of 306° and then a further 11 km on a bearing of 070° to arrive at X. Calculate the distance from P to X.

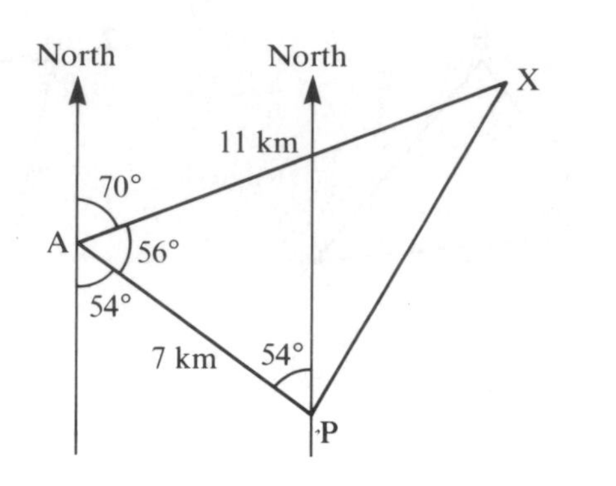

We know two sides and the included angle. Using the cosine rule gives

$$PX^2 = 7^2 + 11^2 - (2 \times 7 \times 11 \times \cos 56°)$$
$$= 49 + 121 - (86{\cdot}12)$$
$$PX^2 = 83{\cdot}88$$
$$PX = 9{\cdot}16\,\text{km} \quad (\text{to 3 S.F.})$$

The distance from P to X is 9·16 km.

Exercise 14

1. Ship B is 58 km south-east of Ship A. Ship C is 70 km due south of Ship A.
   (a) How far is Ship B from Ship C?
   (b) What is the bearing of Ship B from Ship C?

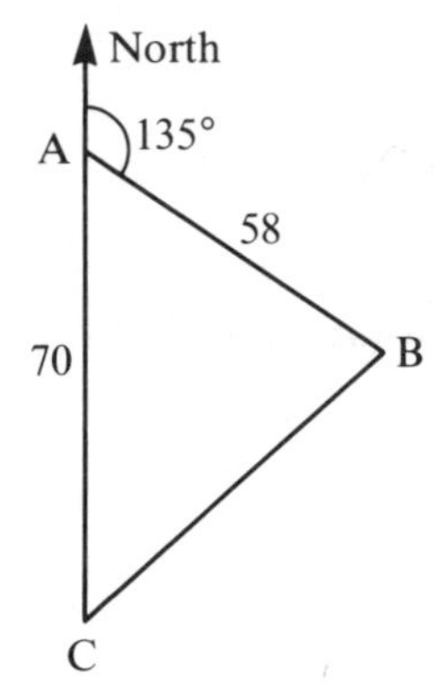

2. A destroyer D and a cruiser C leave port P at the same time. The destroyer sails 25 km on a bearing 040° and the cruiser sails 30 km on a bearing of 320°. How far apart are the ships?
3. Two honeybees A and B leave the hive H at the same time; A flies 27 m due south and B flies 9 m on a bearing of 111°. How far apart are they?
4. The diagram shows a point A which lies 10 km due south of a point B. A straight road AD is such that the bearing of D from A is 043°.
   P and Q are two points on this road which are 8 km from B.
   Calculate the bearing of P from B.
   (Hint: Find angle BPA first)

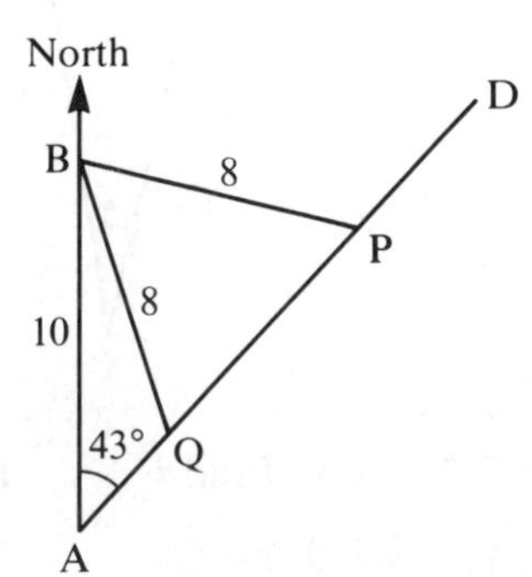

5. Find the largest angle in a triangle in which the sides are in the ratio 5 : 6 : 8.

6. A golfer hits his ball B a distance of 170 m on a hole H which measures 195 m from the tee T to the green. If his shot is directed 10° away from the true line to the hole, find the distance between his ball and the hole.

7. From A, B lies 11 km away on a bearing of 041° and C lies 8 km away on a bearing of 341°. Find
(a) the distance between B and C
(b) the bearing of B from C.

8. From a lighthouse L an aircraft carrier A is 15 km away on a bearing of 112° and a submarine S is 26 km away on a bearing of 200°. Find
(a) the distance between A and S
(b) the bearing of A from S.

9. Find (a) AE
(b) $E\hat{A}C$
If the line BCD is horizontal, find the angle of elevation of E from A.

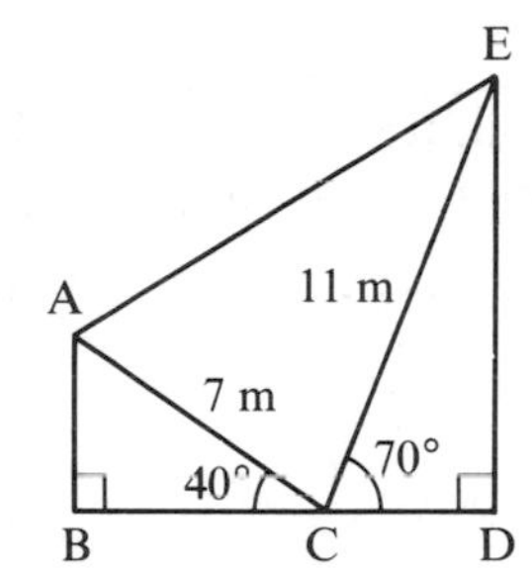

10. An aircraft flies from its base 200 km on a bearing 162°, then 350 km on a bearing 260°, and then returns directly to base. Calculate the length and bearing of the return journey.

11. Town Y is 9 km due north of town Z. Town X is 8 km from Y, 5 km from Z and somewhere to the west of the line YZ.
(a) Draw triangle XYZ and find angle YZX
(b) During an earthquake, town X moves due south until it is due west of Z. Find how far it has moved.

12. Calculate WX, given YZ = 15 m.

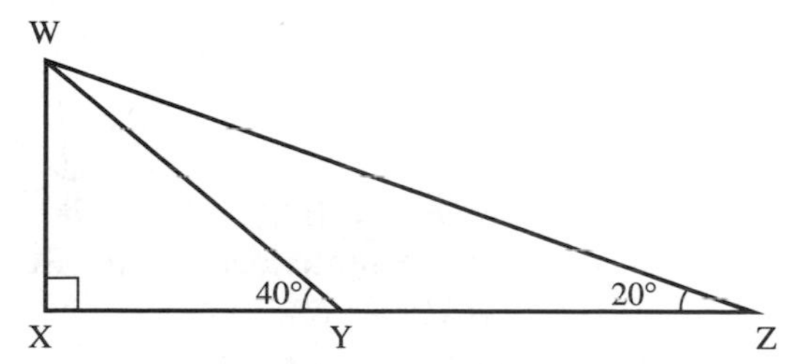

13. A golfer hits her ball a distance of 127 m so that it finishes 31 m from the hole. If the length of the hole is 150 m, calculate the angle between the line of her shot and the direct line to the hole.

14. Find (a) WX (b) WZ (c) WY

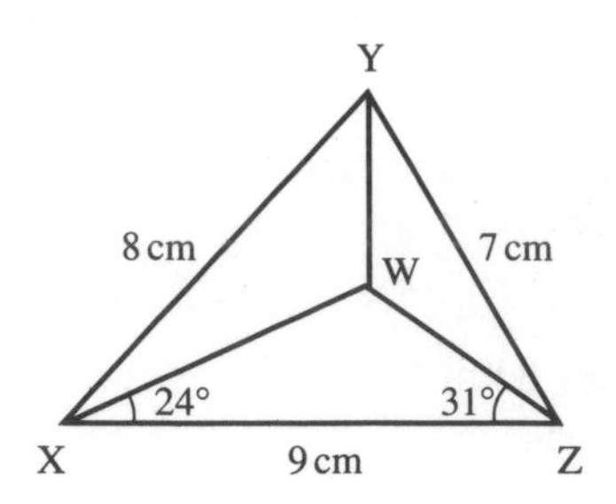

15.† The diagram shows a cube of side 10 cm from which one corner has been cut. Calculate the angle PQR.

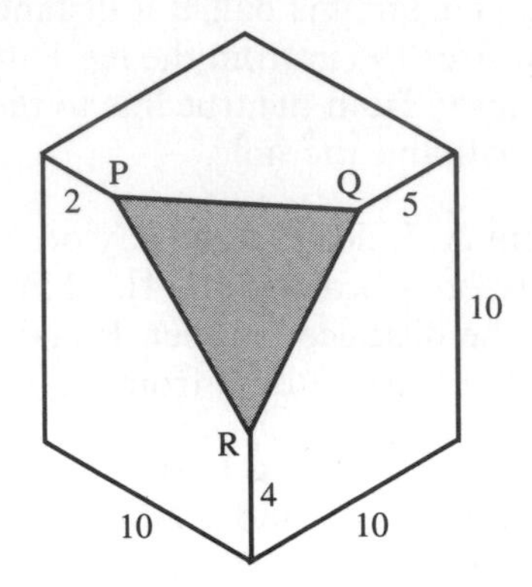

5.8 Circle theorems

1. The angle subtended at the centre of a circle is twice the angle subtended at the circumference.

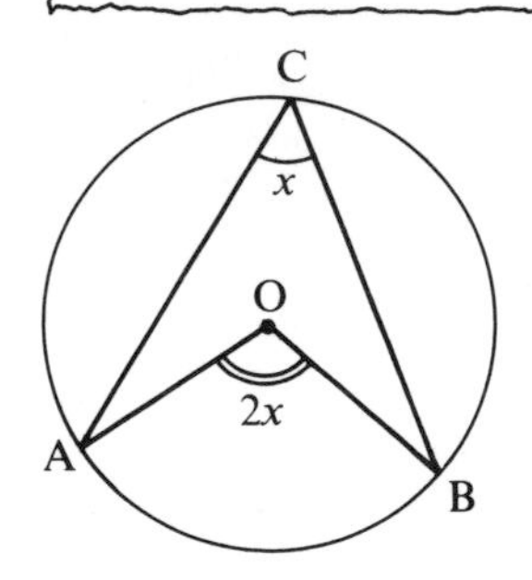

$A\hat{O}B = 2 \times A\hat{C}B$

The proof of this theorem is given on page 377.

2. Angles subtended by an arc in the same segment of a circle are equal.

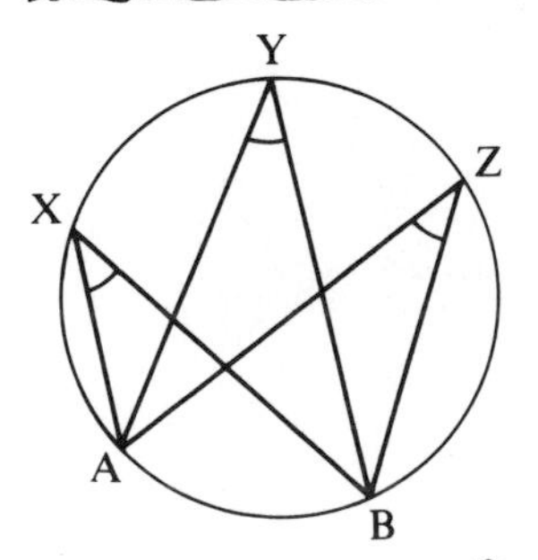

$A\hat{X}B = A\hat{Y}B = A\hat{Z}B$

Example 1

Given $A\hat{B}O = 50°$, find $B\hat{C}A$.

Triangle OBA is isosceles (OA = OB).

$\therefore$ $O\hat{A}B = 50°$

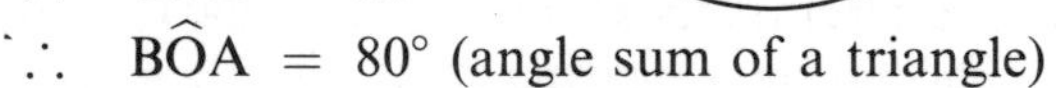

$\therefore$ $B\hat{O}A = 80°$ (angle sum of a triangle)

$\therefore$ $B\hat{C}A = 40°$ (angle at the centre)

Example 2

Given $B\hat{D}C = 62°$ and $D\hat{C}A = 44°$ find $B\hat{A}C$ and $A\hat{B}D$.

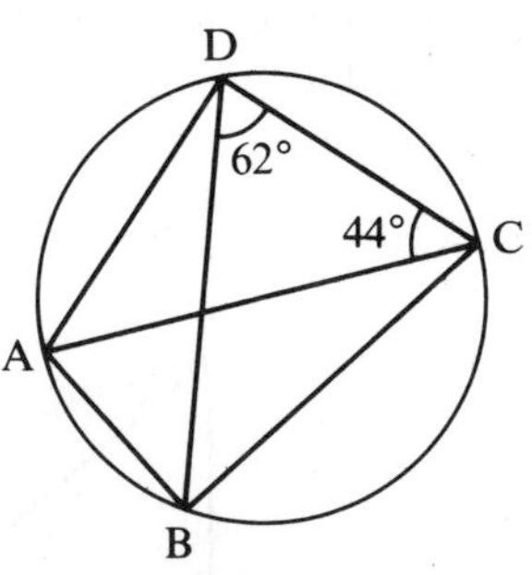

$B\hat{D}C = B\hat{A}C$ (both subtended by arc BC)

$\therefore$ $B\hat{A}C = 62°$

$D\hat{C}A = A\hat{B}D$ (both subtended by arc DA)

$\therefore$ $A\hat{B}D = 44°$.

Exercise 15

Find the angles marked with letters. A line passes through the centre only when point O is shown.

1.
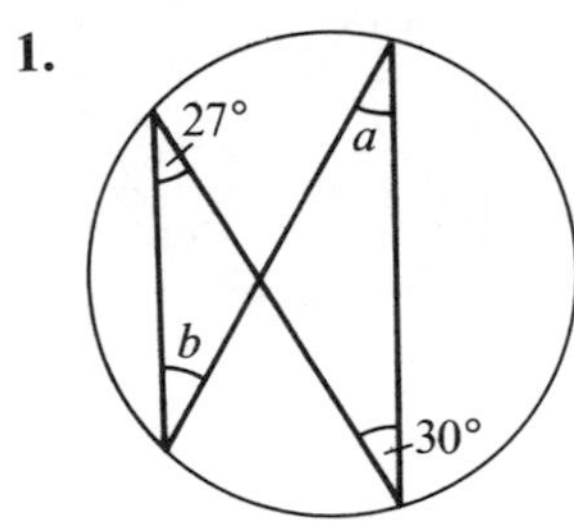

2.
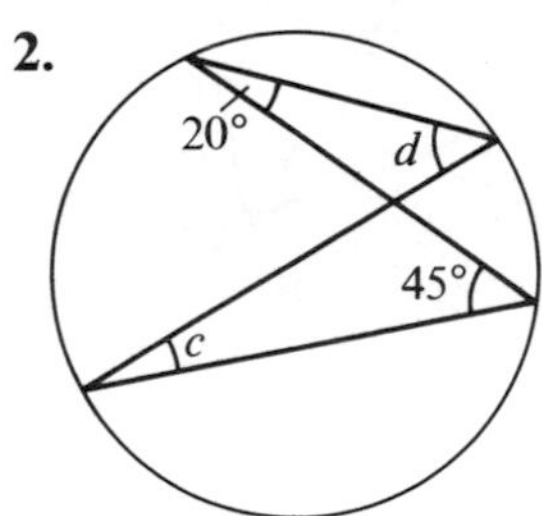

3.
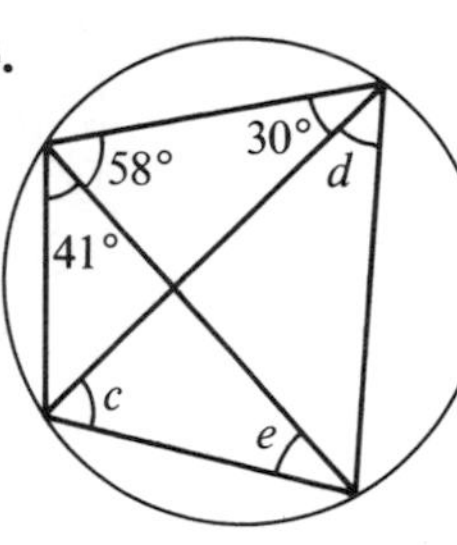

4.
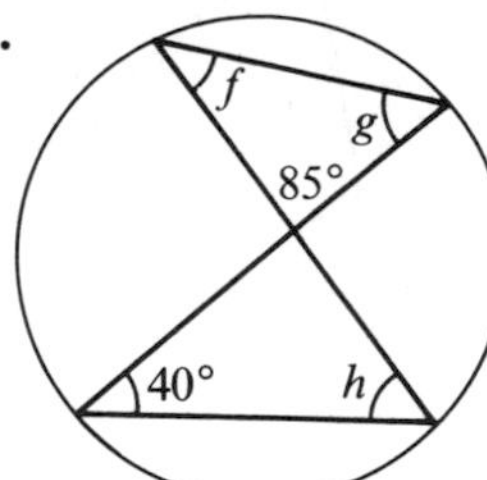

5.
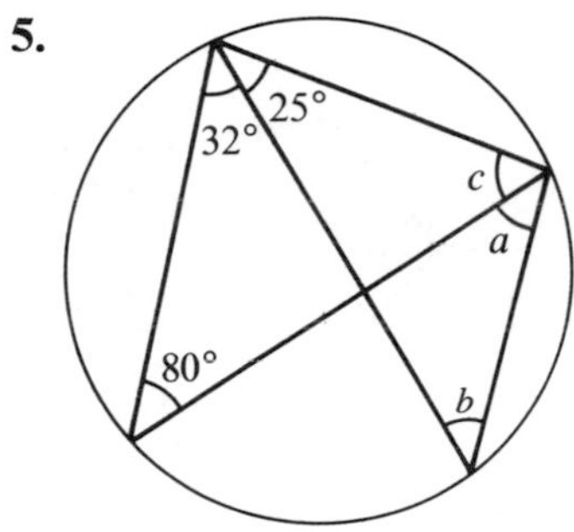

6.
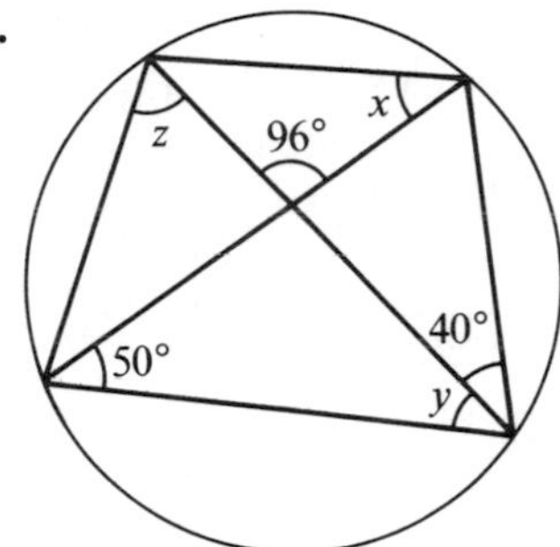

7.
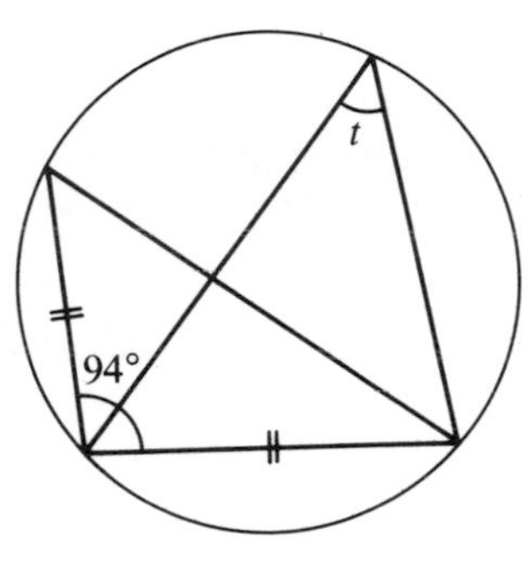

8.
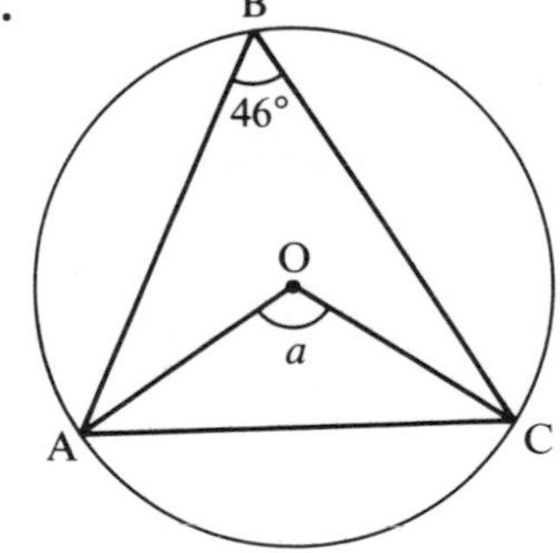

9.
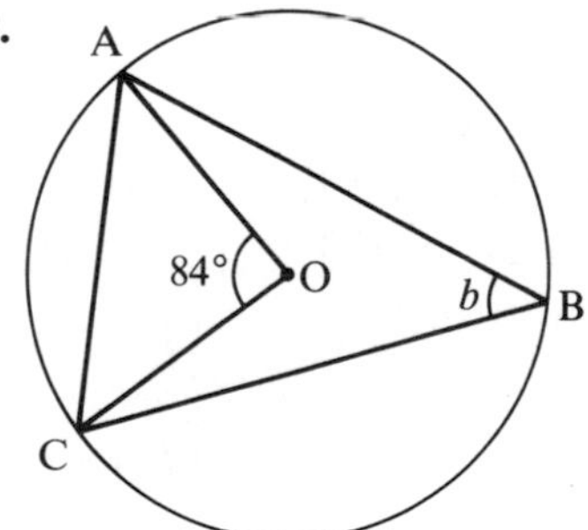

10.
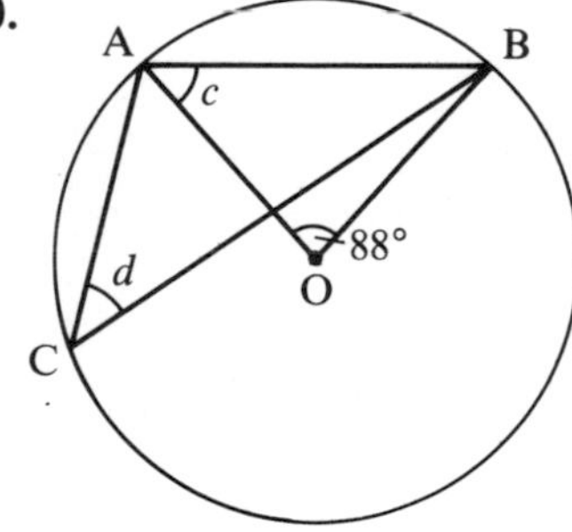

11.
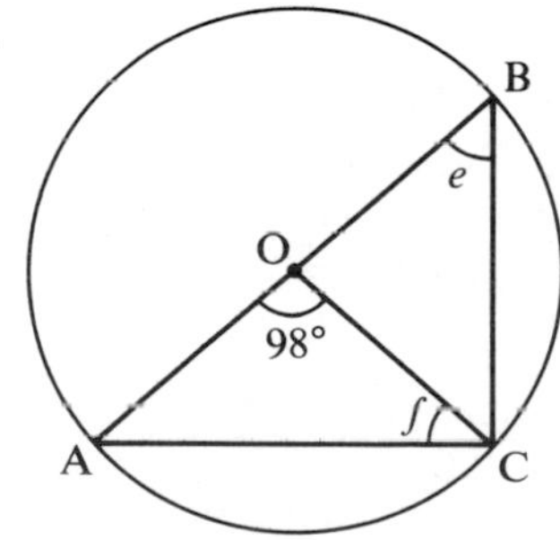

12.
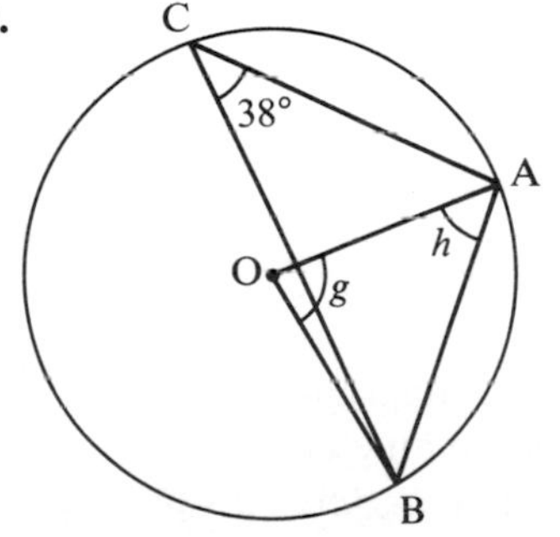

13.
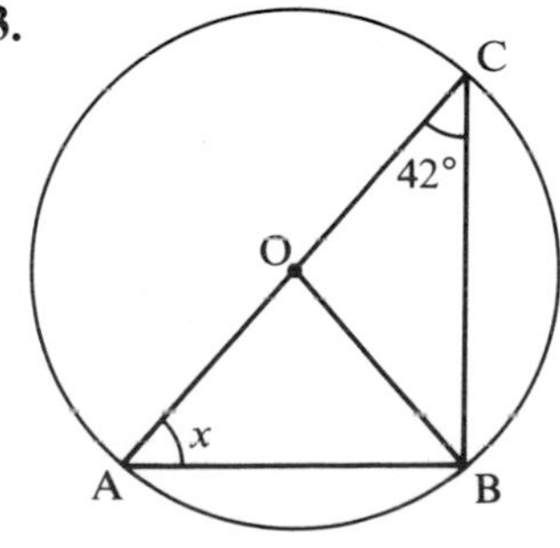

14.

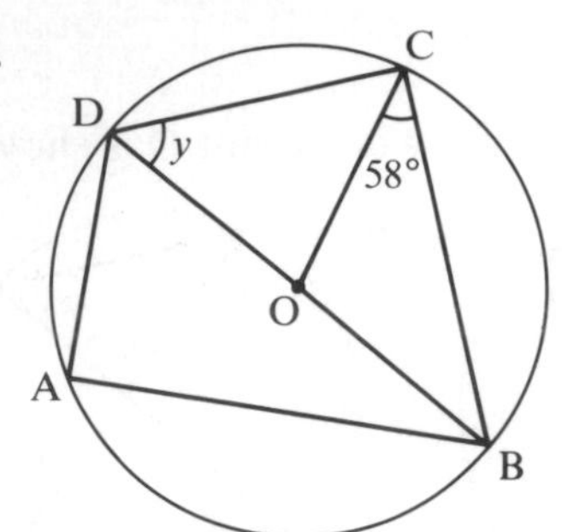

15.

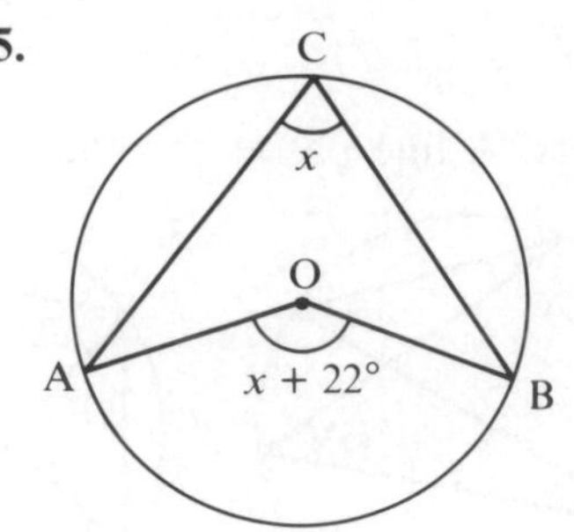

16.

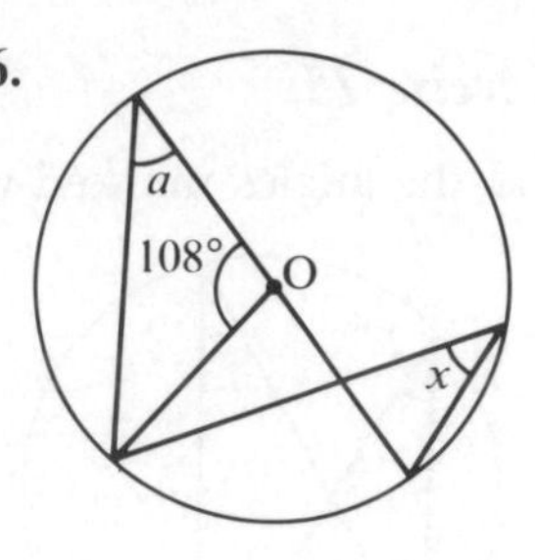

---

3. The opposite angles in a cyclic quadrilateral add up to 180° (the angles are supplementary).

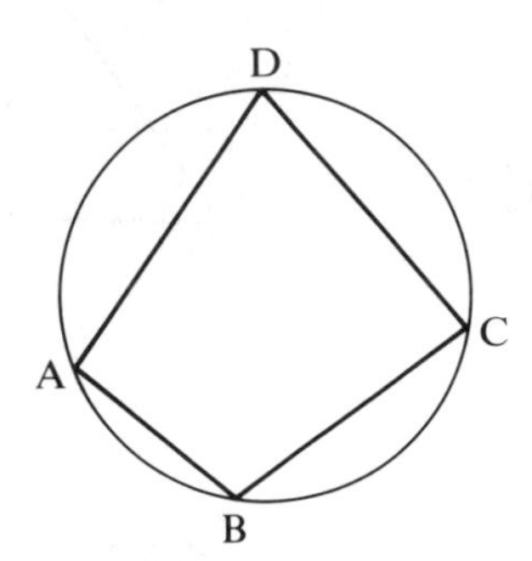

$$\hat{A} + \hat{C} = 180°$$
$$\hat{B} + \hat{D} = 180°$$

Example 3

Find a and x.

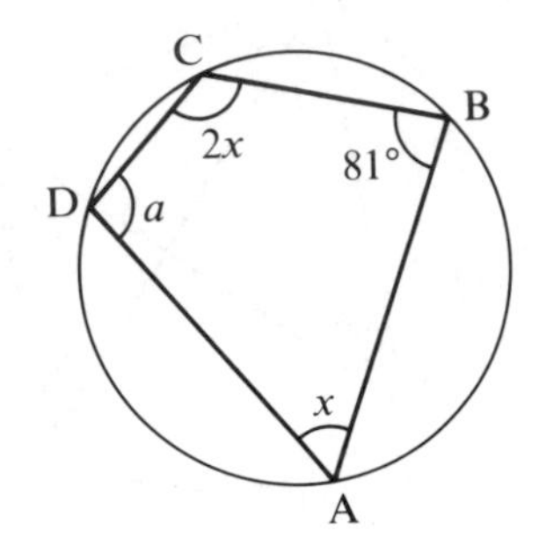

$a = 180° - 81°$
(opposite angles of a cyclic quadrilateral)
$\therefore\ a = 99°$
$x + 2x = 180°$
(opposite angles of a cyclic quadrilateral)
$3x = 180°$
$\therefore\ x = 60°$

4. The angle in a semicircle is a right angle.

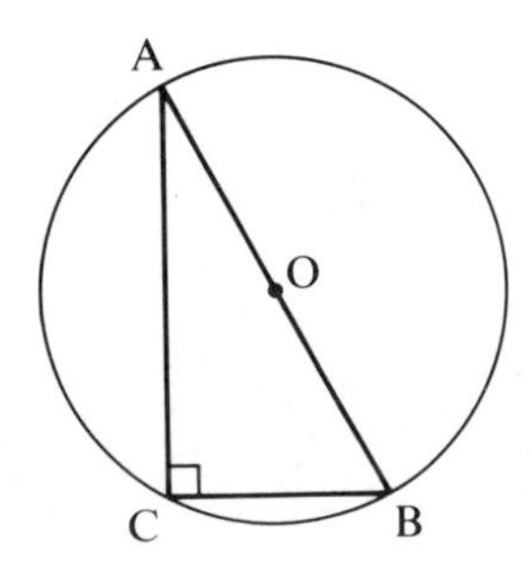

In the diagram,
AB is a diameter.
$A\hat{C}B = 90°$.

Example 4

Find b given that AOB is a diameter.

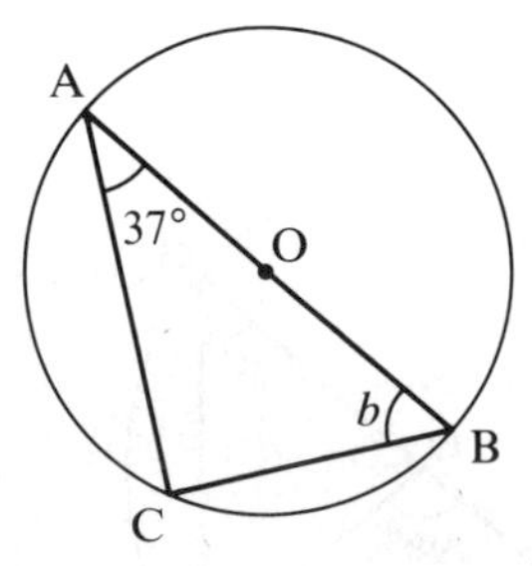

$A\hat{C}B = 90°$ (angle in a semicircle)
$\therefore\ b = 180° - (90 + 37)°$
$= 53°$.

Exercise 16

Find the angles marked with a letter.

1.

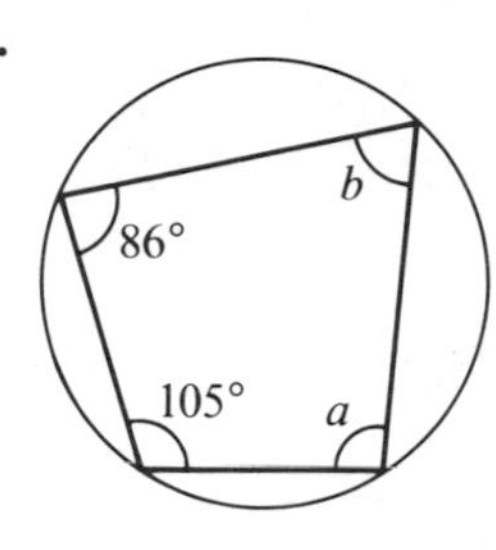

2.

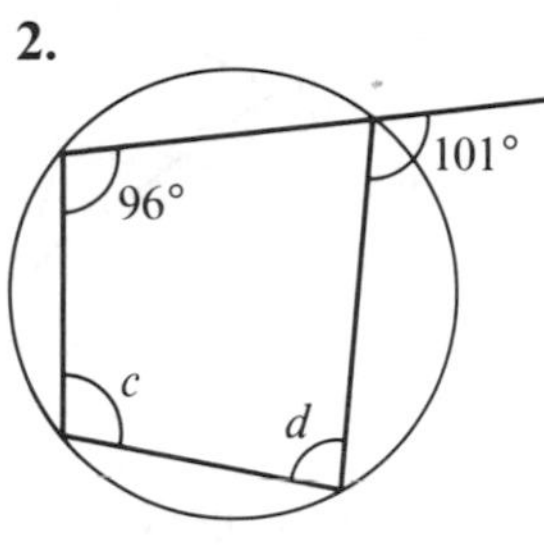

3.

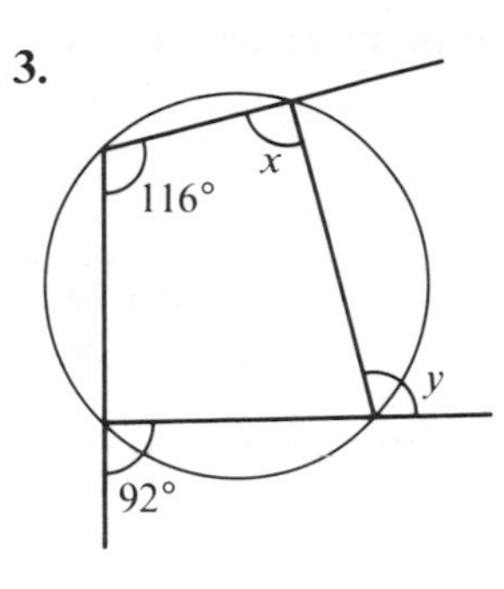

4.

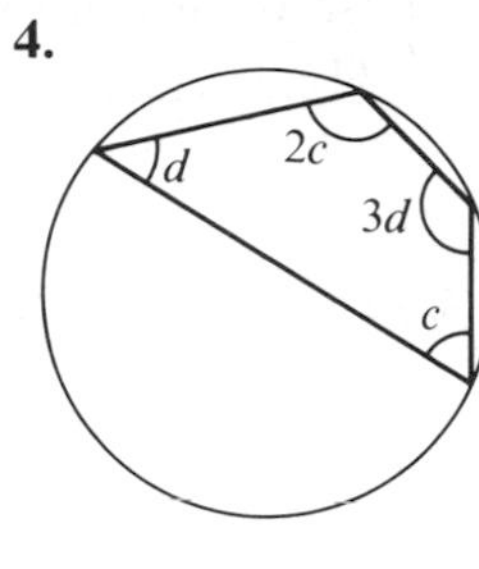

5.

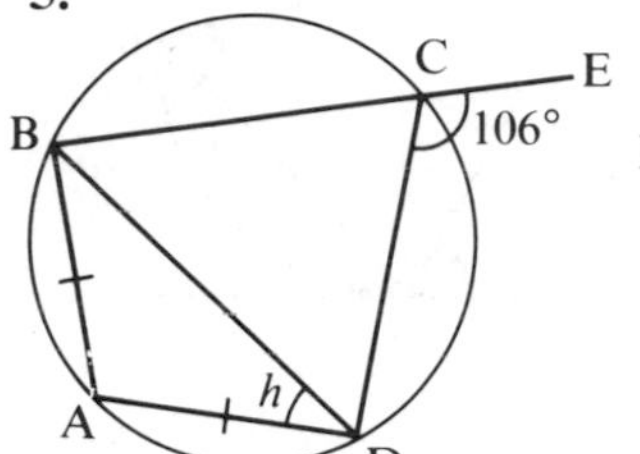

6.

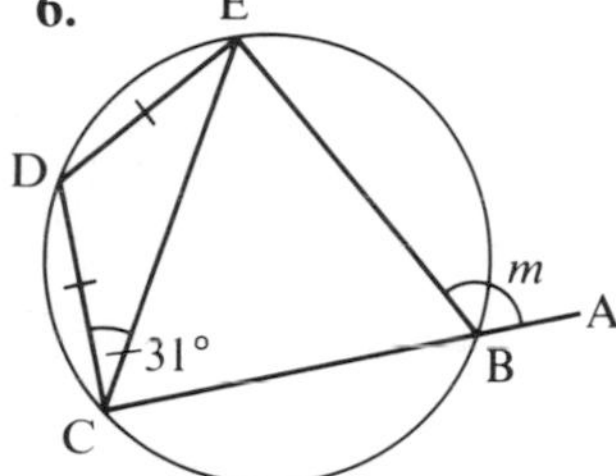

7.

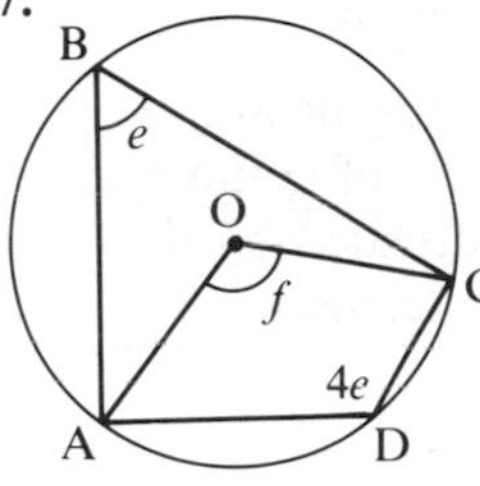

8.

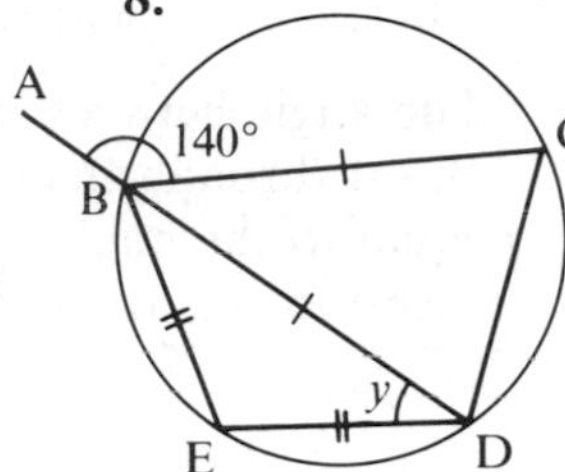

9.

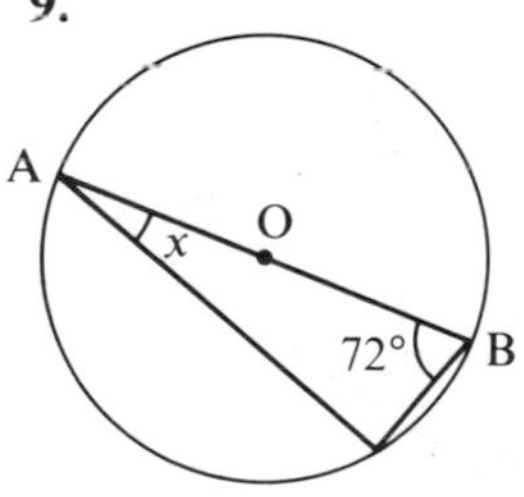

10.

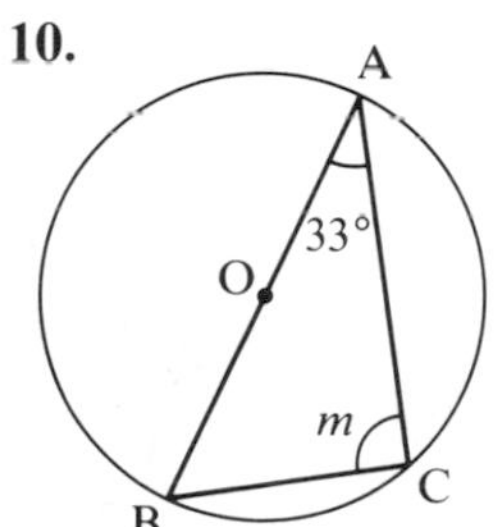

11.

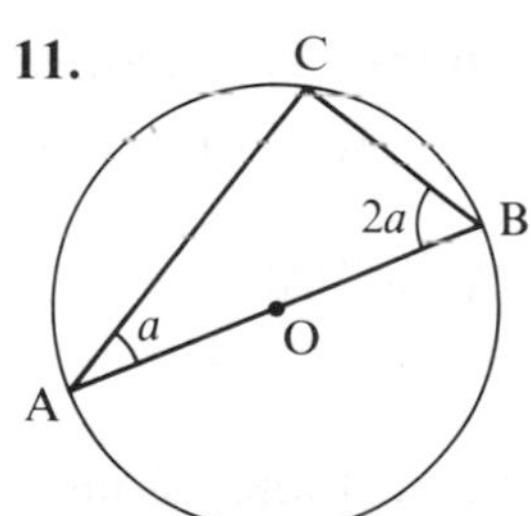

12.

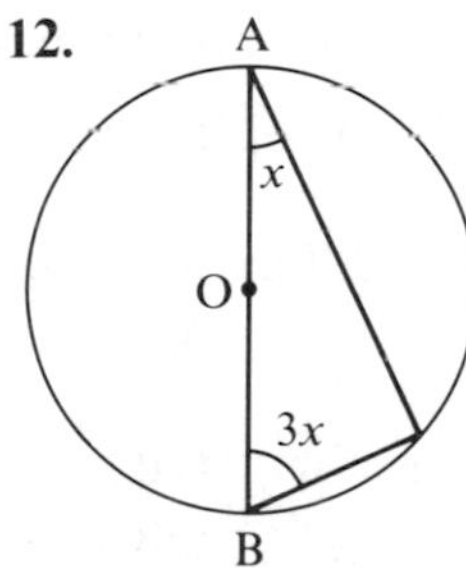

13.

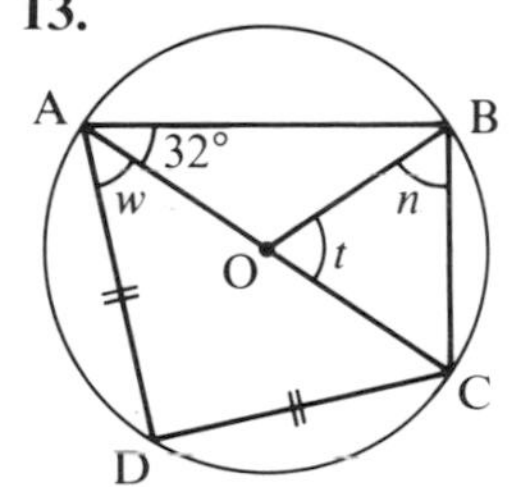

14.

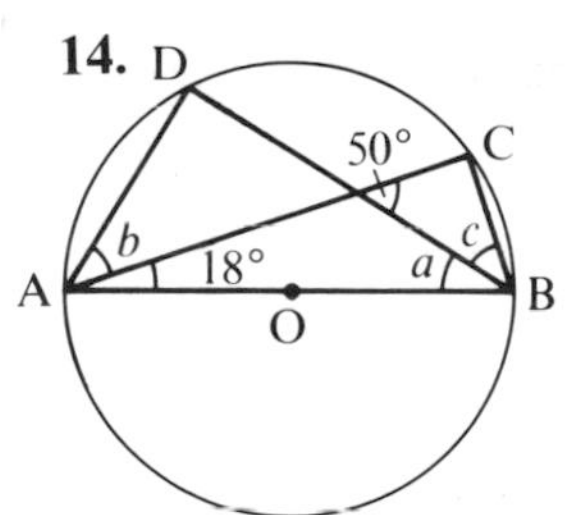

15.

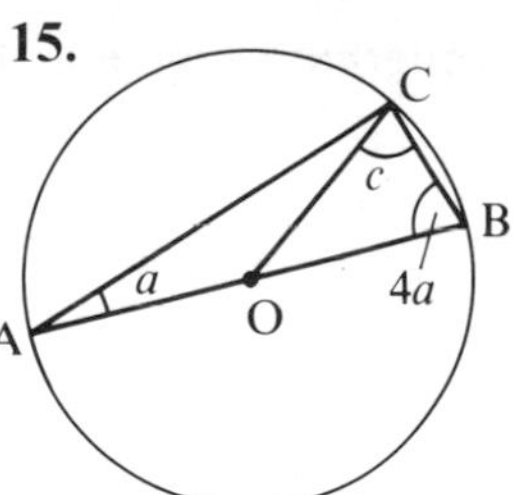

16.

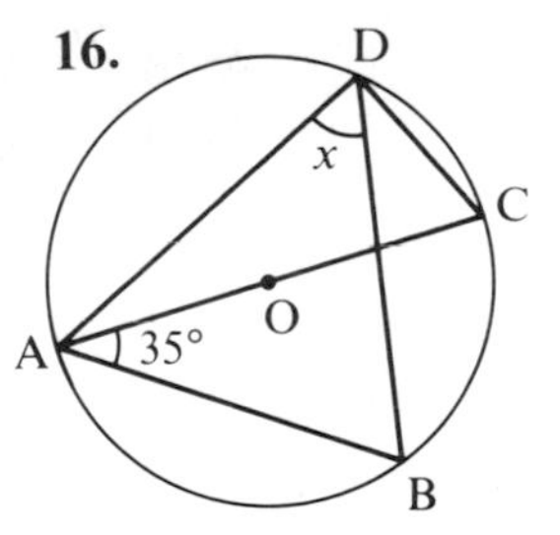

17.

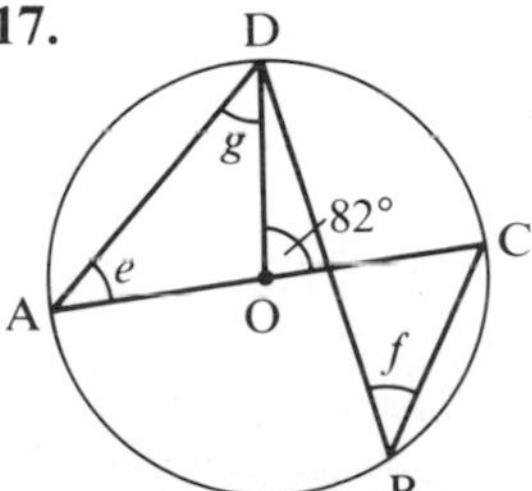

18.

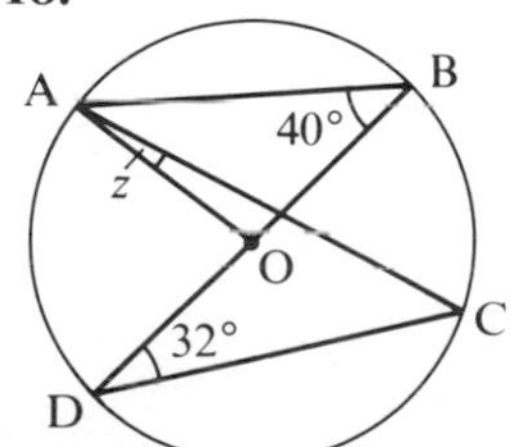

19.

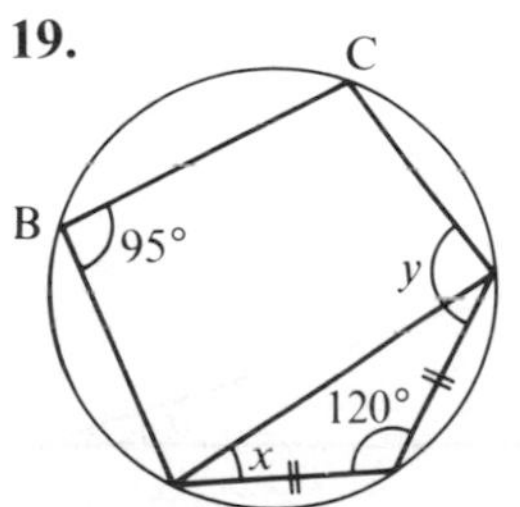

20.

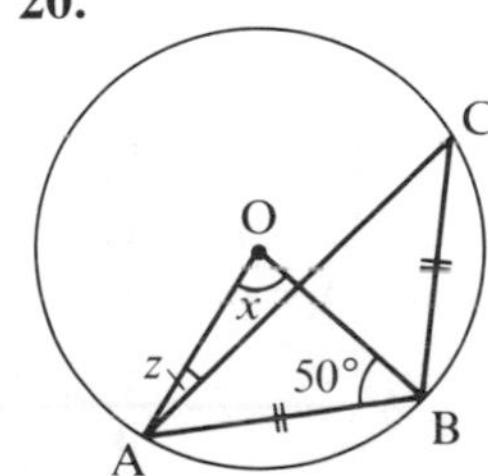

Tangents to circles

1. The angle between a tangent and the radius drawn to the point of contact is 90°.

2. From any point outside a circle just two tangents to the circle may be drawn and they are of equal length.

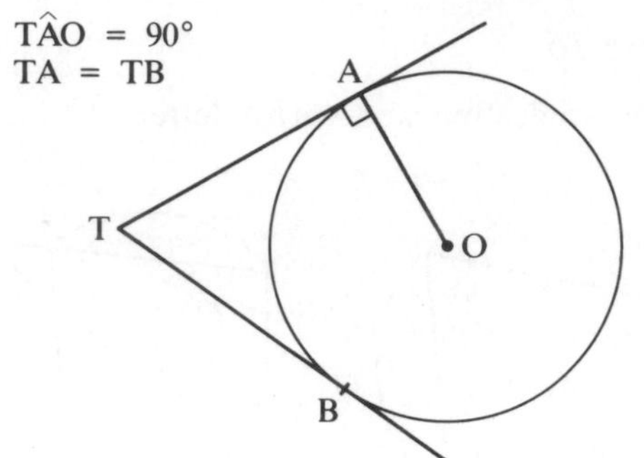

3. Alternate segment theorem.

The angle between a tangent and a chord through the point of contact is equal to the angle subtended by the chord in the alternate segment.

$T\hat{A}B = B\hat{C}A$
$S\hat{A}C = C\hat{B}A$

Example 5

TA and TB are tangents to the circle, centre O.

Given $A\hat{T}B = 50°$, find (a) $A\hat{B}T$ (b) $O\hat{B}A$ (c) $A\hat{C}B$

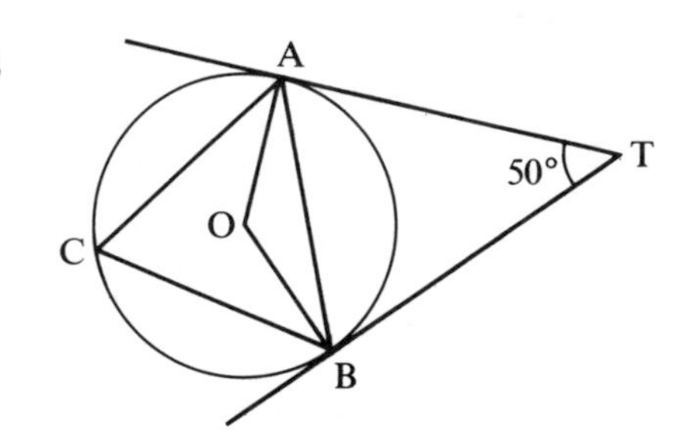

(a) $\triangle$TBA is isosceles (TA = TB)

$\therefore \quad A\hat{B}T = \frac{1}{2}(180 - 50) = 65°$

(b) $\quad O\hat{B}T = 90°$ (tangent and radius)

$\therefore \quad O\hat{B}A = 90 - 65 = 25°$

(c) $\quad A\hat{C}B = A\hat{B}T$ (alternate segment theorem)

$A\hat{C}B = 65°.$

Exercise 17

For Questions **1** to **11**, find the angles marked with a letter.

1.

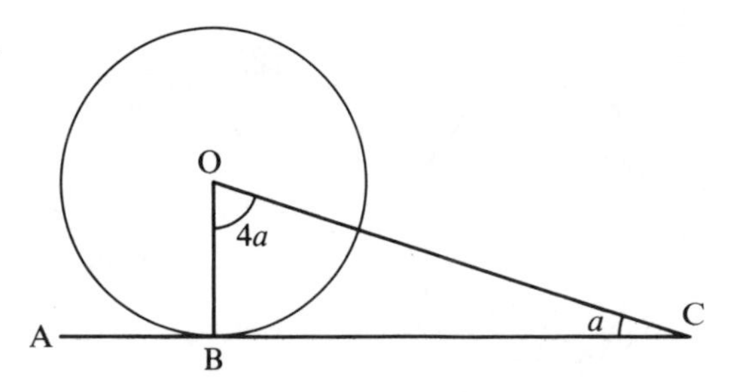

2.

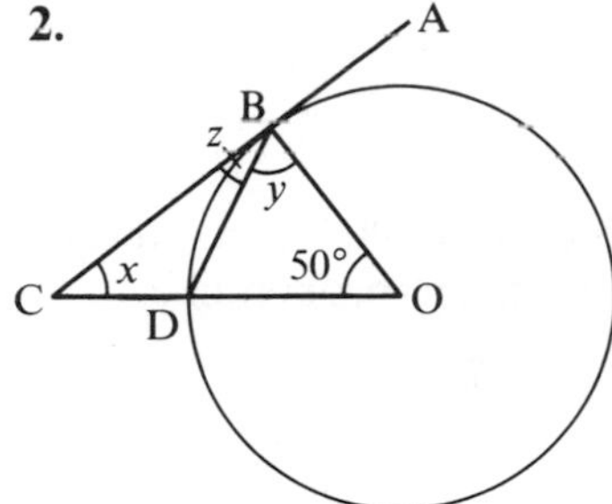

3.

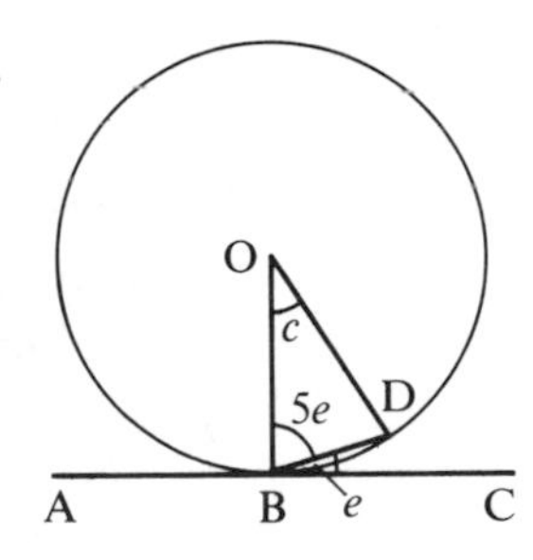

4.

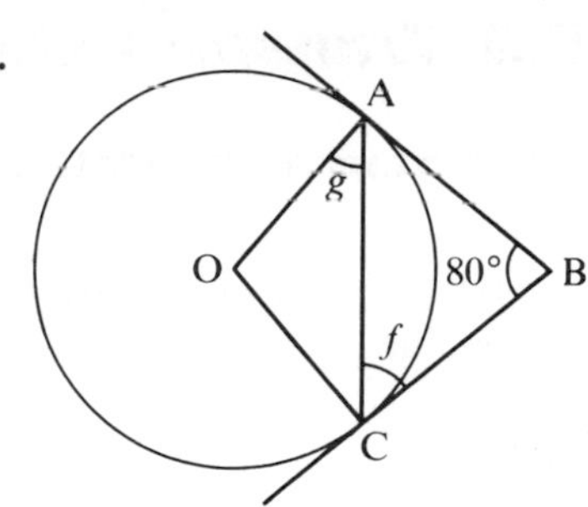

5.

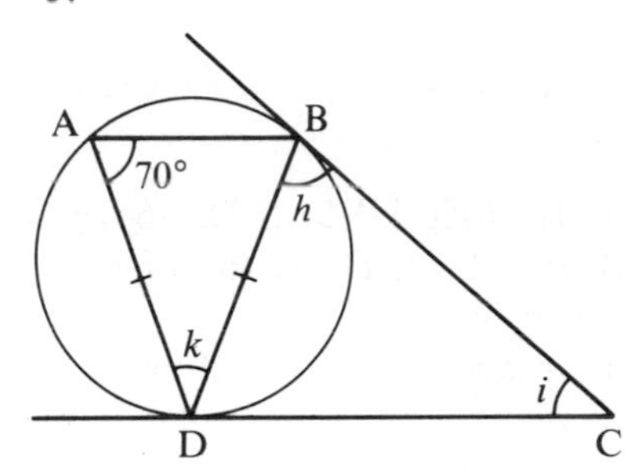

6.

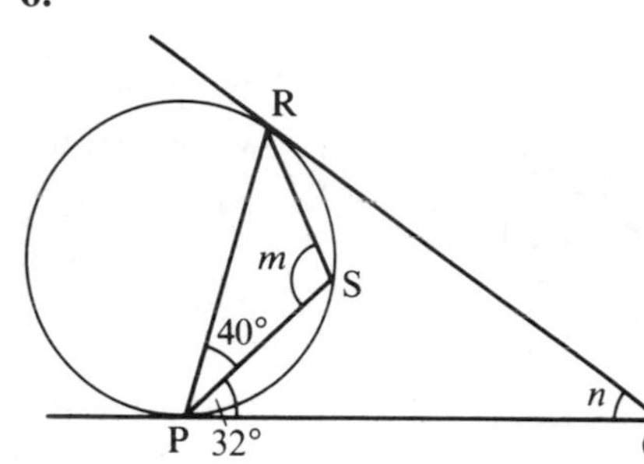

7.

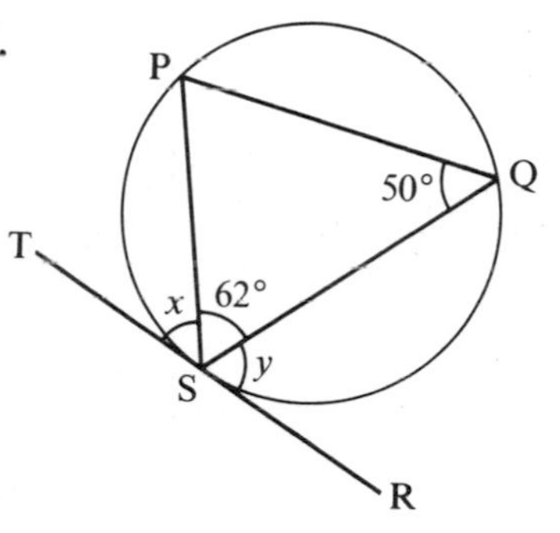

8.

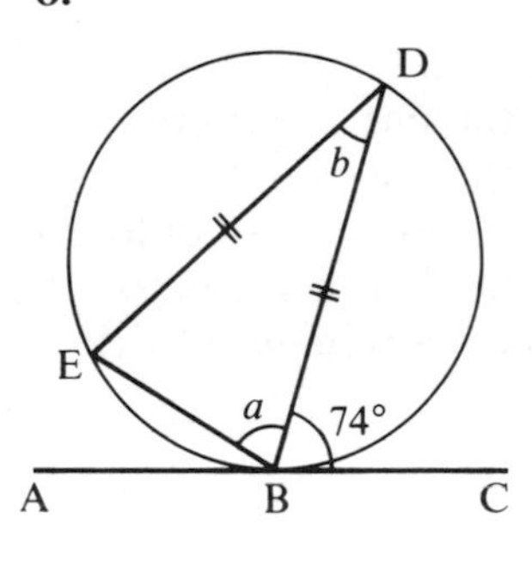

9.

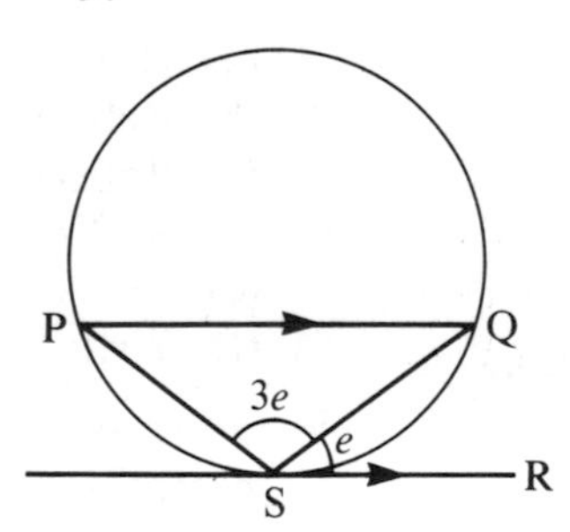

10.

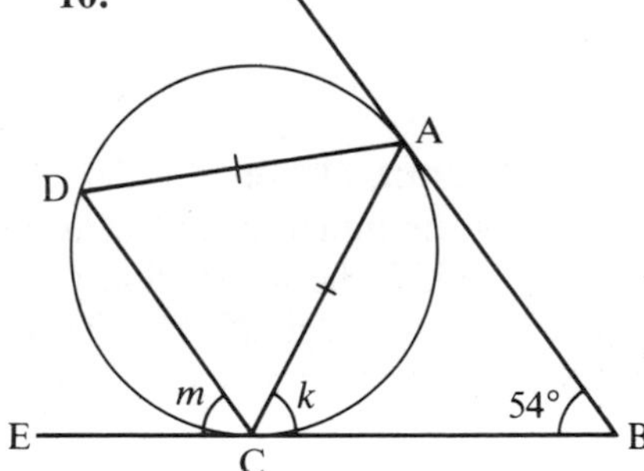

11.

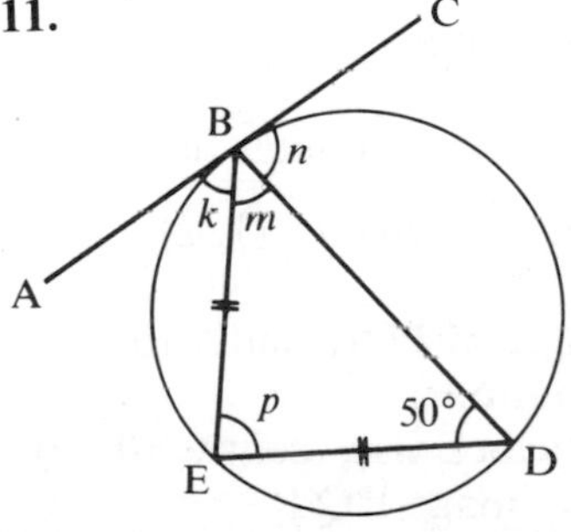

12. Find, in terms of p

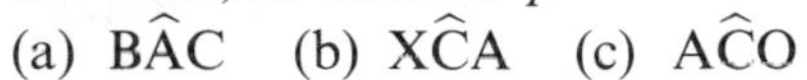

(a) $\mathrm{B\hat{A}C}$ (b) $\mathrm{X\hat{C}A}$ (c) $\mathrm{A\hat{C}O}$

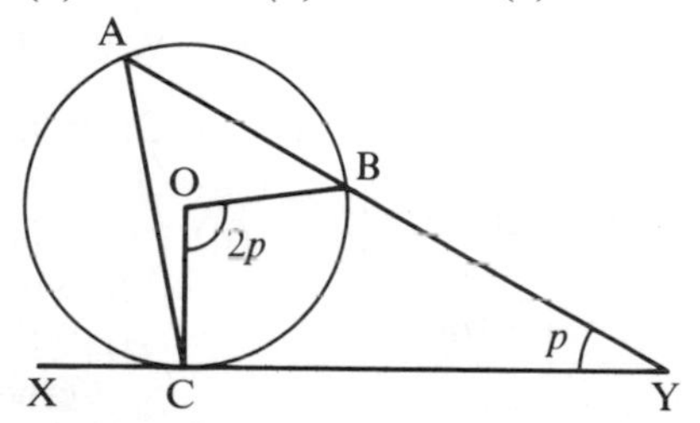

13. Find x, y and z.

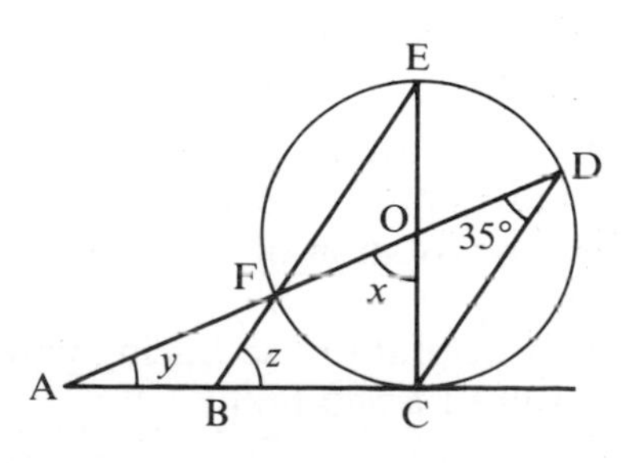

5.9 Transformations

Reflection and rotation

A reflection is specified by the choice of a mirror line.

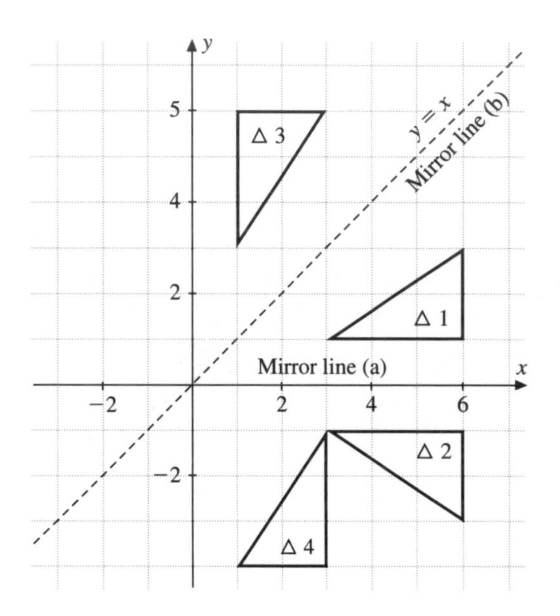

In the diagram above

(a) $\triangle 2$ is the image of $\triangle 1$ after reflection in the x-axis (the mirror line).

(b) $\triangle 3$ is the image of $\triangle 1$ after reflection in the mirror line $y = x$.

A rotation requires specification of three things: angle, direction and centre of rotation.

(c) $\triangle 4$ is the image of $\triangle 2$ after rotation through 90° clockwise about $(3, -1)$.

(d) $\triangle 4$ is the image of $\triangle 3$ after rotation through 180° in either direction about (2,1).

By convention, an anticlockwise rotation is positive and a clockwise rotation is negative. So the rotation of $\triangle 2$ onto $\triangle 4$ is $-90°$, centre $(3, -1)$.
It is helpful to use tracing paper to obtain the result of a rotation.

Exercise 18

In Questions **1** and **2**, draw the object and its image after reflection in the broken line.

1.

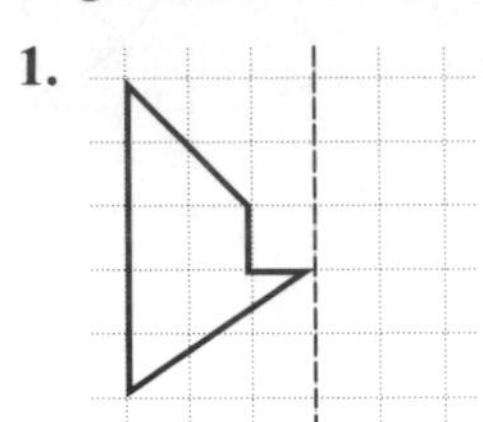

2.

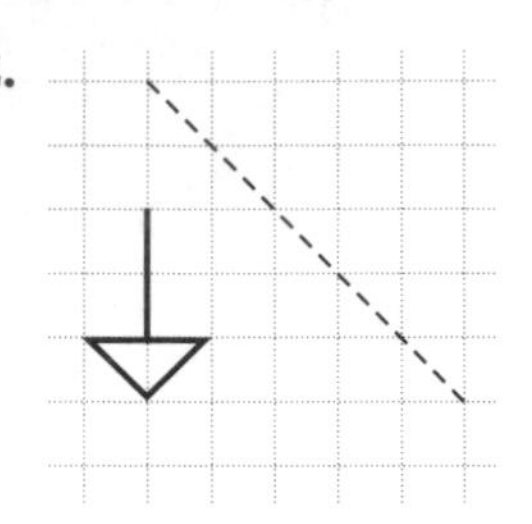

In Questions **3** to **8**, draw x- and y-axes with values from -8 to $+8$.

3. (a) Draw the triangle DEF at D(−6, 8), E(−2, 8), F(−2, 6). Draw the lines $x = 1$, $y = x$, $y = -x$.
(b) Draw the image of $\triangle$DEF after reflection in:
(i) the line $x = 1$. Label it $\triangle 1$.
(ii) the line $y = x$. Label it $\triangle 2$.
(iii) the line $y = -x$. Label it $\triangle 3$.
(c) Write down the coordinates of the image of point D in each case.

4. (a) Draw $\triangle 1$ at (3, 1), (7, 1), (7, 3).
(b) Reflect $\triangle 1$ in the line $y = x$ onto $\triangle 2$.
(c) Reflect $\triangle 2$ in the x-axis onto $\triangle 3$.
(d) Reflect $\triangle 3$ in the line $y = -x$ onto $\triangle 4$.
(e) Reflect $\triangle 4$ in the line $x = 2$ onto $\triangle 5$.
(f) Write down the coordinates of $\triangle 5$.

5. (a) Draw $\triangle 1$ at (2, 6), (2, 8), (6, 6).
(b) Reflect $\triangle 1$ in the line $x + y = 6$ onto $\triangle 2$.
(c) Reflect $\triangle 2$ in the line $x = 3$ onto $\triangle 3$.
(d) Reflect $\triangle 3$ in the line $x + y = 6$ onto $\triangle 4$.
(e) Reflect $\triangle 4$ in the line $y = x - 8$ onto $\triangle 5$.
(f) Write down the coordinates of $\triangle 5$.

6. (a) Draw a triangle PQR at P(1, 2), Q(3, 5), R(6, 2).
(b) Find the image of PQR under the following rotations:
(i) 90° anticlockwise, centre (0, 0); label the image P′Q′R′

(ii) 90° clockwise, centre (−2, 2); label the image P″Q″R″

(iii) 180°, centre, (1, 0); label the image P*Q*R*.

(c) Write down the coordinates of P′, P″, P*.

7. (a) Draw △1 at (1, 2), (1, 6), (3, 5).

(b) Rotate △1 90° clockwise, centre (1, 2) onto △2.

(c) Rotate △2 180°, centre (2, −1) onto △3.

(d) Rotate △3 90° clockwise, centre (2, 3) onto △4.

(e) Write down the coordinates of △4.

8. (a) Draw and label the following triangles:

△1 : (3, 1), (6, 1), (6, 3)
△2 : (−1, 3), (−1, 6), (−3, 6)
△3 : (1, 1), (−2, 1), (−2, −1)
△4 : (3, −1), (3, −4), (5, −4)
△5 : (4, 4), (1, 4), (1, 2)

(b) Describe fully the following rotations:

(i) △1 onto △2 (ii) △1 onto △3
(iii) △1 onto △4 (iv) △1 onto △5
(v) △5 onto △4 (vi) △3 onto △2

Translation and enlargement

A translation can be specified by the choice of a column vector (see page 208).

In the diagram at the top of the next column

(a) △5 is the image of △4 after translation with the column vector $\begin{pmatrix} 4 \\ 2 \end{pmatrix}$.

(b) △4 is the image of △1 after translation with the column vector $\begin{pmatrix} 2 \\ -3 \end{pmatrix}$.

An enlargement requires specification of two things: scale factor and centre of enlargement.

(c) △2 is the image of △1 after enlargement by a scale factor 3 with centre of enlargement (0, 0).

(d) △1 is the image of △2 after enlargement by a scale factor $\frac{1}{3}$ with centre of enlargement (0, 0). Note the construction lines.

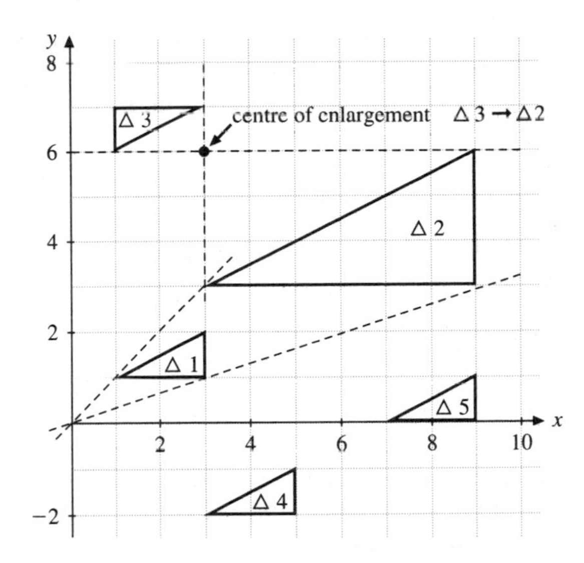

(e) △2 is the image of △3 after enlargement by a scale factor −3 with centre of enlargement (3, 6). Note that the negative scale factor causes the object to be inverted to form the image.

Exercise 19

1. For the diagram below, write down the column vector for each of the following translations.

(a) D onto A (b) B onto F
(c) E onto A (d) A onto C
(e) E onto C (f) C onto B
(g) F onto E (h) B onto C.

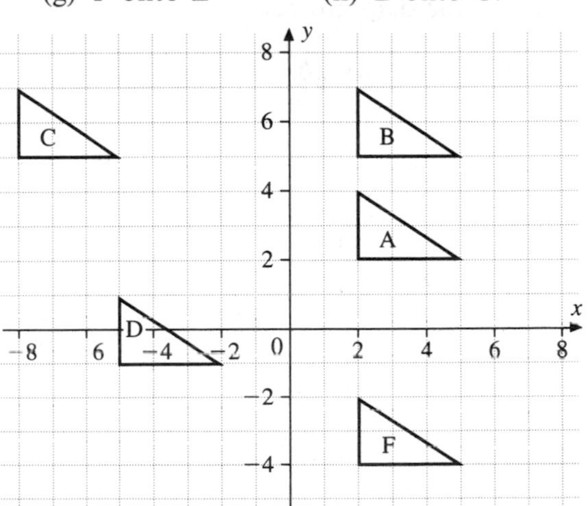

For Questions **2** to **7**, draw x- and y-axes with values from 0 to 15. Enlarge the object using the centre of enlargement and scale factor given.

	object			*centre*	*scale factor*
2.	(2, 4)	(4, 2)	(5, 5)	(1, 2)	$+2$
3.	(1, 1)	(4, 2)	(2, 3)	(1, 1)	$+3$
4.	(1, 2)	(13, 2)	(1, 10)	(0, 0)	$+\frac{1}{2}$
5.	(5, 10)	(5, 7)	(11, 7)	(2, 1)	$+\frac{1}{3}$
6.	(1, 1)	(3, 1)	(3, 2)	(4, 3)	-2
7.	(9, 2)	(14, 2)	(14, 6)	(7, 4)	$-\frac{1}{2}$

For Questions **8** to **11**, copy the diagram and draw an enlargement using the centre O and the scale factor given.

8. Scale factor 2

O

9. Scale factor 3

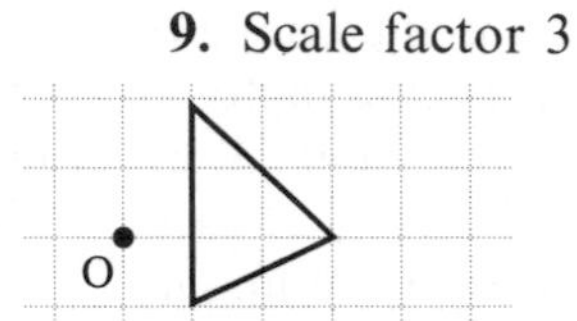

10. Scale factor 3

O

11. Scale factor -2

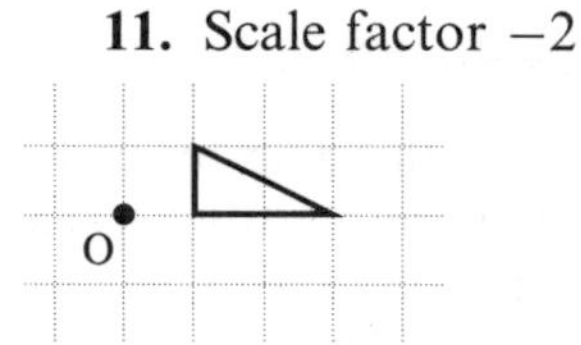

Questions **12** to **14** involve rotation, reflection, enlargement and translation.

12. Copy the diagram below.

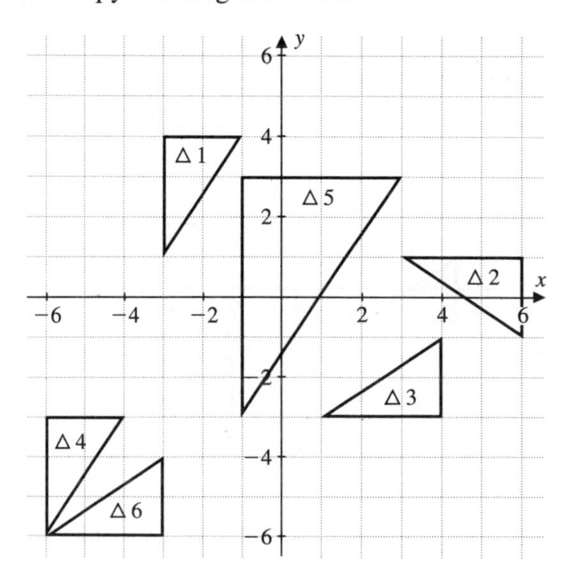

Describe fully the following transformations:

(a) $\Delta 1 \rightarrow \Delta 2$ (b) $\Delta 1 \rightarrow \Delta 3$
(c) $\Delta 4 \rightarrow \Delta 1$ (d) $\Delta 1 \rightarrow \Delta 5$
(e) $\Delta 3 \rightarrow \Delta 6$ (f) $\Delta 6 \rightarrow \Delta 4$

13. Draw x- and y-axes from -8 to $+8$. Plot and label the following triangles:
$\Delta 1: (-5, -5), (-1, -5), (-1, -3)$
$\Delta 2: (1, 7), (1, 3), (3, 3)$
$\Delta 3: (3, -3), (7, -3), (7, -1)$
$\Delta 4: (-5, -5), (-5, -1), (-3, -1)$
$\Delta 5: (1, -6), (3, -6), (3, -5)$
$\Delta 6: (-3, 3), (-3, 7), (-5, 7)$

Describe fully the following transformations:

(a) $\Delta 1 \rightarrow \Delta 2$ (b) $\Delta 1 \rightarrow \Delta 3$
(c) $\Delta 1 \rightarrow \Delta 4$ (d) $\Delta 1 \rightarrow \Delta 5$
(e) $\Delta 1 \rightarrow \Delta 6$ (f) $\Delta 5 \rightarrow \Delta 3$
(g) $\Delta 2 \rightarrow \Delta 3$

14. Draw x- and y-axes from -8 to $+8$. Plot and label the following triangles:
$\Delta 1: (-3, -6), (-3, -2), (-5, -2)$
$\Delta 2: (-5, -1), (-5, -7), (-8, -1)$
$\Delta 3: (-2, -1), (2, -1), (2, 1)$
$\Delta 4: (6, 3), (2, 3), (2, 5)$
$\Delta 5: (8, 4), (8, 8), (6, 8)$
$\Delta 6: (-3, 1), (-3, 3), (-4, 3)$

Describe fully the following transformations:

(a) $\Delta 1 \rightarrow \Delta 2$ (b) $\Delta 1 \rightarrow \Delta 3$
(c) $\Delta 1 \rightarrow \Delta 4$ (d) $\Delta 1 \rightarrow \Delta 5$
(e) $\Delta 1 \rightarrow \Delta 6$ (f) $\Delta 3 \rightarrow \Delta 5$
(g) $\Delta 6 \rightarrow \Delta 2$

Combined transformations

It is convenient to denote transformations by a symbol.
Referring to the diagram at the top of the next page, let **A** denote 'reflection in line $x = 3$' and **B** denote 'translation $\begin{pmatrix} 2 \\ 1 \end{pmatrix}$.'

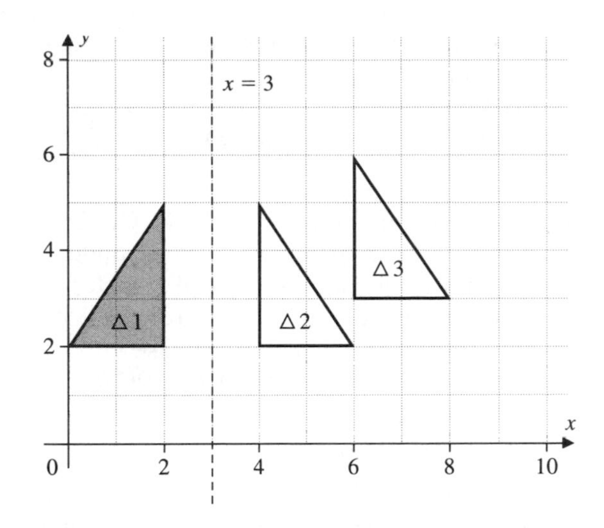

Perform **A** on Triangle 1.
△2 is the image of △1 under the reflection in $x = 3$ i.e. **A** (△1) = △2

A (△1) means
'Perform the transformation **A** on Triangle 1'.

We can see that **B**(△2) = △3. The effect of going from △1 to △3 may be written

BA (△1) = △3

It is very important to note that **BA**(△1) means 'do **A** first and then **B**'.

Repeated transformations

XX(P) means
'Perform transformation **X** on P and then perform **X** on the image'.
This may be written $\mathbf{X}^2$(P)
Similarly **TTT**(P) = $\mathbf{T}^3$(P)

Inverse transformations

If **T** is a translation with column vector $\begin{pmatrix} 3 \\ -2 \end{pmatrix}$, the translation which has the opposite effect is $\begin{pmatrix} -3 \\ 2 \end{pmatrix}$. This is written $\mathbf{T}^{-1}$.

Similarly, if rotation **R** is a 90° clockwise rotation about (0, 0), then the opposite effect, a 90° anticlockwise rotation about (0, 0) is written $\mathbf{R}^{-1}$.

This inverse of a transformation is the transformation which takes the image back to the object.
For example, if **X** is a reflection in $x = 0$, then $\mathbf{X}^{-1}$ is also a reflection in $x = 0$.

The symbol $\mathbf{T}^{-3}$ means $(\mathbf{T}^{-1})^3$, i.e. perform $\mathbf{T}^{-1}$ three times.

Exercise 20

In this exercise, transformations **A**, **B**, ... **H**, are as follows:

A denotes reflection in $x = 2$

B denotes 180° rotation, centre (1, 1)

C denotes translation $\begin{pmatrix} -6 \\ 2 \end{pmatrix}$

D denotes reflection in $y = x$

E denotes reflection in $y = 0$

F denotes translation $\begin{pmatrix} 4 \\ 3 \end{pmatrix}$

G denotes 90° rotation clockwise, centre (0, 0)

H denotes enlargement, scale factor $+\frac{1}{2}$, centre (0, 0)

Draw x- and y-axes with values from −8 to +8.

1. Draw triangle LMN at L(2, 2), M(6, 2), N(6, 4). Find the image of LMN under the following combinations of transformations. Write down the coordinates of the image of point L in each case:
(a) **CA**(LMN) (b) **ED**(LMN)
(c) **DB**(LMN) (d) **BE**(LMN)
(e) **EB**(LMN).

2. Draw triangle PQR at P(2, 2), Q(6, 2), R(6, 4). Find the image of PQR under the following combinations of transformations. Write down the coordinates of the image of point P in each case:
(a) **AF**(PQR) (b) **CG**(PQR)
(c) **AG**(PQR) (d) **HE**(PQR).

3. Draw triangle XYZ at X(−2, 4), Y(−2, 1), Z(−4, 1). Find the image of XYZ under the following combinations of transformations and state the equivalent single transformation in each case.
(a) $\mathbf{G^2E}$(XYZ) (b) **CB**(XYZ)
(c) **DA**(XYZ).

4. Draw triangle OPQ at O(0, 0), P(0, 2), Q(3, 2). Find the image of OPQ under the following combinations of transformations and state the equivalent single transformation in each case.
(a) **DE**(OPQ) (b) **FC**(OPQ)
(c) **DEC**(OPQ) (d) **DFE**(OPQ).

5. Draw triangle RST at R(−4, −1), S($-2\frac{1}{2}$, −2), T(−4, −4). Find the image of RST under the following combinations of transformations and state the equivalent single transformation in each case.
(a) **EAG**(RST) (b) **FH**(RST)
(c) **GF**(RST).

6. Write down the inverses of the transformations **A**, **B**, ... **H**.

7. Draw triangle JKL at J(−2, 2), K(−2, 5), L(−4, 5). Find the image of JKL under the following transformations.
Write down the coordinates of the image of point J in each case.
(a) $\mathbf{C}^{-1}$ (b) $\mathbf{F}^{-1}$
(c) $\mathbf{G}^{-1}$ (d) $\mathbf{D}^{-1}$
(e) $\mathbf{A}^{-1}$.

8. Draw triangle PQR at P(−2, 4), Q(−2, 1), R(−4, 1).Find the image of PQR under the following combinations of transformations. Write down the coordinates of the image of point P in each case.
(a) $\mathbf{DF}^{-1}$(PQR) (b) $\mathbf{EC}^{-1}$(PQR)
(c) $\mathbf{D^2F}$(PQR) (d) **GA**(PQR)
(e) $\mathbf{C}^{-1}\mathbf{G}^{-1}$(PQR).

9. Draw triangle LMN at L(−2, 4), M(−4, 1), N(−2, 1). Find the image of LMN under the following combinations of transformations. Write down the coordinates of the image of point L in each case.
(a) **HE**(LMN) (b) $\mathbf{EAG}^{-1}$(LMN)
(c) **EDA**(LMN) (d) $\mathbf{BG^2E}$(LMN).

10. Draw triangle XYZ at X(1, 2), Y(1, 6), Z(3, 6).
(a) Find the image of XYZ under each of the transformations **BC** and **CB**.
(b) Describe fully the single transformation equivalent to **BC**.
(c) Describe fully the transformation **M** such that **MCB** = **BC**.

5.10 Matrices for transformations

Multiplying matrices

For 2 × 2 matrices, $\begin{pmatrix} a & b \\ c & d \end{pmatrix}\begin{pmatrix} w & x \\ y & z \end{pmatrix} = \begin{pmatrix} aw+by & ax+bz \\ cw+dy & cx+dz \end{pmatrix}$

The same process is used for other orders of matrices.

Example 1

The following show the process of multiplying matrices.

(a) $\begin{pmatrix} 3 & 2 \\ 4 & 1 \end{pmatrix}\begin{pmatrix} 2 & 1 \\ 1 & 5 \end{pmatrix} = \begin{pmatrix} 6+2 & 3+10 \\ 8+1 & 4+5 \end{pmatrix} = \begin{pmatrix} 8 & 13 \\ 9 & 9 \end{pmatrix}$

(b) $\begin{pmatrix} 3 & 0 \\ 0 & 1 \end{pmatrix}\begin{pmatrix} 3 & 2 & 2 \\ 1 & -4 & 2 \end{pmatrix} = \begin{pmatrix} 9 & 6 & 6 \\ 1 & -4 & 2 \end{pmatrix}$

(c) $\begin{pmatrix} 2 & -1 \\ 3 & 0 \end{pmatrix}\begin{pmatrix} 2 \\ 3 \end{pmatrix} = \begin{pmatrix} 1 \\ 6 \end{pmatrix}$

Matrices may be multiplied only if they are *compatible*. The number of *columns* in the left-hand matrix must equal the number of *rows* in the right-hand matrix.

As with combined transformations which are represented by matrices, matrix multiplication is not commutative. So, for square matrices **A** and **B**, the product **AB** does not necessarily equal the product **BA**.

Exercise 21

For Questions **1** to **10**, calculate the resultant value where possible, using:

$\mathbf{A} = \begin{pmatrix} 2 & -1 \\ 3 & 4 \end{pmatrix}$; $\mathbf{B} = \begin{pmatrix} 0 & 5 \\ 1 & -2 \end{pmatrix}$; $\mathbf{C} = \begin{pmatrix} 4 & 3 \\ 1 & -2 \end{pmatrix}$;

$\mathbf{D} = \begin{pmatrix} 1 & 5 & 1 \\ 4 & -6 & 1 \end{pmatrix}$; $\mathbf{E} = \begin{pmatrix} 3 \\ 1 \end{pmatrix}$

1. **AB** **2.** **BA** **3.** **BC** **4.** **CB** **5.** **AE**

6. **BE** **7.** **AD** **8.** **CD** **9.** **(AB)C** **10.** **A(BC)**

In Questions **11** to **13**, find the value of the letters.

11. $\begin{pmatrix} x & 3 \\ -2 & y \end{pmatrix}\begin{pmatrix} 2 \\ 1 \end{pmatrix} = \begin{pmatrix} 5 \\ 0 \end{pmatrix}$

12. $\begin{pmatrix} 2 & 0 \\ 0 & -3 \end{pmatrix}\begin{pmatrix} m \\ n \end{pmatrix} = \begin{pmatrix} 10 \\ 1 \end{pmatrix}$

13. $\begin{pmatrix} 3 & 0 \\ 2 & x \end{pmatrix}\begin{pmatrix} y & z \\ 4 & 0 \end{pmatrix} = \begin{pmatrix} 6 & -3 \\ 8 & w \end{pmatrix}$

14. $\mathbf{A} = \begin{pmatrix} 1 & 0 \\ 3 & 2 \end{pmatrix}$, $\mathbf{B} = \begin{pmatrix} x & 0 \\ 1 & 3 \end{pmatrix}$ and **AB** = **BA**. Find x.

15. $\mathbf{X} = \begin{pmatrix} k & 2 \\ 2 & -k \end{pmatrix}$, and $\mathbf{X}^2 = 5\begin{pmatrix} 1 & 0 \\ 0 & 1 \end{pmatrix}$. Find k.

16. $\mathbf{B} = \begin{pmatrix} 3 & 3 \\ -1 & -1 \end{pmatrix}$.

(a) Find k if $\mathbf{B}^2 = k\mathbf{B}$; (b) Find m if $\mathbf{B}^4 = m\mathbf{B}$

17. $\mathbf{A} = \begin{pmatrix} 5 & 5 \\ -2 & -2 \end{pmatrix}$.

(a) Find n if $\mathbf{A}^2 = n\mathbf{A}$; (b) Find q if $\mathbf{A}^3 = q\mathbf{A}$

Transformations using matrices

Example 2

Find the image of triangle ABC, with A(1, 1), B(3, 1), C(3, 2), under the transformation represented by the matrix

$\mathbf{M} = \begin{pmatrix} 1 & 0 \\ 0 & -1 \end{pmatrix}$.

(a) Write the coordinates of A as a column vector and multiply this vector by **M**.

$$\underset{\mathbf{M}}{\begin{pmatrix} 1 & 0 \\ 0 & -1 \end{pmatrix}} \underset{\text{A}}{\begin{pmatrix} 1 \\ 1 \end{pmatrix}} = \underset{\text{A}'}{\begin{pmatrix} 1 \\ -1 \end{pmatrix}}$$

A′, the image of A, has coordinates (1, −1).

(b) Repeat for B and C.

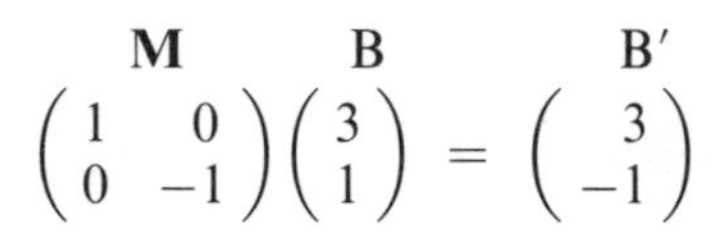

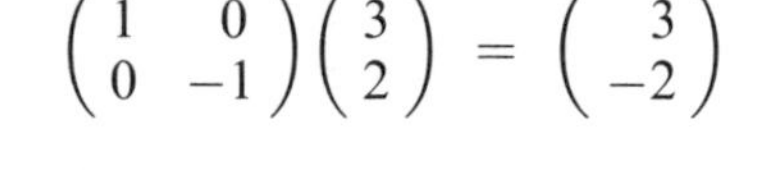

(c) Plot A′(1, −1), B′(3, −1) and C′(3, −2).

The transformation is a reflection in the *x*-axis.

Exercise 22

For Questions **1** and **2**, draw *x*- and *y*-axis with values from −8 to 8. Do all parts of each question on one graph.

1. Draw the triangle A(2, 2), B(6, 2), C(6, 4). Find its image under the transformations represented by the following matrices.

(a) $\begin{pmatrix} 0 & -1 \\ 1 & 0 \end{pmatrix}$, (b) $\begin{pmatrix} -1 & 0 \\ 0 & 1 \end{pmatrix}$, (c) $\begin{pmatrix} \frac{1}{2} & 0 \\ 0 & \frac{1}{2} \end{pmatrix}$.

2. Draw a trapezium at K(2, 2), L(2, 5), M(5, 8), N(6, 6). Find the images of KLMN under the transformations described by the following matrices:

$\mathbf{A} = \begin{pmatrix} 1 & 0 \\ 0 & -1 \end{pmatrix}$ $\mathbf{B} = \begin{pmatrix} -1 & 0 \\ 0 & 1 \end{pmatrix}$ $\mathbf{C} = \begin{pmatrix} 0 & 1 \\ 1 & 0 \end{pmatrix}$

$\mathbf{D} = \begin{pmatrix} 0 & 1 \\ -1 & 0 \end{pmatrix}$ $\mathbf{E} = \begin{pmatrix} 0 & -1 \\ -1 & 0 \end{pmatrix}$ $\mathbf{F} = \begin{pmatrix} -1 & 0 \\ 0 & -1 \end{pmatrix}$

$\mathbf{G} = \begin{pmatrix} 0 & -1 \\ 1 & 0 \end{pmatrix}$ $\mathbf{H} = \begin{pmatrix} 1 & 0 \\ 0 & 1 \end{pmatrix}$

Describe fully each of the eight transformations.

3. (a) Draw axes so that both x and y can take values from -2 to $+8$.
(b) Draw triangle ABC at A(2, 1), B(7, 1), C(2, 4).
(c) Find the image of ABC under the transformation represented by the matrix $\begin{pmatrix} 1 & -1 \\ 1 & 1 \end{pmatrix}$ and plot the image on the graph.
(d) The transformation is a rotation followed by an enlargement. Calculate the angle of the rotation and the scale factor of the enlargement.

4. (a) Draw axes so that x can take values from 0 to 15 and y can take values from -6 to $+6$.
(b) Draw triangle PQR at P(2, 1), Q(7, 1), R(2, 4).
(c) Find the image of PQR under the transformation represented by the matrix $\begin{pmatrix} 2 & 1 \\ -1 & 2 \end{pmatrix}$ and plot the image on the graph.
(d) The transformation is a rotation followed by an enlargement. Calculate the angle of the rotation and the scale factor of the enlargement.

5. (a) On graph paper, draw the triangle T whose vertices are (2, 2), (6, 2) and (6, 4).
(b) Draw the image U of T under the transformation whose matrix is $\begin{pmatrix} 0 & 1 \\ 1 & 0 \end{pmatrix}$.
(c) Draw the image V of T under the transformation whose matrix is $\begin{pmatrix} 1 & 0 \\ 0 & -1 \end{pmatrix}$.
(d) Describe the single transformation which would map U onto V.

6. Using a scale of 1 cm to one unit in each case, draw x- and y-axes, taking values of x from -4 to $+6$ and values of y from 0 to 12.
(a) Draw and label the quadrilateral OABC with O(0, 0), A(2, 0), B(4, 2), C(0, 2).
(b) Find and draw the image of OABC under the transformation whose matrix is $\mathbf{R}$, where $\mathbf{R} = \begin{pmatrix} 2{\cdot}4 & -1{\cdot}8 \\ 1{\cdot}8 & 2{\cdot}4 \end{pmatrix}$.
(c) Calculate, in surd form, the lengths OB and O′B′.
(d) Calculate the angle AOA′.
(e) Given that the transformation $\mathbf{R}$ consists of a rotation about O followed by an enlargement, state the angle of the rotation and the scale factor of the enlargement.

7. Draw the rectangle (0, 0), (0, 1), (2, 1), (2, 0) and its image under the following transformations and describe the *single* transformation which each represents.

(a) $\begin{pmatrix} 0 & 1 \\ 1 & 0 \end{pmatrix}\begin{pmatrix} x \\ y \end{pmatrix} + \begin{pmatrix} 1 \\ -1 \end{pmatrix}$ (b) $\begin{pmatrix} 1 & 0 \\ 0 & -1 \end{pmatrix}\begin{pmatrix} x \\ y \end{pmatrix} + \begin{pmatrix} 0 \\ 2 \end{pmatrix}$

(c) $\begin{pmatrix} 0 & 1 \\ -1 & 0 \end{pmatrix}\begin{pmatrix} x \\ y \end{pmatrix} + \begin{pmatrix} 4 \\ 0 \end{pmatrix}$ (d) $\begin{pmatrix} 3 & 0 \\ 0 & 3 \end{pmatrix}\begin{pmatrix} x \\ y \end{pmatrix} + \begin{pmatrix} -4 \\ 2 \end{pmatrix}$

8. (a) Draw L(1, 1), M(3, 3), N(4, 1) and its image L′M′N′ under the matrix $\mathbf{A} = \begin{pmatrix} 1 & 0 \\ -1 & -1 \end{pmatrix}$.

(b) Find and draw the image of L′M′N′ under matrix $\mathbf{B} = \begin{pmatrix} 0 & 1 \\ -1 & 0 \end{pmatrix}$ and label it L″M″N″.

(c) Calculate the matrix product **BA**.

(d) Find the image of LMN under the matrix **BA**, and compare with the result of performing **A** and then **B**.

Examination exercise 5

1. There is an athletics competition at Suki's school. Suki is told to mark out an area of the sports field that is to be used for the shot putt competition. The area consists of two parts:
A circle of radius 1 metre from where the shot is thrown.
A sector of angle 30° and radius 26 metres which overlaps the throwing circle as shown.

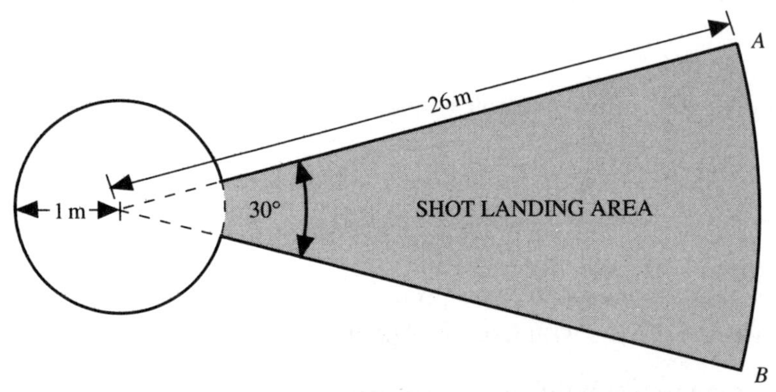

(a) Find the area of the circle from which the shot is thrown. Take π to be 3·14 or use the π key on your calculator.

The throwing circle is made of concrete to a depth of 5 cm.

(b) Calculate the volume of concrete needed in cubic metres.

So that no spectators are within range of the shot when it is thrown, Suki paints the arc AB of the shot landing area.

(c) What is the length of the arc she paints?

(d) Find the shaded area into which the shot can land. [S]

2. A rectangular block of steel measures 1 m by 20 cm by 4 cm. It is melted down to make ball bearings as spheres of radius 4 mm.

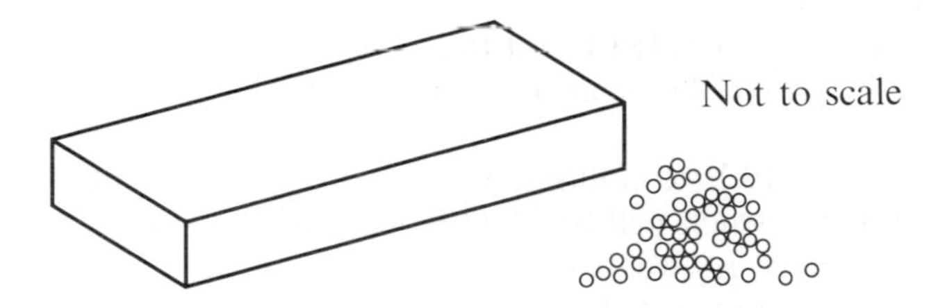

(a) Find the volume of the block of steel in cubic centimetres.
(b) Find the volume of a ball bearing in cubic millimetres.
(c) How many ball bearings can be made from the block of steel?
[S]

3. The diagram shows a circle, centre O, of radius 12 metres. The chord AB is 10 metres long.

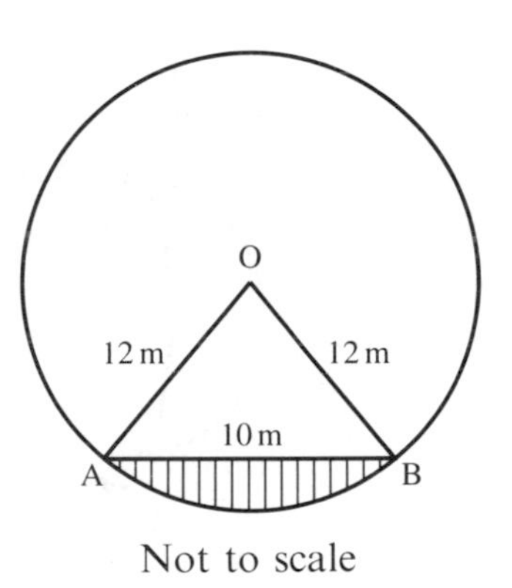

(a) Calculate
  (i) the size, correct to the nearest degree, of the angle AOB,
  (ii) the area of the shaded segment.
  (You may take π to be 3·142.)
(b) The cross section of a river may be taken to be the same as the shaded segment.

If the river is flowing at 0·8 metres per second, calculate the volume of water, in cubic metres, that will flow past AB in one hour.
[M]

4. When a new car came onto the market recently, a scale model of it, similar in every respect to the actual car, was made for sale in toy shops.
The table below compares certain details of the actual car with the scale model. Find the values of x, y and z.

	Car	Model
Length	420 cm	12 cm
Width	x cm	5 cm
Area of windscreen	8330 cm^2	y cm^2
Capacity of boot	z cm^3	19·2 cm^3

[L]

5. An ice-cream firm is to introduce a new ice cream cone called the *Concerto*. The ice cream completely fills the cone which has a base diameter of 7 cm and a height of 14 cm.

(a) Calculate the volume of the ice cream needed to fill the *Concerto*. (Give your answer in cm^3 to the nearest whole number.)

The ice cream used is manufactured in 10 litre containers.

(b) Calculate the number of *Concertos* which will be filled from each container assuming that there is no wastage.

The firm is also to produce a smaller version of the *Concerto* called the *Concertino*. The *Concertino* is geometrically similar to the *Concerto* but only $\frac{4}{5}$ of its height.

(c) Calculate the volume of ice cream in one *Concertino*. [L]

6. The list below contains an expression for the volume and an expression for the surface area of the object shown.

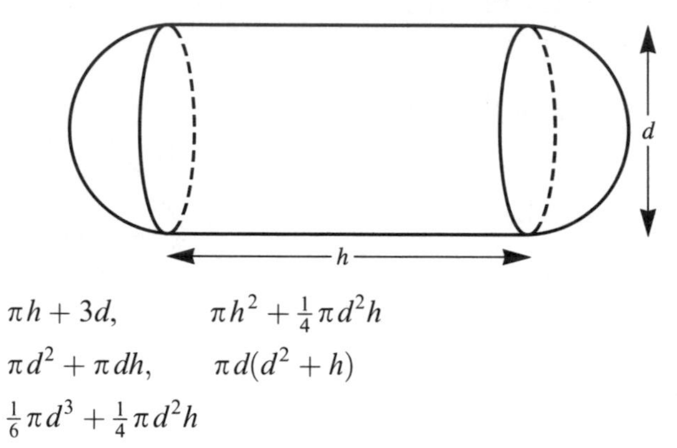

$\pi h + 3d, \qquad \pi h^2 + \frac{1}{4}\pi d^2 h$

$\pi d^2 + \pi dh, \qquad \pi d(d^2 + h)$

$\frac{1}{6}\pi d^3 + \frac{1}{4}\pi d^2 h$

(a) Write down the expression for the volume.

(b) Write down the expression for the surface area.

7. A ship has a speed relative to the water of 12 m/s and sets course due East. The ship is driven off course by a current flowing due South at 3 m/s.
Calculate the resultant velocity of the ship, giving the speed and the bearing.

8. If $\cos x = 0{\cdot}5$, find three possible values of x.

9. A wedge of cheese is in the shape of a prism as shown. It is cut into two equal volumes by a single cut, parallel to one face. Calculate the value of x. [N]

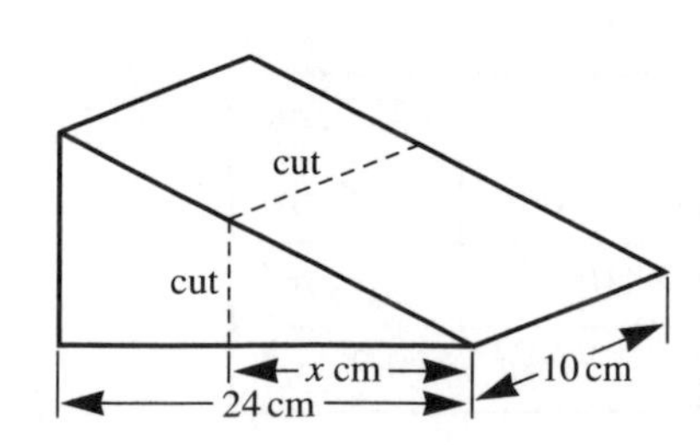

10. The diagram represents a car bonnet, AB, hinged at A and propped open by a stay, CD.

AC = 23 cm, AB = 47 cm, AD = 18 cm and CD = 20 cm.

(a) Calculate the size of angle CAD correct to the nearest degree.
(b) Find the height of B above the level of AC, giving your answer correct to the nearest centimetre. [M]

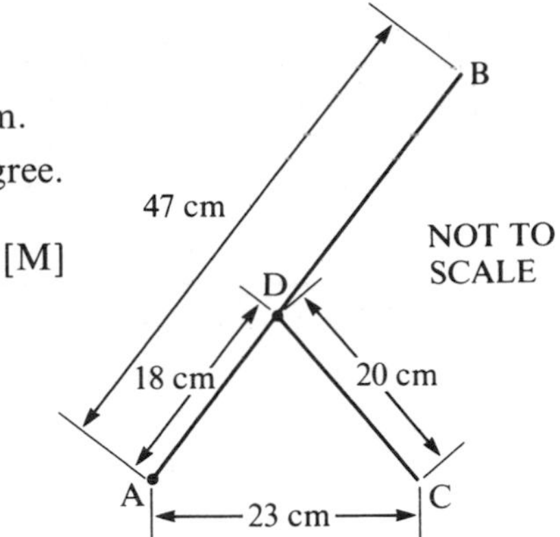

11.

The diagram shows part of the coast of the Isle of Wight and the Solent. The point R represents the position of Ryde, point P the position of Puckpool Point and point N the position of No Man's Land Fort.

RP = 2200 metres, angle NRP = 29° and angle RPN = 131°.

(a) Calculate the distance RN, correct to the nearest 10 metres.
(b) A boat sails directly from Ryde to No Man's Land Fort. Calculate, correct to the nearest 10 metres, its closest distance to Puckpool Point. [M]

12. In the diagram, ABCD is a cyclic quadrilateral. The line SBT is the tangent at B to the circle. The lines BA produced and CD produced meet at E and EA = AC. Angle CBT = 42° and angle EDA = 53°.
(To obtain full marks in this question you must give reasons for your answers.)
Calculate the sizes of the angles.
(a) angle CAB
(b) angle ABC
(c) angle ECA
(d) angle EAD [L]

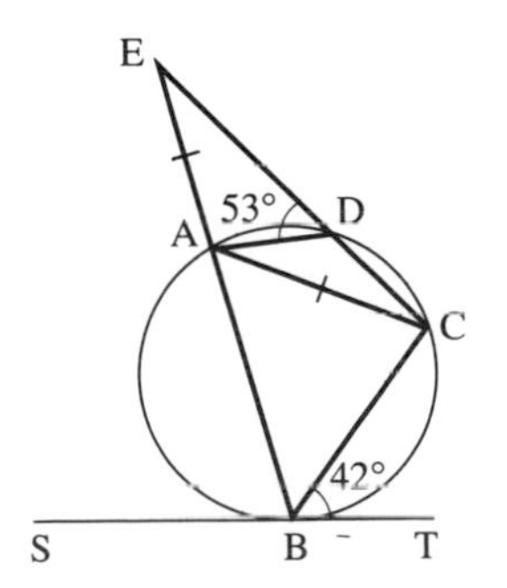

13. The depth d metres of water in Portsea Harbour is given by the formula $d = 5 + 3\cos(30t)°$ at time t hours after midnight.

(a) Copy and complete the table, giving d to 1 decimal place.

t	0	1	2	3	4	5	6	7	8	9	10	11	12
d	8·0		6·5										

(b) Draw a graph of the depth of water between midnight and noon. Use a scale of 1 cm to 1 hour on the t-axis and 1 cm to 1 metre on the d-axis.

(c) Use your graph to estimate
  (i) the depth of water at 2.30 a.m.,
  (ii) at what times the water is 4 m deep.

(d) Estimate the rate at which the depth of water is increasing at 7 a.m.

(e) At what time is the rate of increase of the depth of water a maximum? [M]

14.

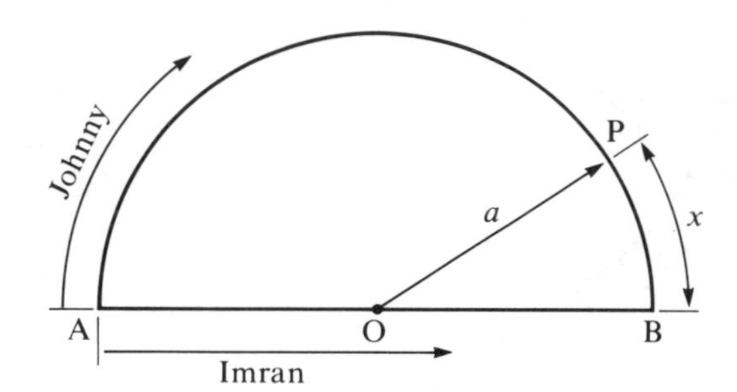

(a) Part of the road system of a modern town was laid out in the form of a semicircle, radius a, as shown.
B is due East of A.
Imran and Johnny cycled in opposite directions around the semicircle, starting together at A and arriving at P at the same time.

The arc length BP $= x$.

Find, in terms of a and x,
  (i) the distance cycled by Imran,
  (ii) the distance cycled by Johnny.

(b) If they both rode at the same speed, show that

$$x = \frac{a(\pi - 2)}{2}.$$

(c) Calculate the size of the angle POB in degrees.

(d) When the boys met at P, what was their bearing from the start at A? [N]

6 Algebra 3

6.1 Growth and decay rates

Under certain conditions, the number of bacteria in a particular culture doubles every 10 minutes.

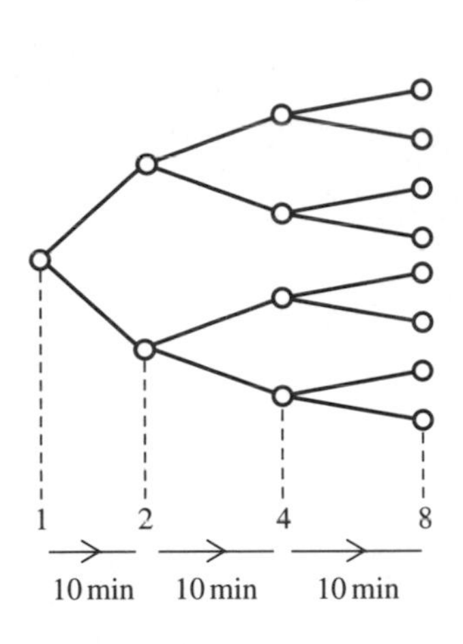

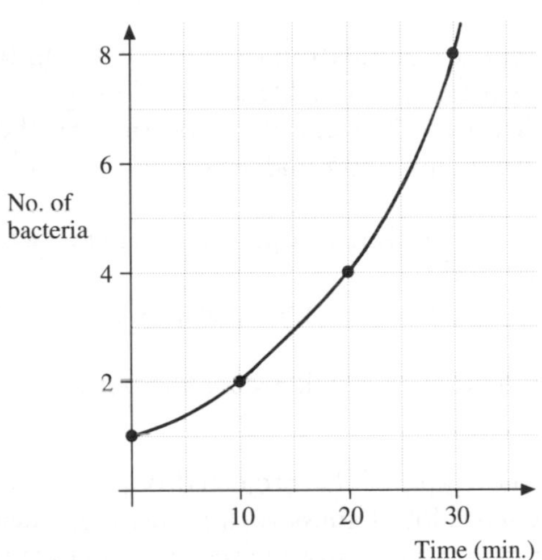

This is an example of *exponential growth*. The number of bacteria is multiplied by the same number in equal periods of time.

Radioactive materials like uranium or plutonium decay very, very slowly. The *half life* is the length of time for half the material to decay. Suppose uranium has a half life of 40 years and there is 100 g of it now. In 40 years there would be 50 g. After another 40 years there would still be 25 g and so on.

Exercise 1

1. At time $t = 0$, one bacteria is placed in a culture in a laboratory. The number of bacteria doubles every 10 minutes.

(a) Draw a graph to show the growth of the bacteria from $t = 0$ to $t = 120$ mins.
Use a scale of 1 cm to 10 s across the page and 5 cm to 1000 units up the page.

(b) Use your graph to estimate the time taken to reach 800 bacteria.

2. Here are two population graphs.

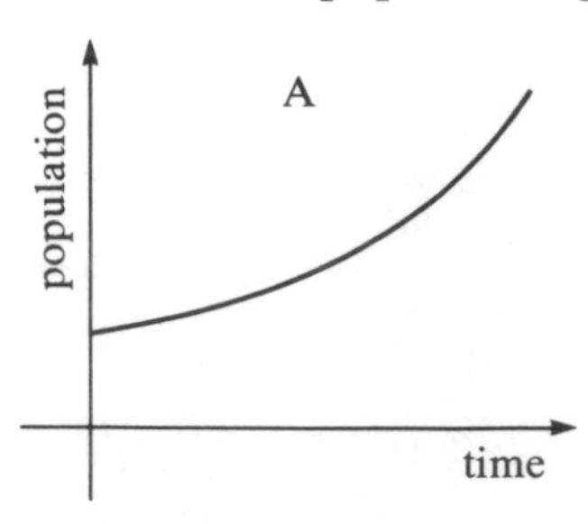

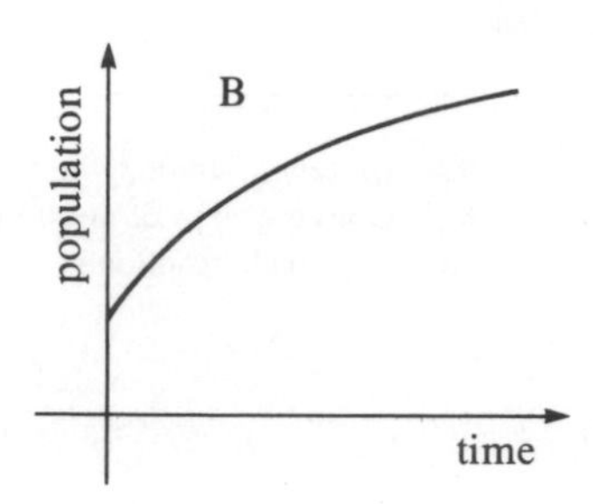

In which graph is
(a) the growth rate decreasing?
(b) the growth rate increasing?

3. Sketch a population graph in which the growth rate is constant.

4. An economist estimates that the population of country A will be multiplied by 1·2 every 10 years and that the population of country B will be multiplied by 1·05 every 10 years. In 1980 the populations of A and B were 36 million and 100 million respectively.
(a) Draw a graph to show the projected populations of the two countries from 1980 to 2060.
Use a scale of 2 cm to 10 years across the page and 2 cm to 20 million up the page.
(b) Estimate when the population of A will exceed the population of B for the first time.

5. In Question **4**, you drew the graph of the projected population growth over 80 years. There is a lot of guesswork involved in such predictions. In practice, populations of any species (man included) do not grow exponentially for ever.
Suppose a virus infects a plant. Initially the virus might double in number every day, but after a while, some viruses will die and a reduced supply of food will mean that the population will grow more and more slowly. Eventually the number will reach a peak.
(a) Draw a sketch graph to show how the number of viruses increases over this period.
(b) After a while, the plant's defence mechanisms will react to the virus and (hopefully for the plant) the number of viruses will be reduced to a very small number.
Extend your sketch graph to show this.

6. Radioactive waste from a nuclear power station has a half life of 50 years. One day 360 kg of the waste was sealed in a concrete bunker.
(a) Draw a graph to show how the waste decays over the next 200 years.
(b) From your graph estimate the number of years until there is only 60 kg of radioactive material.

6.2 Gradient of a curve

Mathematicians have been interested in finding the gradient of a curve for hundreds of years. When we say 'gradient of the curve', we really mean the 'gradient of the tangent to the curve' at that point.

In the sketch, the gradient of the curve at B is greater than the gradient of the curve at A.

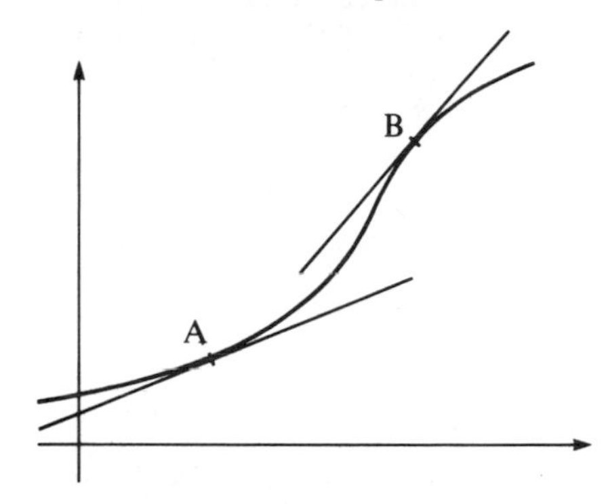

Newton and Leibnitz independently discovered a method for *calculating* the gradient of a curve in about 1680. This branch of mathematics is beyond the scope of this book. We will find the gradient of a curve by drawing a tangent to the curve.

The gradient of a curve is used to find speed or acceleration and such questions appear later in this chapter.

Example 1

The graph of $y = \frac{12}{x} + x - 6$ is drawn on the right.

Find the gradient of the tangent to the curve drawn at the point where $x = 5$.

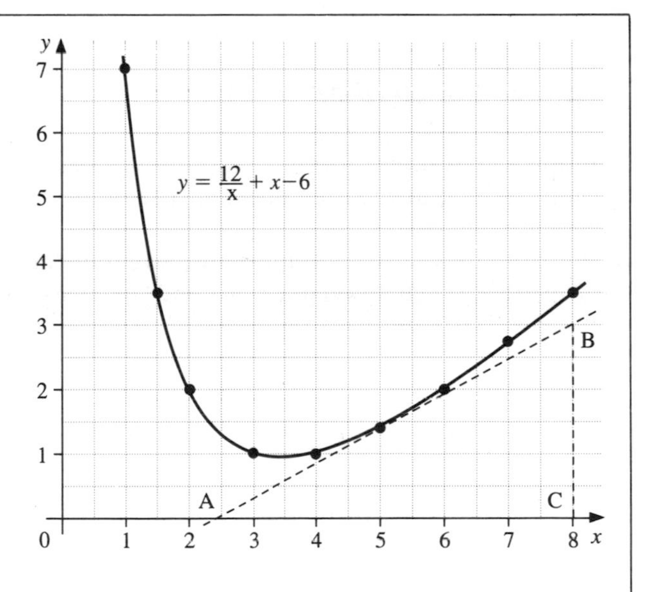

The tangent AB is drawn to touch the curve at $x = 5$.

The gradient of AB $= \frac{BC}{AC}$.

gradient $= \frac{3}{8 - 2{\cdot}4} = \frac{3}{5{\cdot}6} \approx 0{\cdot}54$

It is difficult to obtain an accurate value for the gradient of a tangent so the above result is more realistically 'approximately 0·5'.

Exercise 2

1. Draw the graph of $y = x^2 - 3x$, for $-2 \leqslant x \leqslant 5$.
(Scales: 2 cm = 1 unit for x; 1 cm = 1 unit for y). Find
(a) the gradient of the tangent to the curve at $x = 3$,
(b) the gradient of the tangent to the curve at $x = -1$,
(c) the value of x where the gradient of the curve is zero.

2. Draw the graph of $y = 3^x$, for $-3 \leqslant x \leqslant 3$.
(Scales: 2 cm = 1 unit for x; 1 cm = 2 units for y).
Find the gradient of the curve at $x = 1$.

3. Draw the graph of $y = x^2$, for $0 \leqslant x \leqslant 6$.
(Scales: 2 cm = 1 unit for x; 1 cm = 2 units for y).
(a) Find
(i) the gradient of the curve at $x = 2$,
(ii) the gradient of the curve at $x = 4$.
Give both answers to the nearest whole number.
(b) Finding the gradient of a tangent to a curve is notoriously inaccurate. You will get more reliable results if you take the 'class average' of your answers to (i) and (ii) above.
(c) Use your answers to *predict* the gradient of the curve at $x = 5$.
(d) Draw a tangent at $x = 5$ and find its gradient to check if your prediction was correct.

6.3 Area under a curve

It is sometimes useful to know the area under a curve. In the next section we look at the area under a speed-time graph to calculate distance travelled.
The method below, using trapeziums, gives only an approximate value for the area. When a computer is used, the answer can be found to a high degree of accuracy.

Example 1

Find an approximate value for the area under the curve $y = \dfrac{12}{x}$ between $x = 2$ and $x = 5$.

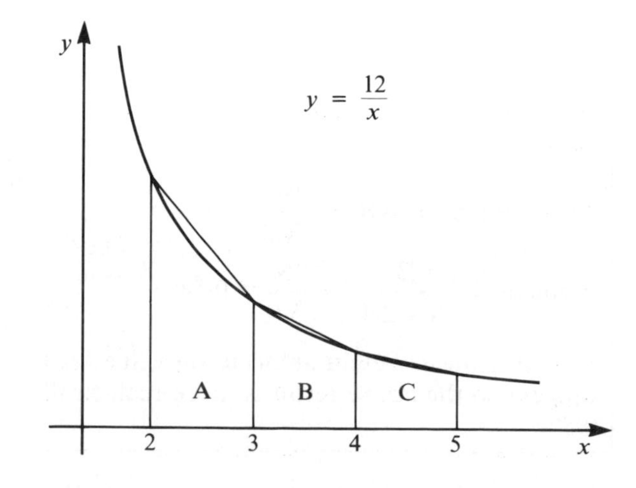

(a) Draw lines parallel to the y-axis at $x = 2, 3, 4, 5$ to form three trapeziums. The area under the curve is approximately equal to the area of the three trapeziums A, B and C.

(b) Calculate the values of y at $x = 2, 3, 4, 5$

$$\text{at } x = 2, y = \tfrac{12}{2} = 6$$

$$\text{at } x = 3, y = \tfrac{12}{3} = 4$$

$$\text{at } x = 4, y = \tfrac{12}{4} = 3$$

$$\text{at } x = 5, y = \tfrac{12}{5} = 2{\cdot}4$$

(c) Calculate the area of each of the trapeziums using the formula area $= \frac{1}{2}(a + b)h$, where a and b are the parallel sides and h is the distance in between.

Area A $= \frac{1}{2}(6 + 4) \times 1 = 5$ sq. units
Area B $= \frac{1}{2}(4 + 3) \times 1 = 3{\cdot}5$ sq. units
Area C $= \frac{1}{2}(3 + 2{\cdot}4) \times 1 = 2{\cdot}7$ sq. units

Total area under the curve $\approx (5 + 3{\cdot}5 + 2{\cdot}7)$
$\approx 11{\cdot}2$ sq. units

Exercise 3

1. Find an approximate value for the area under the curve $y = \dfrac{8}{x}$ between $x = 2$ and $x = 5$.

Divide the area into three trapeziums as shown.

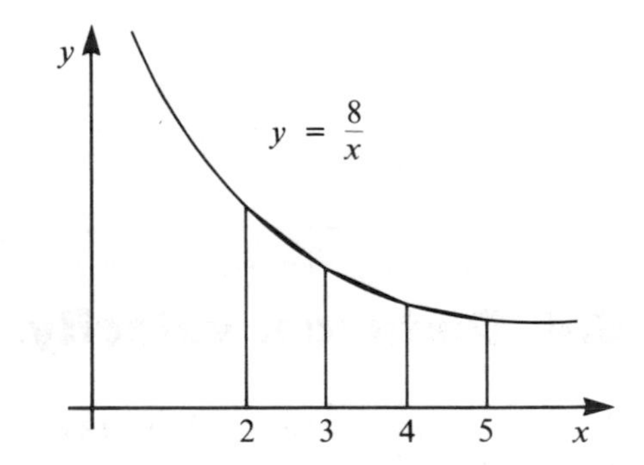

2. Find an approximate value for the area under the curve $y = \dfrac{12}{(x + 2)}$ between $x = 0$ and $x = 4$.

Divide the area into four trapeziums as shown.

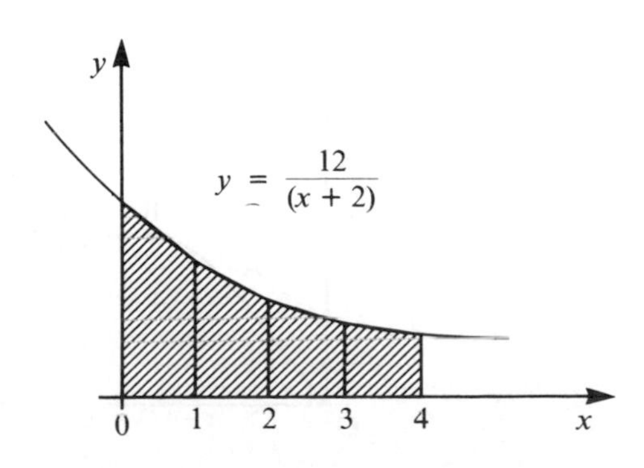

3. Find an approximate value for the area under the curve $y = 6x - x^2$ between $x = 1$ and $x = 4$. Divide the area into three trapeziums of equal width.

4. (a) Sketch the curve $y = 16 - x^2$ for $-4 \leqslant x \leqslant 4$.
(b) Find an approximate value for the area under the curve $y = 16 - x^2$ between $x = 0$ and $x = 3$.
(Divide the area into three trapeziums of equal width.)
(c) State whether your approximate value is greater than or less than the actual value for the area.

5. (a) Find an approximate value for the area under the curve $y = x^2 + 2x + 5$ between $x = 0$ and $x = 4$.
Divide the area into four trapeziums.
(b) State whether your approximate value is greater than or less than the actual value for the area.

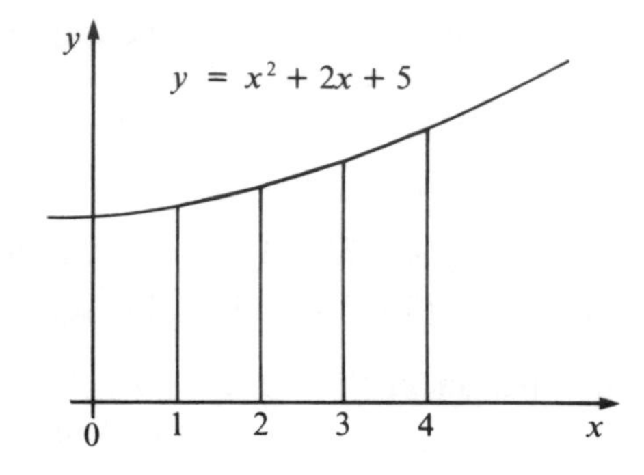

6.† The sketch shows the curve $y = x^2 - 5x + 8$ and the line $y = 4$.
(a) Calculate the x-values at the points A and B.
(b) Find an approximate value for the area shaded.

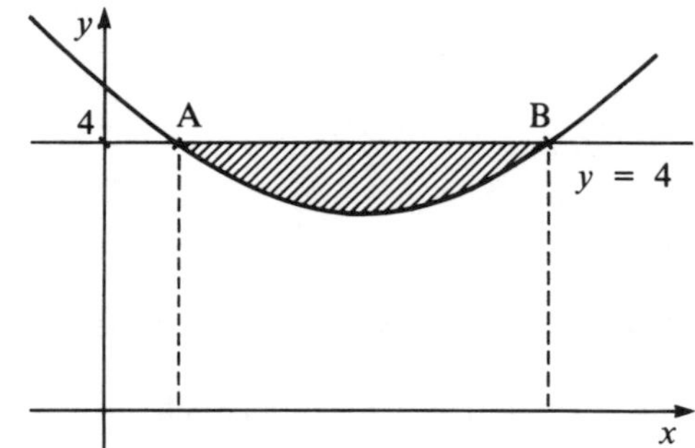

6.4 Distance, velocity, acceleration

When a *distance-time* graph is drawn, the gradient of the graph gives the speed of the object.

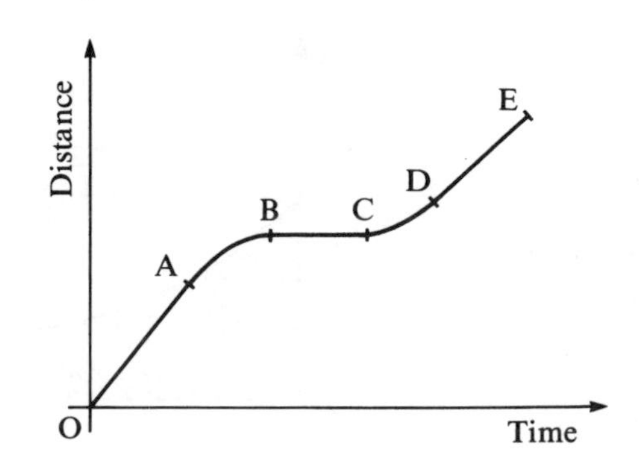

From O to A : constant speed
A to B : speed goes down to zero
B to C : at rest
C to D : speed increases
D to E : constant speed (not as fast as O to A)

When a *velocity-time* graph is drawn two quantities can be found.

(i) acceleration = gradient of graph.
(ii) distance travelled = area under graph.

Example 1

The diagram is the speed-time graph of the first 30 seconds of a car journey.

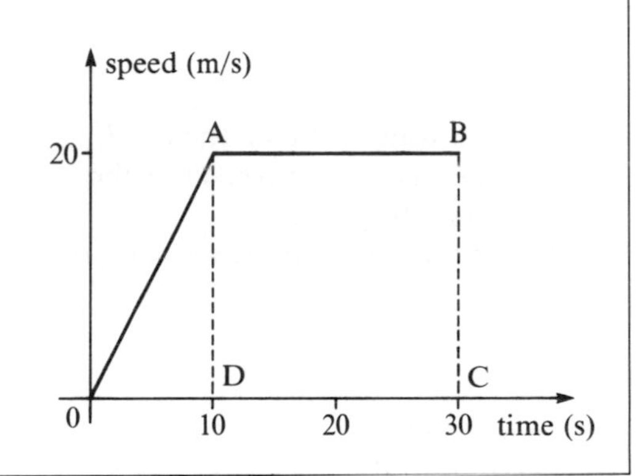

(a) The gradient of line OA $= \frac{20}{10} = 2$.

$\therefore$ The acceleration in the first 10 seconds is $2\,\text{m/s}^2$.

(b) The distance travelled in the first 30 seconds is given by the area of OAD plus the area of ABCD.

Distance $= (\frac{1}{2} \times 10 \times 20) + (20 \times 20) = 500\,\text{m}$

Example 2

The sketch shows the velocity-time graph of a rocket as it takes off. The equation of the curve is $v = 11t^2$

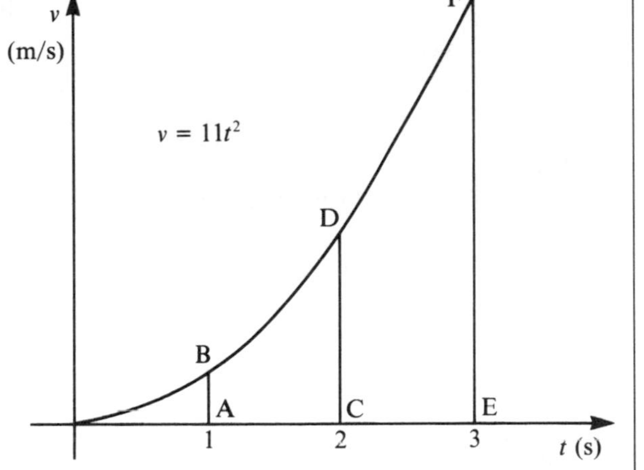

Find an approximate value for the distance travelled by the rocket in the first 3 seconds.

We use 'distance travelled = area under graph'.
Divide the area into 3 trapeziums (one of which is a triangle).

AB $= 11$, CD $= 44$, EF $= 99$
(from $v = 11t^2$)

Area $\approx \frac{1}{2}(0 + 11) \times 1 + \frac{1}{2}(11 + 44) \times 1 + \frac{1}{2}(44 + 99) \times 1$
$\approx 104{\cdot}75$
Distance travelled in the first 3 seconds is about 105 m.

Note: To find the acceleration of the rocket at $t = 2$, we could draw a tangent to the curve at this point and so find its gradient.

Exercise 4

The graphs show speed v in m/s and time t in seconds.

1. Find
   (a) the acceleration when $t = 4$,
   (b) the total distance travelled.

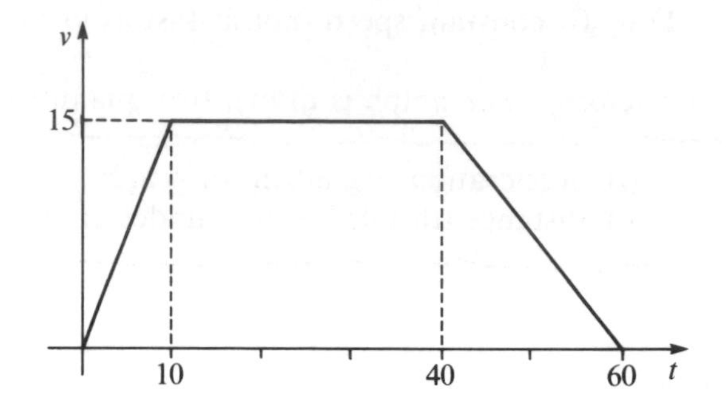

2. Find
   (a) the total distance travelled,
   (b) the distance travelled in the first 10 seconds,
   (c) the acceleration when $t = 20$.

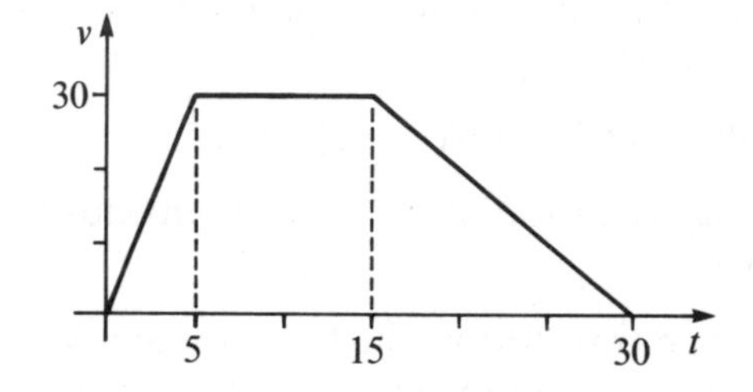

3. Find
   (a) the total distance travelled,
   (b) the distance travelled in the first 40 seconds.
   (c) the acceleration when $t = 15$.

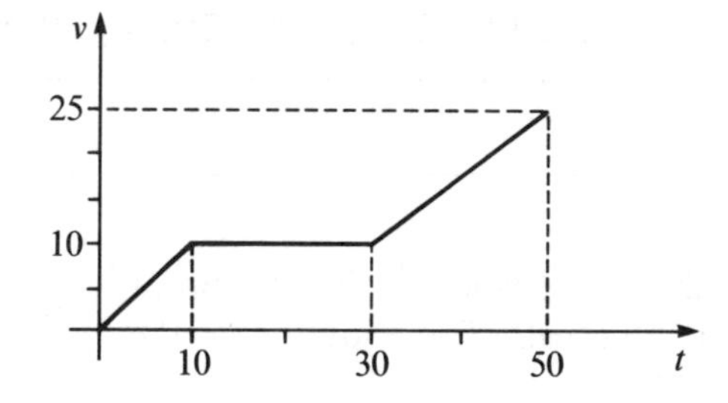

4. Find
   (a) V if the total distance travelled is 900 m,
   (b) the distance travelled in the first 60 seconds.

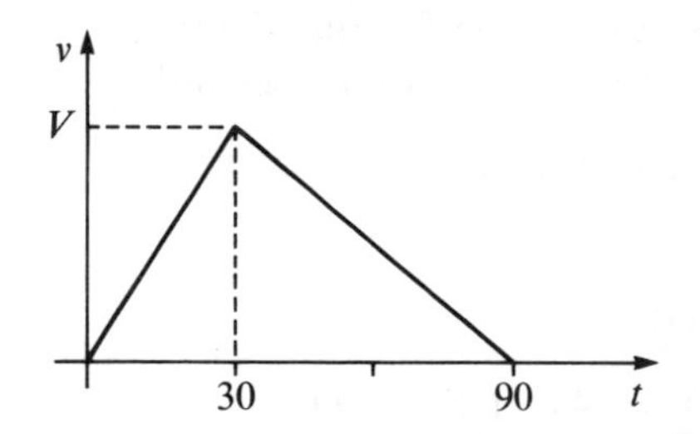

5. Find
   (a) T if the initial acceleration is $2\,\text{m/s}^2$,
   (b) the total distance travelled,
   (c) the average speed for the whole journey.

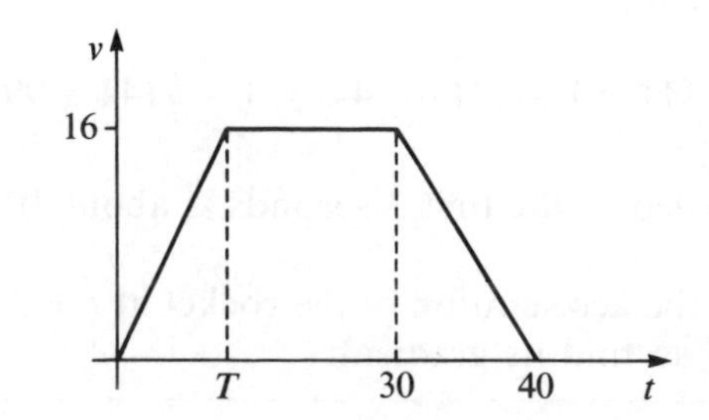

6. The speed of a train is measured at regular intervals of time from $t = 0$ to $t = 60$ s, as shown below.

t s	0	10	20	30	40	50	60
v m/s	0	10	16	19·7	22·2	23·8	24·7

Draw a speed-time graph to illustrate the motion. Plot t on the horizontal axis with a scale of 1 cm to 5 s and plot v on the vertical axis with a scale of 2 cm to 5 m/s.
Use the graph to estimate:
(a) the acceleration at $t = 10$,
(b) the distance travelled by the train from $t = 30$ to $t = 60$.

7. The speed of a car is measured at regular intervals of time from $t = 0$ to $t = 60$ s, as shown below.

t s	0	10	20	30	40	50	60
v m/s	0	1·3	3·2	6	10·1	16·5	30

Draw a speed-time graph using the same scales as in Question **6**.
Use the graph to estimate:
(a) the acceleration at $t = 30$,
(b) the distance travelled by the car from $t = 20$ to $t = 50$.

8. The sketch graph shows the distance-time graph of an object in an experiment. [N.B. Not a velocity-time graph.]

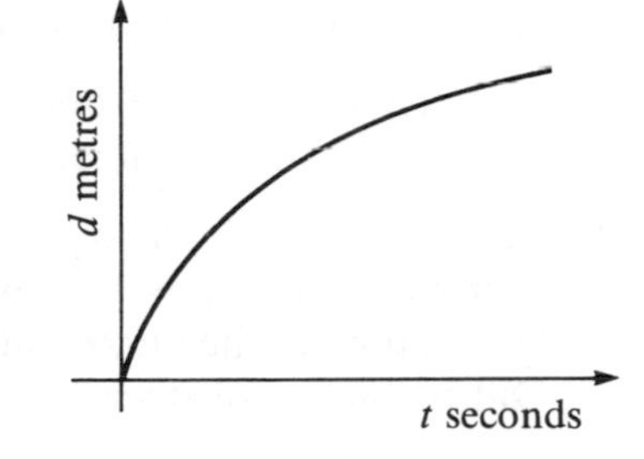

The graph passes through (0, 0), (0·2, 1·2), (0·5, 1·6), (1, 2), (2, 2·5), (3, 2·9).

Draw your own accurate graph and then draw a suitable tangent to find the speed of the object at $t = 0·5$ seconds. Use a scale of 5 cm to 1 second across the page and 5 cm to 1 unit up the page.

9.† Given that the average speed for the whole journey is 37·5 m/s and that the deceleration between T and $2T$ is 2·5 m/s^2, find
(a) the value of V, (b) the value of T.

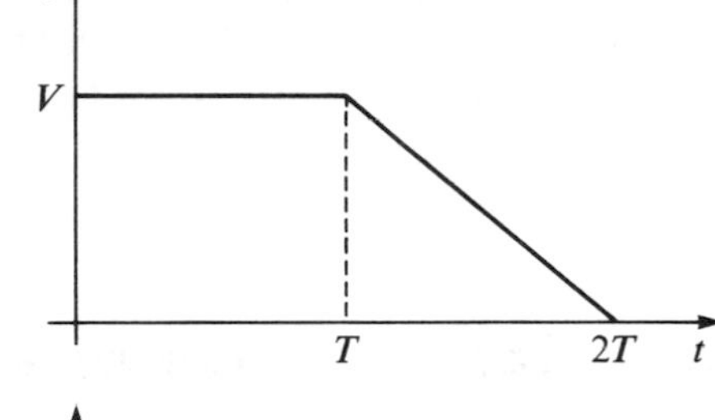

10.† Given that the total distance travelled is 4 km and that the initial deceleration is 4 m/s^2, find
(a) the value of V, (b) the value of T.

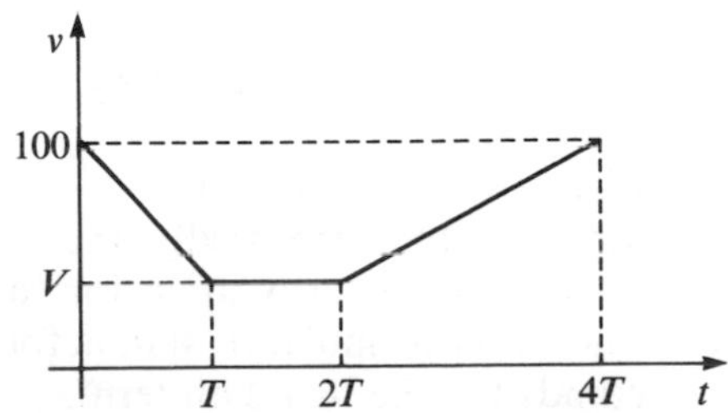

Exercise 5

Sketch a speed-time graph for each question. All accelerations are taken to be uniform.

1. A car accelerated from 0 to 50 m/s in 9 s. How far did it travel in this time?

2. A motor cycle accelerated from 10 m/s to 30 m/s in 6 s. How far did it travel in this time?

3. A train slowed down from 50 km/h to 10 km/h in 2 minutes. How far did it travel in this time?

4. When taking off, an aircraft accelerates from 0 to 100 m/s in a distance of 500 m. How long did it take to travel this distance?

5. An earthworm accelerates from a speed of 0·01 m/s to 0·02 m/s over a distance of 0·9 m. How long did it take?

6. A car travelling at 60 m.p.h. is stopped in 6 seconds. How far does it travel in this time?
Give the answer in yards. [1 mile = 1760 yards].

7. A car accelerates from 15 m.p.h. to 60 m.p.h. in 3 seconds. How many yards does it travel in this time?

8. At lift-off a rocket accelerates from 0 to 1000 km/h in just 10 s. How far does it travel in this time?

9. A coach accelerated from 0 to 60 km/h in 30 s. How many metres did it travel in this time?

10. Mr Wheeler was driving a car at 30 m/s when he saw an obstacle 45 m in front of him. It took a reaction time of 0·3 seconds before he could press the brakes and a further 2·5 seconds to stop the car. Did he hit the obstacle?

11. An aircraft is cruising at a speed of 200 m/s. When it lands it must be travelling at a speed of 50 m/s. In the air it can slow down at a rate of 0·2 m/s^2. On the ground it slows down at a rate of 2 m/s^2. Draw a velocity time graph for the aircraft as it reduces its speed from 200 m/s to 50 m/s and then to 0 m/s.
How far does it travel in this time?

6.5 Convergence of sequences

Most sequences are generated by iterative procedures.
In an iterative procedure, a calculation is done repeatedly; the output becomes the new input for further use.

Input → Calculation → Output (output fed back to input)

In an iteration formula, u_n stands for the nth term of a sequence and u_{n+1} stands for the $(n+1)$th term.

Example 1

Consider the iteration formula $u_{n+1} = \dfrac{u_n}{3} + 4$

This can be thought of as: next term $= \dfrac{\text{last term}}{3} + 4$

Take any value for u_1, say 7, to begin.
Then the following terms occur:

$u_1 = 7$ $\quad u_5 = 6{\cdot}012345679$
$u_2 = 6{\cdot}3$ $\quad u_6 = 6{\cdot}004115226$
$u_3 = 6{\cdot}1$ $\quad u_7 = 6{\cdot}001371742$
$u_4 = 6{\cdot}037037037$

Each term of the sequence is getting nearer and nearer to a fixed number, in this case 6.
We say this sequence *converges* towards 6 and we call 6 the *limit* of the sequence. Whatever the number we start with for u_1, the sequence will always converge towards 6.

In some sequences, the terms do not get nearer to a fixed number. We say the sequence *diverges* and it has no limit.

In mathematics, convergent sequences are generally useful but divergent sequences are not.

Exercise 6

1. The iteration formula for a sequence is $a_{n+1} = \dfrac{a_n}{3} + 7$.

(a) Take any value for a_1 and work out $a_2, a_3, \ldots a_{12}$. Does the sequence appear to converge?
(b) If so, what is the limit of the sequence?

2. The iteration formula for a sequence is $a_{n+1} = \dfrac{a_n}{5} + 2$.

Take any value for a_1 and find the sequence $a_2, a_3, a_4 \ldots$.
What is the limit of the sequence?

3. Here are some more iteration formulas. Take $a_1 = 2$ and work out $a_2, a_3, \ldots$. State whether the sequence converges or diverges. If it does converge state the limit of the sequence. Sometimes the limit is not an obvious whole number or fraction. If necessary give the limit correct to 3 s.f.

(a) $a_{n+1} = \dfrac{a_n}{5} + 8$ (b) $u_{n+1} = \dfrac{5}{6 + u_n}$

(c) $a_{n+1} = \dfrac{3}{a_n^3 - 1}$ (d) $u_{n+1} = \sqrt{(u_n + 8)}$

(e) $u_{n+1} = \sqrt{2 + \sqrt{1 + u_n}}$

4. A calculator can be used ` ficiently using the reciprocal button $\boxed{\frac{1}{x}}$ as follows for the formula

$$a_{n+1} = \frac{11}{a_n + 3}$$

Take $a_1 = 5$.

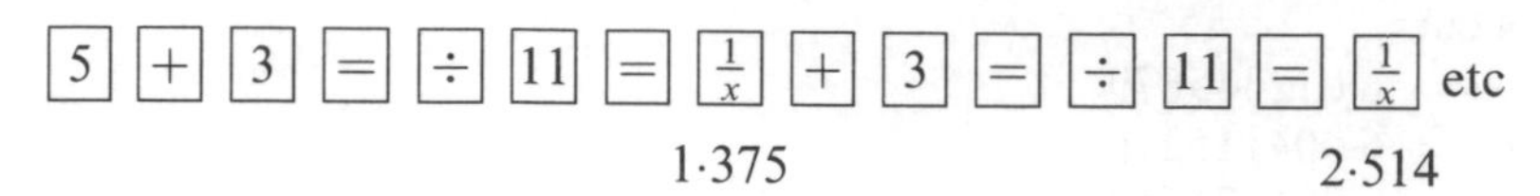

Find the limit of this sequence using the reciprocal button as suggested.

5. Some sequences converge rather slowly. Try this simple program written in BASIC to print out a_2, a_3, a_4 ... for the formula

$$a_{n+1} = \frac{11}{a_n + 3}.$$

```
10    A = 1
20    FOR N = 1 TO 25
30    A = 11/(A + 3)
40    PRINT A
50    NEXT N
RUN
```

In line 10, we took $a_1 = 1$. You could take any value. In line 20 we asked for 25 terms of the sequence. You could ask for more terms or less terms if you wish.

6. Use this program to work out terms of the sequence

$$u_{n+1} = \frac{11}{u_n} - 3.$$

```
10    U = 5                                   [i.e. u1 = 5]
20    FOR N = 1 TO 25
30    U = 11/U - 3
40    PRINT U
50    NEXT N
RUN
```

Run the program on a computer and write down the limit of the sequence correct to 4 s.f.

7. Write your own program for the sequence

$$u_{n+1} = \frac{7}{u_n + 1}$$

Write down the limit of the sequence correct to 5 s.f.

8. Consider the iteration formula

$$u_{n+1} = 3 - u_n^2$$

(a) Describe the sequence if you take $u_1 = 3$.
(b) Describe the sequence if you take $u_1 = 1$.

9. This is a rather nice sequence. Start with any two (different) numbers. After these, each new term is the mean of the two preceding terms.
(a) Here we started with 3 and 6.

$$3, \quad 6, \quad \underset{\left(\frac{3+6}{2}\right)}{4{\cdot}5}, \quad \underset{\left(\frac{6+4{\cdot}5}{2}\right)}{5{\cdot}25}, \quad \underset{\left(\frac{4{\cdot}5+5{\cdot}25}{2}\right)}{4{\cdot}875}, \quad \ldots$$

Carry on this sequence for another 16 terms and write down the limit.
(b) Now start with another two numbers, say 5 and 8. What is the limit of this sequence?
(c) Can you see a connection between the limit and the two numbers you start with?
[If not, try the sequence starting 6, 12, ...]
(d) (i) Predict the limit for the sequence which starts 4, 13, ...
(ii) Check to see if your prediction was correct.
(e)† If you have time, investigate the sequence starting with any *three* numbers. After these each new term is the mean of the preceding *three* terms.
Can you find a connection between the limit and the three numbers you start with?

10. Here is an iteration formula in which K is any positive integer.

$$a_{n+1} = \frac{1}{2}\left(a_n + \frac{K}{a_n}\right)$$

(a) Choose a value for K, say 8.
(b) Take any value for a_1 and find the sequence $a_2, a_3, a_4 \ldots$
(c) Write down the apparent limit of the sequence.
(d) Can you find any connection between the number K and the limit of the sequence?
(e) Repeat for other values of K.

11.† This sequence is known as the Fibonacci Sequence.
1, 1, 2, 3, 5, 8, 13, 21.

We can see that the nth term = $(n-1)$th term + $(n-2)$th term.
or $u_n = u_{n-1} + u_{n-2}$
Here is an expression for the general term, u_n

$$u_n = \frac{1}{\sqrt{5}}\left(\frac{1+\sqrt{5}}{2}\right)^n - \frac{1}{\sqrt{5}}\left(\frac{1-\sqrt{5}}{2}\right)^n$$

Put $n = 4$, $n = 6$, $n = 9$ to check if this formula for u_n gives the expected values.

Solving equations by iterative methods

We have seen that many sequences converge to a limit. Consider the iterative formula $u_{n+1} = \frac{u_n}{4} + 6$

After many cycles, u_{n+1} and u_n will be approximately the same and each could be called x.

$$\therefore \quad x = \frac{x}{4} + 6$$

$$\frac{3x}{4} = 6$$

$$x = \frac{24}{3} = 8 \qquad \text{(which } is \text{ the limit).}$$

This idea can be used 'in reverse' to help solve equations.

Consider solving the equation $x^2 + x - 5 = 0$ by iterative methods.

(a) As a first attempt, we might put $x = 5 - x^2$ and use this iterative formula $x_{n+1} = 5 - x_n^2$

If $x_1 = 1$
$x_2 = 4,\ x_3 = -9,\ x_4 = -75.$

Unfortunately, the sequence diverges which is not helpful.

(b) Try rearranging $x^2 + x - 5 = 0$ differently.

$$x^2 + x = 5$$

$$x(x + 1) = 5 \qquad \ldots [A]$$

$$x = \frac{5}{x + 1}$$

This suggests the iterative formula $x_{n+1} = \frac{5}{x_n + 1}$. Try this formula several times, starting with $x_1 = 2$. Here are the next terms (rounded off).

2, 1·666, 1·875, 1·739, 1·825, 1·769, 1·805, 1·782, 1·797, 1·787, 1·794, 1·790, 1·792.

The terms are converging (rather slowly) towards 1·79 correct to 2 d.p.
So $x = 1{\cdot}79$ is a solution (to 2 d.p.) of the equation $x^2 + x - 5 = 0$

(c) But this is a *quadratic* equation and it should have *two* solutions. Go back to line [A]

$$x(x + 1) = 5$$

This time divide by x first.

$$x + 1 = \frac{5}{x}$$

$$x = \frac{5}{x} - 1$$

This suggests the iterative formula $x_{n+1} = \dfrac{5}{x_n} - 1$

If we take any value for x_1 ($\neq 0$), the sequence does in fact have a limit of $-2{\cdot}79$ (to 2 d.p.)
So we have found the two solutions of the equation
$x^2 + x - 5 = 0$

They are $x = 1{\cdot}79$ and $x = -2{\cdot}79$ (to 2 d.p.)

The method of factorising and dividing (as in line [A]) works for quadratic equations.

Example 2

Find solutions of the quadratic equation $x^2 - 3x - 1 = 0$ by iterative methods.

We write
$$x^2 - 3x = 1$$
$$x(x - 3) = 1$$

either $x = \dfrac{1}{x - 3}$ or $x - 3 = \dfrac{1}{x}$

$$x = \frac{1}{x} + 3$$

To find the two solutions, we use the two iterative formulas

$$x_{n+1} = \frac{1}{x_n - 3} \quad \text{and} \quad x_{n+1} = \frac{1}{x_n} + 3$$

Taking any value for x_1 to start with, the solutions we obtain are $x = -0{\cdot}30$ and $x = 3{\cdot}30$ (both to 2 d.p.)

Note: These sequences converge rather slowly, but the methods are useful when used with a computer which gives results much more quickly.

Exercise 7

1. Find solutions of the following quadratic equations. For each equation, find two iterative formulas as in the example above. Give solutions correct to 2 d.p.

(a) $x^2 - 5x - 1 = 0$ (b) $x^2 + x - 4 = 0$
(c) $x^2 = x + 3$

2. Show that the iterative formula $x_{n+1} = \frac{1}{2}\left(\frac{1}{x_n} + 5\right)$ can be obtained to solve the equation $2x^2 - 5x - 1 = 0$. Find the other iterative formula and hence find both solutions correct to 2 d.p.

3. Find two iterative formulas to solve the equation $x^2 + 9 = 6x$. What do you notice?

4. Show that the equation $x^2 - x + 4 = 0$ can be rearranged to give either $x = \frac{4}{1 - x}$ or $x = 1 - \frac{4}{x}$.

Use these iterative formulas to solve the equation. What do you notice?

5. The sketch shows $y = x^4$ and $y = 3x^3 + 10$

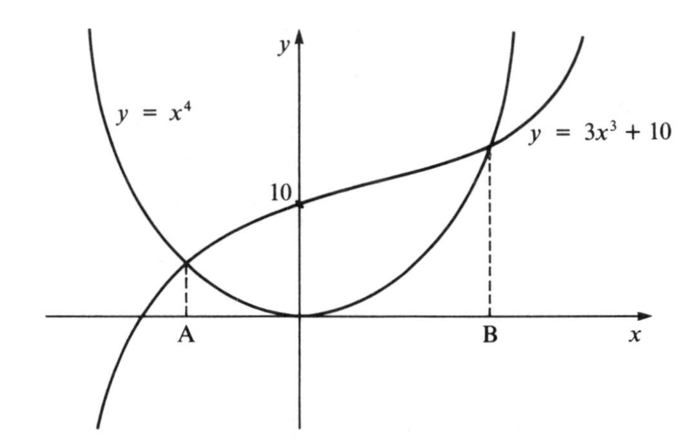

The equation $x^4 = 3x^3 + 10$ will have two solutions at the points marked A, B. Show that the equation can be rearranged to give

either $x = \frac{10}{x^3} + 3$ or $x = \sqrt[3]{\frac{10}{x - 3}}$.

Use these iterative formulas to find the two solutions. Start with $x_1 = 3$ in the first formula and with $x_1 = -1$ in the second formula.

6. Show that the equation $x^3 = 9 - x^2$ can be rearranged as either

$x = \frac{9}{x^2} - 1$ or $x = \sqrt{\left(\frac{9}{x + 1}\right)}$.

On the same axes, draw a sketch graph of $y = x^3$ and $y = 9 - x^2$ to show that there is only one solution.
Find which of the above iterative formulas converges and give the solution correct to 2 d.p.

6.6 Algebraic fractions

Simplifying fractions

Example 1

Simplify the following (a) $\frac{3a}{5a^2}$ (b) $\frac{3y+y^2}{6y}$

(a) $\frac{3a}{5a^2} = \frac{3 \times \not{a}}{5 \times a \times \not{a}} = \frac{3}{5a}$ (b) $\frac{y(3+y)}{6y} = \frac{3+y}{6}$

Exercise 8

Simplify as far as possible, where you can.

1. $\frac{25}{35}$ **2.** $\frac{84}{96}$ **3.** $\frac{5y^2}{y}$

4. $\frac{y}{2y}$ **5.** $\frac{8x^2}{2x^2}$ **6.** $\frac{2x}{4y}$

7. $\frac{6y}{3y}$ **8.** $\frac{5ab}{10b}$ **9.** $\frac{8ab^2}{12ab}$

10. $\frac{7a^2b}{35ab^2}$ **11.** $\frac{(2a)^2}{4a}$ **12.** $\frac{7yx}{8xy}$

13. $\frac{5x+2x^2}{3x}$ **14.** $\frac{9x+3}{3x}$ **15.** $\frac{25+7}{25}$

16. $\frac{4a+5a^2}{5a}$ **17.** $\frac{3x}{4x-x^2}$ **18.** $\frac{5ab}{15a+10a^2}$

19. $\frac{5x+4}{8x}$ **20.** $\frac{12x+6}{6y}$ **21.** $\frac{5x+10y}{15xy}$

22. $\frac{18a-3ab}{6a^2}$ **23.** $\frac{4ab+8a^2}{2ab}$ **24.** $\frac{(2x)^2-8x}{4x}$

25. $\frac{x^2+2x}{x^2-3x}$ **26.** $\frac{x^2-3x}{x^2-2x-3}$ **27.** $\frac{x^2+4x}{2x^2-10x}$

28. $\frac{x^2+6x+5}{x^2-x-2}$ **29.** $\frac{x^2-4x-21}{x^2-5x-14}$ **30.** $\frac{x^2+7x+10}{x^2-4}$

Addition and subtraction of algebraic fractions

Example 2

Write as a single fraction $\frac{2}{x}+\frac{3}{y}$

The L.C.M. of x and y is xy.

$$\therefore \quad \frac{2}{x}+\frac{3}{y}=\frac{2y}{xy}+\frac{3x}{xy}=\frac{2y+3x}{xy}$$

Exercise 9

Simplify the following.

1. $\frac{2}{5}+\frac{1}{5}$ **2.** $\frac{2x}{5}+\frac{x}{5}$ **3.** $\frac{2}{x}+\frac{1}{x}$

4. $\frac{1}{7}+\frac{3}{7}$ **5.** $\frac{x}{7}+\frac{3x}{7}$ **6.** $\frac{1}{7x}+\frac{3}{7x}$

7. $\frac{5}{8}+\frac{1}{4}$ **8.** $\frac{5x}{8}+\frac{x}{4}$ **9.** $\frac{5}{8x}+\frac{1}{4x}$

10. $\frac{2}{3}+\frac{1}{6}$ **11.** $\frac{2x}{3}+\frac{x}{6}$ **12.** $\frac{2}{3x}+\frac{1}{6x}$

13. $\frac{3}{4}+\frac{2}{5}$ **14.** $\frac{3x}{4}+\frac{2x}{5}$ **15.** $\frac{3}{4x}+\frac{2}{5x}$

16. $\frac{3}{4}-\frac{2}{3}$ **17.** $\frac{3x}{4}-\frac{2x}{3}$ **18.** $\frac{3}{4x}-\frac{2}{3x}$

19. $\frac{x}{2}+\frac{x+1}{3}$ **20.** $\frac{x-1}{3}+\frac{x+2}{4}$ **21.** $\frac{2x-1}{5}+\frac{x+3}{2}$

22. $\frac{x+1}{3}-\frac{(2x+1)}{4}$ **23.** $\frac{x-3}{3}-\frac{(x-2)}{5}$ **24.** $\frac{2x+1}{7}-\frac{(x+2)}{2}$

25. $\frac{1}{x}+\frac{2}{x+1}$ **26.** $\frac{3}{x-2}+\frac{4}{x}$ **27.** $\frac{5}{x-2}+\frac{3}{x+3}$

28. $\frac{7}{x+1}-\frac{3}{x+2}$ **29.** $\frac{2}{x+3}-\frac{5}{x-1}$ **30.** $\frac{3}{x-2}-\frac{4}{x+1}$

6.7 Quadratic equations

Factorising quadratic expressions

Earlier in the book a pair of brackets like $(x+4)(x-3)$ were multiplied to give x^2+x-12.

The reverse of this process is called factorising.

Example 1

Factorise x^2+6x+8

(a) Find two numbers which multiply to give 8 and add up to 6.

(b) Put these numbers into brackets.
So $x^2+6x+8 = (x+4)(x+2)$.

Example 2

Factorise (a) $x^2+2x-15$
(b) x^2-6x+8

(a) Two numbers which multiply to give -15 and add up to $+2$ are -3 and 5.
$\therefore\ x^2+2x-15 = (x-3)(x+5)$.

(b) Two numbers which multiply to give $+8$ and add up to -6 are -2 and -4.
$\therefore\ x^2-6x+8 = (x-2)(x-4)$.

Exercise 10

Factorise the following:

1. $x^2+7x+10$ **2.** $x^2+7x+12$ **3.** $x^2+8x+15$

4. $x^2+10x+21$ **5.** $x^2+8x+12$ **6.** $y^2+12y+35$

7. $y^2+11y+24$ **8.** $y^2+10y+25$ **9.** $y^2+15y+36$

10. $a^2-3a-10$ **11.** a^2-a-12 **12.** z^2+z-6

13. $x^2-2x-35$ **14.** $x^2-5x-24$ **15.** x^2-6x+8

16. y^2-5y+6 **17.** $x^2-8x+15$ **18.** a^2-a-6

19. $a^2+14a+45$ **20.** $b^2-4b-21$ **21.** $x^2-8x+16$

22. y^2+2y+1 **23.** $y^2-3y-28$ **24.** x^2-x-20

25. $x^2-8x-240$ **26.** $x^2-26x+165$ **27.** $y^2+3y-108$

28. x^2-49 **29.** x^2-9 **30.** x^2-16

Example 3

Factorise $3x^2 + 13x + 4$

(a) Find two numbers which multiply to give (3×4), i.e. 12, and add up to 13. In this case the numbers are 1 and 12.

(b) Split the '$13x$' term, $3x^2 + x + 12x + 4$

(c) Factorise in pairs, $x(3x + 1) + 4(3x + 1)$

(d) $(3x + 1)$ is common, $(3x + 1)(x + 4)$

Exercise 11

Factorise the following:

1. $2x^2 + 5x + 3$ **2.** $2x^2 + 7x + 3$ **3.** $3x^2 + 7x + 2$
4. $2x^2 + 11x + 12$ **5.** $3x^2 + 8x + 4$ **6.** $2x^2 + 7x + 5$
7. $3x^2 - 5x - 2$ **8.** $2x^2 - x - 15$ **9.** $2x^2 + x - 21$
10. $3x^2 - 17x - 28$ **11.** $6x^2 + 7x + 2$ **12.** $3x^2 - 11x + 6$
13. $3y^2 - 11y + 10$ **14.** $6y^2 + 7y - 3$ **15.** $10x^2 + 9x + 2$
16. $6x^2 - 19x + 3$ **17.** $8x^2 - 10x - 3$ **18.** $12x^2 + 23x + 10$
19. $4y^2 - 23y + 15$ **20.** $6x^2 - 27x + 30$

The difference of two squares

$x^2 - y^2 = (x - y)(x + y)$
Remember this result.

Example 4

Factorise (a) $y^2 - 16$ (b) $4a^2 - b^2$

(a) $y^2 - 16 = (y - 4)(y + 4)$ (b) $4a^2 - b^2 = (2a - b)(2a + b)$

Exercise 12

Factorise the following:

1. $y^2 - a^2$ **2.** $m^2 - n^2$ **3.** $x^2 - t^2$
4. $y^2 - 1$ **5.** $x^2 - 9$ **6.** $a^2 - 25$
7. $x^2 - \frac{1}{4}$ **8.** $x^2 - \frac{1}{9}$ **9.** $4x^2 - y^2$
10. $a^2 - 4b^2$ **11.** $25x^2 - 4y^2$ **12.** $9x^2 - 16y^2$
13. $x^2 - \frac{y^2}{4}$ **14.** $9m^2 - \frac{4}{9}n^2$ **15.** $16t^2 - \frac{4}{25}s^2$
16. $4x^2 - \frac{z^2}{100}$ **17.** $x^3 - x$ **18.** $a^3 - ab^2$
19. $4x^3 - x$ **20.** $8x^3 - 2xy^2$

Methods of solving quadratic equations

The last three exercises have been about factorising expressions. These contain no 'equals' sign, but equations do.
Quadratic equations always have an x^2 term, and often an x term as well as a number term. They generally have two different solutions. Here three different methods of solution are considered.

(a) Solution by factors

Consider the equation $a \times b = 0$, where a and b are numbers. The product $a \times b$ can only be zero if either a or b (or both) is equal to zero. Can you think of other possible pairs of numbers which multiply together to give zero?

Example 5

Solve the equation $x^2 + x - 12 = 0$

Factorising, $(x - 3)(x + 4) = 0$

either $x - 3 = 0$ or $x + 4 = 0$
$x = 3$ $x = -4$

Example 6

Solve the equation $6x^2 + x - 2 = 0$

Factorising, $(2x - 1)(3x + 2) = 0$

either $2x - 1 = 0$ or $3x + 2 = 0$
$2x = 1$ $3x = -2$
$x = \frac{1}{2}$ $x = -\frac{2}{3}$

Exercise 13

Solve the following equations

1. $x^2 + 7x + 12 = 0$
2. $x^2 + 7x + 10 = 0$
3. $x^2 + 2x - 15 = 0$
4. $x^2 + x - 6 = 0$
5. $x^2 - 8x + 12 = 0$
6. $x^2 + 10x + 21 = 0$
7. $x^2 - 5x + 6 = 0$
8. $x^2 - 4x - 5 = 0$
9. $x^2 + 5x - 14 = 0$
10. $2x^2 - 3x - 2 = 0$
11. $3x^2 + 10x - 8 = 0$
12. $2x^2 + 7x - 15 = 0$
13. $6x^2 - 13x + 6 = 0$
14. $4x^2 - 29x + 7 = 0$
15. $10x^2 - x - 3 = 0$
16. $y^2 - 15y + 56 = 0$
17. $12y^2 - 16y + 5 = 0$
18. $y^2 + 2y - 63 = 0$
19. $x^2 + 2x + 1 = 0$
20. $x^2 - 6x + 9 = 0$
21. $x^2 + 10x + 25 = 0$
22. $x^2 - 14x + 49 = 0$
23. $6a^2 - a - 1 = 0$
24. $4a^2 - 3a - 10 = 0$

25. $z^2 - 8z - 65 = 0$
26. $6x^2 + 17x - 3 = 0$
27. $10k^2 + 19k - 2 = 0$
28. $y^2 - 2y + 1 = 0$
29. $36x^2 + x - 2 = 0$
30. $20x^2 - 7x - 3 = 0$

Example 7

Solve the equation $x^2 - 7x = 0$

Factorising, $x(x-7) = 0$

either $x = 0$ or $x - 7 = 0$

$x = 7$

The solutions are $x = 0$ and $x = 7$.

Exercise 14

Solve the following equations.

1. $x^2 - 3x = 0$
2. $x^2 + 7x = 0$
3. $2x^2 - 2x = 0$
4. $3x^2 - x = 0$
5. $x^2 - 16 = 0$
6. $x^2 - 49 = 0$
7. $4x^2 - 1 = 0$
8. $9x^2 - 4 = 0$
9. $6y^2 + 9y = 0$
10. $6a^2 - 9a = 0$
11. $10x^2 - 55x = 0$
12. $16x^2 - 1 = 0$
13. $y^2 - \frac{1}{4} = 0$
14. $56x^2 - 35x = 0$
15. $36x^2 - 3x = 0$
16. $x^2 = 6x$
17. $x^2 = 11x$
18. $2x^2 = 3x$
19. $x^2 = x$
20. $4x = x^2$

(b) Solution by formula

The solutions of the quadratic equation $ax^2 + bx + c = 0$ are given by the formula $x = \dfrac{-b \pm \sqrt{(b^2 - 4ac)}}{2a}$.

Use this formula only after trying (and failing) to factorise.

Example 8

Solve the equation $2x^2 - 3x - 4 = 0$.

In this case $a = 2,\ b = -3,\ c = -4$.

$$x = \frac{-(-3) \pm \sqrt{[(-3)^2 - (4 \times 2 \times -4)]}}{2 \times 2}$$

$$x = \frac{3 \pm \sqrt{[9 + 32]}}{4} = \frac{3 \pm \sqrt{41}}{4}$$

$$x = \frac{3 \pm 6{\cdot}403}{4}$$

either $x = \dfrac{3 + 6{\cdot}403}{4} = 2{\cdot}35$ (2 decimal places)

or $x = \dfrac{3 - 6{\cdot}403}{4} = \dfrac{-3{\cdot}403}{4} = -0{\cdot}85$ (2 decimal places).

Exercise 15

Solve the following, giving answers to two decimal places where necessary.

1. $2x^2 + 11x + 5 = 0$
2. $3x^2 + 11x + 6 = 0$
3. $6x^2 + 7x + 2 = 0$
4. $3x^2 - 10x + 3 = 0$
5. $5x^2 - 7x + 2 = 0$
6. $6x^2 - 11x + 3 = 0$
7. $2x^2 + 6x + 3 = 0$
8. $x^2 + 4x + 1 = 0$
9. $5x^2 - 5x + 1 = 0$
10. $x^2 - 7x + 2 = 0$
11. $2x^2 + 5x - 1 = 0$
12. $3x^2 + x - 3 = 0$
13. $3x^2 + 8x - 6 = 0$
14. $3x^2 - 7x - 20 = 0$
15. $2x^2 - 7x - 15 = 0$
16. $x^2 - 3x - 2 = 0$
17. $2x^2 + 6x - 1 = 0$
18. $6x^2 - 11x - 7 = 0$
19. $3x^2 + 25x + 8 = 0$
20. $3y^2 - 2y - 5 = 0$
21. $2y^2 - 5y + 1 = 0$
22. $\frac{1}{2}y^2 + 3y + 1 = 0$
23. $2 - x - 6x^2 = 0$
24. $3 + 4x - 2x^2 = 0$

The solution to a problem can involve an equation which does not at first appear to be quadratic. The terms in the equation may need to be rearranged as shown below.

Example 9

Solve:

$$\begin{aligned} 2x(x-1) &= (x+1)^2 - 5 \\ 2x^2 - 2x &= x^2 + 2x + 1 - 5 \\ 2x^2 - 2x - x^2 - 2x - 1 + 5 &= 0 \\ x^2 - 4x + 4 &= 0 \\ (x-2)(x-2) &= 0 \\ x &= 2 \end{aligned}$$

In this example the quadratic has a repeated root of $x = 2$.

Exercise 16

Solve the following equations, giving answers to two decimal places where necessary.

1. $x^2 = 6 - x$
2. $x(x + 10) = -21$
3. $3x + 2 = 2x^2$
4. $x^2 + 4 = 5x$
5. $6x(x + 1) = 5 - x$
6. $(2x)^2 = x(x - 14) - 5$
7. $(x - 3)^2 = 10$
8. $(x + 1)^2 - 10 = 2x(x - 2)$
9. $(2x - 1)^2 = (x - 1)^2 + 8$
10. $3x(x + 2) - x(x - 2) + 6 = 0$
11. $x = \frac{15}{x} - 22$
12. $x + 5 = \frac{14}{x}$
13. $4x + \frac{7}{x} = 29$
14. $10x = 1 + \frac{3}{x}$

15. $2x^2 = 7x$

16. $16 = \dfrac{1}{x^2}$

17. $2x + 2 = \dfrac{7}{x} - 1$

18. $\dfrac{2}{x} + \dfrac{2}{x+1} = 3$

19. $\dfrac{3}{x-1} + \dfrac{3}{x+1} = 4$

20. $\dfrac{2}{x-2} + \dfrac{4}{x+1} = 3$

21. One of the solutions published by Cardan in 1545 for the solution of cubic equations is given below. For an equation in the form $x^3 + px = q$

$$x = \sqrt[3]{\left[\sqrt{\left(\frac{p}{3}\right)^3 + \left(\frac{q}{2}\right)^2} + \frac{q}{2}\right]} - \sqrt[3]{\left[\sqrt{\left(\frac{p}{3}\right)^3 + \left(\frac{q}{2}\right)^2} - \frac{q}{2}\right]}$$

Use the formula to solve the following equations, giving answers to 4 sig. fig. where necessary.

(a) $x^3 + 7x = -8$

(b) $x^3 + 6x = 4$

(c) $x^3 + 3x = 2$

(d) $x^3 + 9x - 2 = 0$

(c) Solution by completing the square

Look at the function $f(x) = x^2 + 6x$

Completing the square, this becomes $f(x) = (x + 3)^2 - 9$

This is done as follows.

1. 3 is half of 6 and gives $6x$.
2. Having added 3 to the square term, 9 needs to be subtracted from the expression to cancel the $+9$ obtained.

Here are some more examples.

(a) $x^2 - 12x = (x - 6)^2 - 36$

(b) $x^2 + 3x = (x + \frac{3}{2})^2 - \frac{9}{4}$

(c) $x^2 + 6x + 1 = (x + 3)^2 - 9 + 1$
$\quad = (x + 3)^2 - 8$

(d) $x^2 - 10x - 17 = (x - 5)^2 - 25 - 17$
$\quad = (x - 5)^2 - 42$

(e) $2x^2 - 12x + 7 = 2[x^2 - 6x + \frac{7}{2}]$
$\quad = 2[(x - 3)^2 - 9 + \frac{7}{2}]$
$\quad = 2[(x - 3)^2 - \frac{11}{2}]$

Example 10

Solve the quadratic equation $x^2 - 6x + 7 = 0$ by completing the square.

$$(x-3)^2 - 9 + 7 = 0$$
$$(x-3)^2 = 2$$
$$\therefore \quad x - 3 = +\sqrt{2} \text{ or } -\sqrt{2}$$
$$x = 3 + \sqrt{2} \text{ or } 3 - \sqrt{2}$$

So, $\quad x = 4{\cdot}41$ or $1{\cdot}59$ to 2 d.p.

Example 11

Given $f(x) = x^2 - 8x + 18$, show that $f(x) \geqslant 2$ for all values of x.

Completing the square, $f(x) = (x-4)^2 - 16 + 18$
$$f(x) = (x-4)^2 + 2.$$

Now $(x-4)^2$ is always greater than or equal to zero because it is 'something squared.'

$$\therefore \quad f(x) \geqslant 2$$

Exercise 17

In Questions **1** to **10**, complete the square for each expression by writing each one in the form $(x+a)^2 + b$ where a and b can be positive or negative.

1. $x^2 + 8x$ **2.** $x^2 - 12x$ **3.** $x^2 + x$

4. $x^2 + 4x + 1$ **5.** $x^2 - 6x + 9$ **6.** $x^2 + 2x - 15$

7. $2x^2 + 16x + 5$ **8.** $2x^2 - 10x$ **9.** $6 + 4x - x^2$

10. $3 - 2x - x^2$

11. Solve these equations by completing the square
(a) $x^2 + 4x - 3 = 0$ (b) $x^2 - 3x - 2 = 0$
(c) $x^2 + 12x = 1$

12. Try to solve the equation $x^2 + 6x + 10 = 0$, by completing the square. Explain why you can find no solutions.

13. Given $f(x) = x^2 + 6x + 12$, show that $f(x) \geqslant 3$ for all values of x.

14. Given $g(x) = x^2 - 7x + \frac{1}{4}$, show that the least possible value of $g(x)$ is -12.

15. If $f(x) = x^2 + 4x + 7$ find
(a) the smallest possible value of $f(x)$
(b) the value of x for which this smallest value occurs.
(c) the greatest possible value of $\dfrac{1}{(x^2 + 4x + 7)}$

16. Given $y = x^2 - x + 1$, find
(a) the lowest value of y,
(b) the value of x for which this lowest value occurs,
(c) the greatest possible value of $\frac{1}{(x^2 - x + 1)}$

17. Simplify $(x + 4)(x + 2) - 2(x + 2)$, and explain why the expression can never be negative, whatever the value of x.

Using quadratic equations to solve problems

Example 12

The perimeter of a rectangle is 42 cm. If the diagonal is 15 cm, find the width of the rectangle.

Let the width of the rectangle be x cm.
Since the perimeter is 42 cm, the sum of the length and the width is 21 cm.
$\therefore$ length of rectangle $= (21 - x)$ cm

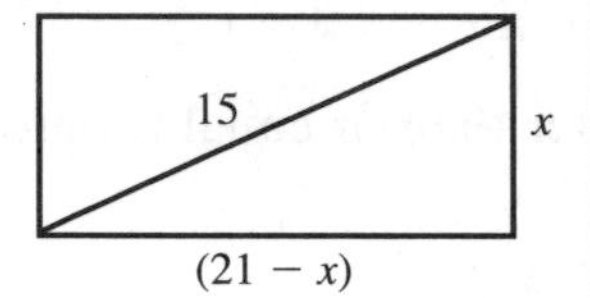

By Pythagoras' theorem

$$\begin{aligned} x^2 + (21 - x)^2 &= 15^2 \\ x^2 + (21 - x)(21 - x) &= 15^2 \\ x^2 + 441 - 42x + x^2 &= 225 \\ 2x^2 - 42x + 216 &= 0 \\ x^2 - 21x + 108 &= 0 \\ (x - 12)(x - 9) &= 0 \\ x &= 12 \\ \text{or} \quad x &= 9 \end{aligned}$$

Note that the dimensions of the rectangle are 9 cm and 12 cm, whichever value of x is taken.
$\therefore$ The width of the rectangle is 9 cm.

Exercise 18

Solve by forming a quadratic equation.

1. Two numbers, which differ by 3, have a product of 88. Find them.

2. The product of two consecutive odd numbers is 143. Find the numbers. (Hint: If the first odd number is x, what is the next odd number?)

3. The length of a rectangle exceeds the width by 7 cm. If the area is 60 cm^2, find the length of the rectangle.

4. The length of a rectangle exceeds the width by 2 cm. If the diagonal is 10 cm long, find the width of the rectangle.

5. The area of the rectangle exceeds the area of the square by $24\,m^2$. Find x.

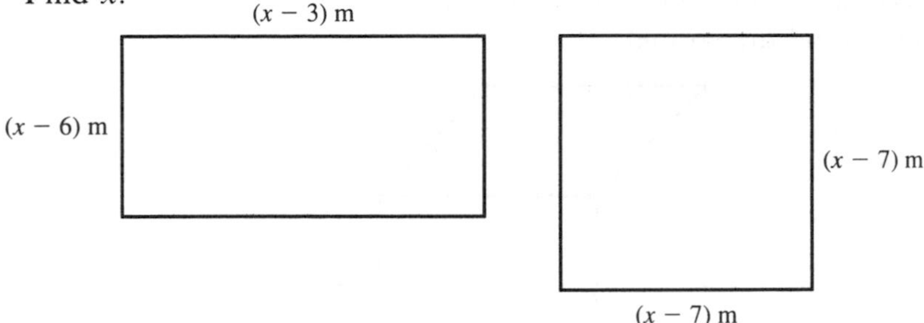

6. The perimeter of a rectangle is 68 cm. If the diagonal is 26 cm, find the dimensions of the rectangle.

7. A man walks a certain distance due North and then the same distance plus a further 7 km due East. If the final distance from the starting point is 17 km, find the distances he walks North and East.

8. A farmer makes a profit of x pence on each of the $(x + 5)$ eggs her hen lays. If her total profit was 84 pence, find the number of eggs the hen lays.

9. Rectangle ABCD measures 10 cm by 6 cm.

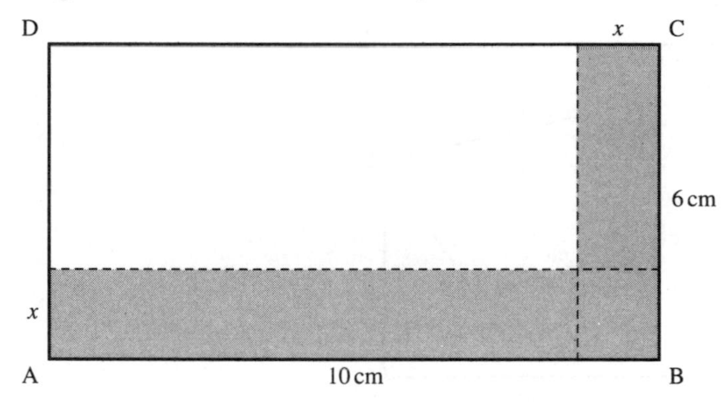

The two shaded strips of width x cm are cut away. The area of the remaining rectangle is three quarters of the area of ABCD. Find x.

10. A positive number exceeds four times its reciprocal by 3. Find the number.

11. Two numbers differ by 3. The sum of their reciprocals is $\frac{7}{10}$; find the numbers.

12. A cyclist travels 40 km at a speed x km/h. Find the time taken in terms of x. Find the time taken when his speed is reduced by 2 km/h. If the difference between the times is 1 hour, find the original speed.

13. An increase of speed of 4 km/h on a journey of 32 km reduces the time taken by 4 hours. Find the original speed.

14. A train normally travels 60 miles at a certain speed. One day, due to bad weather, the train's speed is reduced by 10 mph so that the journey takes 3 hours longer. Find the normal speed.

15. When one edge of a cube is decreased by 1 cm, its volume is decreased by 91 cm^3. Find the length of a side of the original cube.

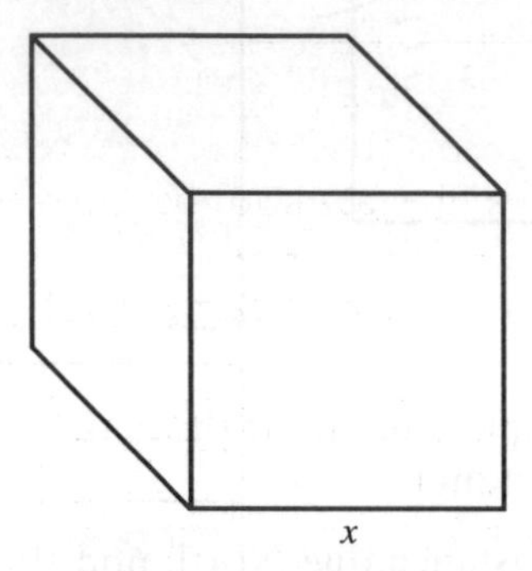

16. An aircraft flies a certain distance on a bearing of 135° and then twice the distance on a bearing of 225°. Its distance from the starting point is then 350 km. Find the length of the first part of the journey.

17. In Fig. 1, ABCD is a rectangle with AB = 12 cm and BC = 7 cm. AK = BL = CM = DN = x cm. If the area of KLMN is 54 cm^2 find x.

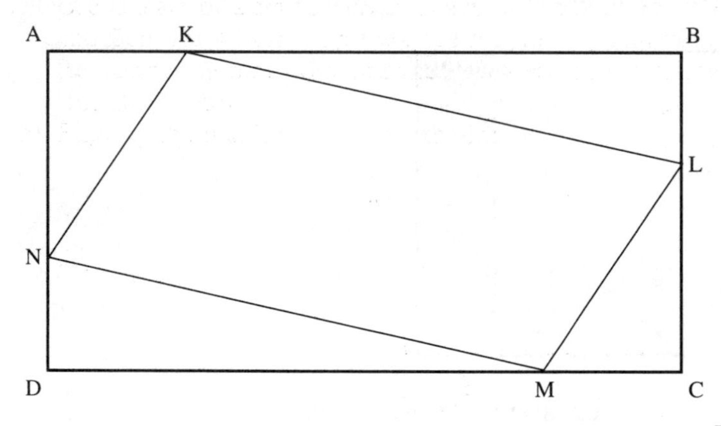

Fig. 1

18. In Fig. 1, AB = 14 cm BC = 11 cm and AK = BL = CM = DN = x cm. If the area of KLMN is now 97 cm^2, find x.

19. The numerator of a fraction is 1 less than the denominator. When both numerator and denominator are increased by 2, the fraction is increased by $\frac{1}{12}$. Find the original fraction.

20. The perimeters of a square and a rectangle are equal. The length of the rectangle is 11 cm and the area of the square is 4 cm^2 more than the area of the rectangle. Find the side of the square.

21. A lot of paper comes in standard sizes A0, A1, A2, A3 etc. The large sheet shown below is size A0; A0 can be cut in half to give two sheets of A1; A1 can be cut in half to give two sheets of A2 and so on.

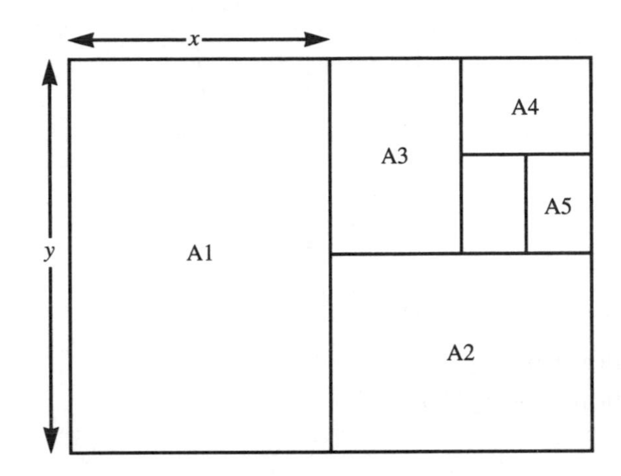

Call the sides of A1 x and y, as shown.

(a) What are the sides of A0 in terms of x and y?

(b) All the sizes of paper are similar. Form an equation involving x and y.

(c) Size A0 has an area of $1\,\text{m}^2$. Calculate the values of x and y correct to the nearest mm.

(d) Hence calculate the long side of a sheet of A4. Check your answer by measuring a sheet of A4.

6.8 Transformation of curves

The notation f(x) means 'function of x'. A function of x is an expression which (usually) varies, depending on the value of x. Examples of functions are:

$f(x) = x^2 + 3$; $f(x) = \frac{1}{x} + 7$; $f(x) = \sin x$.

We can imagine a box which performs the function f on any input.

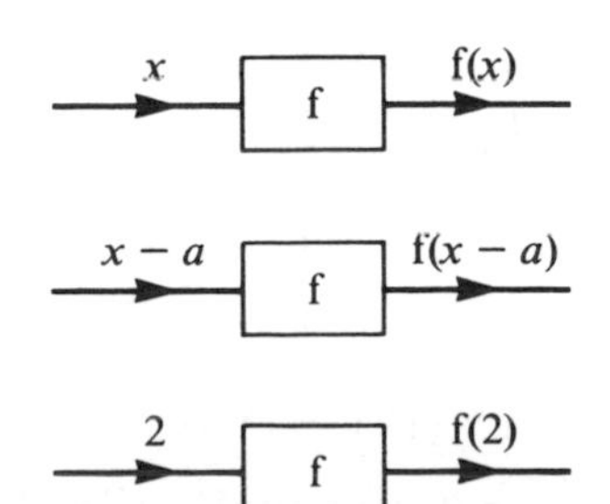

If $f(x) = x^2 + 7x + 2$, $f(3) = 3^2 + 7 \times 3 + 2 = 32$

Here are *five* ways in which any function f(x) can be transformed.

$$f(x) + a; \quad f(x - a); \quad -f(x); \quad f(ax); \quad \frac{1}{f(x)}$$

(a) $y = f(x) + a$

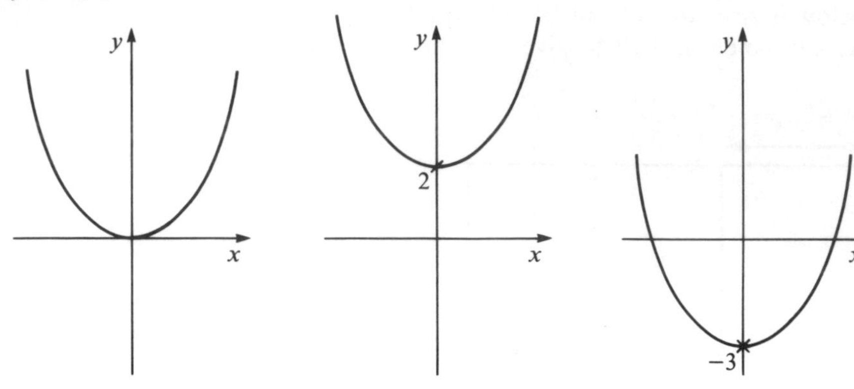

① $y = f(x)$ ② $y = f(x) + 2$ ③ $y = f(x) - 3$

- Can you describe the transformation from ① to ②?
- Can you describe the transformation from ① to ③?

(b) $y = f(x - a)$

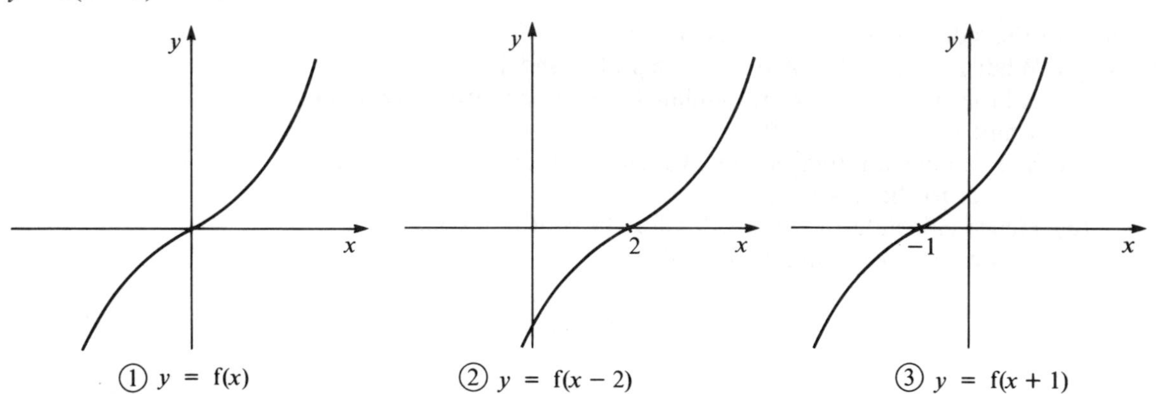

① $y = f(x)$ ② $y = f(x - 2)$ ③ $y = f(x + 1)$

- Can you describe the transformation from ① to ②?
- Can you describe the transformation from ① to ③?

(c) $y = -f(x)$

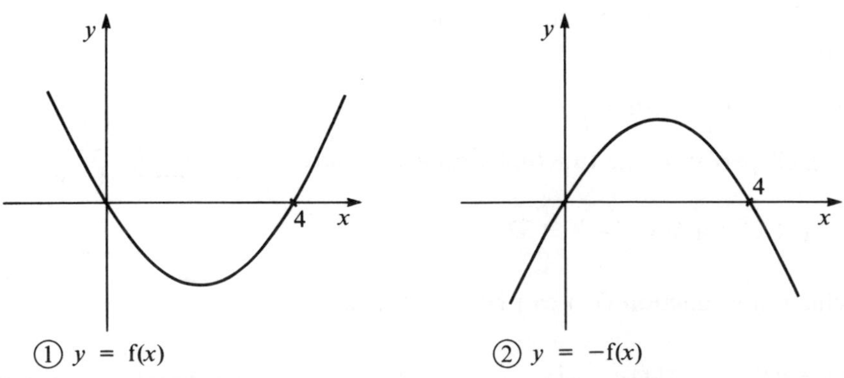

① $y = f(x)$ ② $y = -f(x)$

- Can you describe the transformation from ① to ②?

(d) $y = f(ax)$

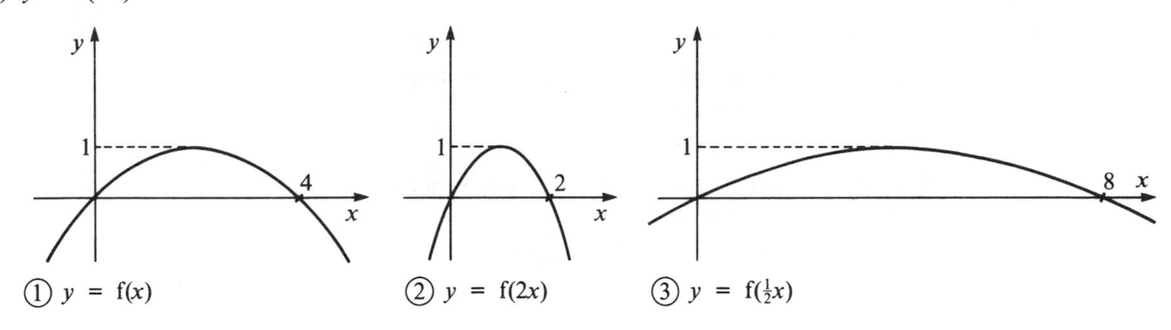

- Can you describe the transformation from ① to ②?
- Can you describe the transformation from ① to ③?

This is the most difficult one.

From ① to ② the transformation is a stretch parallel to the x-axis by a scale factor $\frac{1}{2}$.

From ② to ① the transformation is a stretch parallel to the x-axis by a scale factor 2.

Similarly $f(3x)$ would be a stretch with scale factor $\frac{1}{3}$ and $f(\frac{1}{5}x)$ would be a stretch with scale factor 5.

(e) $y = \dfrac{1}{f(x)}$

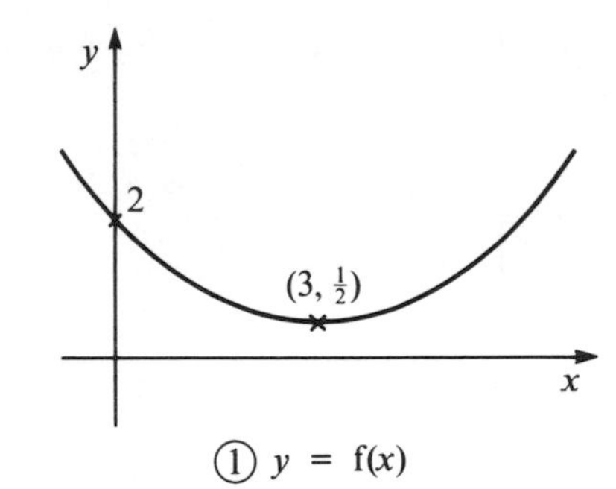

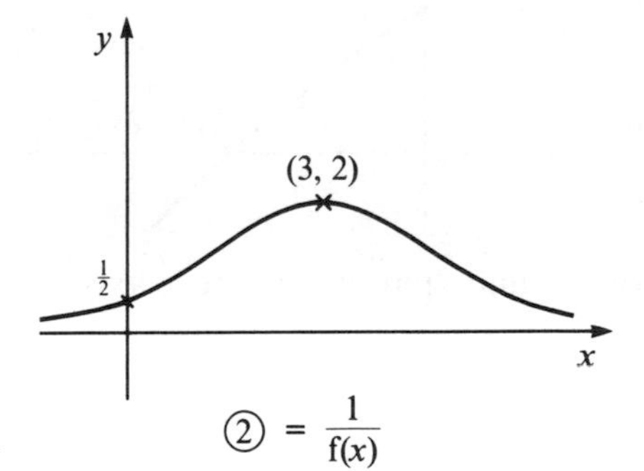

When $f(x)$ is large, $\dfrac{1}{f(x)}$ is small.

When $f(x)$ has a minimum value, $\dfrac{1}{f(x)}$ has a maximum value.

When $f(x)$ has a maximum value, $\dfrac{1}{f(x)}$ has a minimum value.

$\dfrac{1}{f(x)}$ has the same sign as $f(x)$.

Summary rules for curve transformations

(a) $y = \mathrm{f}(x) + a$: Translation by a units parallel to the y-axis.

(b) $y = \mathrm{f}(x - a)$: Translation by a units parallel to the x-axis. (note the negative sign).

(c) $y = -\mathrm{f}(x)$: Reflection in the x-axis.

(d) $y = \mathrm{f}(ax)$: Stretch parallel to the x-axis by a scale factor $\frac{1}{a}$ (note the inverse of a).

(e) $y = \frac{1}{\mathrm{f}(x)}$: When $\mathrm{f}(x)$ is large, $y = \frac{1}{\mathrm{f}(x)}$ is small.

Example 1

From the sketch of $y = \mathrm{f}(x)$, draw

(a) $y = \mathrm{f}(x + 3)$,

(b) $y = \mathrm{f}(3x)$

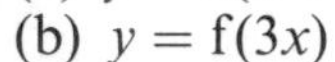

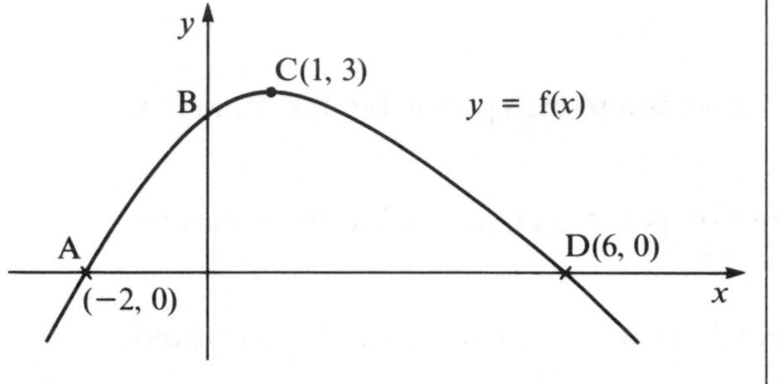

(a)

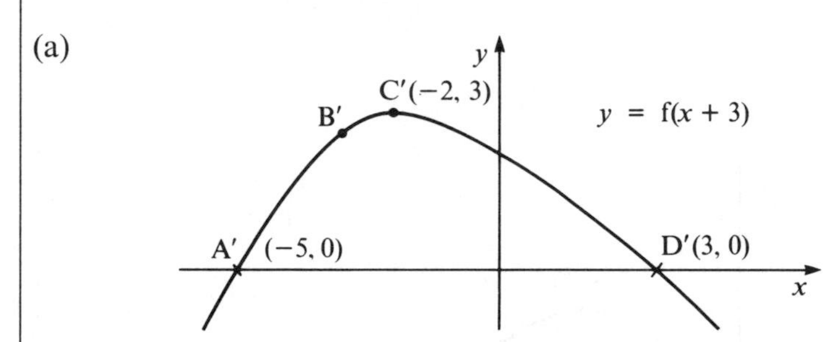

This is a translation of −3 units parallel to the x-axis.

(b)

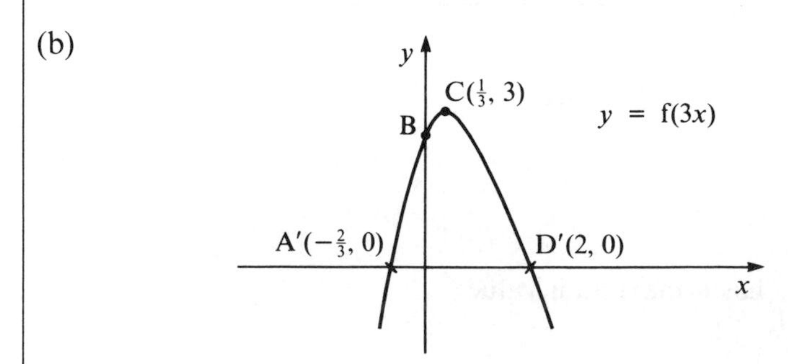

This is a stretch parallel to the x-axis by a scale factor $\frac{1}{3}$. Notice that B remains in the same place and that the y-coordinate of C remains unchanged.

Exercise 19

A computer or calculator which sketches curves can be used effectively in this exercise, although it is not essential.

1. (a) Draw an accurate graph of $y = x^2$ for values of x from -3 to 3.
(b) On the same axes draw an accurate graph of $y = x^2 + 4$ and $y = (x - 1)^2$.
Scales: x from -3 to $+3$, 2 cm = 1 unit
y from 0 to 14, 1 cm = 1 unit.

2. (a) Draw an accurate graph of $y = (x + 2)^2$ for values of x from -5 to 1.
(b) Draw an accurate graph of $y = (\frac{1}{2}x + 2)^2$ for values of x from -9 to 1.
Scales: x from -10 to 1, 1 cm = 1 unit
y from 0 to 10, 1 cm = 1 unit.

This is an example of result (d) above. You have drawn $y = f(x)$ and $y = f(\frac{1}{2}x)$.
Describe the transformation from $y = (x + 2)^2$ onto $y = (\frac{1}{2}x + 2)^2$.

3. This is the sketch graph of $y = f(x)$ where $f(x) = x(x + 2)$.

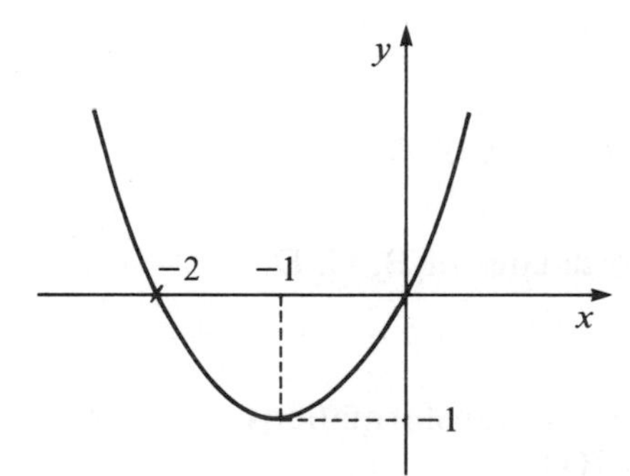

(a) Sketch the graph of $y = f(x) + 3$
(b) Sketch the graph of $y = f(x + 1)$ [i.e. $y = (x + 1)(x + 3)$]
(c) Sketch the graph of $y = -f(x)$.

4. This is the sketch graph of $y = f(x)$ where $f(x) = x(x + 2)(x - 1)$.

(a) Sketch $y = f(x) - 2$
(b) Sketch $y = f(x - 7)$

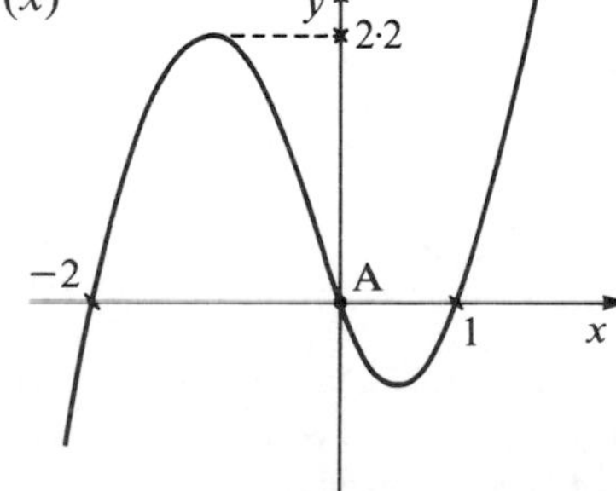

Give the new coordinates of the point A on the two sketches.

5. This is the sketch of $y = \mathrm{f}(x)$ which passes through A,B,C.

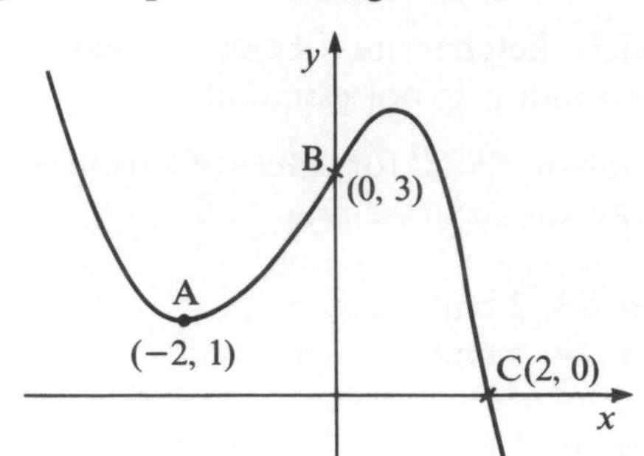

Sketch the following curves, giving the new coordinates of A,B,C in each case.

(a) $y = -\mathrm{f}(x)$
(b) $y = \mathrm{f}(x - 2)$
(c) $y = \mathrm{f}(2x)$

6. Here are the sketches of $y = \mathrm{f}(x)$ and $y = \mathrm{g}(x)$

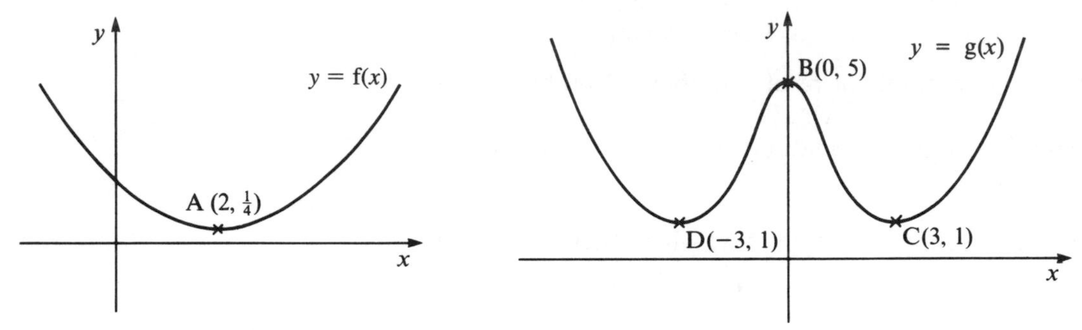

(a) Sketch $y = \dfrac{1}{\mathrm{f(x)}}$, showing the new coordinates of A.

(b) Sketch $y = \dfrac{1}{\mathrm{g(x)}}$, showing the new coordinates of B, C, D.

7. It is possible to perform two or more successive transformations on the same curve. This is a sketch of $y = \mathrm{f}(x)$.

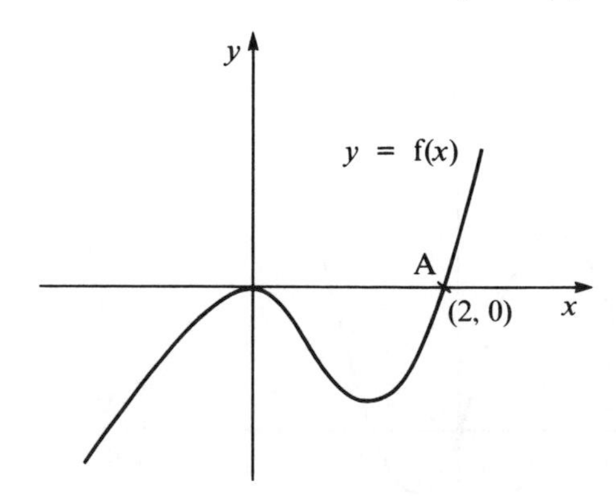

(a) Sketch $y = \mathrm{f}(x + 1) + 5$
(b) Sketch $y = \mathrm{f}(x - 3) - 4$

Show the new coordinates of the point A on each sketch.

8.† Find the equation of the curve obtained when the graph of $y = x^2 + 3x$ is:

(a) translated 5 units in the direction ↑
(b) translated 2 units in the direction →
(c) reflected in the x-axis.

9.† $f(x) = x^2$ and $g(x) = x^2 - 4x + 7$

(a) If $g(x) = f(x - a) + b$, find the values of a and b.
(b) Hence sketch the graphs of $y = f(x)$ and $y = g(x)$ showing the transformation from f to g.

Examination exercise 6

1. The perimeter of the rectangle ABCD is 42 cm. The width BC is x cm.

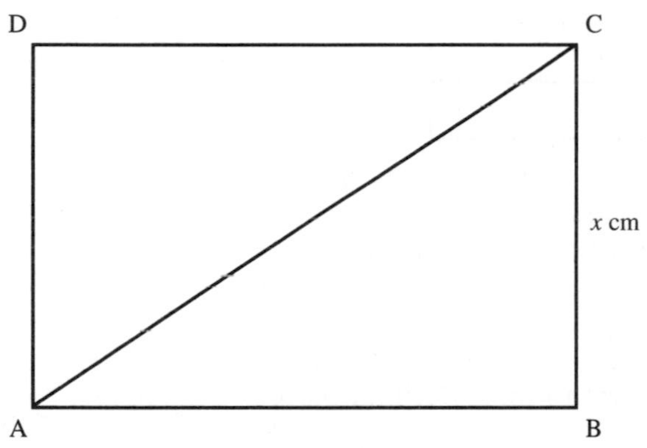

(a) Write, in terms of x, an expression for the length of AB.
(b) The diagonal AC is 15 cm. Calculate the length and width of the rectangle. [S]

2.

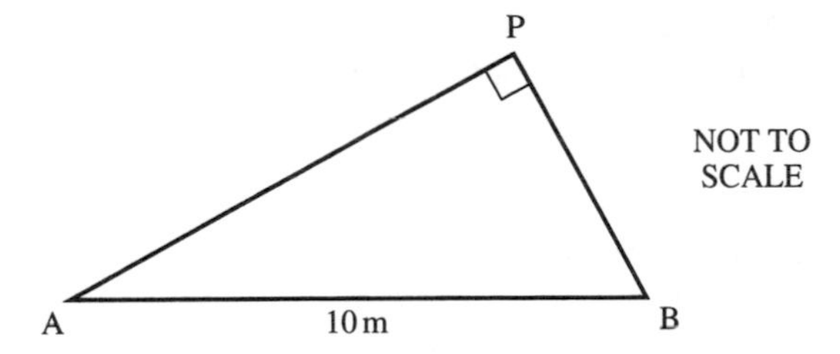

Derrick has to mark out a triangular flower bed ABP, as shown in the diagram above. The distance AB must be 10 m and the angle APB must be 90°. The lengths of the other two sides, AP and BP, must total 13 m.

(a) Taking the length of AP as x metres, form an equation in x, and show that it simplifies to $2x^2 - 26x + 69 = 0$.
(b) Solve the equation to find the two possible values of x, correct to 2 decimal places. [M]

3. The sum of the squares of three consecutive whole numbers is 6914. The middle number is n.
(a) Write down expressions for the other two numbers.
(b) By squaring the algebraic expressions for the three numbers and adding them together, create an equation in n.
(c) (i) Solve your equation to find n.
(ii) Write down the three consecutive numbers. [L]

4. The mass, m grams, of a radioactive chemical after t years is given by $m = 80 \times 0{\cdot}5^t$.
(a) Find the mass after 3 years.
(b) Find the mass after 6 months.
(c) What was the initial mass? [M]

5. (a) Solve the equation $x^5 = 0{\cdot}00243$
(b) Once a year a scientist measured the mass of a certain piece of a decaying radioactive element. His results are shown.

Time (years)	0	1	2	3
Mass (kg)	20	18	16·2	14·58

(i) Calculate the annual percentage decrease of the mass.
(ii) Calculate the mass after 10 years.
(iii) Estimate the half life of the element (ie the time it takes to lose half of its mass). [S]

6. Given that $u_1 = 1$ and $u_{n+1} = \sqrt{1 + \sqrt{2 + u_n}}$, calculate the value of u_2 and u_3, giving your answers correct to three decimal places. [N]

7. (a) Given that $f(x) = 2x^2 - 4x - 1$, calculate the value of
(i) $f(2)$, (ii) $f(3)$.
(b) Explain briefly what the results obtained in (a) show about one of the roots of the equation $2x^2 - 4x - 1 = 0$.
(c) Show that the quadratic equation $2x^2 - 4x - 1 = 0$ can be written in the form $x = 2 + \dfrac{1}{2x}$.
(d) Use the iterative formula $x_{n+1} = 2 + \dfrac{1}{2x_n}$ to find,
(i) the value of x_2 if $x_1 = 2{\cdot}5$,
(ii) the value of x_3, giving your answer to three places of decimals.
(e) The solutions of the quadratic equation $ax^2 + bx + c = 0$ are

$$x = \frac{-b \pm \sqrt{b^2 - 4ac}}{2a}.$$

Use this formula to obtain the negative solution of the equation $2x^2 - 4x - 1 = 0$, giving your answer correct to three decimal places. [N]

8. A sequence of numbers, $x_1, x_2, x_3, x_4, \ldots$ is defined by the iterative equation $x_{n+1} = 4 + \dfrac{12}{x_n}$.

(a) Starting with $x_1 = 5$,
   (i) calculate the value of x_2,
   (ii) show that $x_3 = 5{\cdot}875$.

(b) By continuing this process and observing how the values of x_1, x_2, x_3, x_4, etc. are converging, predict the value of x_n as n tends to infinity.

(c) (i) Rearrange the equation $x = 4 + \dfrac{12}{x}$ into the form $px^2 + qx + r = 0$ where p, q and r are constants.
   (ii) Write down, or calculate, the negative solution of this equation. [N]

9.

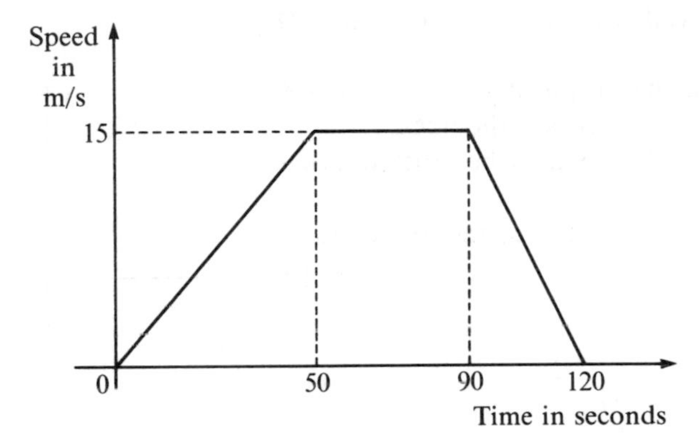

The diagram is the speed-time graph of a bus on a short journey between two bus stops.

Calculate

(a) the acceleration, in m/s^2, of the bus during the first 50 seconds.

(b) the total distance, in m, travelled in the journey,

(c) the average speed, in m/s, of the bus for the whole journey. [L]

10. A car accelerates from rest at $4\,m/s^2$ for 5 seconds, then moves with constant speed for 10 seconds, then decelerates at $2\,m/s^2$ back to rest.

(a) Draw a speed-time graph for the journey of the car.

(b) Calculate the total distance travelled by the car on this journey. [L]

11. (a) Express $x^2 + 5x - 7$ in the form $(x + a)^2 + b$.

(b) Hence, solve $x^2 + 5x - 7 = 0$. [N]

12. At time t seconds, the velocity, v metres per second, of a particle moving along a straight line is given by $v = 2t^2 - 9t + 8$.
Some values of v and t are given in the table

t	0	1	2	3	4	5	6
v	8		−2	−1	4		26

(a) Calculate the value of v when
(i) $t = 1$, (ii) $t = 5$.

(b) Taking 2 cm to represent 1 unit on the t axis and 1 cm to represent 2 units on the v axis, draw the graph of $v = 2t^2 - 9t + 8$ for values of t from 0 to 6.

(c) Use your graph to estimate
(i) the values of t for which the velocity is zero,
(ii) the value of t for which the acceleration is zero.

(d) Use the trapezium rule, with four equal intervals, to calculate the approximate distance travelled from $t = 4$ to $t = 6$. [L]

13. (a) Using the axes shown, sketch the graph of $y = x^2 - 2x - 8$, indicating clearly where the graph crosses the axes.

(b) Show that the expression $x^2 - 2x - 8$ may be written as $(x - 1)^2 - 9$.

(c) Hence, or otherwise, find the least value of the expression $x^2 - 2x - 8$. [M]

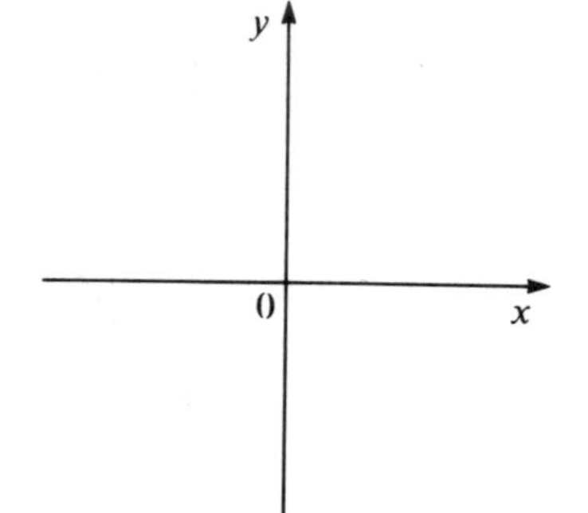

14. Two graphs are drawn below.

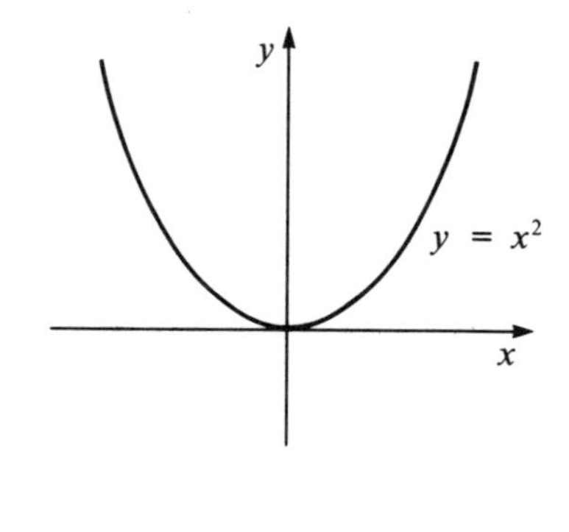

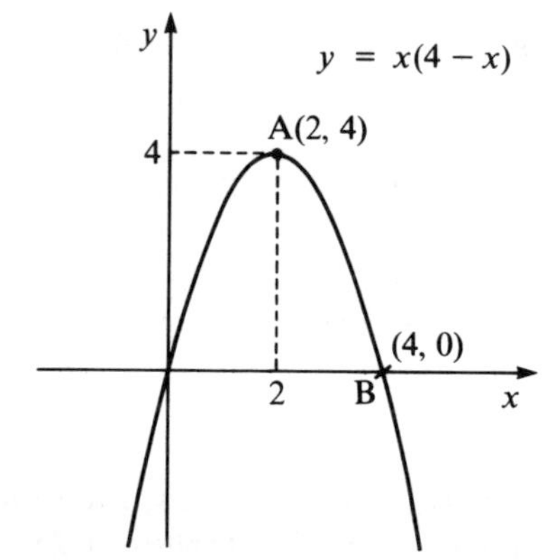

(a) Sketch the graph of $y = x^2 - 2$

(b) Sketch the graph of $y = (x - 2)^2$

(c) Sketch the graph of $y = 2x(4 - 2x)$, showing the new coordinates of points A and B.

7 Handling data 1

7.1 Data presentation

The results of a statistical investigation are known as data. Data can be presented in a variety of different ways.

(a) Raw data

This is data in the form that it was collected, for example, the number of peas in 40 pods.
Data in this form is difficult to interpret.

5	3	6	5	4	6	6	7	4	6
4	7	7	3	7	4	7	5	7	5
6	7	6	7	5	6	6	7	6	6
5	3	6	4	6	5	7	3	6	4

(b) Frequency tables

This presentation is in the form of a tally; the tally provides a numerical value to the frequency of occurrence.
From the example above, there are 4 pea pods that contain 3 peas, etc.

Number of peas	Tally	Frequency (f)
3						4
4	~~				~~		6
5	~~||||~~ ||	7
6	~~||||~~ ~~||||~~ |||	13
7	~~||||~~ ~~||||~~	10

(c) Display charts

Three types of display chart are shown below; each refers to the initial data about the number of peas in a pod.

Number of peas	Angle on pie chart
3	$\frac{4}{40} \times 360 = 36°$
4	$\frac{6}{40} \times 360 = 54°$
5	$\frac{7}{40} \times 360 = 63°$
6	$\frac{13}{40} \times 360 = 117°$
7	$\frac{10}{40} \times 360 = 90°$

Pie chart

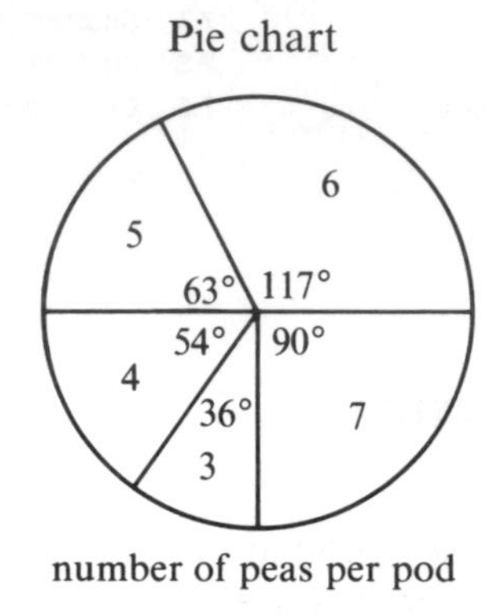

number of peas per pod

Bar chart

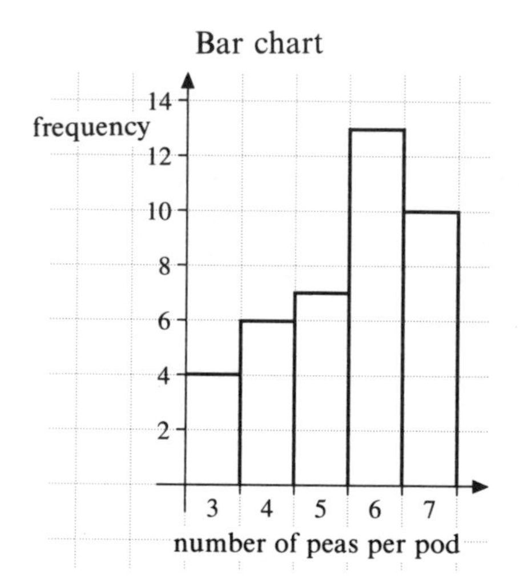

Frequency polygon

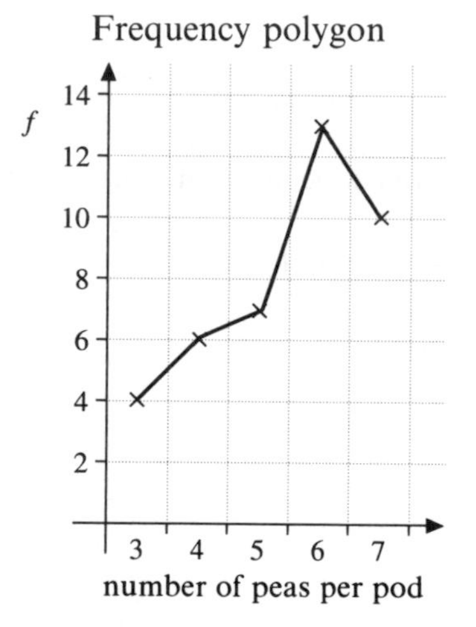

Note that in the bar chart and the frequency polygon, the vertical axis is always used to show frequency. It is the frequency that most clearly shows the mode, i.e. that more pods contained 6 peas than any other number of peas.
In the frequency polygon, the boxes of the bar chart are replaced by a line joining the tops of their mid-points.

All the above is discrete data; this is concerned solely with individually distinct number quantities. Continuous data is derived from measurements, e.g. height, weight, age, time. It can be handled as follows.

(a) Rounded data

Each measurement can be rounded and the frequency of this rounded quantity then recorded as for discrete data. For example, the lengths of 36 pea pods can be rounded to the nearest mm and then recorded as raw data — a pea pod measuring 59·2 mm is recorded as one that has a length of 59 mm.

Rounded data

52	80	65	82	77	60	72	83	63
78	84	75	53	73	70	86	55	88
85	59	76	86	73	89	91	76	92
66	93	84	62	79	90	73	68	71

(b) Grouped data

Each measurement can be grouped into classes with defined class boundaries. In the table below, the same data is grouped into the classes shown in the left-hand column.

Grouped frequency table

Length (mm)	Tally	Frequency
50–59	||||	4
60–69	~~||||~~ |	6
70–79	~~||||~~ ~~||||~~ ||	12
80–89	~~||||~~ ~~||||~~	10
90–99	||||	4

So, the pea pod measuring 59·2 mm is one of 4 recorded in the class 50–59. Because of the rounding of the data, the actual boundary of this first class is from 49·5 to 59·5 (a length of 59·5 mm is rounded up to 60 mm). This is shown below on the horizontal axis of the bar chart for these class boundaries.

Bar chart of grouped data

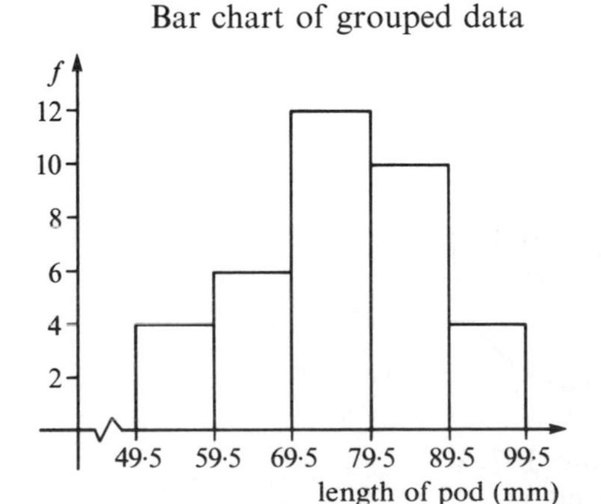

Frequency polygon

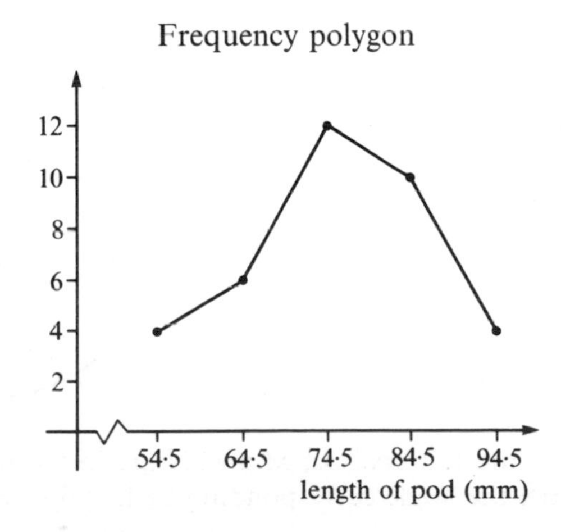

The frequency polygon of the same data is also shown. Here the horizontal axis records the mid-points of the classes where the mid-point of the class 50–59 is calculated as $\frac{49 \cdot 5 + 59 \cdot 5}{2} = 54 \cdot 5$.

In some cases where there are many small classes, the mid-points of the frequency polygon are joined with a curve. The diagram is then known as a frequency curve.

Exercise 1

1. The pie chart illustrates the values of various goods sold by a certain shop. If the total value of the sales was £24 000, find the sales value of
(a) toys
(b) grass seed
(c) records
(d) food.

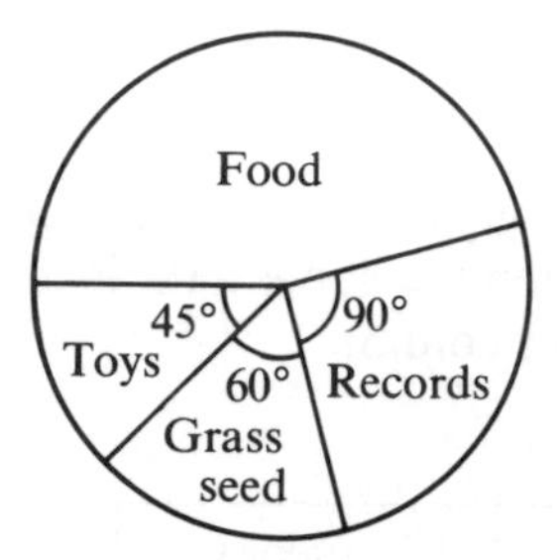

2. A quantity of scrambled eggs is made using the following recipe:

Ingredient	eggs	milk	butter	cheese	salt/pepper
Mass	450 g	20 g	39 g	90 g	1 g

Calculate the angles on a pie chart corresponding to each ingredient.

3. A firm making artificial sand sold its products in four countries.
5% were sold in Spain
15% were sold in France
15% were sold in Germany
65% were sold in U.K.
What would be the angles on a pie chart drawn to represent this information?

4. The pie chart illustrates the sales of various makes of petrol.
(a) What percentage of sales does 'Esso' have?
(b) If 'Jet' accounts for $12\frac{1}{2}\%$ of total sales, calculate the angles x and y.

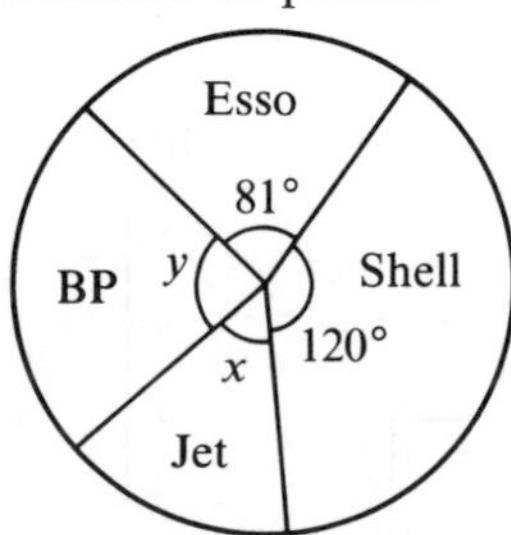

5. The cooking times for meals L, M and N are in the ratio $3 : 7 : x$. On a pie chart, the angle corresponding to L is 60°. Find x.

6. In a survey, the number of people in 100 cars passing a set of traffic lights was counted. Here are the results:

Number of people in car	0	1	2	3	4	5	6
Frequency	0	10	35	25	20	10	0

(a) Draw a bar chart to illustrate this data.
(b) On the same graph draw the frequency polygon.

Here the bar chart has been started. For frequency, use a scale of 1 cm for 5 units.

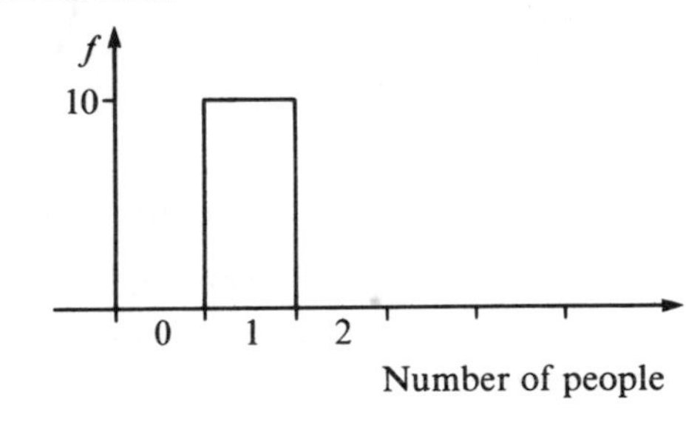

7. In an international survey of maths ability one hundred 14-year olds from several countries were given the same test. Here are the results for USA and Singapore.

Marks	30–32	33–35	36–38	39–41	42–44	45–47	48–50
f (USA)	5	5	10	15	40	15	10
f (Singapore)	0	0	5	15	25	35	20

(a) Find the midpoints for each class interval.
(b) Using the same axes draw a frequency polygon for the results of each country using different colours. [Scales: 1 cm to 10 units for frequency; 1 cm for 3 marks across the page].
(c) What is the main difference between the two sets of results?

8. Here is the frequency table of the weights in kg of 50 Sumo wrestlers.

Class interval	140–144	145–149	150–154	155–159	160–164	165–169
f	5	7	10	16	8	4

This is continuous data so for the 140–144 interval the class boundaries are 139·5 to 144·5.
(a) Find the midpoint of each class interval.
(b) Draw a bar chart and a frequency polygon to illustrate this data.

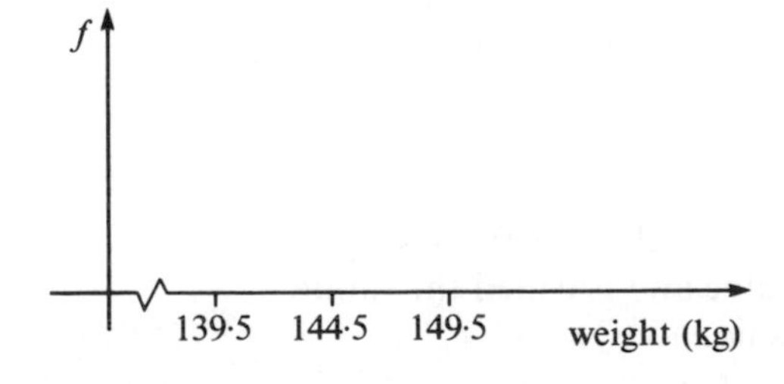

9. The diagram illustrates the production of apples in two countries.

In what way could the pictorial display be regarded as misleading?

10. The graph shows the performance of a company in the year in which a new manager was appointed. In what way is the graph misleading?

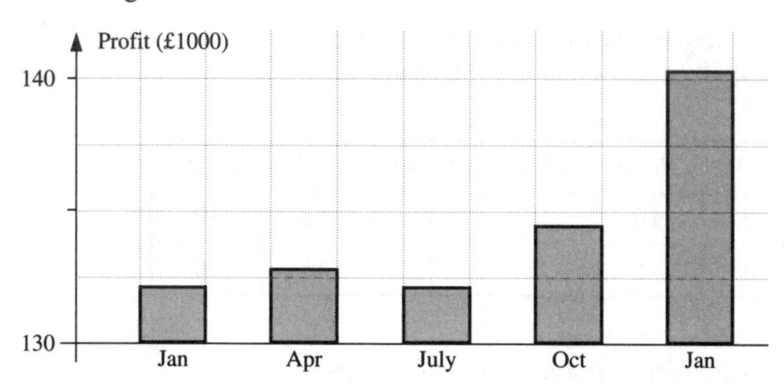

7.2 Averages

A single number called an average can be used to represent the whole range of a set of data, whether the data is discrete (e.g. exam marks) or continuous (e.g. heights).

(a) The median

The data is arranged in order from the smallest to the largest; the middle number is then selected. This is really the central number of the range and is called the median.
If there are two 'middle' numbers, the median is in the middle of these two numbers.

(b) The mean

All the data is added up and the total divided by the number of items. This is called the mean and is equivalent to sharing out all the data evenly.

(c) The mode

The number of items which occurs most frequently in a frequency table is selected. This is the most popular value and is called the mode (from the French 'a la mode' meaning 'fashionable').

Each ‘average’ has its purpose and sometimes one is preferable to the other.

The median is fairly casy to find and has an advantage in being hardly affected by untypical values such as very large or very small values that occur at the ends of a distribution.
Consider these examination marks.

20, 21, 21, 22, 23, 23, 25, 27, 27, 27, 29, 98, 98
↑

Clearly the median value is 25.
The mean of the above data is 35.5. It is easier to use in further work such as standard deviation (page 350), but clearly it does not give a true picture of the centre of distribution of the data.
The mode of this data is 27. It is easy to calculate and it eliminates some of the effects of extreme values. However it does have disadvantages, particularly in data which has two ‘most popular’ values, and it is not widely used.

Range

In addition to knowing the centre of a distribution, it is useful to know the range or spread of the data.

range = (largest value) – (smallest value)

For the examination marks, range = 98 – 20 = 78.

Example 1

Find the median, the mean, the mode and the range of this set of 10 numbers: 5, 4, 10, 3, 3, 4, 7, 4, 6, 5.

(a) Arrange the numbers in order of size to find the median.

3, 3, 4, 4, 4, 5, 5, 6, 7, 10
↑

the median is the ‘average’ of 4 and 5.
$\therefore$ median = 4·5

(b) mean $= \dfrac{(5+4+10+3+3+4+7+4+6+3)}{10} = \dfrac{51}{10} = 5{\cdot}1$

(c) mode = 4 because there are more 4’s then any other number

(d) range = 10 – 3 = 7

Exercise 2

1. Find the median, the mean, the mode and the range of the following sets of numbers:
(a) 3, 12, 4, 6, 8, 5, 4
(b) 7, 21, 2, 17, 3, 13, 7, 4, 9, 7, 9
(c) 12, 1, 10, 1, 9, 3, 4, 9, 7, 9
(d) 8, 0, 3, 3, 1, 7, 4, 1, 4, 4.

2. The following are the salaries of 5 employees in a small business:

Mr A : £22,500 Mr B : £17,900 Mr C : £21,400
Mr D : £22,500 Mr E : £85,300.

(a) Find the mean, median and mode of their salaries.
(b) Which does *not* give a fair 'average'? Explain why in one sentence.

3. A farmer has 32 cattle to sell. Their weights in kg are

81	81	82	82	83	84	84	85
85	86	86	87	87	88	89	91
91	92	93	94	96	150	152	153
154	320	370	375	376	380	381	390

[Total weight = 5028 kg].

On the telephone to a potential buyer, the farmer describes the cattle and says the 'average' weight is 'over 157 kg'.
(a) Find the mean weight and the median weight.
(b) Which 'average' has the farmer used to describe his animals? Does this average describe the cattle fairly?

4. A gardening magazine sells seedlings of a plant through the post and claims that the average height of the plants after one year's growth will be 85 cm. A sample of 24 of the plants were measured after one year with the following results (in cm)

6	7	7	9	34	56	85	89
89	90	90	91	91	92	93	93
93	94	95	95	96	97	97	99

[The sum of the heights is 1788 cm].

(a) Find the mean and the median height of the sample.
(b) Is the magazine's claim about average height justified?

5. The mean weight of five men is 76 kg. The weights of four of the men are 72 kg, 74 kg, 75 kg and 81 kg. What is the weight of the fifth man?

6. The mean length of 6 rods is 44·2 cm. The mean length of 5 of them is 46 cm. How long is the sixth rod?

7. (a) The mean of 3, 7, 8, 10 and x is 6. Find x.
(b) The mean of 3, 3, 7, 8, 10, x and x is 7. Find x.

8. The mean height of 12 men is 1·70 m, and the mean height of 8 women is 1·60 m. Find
(a) the total height of the 12 men,
(b) the total height of the 8 women,
(c) the mean height of the 20 men and women.

9. The total weight of 6 rugby players is 540 kg and the mean weight of 14 ballet dancers is 40 kg. Find the mean weight of the group of 20 rugby players and ballet dancers.

Frequency tables

A frequency table shows a number x such as a mark or a score, against the frequency f or number of times that x occurs.
The next examples show how these symbols are used in calculating the mean, the median and the mode.
The symbol Σ (or sigma) means 'the sum of'.

Example 2 Discrete data

The marks obtained by 100 students in a test were as follows:

Mark (x)	0	1	2	3	4
Frequency (f)	4	19	25	29	23

Find
(a) the mean mark (b) the median mark (c) the modal mark

(a) Mean $= \dfrac{\Sigma xf}{\Sigma f}$

where Σxf means 'the sum of the products xf'
i.e. Σ(number $\times$ frequency)
and Σf means 'the sum of the frequencies'

$$\text{Mean} = \frac{(0 \times 4) + (1 \times 19) + (2 \times 25) + (3 \times 29) + (4 \times 23)}{100}$$

$$= \frac{248}{100} = 2{\cdot}48$$

(b) The median mark is the number between the 50th and 51st numbers. By inspection, both the 50th and 51st numbers are 3.

$\therefore$ Median = 3 marks.

(c) The modal mark = 3.

For grouped data, each group can be represented approximately by its mid-point. Suppose the marks of 51 students in a test were:

Mark	30–39	40–49	50–59	60–69
Frequency	7	14	21	9

We say that, for the 30–39 interval, there are 7 marks of 34·5.

Note that the mean calculated by this method is only an estimate because the raw data is not available and an assumption has been made in regard to the mid-point of each interval whereas in Example 2 a true mean could be found.

Exercise 3

1. A group of 50 people were asked how many books they had read in the previous year; the results are shown in the frequency table below. Calculate the mean number of books read per person.

Number of books	0	1	2	3	4	5	6	7	8
Frequency	5	5	6	9	11	7	4	2	1

2. A teacher conducted a mental arithmetic test for 26 pupils and the marks out of 10 were as follows.

Mark	3	4	5	6	7	8	9	10
Frequency	6	3	1	2	0	5	5	4

(a) Find the mean, median and mode.
(b) The teacher congratulated the class saying that "over three quarters were above average". Which 'average' justifies this statement?

3. The following tables give the distribution of marks obtained by different classes in various tests. For each table, find the mean, median and mode.

(a)

Mark	0	1	2	3	4	5	6
Frequency	3	5	8	9	5	7	3

(b)

Mark	15	16	17	18	19	20
Frequency	1	3	7	1	5	3

4. The results of 24 students in a test are given.

(a) Find the mid-point of each group of marks and calculate an estimate of the mean mark.
(b) Find the modal group.

Mark	Frequency
85–99	4
70–84	7
55–69	8
40–54	5

5. The number of letters delivered to the 26 houses in a street was as follows:

Calculate an estimate of the mean number of letters delivered per house.

Number of letters delivered	Number of houses (i.e. frequency)
0– 2	10
3– 4	8
5– 7	5
8–12	3

6. The number of goals scored in a series of football matches was as follows:

Number of goals	1	2	3
Number of matches	8	8	x

(a) If the mean number of goals is 2·04, find x.
(b) If the modal number of goals is 3, find the smallest possible value of x.
(c) If the median number of goals is 2, find the largest possible value of x.

7. In a survey of the number of occupants in a number of cars, the following data resulted.

Number of occupants	1	2	3	4
Number of cars	7	11	7	x

(a) If the mean number of occupants is $2\frac{1}{3}$, find x.
(b) If the mode is 2, find the largest possible value of x.
(c) If the median is 2, find the largest possible value of x.

8. The numbers 3, 5, 7, 8 and N are arranged in ascending order. If the mean of the numbers is equal to the median, find N.

9. The mean of 5 numbers is 11. The numbers are in the ratio 1:2:3:4:5. Find the smallest number.

10. The mean of a set of 7 numbers is 3·6 and the mean of a different set of 18 numbers is 5·1. Calculate the mean of the 25 numbers.

11.† The median of five consecutive integers is N.
(a) Find the mean of the five numbers.
(b) Find the mean and the median of the squares of the integers.
(c) Find the difference between these values.

12.† The marks obtained by the members of a class are summarised in the table.

Mark	x	y	z
Frequency	a	b	c

Calculate the mean mark in terms of a, b, c, x, y.

7.3 Hypotheses and questionnaires

- Cats prefer Kit-e-Kat.
- Boys are better at Mathematics than girls.
- Smoking damages your health.
- A university degree guarantees a higher standard of living.

All these are examples of hypotheses. A hypothesis is generally considered to be a statement which may be true but for which no proof has yet been found.
Statisticians are employed to collect and analyse data in order to obtain information from it which will prove or disprove a hypothesis. Here again are the hypotheses put in the form of questions.

- Is more Kit-e-Kat sold than other brands? — after all, you can't ask the cats!
- Do boys score higher than girls in maths tests? — you might want to write your own test for this.
- Do smokers have a shorter lifespan than non-smokers?
- Do people with a degree earn more than those without a degree?

Several factors need considering when choosing a hypothesis.

Can you test it?

In the smoker's problem, a wide variety of influences affect the lifespan of people: diet, fitness, stress, heredity, etc. How can these be eliminated so that *only* the smoking counts?

Can enough data be collected to give a reasonable result?

Think of where the data will come from. For example in the cat food problem, can you ask your local shop? Will one shop be enough? In the case of the university graduates, how can a lot of graduates (and of course non-graduates) be asked for comparison? Will they each be willing to tell you their income?

How can you know if the hypothesis has been proved or disproved?

Before collecting data, consider the criteria for proof (or disproof). In the mathematics question, what is meant by better? Do the marks of the boys have to be 5% higher (or more) on average than those of the girls?

Can the type of data to be collected be analysed?

Consider the techniques available to you: mean, median, mode, range, scatter diagrams, pie charts, frequency polygons. There are other techniques like histograms, percentiles and standard deviation which will be discussed later.

Is it interesting?

If not, the whole piece of work will be dull and tedious both for you and for others to study.
An excellent way to collect data is by means of a questionnaire.
Having decided what you want to know, it is just a matter of asking the right questions.

Questionnaire design

1. Provide an introduction to the problem so that your subject knows the purpose of the questionnaire.

2. Makc thc qucstions easy to understand and straightforward so that they are specific to answer.
   Do not ask vague questions like 'Did you see much of the Olympics on TV?' The answers could be 'Yes, a lot', 'Not much', 'Only bits', 'Once or twice a day'. Such answers are hard to analyse.
   A better question is
   'How much of the Olympics coverage did you watch?'
   Then give boxes to be ticked.
   Not at all ☐
   Up to 1 hour per day ☐
   Up to 2 hours per day ☐
   More than 2 hours per day ☐

3. Make sure that the questions are not leading questions. It is human nature not to contradict the questioner, but remember the survey is to find other people's opinions, not to support your own.
   Do not ask:
   'Do you agree that BBC has the best sports coverage?'
   A better question is:
   'Which of the following has the best sports coverage?'

BBC	ITV	Ch4	Satellite TV
☐	☐	☐	☐

   You might ask for one tick, or possibly numbers 1, 2, 3, 4 to show an order of preference.

4. To avoid asking sensitive questions about age or income for example, design the questions with care so as not to offend or embarrass. Do not ask 'How old are you?' or 'Give your date of birth'.
   A better question is:
   'Tick one box for your age group'.

15–17	18–20	21–30	31–50
☐	☐	☐	☐

5. Put the easy questions first and do not ask more questions than necessary.

Exercise 4

Criticise the following questions and suggest a better question which overcomes the problem involved. Write some questions with 'yes/no' answers and some questions which involve multiple responses. Remember to word your questions simply.

1. Do you think it is ridiculous to spend money on food 'mountains' in Europe while people in Africa are starving?
2. What do you think of the new headteacher?
3. How dangerous do you think it is to fly in a single-engined aeroplane?
4. What is your weekly pay?
5. Do you agree that English and Maths are the most important subjects at school?
6. Do you or your parents often hire videos from a shop?
7. Do you think too much homework is set?
8. Do you think you would still eat meat if you had been to see the animals killed?

Pilot survey

Stop and consider two points.
(a) Will the questionnaire help to prove or disprove the hypothesis?
(b) How many people are needed for the survey?

Both of these questions can be answered by first doing a pilot survey, that is, a quick mini-survey of 5–10 people. Some of the questions may need changing straightaway.

Analysis

Having stated clearly what was the hypothesis being tested, display the results clearly. Diagrams like pie charts or bar charts, possibly showing percentages, are a good idea, particularly if drawn using colours. A data base or spreadsheet program on a computer may help. Draw conclusions from the results, but make sure they are justified by the evidence.

Your own work

Almost certainly the best hypotheses are your own because you are personally interested.
Here is a list of hypotheses which some students have enjoyed investigating.
(a) Young people are more superstitious than old people.
(b) Given a free choice, most girls would hardly ever choose to wear a dress in preference to something else.
(c) More babies are born in the Winter than in the Summer.
(d) First years watch more TV than fifth years.
(e) Most cars these days use unleaded petrol.

7.4 Scatter diagrams

Sometimes it is helpful to discover if there is a relationship or correlation between two sets of data to answer such questions as the following.

- Do tall people weigh more than short people?
- Are pupils who are good at Science also good at Mathematics?
- Do tall parents have tall children?
- Do heavy people have a slower pulse rate?
- Does the number of Olympic gold medals won by British athletes affect the rate of inflation?

If there is a relationship, it will be easy to see from a plot of the data on a scatter diagram. This is a graph in which one set of data is plotted on the horizontal axis and the other on the vertical axis. It is then possible to see if a satisfactory 'line of best fit' can be drawn through the points.

Example 1

Each month, the average outdoors temperature was recorded together with the number of therms of gas used to heat the house. The results were plotted on the scatter diagram as shown.
Is there a correlation between the outdoors temperature and the number of therms of gas used?

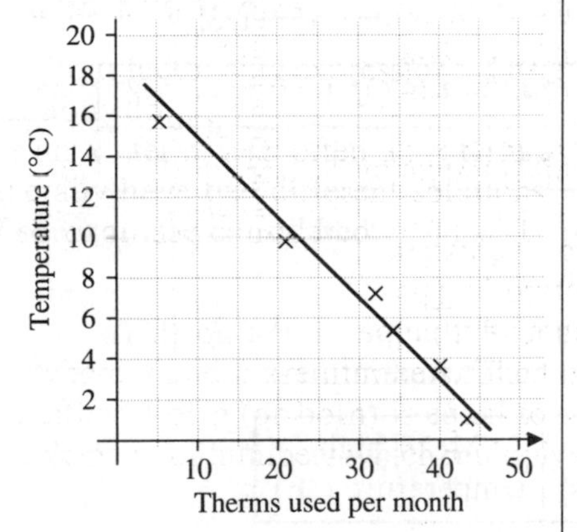

Clearly, there is a high degree of correlation between these two figures. In fact, British Gas do use weather forecasts as their main short-term predictor of future gas consumption over the whole country.
The 'line of best fit' has been drawn 'by eye'.
From the graph, it can be predicted that an outdoor temperature of 12°C will require about 17 therms of gas to be used.

Note that it is only possible to predict with any certainty within the range of values given. Extension of the line of best fit below zero suggests that 60 therms would be used when the temperature is –4°C, but perhaps a lot of people might stay in bed when the temperature is this cold so that the gas consumption would not increase to this extent. The point is that nobody knows!

The line in Example 1 has a negative gradient so there is a negative correlation. If the line of best fit has a positive gradient, there is said to be a positive correlation between the two sets of data.
Where there does not seem to be any fit between the line and the data, there is no correlation. For example, there is no correlation between the number of Olympic gold medals won and the rate of inflation.

Exercise 5

In Questions **1** and **2**, draw a scatter diagram and put in the line of best fit.

1. The marks of 7 pupils in the two papers of a physics examination were as follows

Paper 1	20	32	40	60	71	80	91
Paper 2	15	25	40	50	64	75	84

A pupil scored a mark of 50 on paper 1. What would you expect her to get on paper 2?

2. The table shows (a) the engine size in litres of various cars and (b) the distance travelled in km on one litre of petrol.

Engine	0·8	1·6	2·6	1·0	2·1	1·3	1·8
Distance	13	10·2	5·4	12	7·8	11·2	8·5

A car has a 3 litre engine. How far would you expect it to go on one litre of petrol?

3. The table shows the latitude of 10 cities in the northern hemisphere and the mean high temperatures.

City	Latitude (degrees)	Mean high temperature (°F)
Bogota	5	66
Bombay	19	87
Casablanca	34	72
Dublin	53	56
Hong Kong	22	77
Istanbul	41	64
St Petersburg	60	46
Manila	15	89
Oslo	60	50
Paris	49	59

(a) Draw a scatter diagram and draw a line of best fit.
(b) Which city lies well off the line? What factor might cause this discrepancy.
(c) The latitude of Shanghai is 31°N. What is the likely mean high temperature?

4. What sort of pattern would you expect if you took readings of the following and drew a scatter diagram?
(a) cars on roads; accident rate.
(b) sales of perfume; advertising costs.
(c) birth rate; rate of inflation.
(d) petrol consumption of car; price of petrol.

5. Some measurements of people in your class will be needed here.
(a) Measure the height and armspan of each person to the nearest centimetre.

Plot the measurements on a scatter graph. Is there any correlation?
(b) Now measure the head circumference of each person just above the eyes.
Plot head circumference against height on a scatter graph. Is there any correlation?
(c) Decide as a class which other measurements [e.g. pulse rate] you can make easily and plot these to see if any correlation exists.
(d) Which pair of measurements gave the best correlation?

7.5 Flow diagrams

A flow diagram breaks down a calculation or operation into a series of simple steps connected by direction arrows.
Flow diagrams are sometimes used as a first step in writing a computer program.

Exercise 6

In Questions **1** and **2**, put each of the numbers 1, 2, 3, 4, 5, 6, 7 in at the box marked INPUT N. Work out what number would be printed in each case.
Record the results in a table:

Input	1	2	3	4	5	6	7
Output							

1.

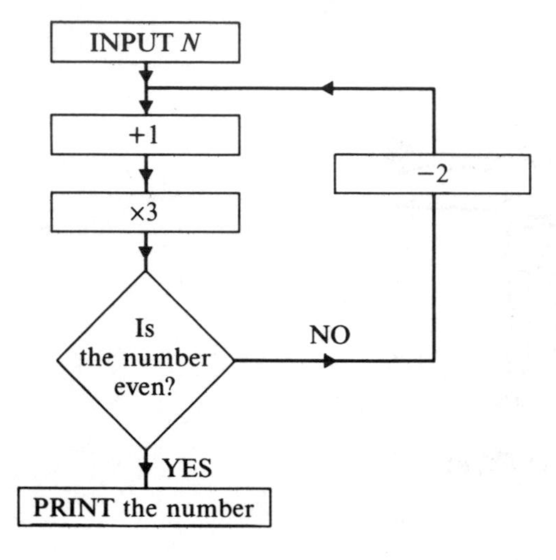

2.

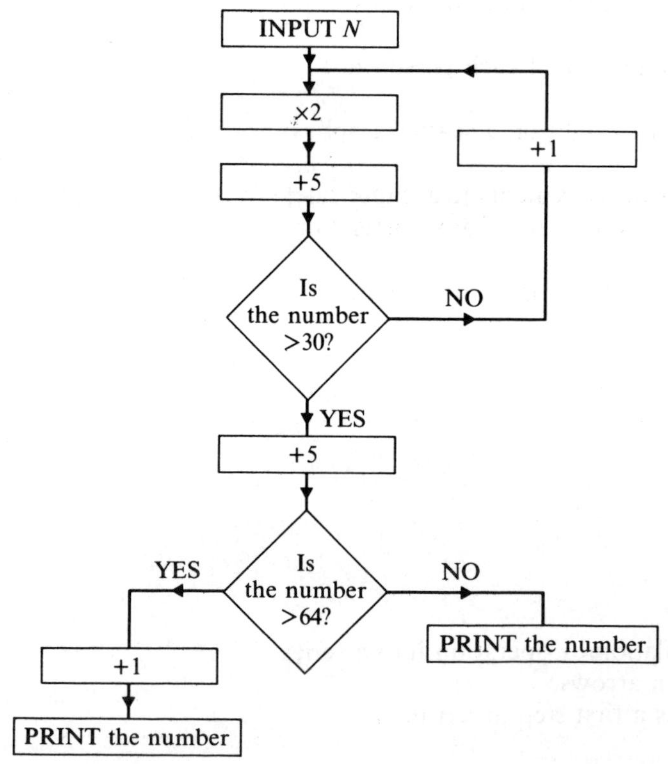

3. Work through this flow diagram several times. Take any value of Z that you like, say 7, 11, 197, -17. What do you notice?

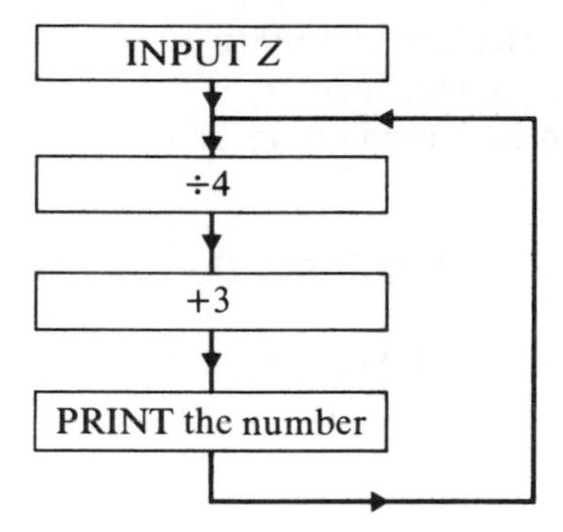

4. Try this for several values of N (say 68, 350, ...). What does the flow chart find?

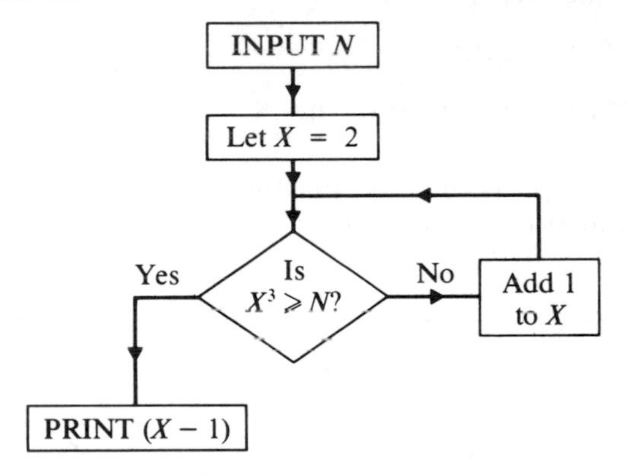

5. In this flow diagram the statement '$N = N - 5$' makes the new value of N five less than the old value. Similarly $K = K + 1$ makes the new value of K one more than the old value.

(a) Find the output for inputs 21, 22, 31, 35.
(b) How many different inputs give an output of 4?
(c) What does the flow chart find?
(d) Find the output if the input is 280.

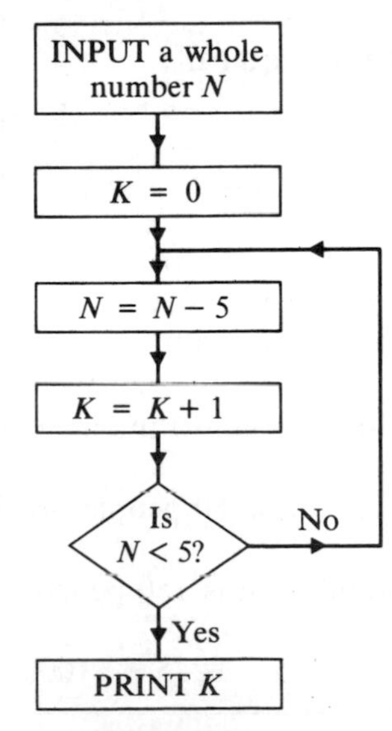

7.6 Cumulative frequency

Data given in a frequency table can be used to calculate cumulative frequencies. These new values, when plotted and joined, form a cumulative frequency curve, sometimes called an S-shaped curve or ogive.

It is a simple matter to find the median from the halfway point of a cumulative frequency curve.
Other points of location can also be found from this curve. The cumulative frequency axis can be divided into 100 parts.

- The upper quartile is at the 75% point
- The lower quartile is at the 25% point

The quartiles are particularly useful in finding the central 50% of the range of the distribution; this is known as the interquartile range.

interquartile range = (upper quartile) − (lower quartile)

The interquartile range is an important measure of spread in that it shows how widely the data is spread.
Half the distribution is in the interquartile range. If the interquartile range is small, then the middle half of the distribution is bunched together.

Example 1

In a survey, 200 people were asked to state their weekly earnings. The results were plotted on the cumulative frequency curve.

(a) How many people earned up to £350?
(b) How many people earned more than £200 per week?
(c) Find the median and the interquartile range.

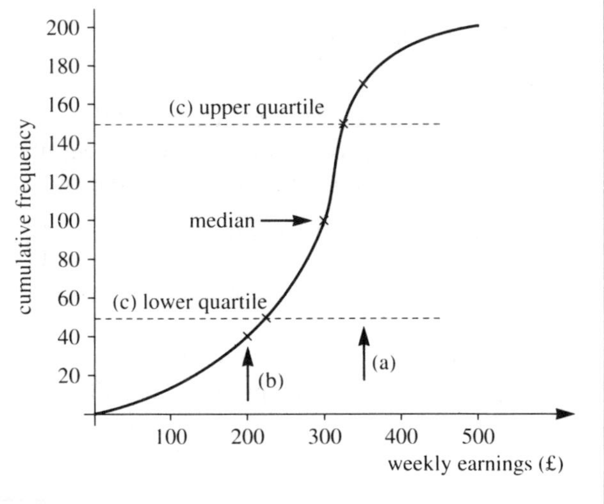

(a) From the curve, about 170 people earned up to £350 per week.
(b) About 40 people earned up to £200 per week. There are 200 people in the survey, so 160 people earned more than £200 per week.

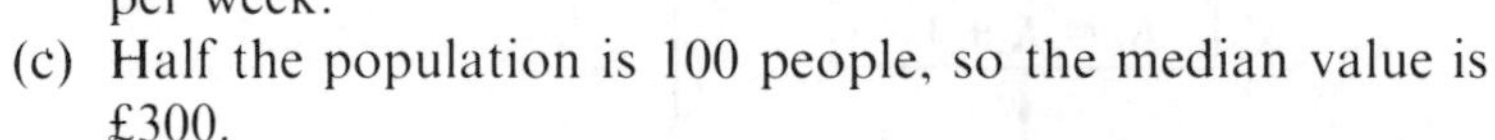

(c) Half the population is 100 people, so the median value is £300.
One quarter of the population is 50 people, so the lower quartile is £225.
Three quarters of the population is 150 people, so the upper quartile is £325.
The interquartile range = £325 − £225 = £100.

Example 2

A pet shop owner likes to weigh all his mice every week as a check on their state of health. The weights of the 80 mice are shown below.

weight (g)	frequency	cumulative frequency	upper limits
0–10	3	3	⩽ 10
10–20	5	8	⩽ 20
20–30	5	13	⩽ 30
30–40	9	22	⩽ 40
40–50	11	33	⩽ 50
50–60	15	48	⩽ 60
60–70	14	62	⩽ 70
70–80	8	70	⩽ 80
80–90	6	76	⩽ 90
90–100	4	80	⩽ 100

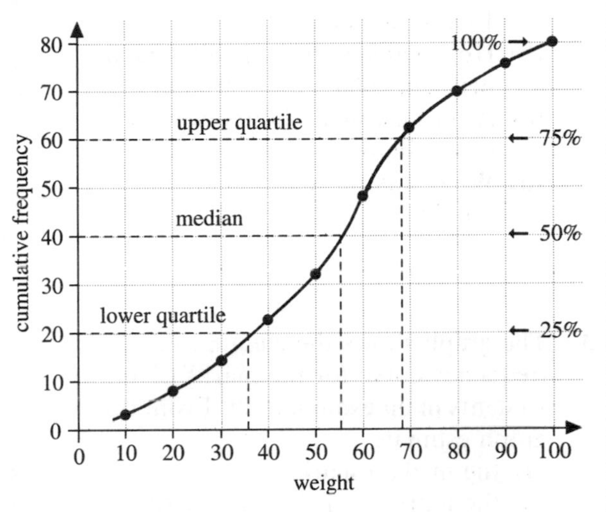

The table also shows the cumulative frequency. On the right above is a plot of the curve. Estimate (a) the median (b) the interquartile range.

Note that the points on the graph are plotted at the upper limit of each group of weights.

From the cumulative frequency curve,

median = 55 g
lower quartile = 37·5 g
upper quartile = 68 g
interquartile range = (68 – 37·5) g = 30·5 g

Exercise 7

1. The graph shows the cumulative frequency curve for the marks of 60 students in an examination. From the graph estimate

(a) the median mark
(b) the mark at the lower quartile and at the upper quartile
(c) the interquartile range
(d) the pass mark if two thirds of the students passed.

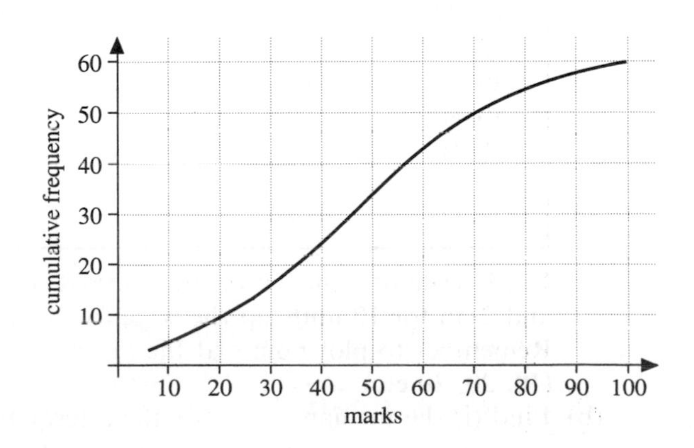

2. The lifetime of 500 electric light bulbs was measured in a laboratory. The results are shown in the cumulative frequency diagram.
   (a) How many bulbs had a lifetime of 1500 hours or less?
   (b) How many bulbs had a lifetime of between 2000 and 3000 hours?
   (c) After how many hours were 70% of the bulbs dead?
   (d) What was the shortest lifetime of a bulb?

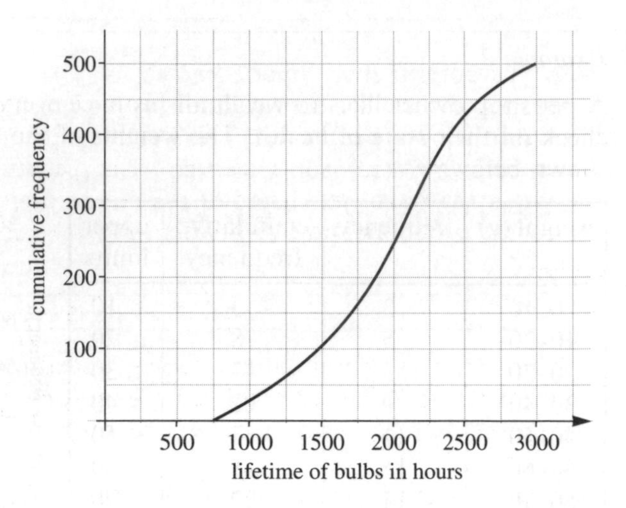

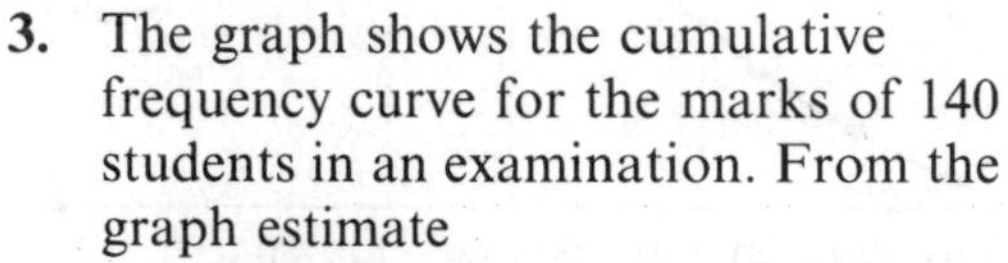

3. The graph shows the cumulative frequency curve for the marks of 140 students in an examination. From the graph estimate
   (a) the median mark
   (b) the mark at the lower quartile and at the upper quartile
   (c) the interquartile range
   (d) the pass mark if three fifths of the students passed
   (e) the number of students achieving more than 30 marks.

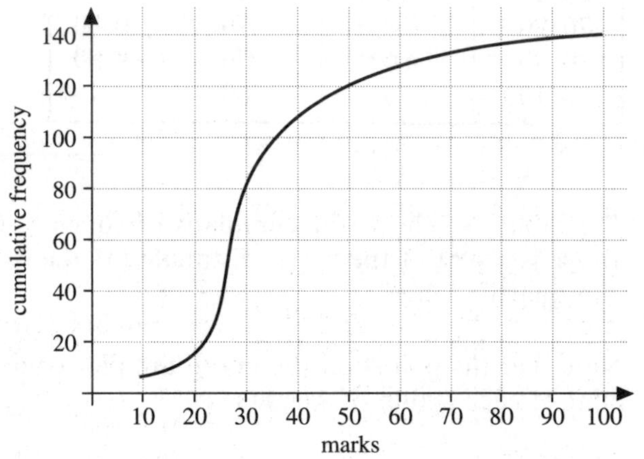

4. A photographer measures all the snakes required for a scene in a film involving a snake pit.
   (a) Draw a cumulative frequency curve for the results below.

length (cm)	frequency	cumulative frequency	upper limit
0–10	0	0	$\leqslant 10$
10–20	2	2	$\leqslant 20$
20–30	4	6	$\leqslant 30$
30–40	10	16	$\leqslant 40$
40–50	17		⋮
50–60	11		⋮
60–70	3		⋮
70–80	3		⋮

Use a scale of 2 cm for 10 units across the page for the lengths and 2 cm for 10 units up the page for the cumulative frequency. Remember to plot points at the *upper* end of the classes (10, 20, 30 etc).

   (b) Find (i) the median (ii) the interquartile range.

5. In an international competition 60 children from Britain and France did the same science test.

Marks	Britain frequency	France frequency
1–5	1	2
6–10	2	5
11–15	4	11
16–20	8	16
21–25	16	10
26–30	19	8
31–35	10	8

Note: The upper class boundaries for the marks are 5·5, 10·5, 15·5 etc.
The cumulative frequency graph should be plotted for values $\leqslant 5{\cdot}5$, $\leqslant 10{\cdot}5$, $\leqslant 15{\cdot}5$ and so on.

(a) Using the same axes draw the cumulative frequency curves for the British and French results.
Use a scale of 2 cm for 5 marks across the page and 2 cm for 10 people up the page.
(b) Find the median mark for each country.
(c) Find the interquartile range for the British results.
(d) Describe in one sentence the main difference between the two sets of results.

6. The weights of the fish in an aquarium are as follows.

weight (g)	41–50	51–60	61–70	71–80	81–90	91–100	101–110	111–120
frequency	6	8	14	21	26	14	7	4

Note that this is continuous data. The class boundaries for the '41–50' group are 40·5 and 50·5.

(a) Copy and complete the cumulative frequency table

weight less than	50·5	60·5	70·5	80·5	90·5	100·5	110·5	120·5
cum. frequency	6	14						

(b) Draw a cumulative frequency curve and find
(i) the median (ii) the interquartile range.
(c) What does the interquartile range tell you?
(d) What is the weight of the 30th percentile?

7. A new variety of plant is tested by planting a sample of seeds and measuring the heights of the plants after six weeks. The results were:

height (cm)	0–	2–	4–	6–	8–	10–	12–	14–	16 and over
frequency	12	0	2	4	10	12	10	8	0

Note: The height interval '0–' means $0 \leqslant \text{height} < 2$. The class boundaries are 0 and 2.

(a) Draw a cumulative frequency curve for the data.
Use a scale of 1 cm to 1 cm across the page and 2 cm to 10 units up the page.

(b) Find the median height of the plants.

(c) Why do you think there are so many plants in the '0–' interval?

(d) The 'average' height of the plants after six weeks is to be printed on the packet.
The actual mean height of the plants is 9·0 cm.
Which 'average' height gives the better indication of the likely performance of the plants: the mean or the median? Explain your answer.

8. In an experiment, 50 people were asked to guess the weight of a bunch of daffodils in grams. The guesses were as follows:

47 39 21 30 42 35 44 36 19 52 17 25 28
23 32 66 29 5 40 33 11 44 22 45 40 64
27 58 38 37 48 63 23 40 53 24 38 51 56
47 22 44 33 13 59 33 49 57 30 33

Construct a frequency table using class intervals 0–9, 10–19, 20–29 etc.
Draw a cumulative frequency curve and estimate

(a) the median weight

(b) the interquartile range

(c) the number of people who guessed a weight within 10 grams of the median.

9.† The age distribution of the populations of two countries, A and B, is shown below

Age	Number of people in A (millions)	Number of people in B (millions)
Under 10	15	2
10–19	11	3
20–39	18	5
40–59	7	13
60–79	3	14
80–99	1	7

(a) Copy and complete the cumulative frequency tables below

Age	Country A	Country B
Under 10	15	
Under 20	26	
Under 40		
Under 60		
Under 80		
Under 100		

(b) Using the same axes draw cumulative frequency curves for countries A and B.
Use a scale of 2 cm to 20 years across the page and 2 cm to 10 million up the page.
(c) State the population of country B.
(d) State the median age for the two countries.
(e) Describe the main difference in the age distribution of the two countries.

Examination exercise 7

1. The annual salaries of the workers in a factory were as follows:

20 apprentices........................ £ 4 500 each
15 semi-skilled workers .... £ 9 000 each
10 skilled workers .... £13 500 each
2 foremen .... £15 000 each
1 manager............................. £25 000

At discussions on pay rises, average salaries for everyone in the factory were discussed and quoted.
(a) Jack, the union representative, quoted the modal salary as the average. What was the modal salary?
(b) Wilma, one of the foremen, quoted the median as the average. What was the median salary?
(c) Rashid, the manager, quoted the mean as the average. What was the mean salary? [N]

2. The table below shows how each £1 of expenditure of Derbyshire County Council was divided up in 1987/8.

How spent	Amount
day-to-day business	47p
repaying loans and interest	23p
investing in assets	30p

Draw a pie chart to illustrate this information, showing clearly how you calculated the angles required. [M]

3. In May 1990 an estate agent sold nine, 3-bedroomed houses. The sale prices were

£59 000	£65 000	£52 000	£129 500	£52 000
£62 500	£54 500	£57 000	£56 000	

(a) What is (i) the modal sale price, (ii) the median sale price?
(b) The mean of these sale prices is approximately £65 300.
On the basis of these sale prices, which of the averages mode, median or mean would give the fairest idea of the "average" price paid for a 3-bedroom house in this area in May 1990? Give TWO reasons to justify your answer. [S]

4. A poultry farmer recorded the weight (w kg) of 50 chickens at the age of seven weeks. The results are shown in the table below.

Weight (w kg)	Number of chickens
$1 \leqslant w < 1{\cdot}2$	12
$1{\cdot}2 \leqslant w < 1{\cdot}4$	14
$1{\cdot}4 \leqslant w < 1{\cdot}6$	5
$1{\cdot}6 \leqslant w < 1{\cdot}8$	15
$1{\cdot}8 \leqslant w < 2{\cdot}0$	4

(a) What is the modal class?
(b) In which class does the median weight lie?
(c) Calculate an estimate of the mean weight of these 50 chickens. [M]

5. A teacher thinks that the nearer her pupils sit to the front of the class, the higher their test results will be. The scatter diagrams show the results for her second and fourth year classes.

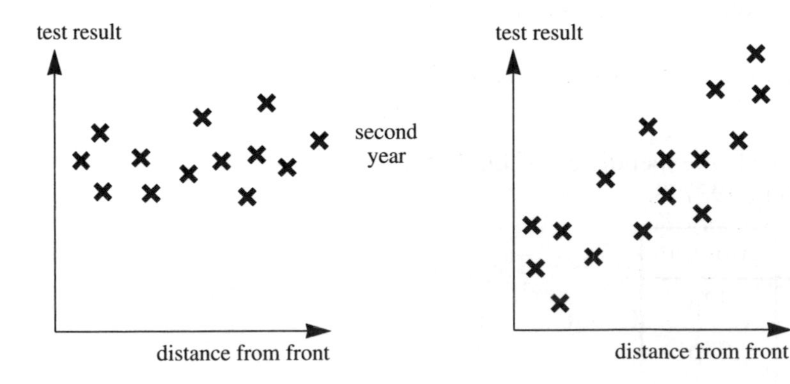

For each class, say whether the teacher's ideas are right or wrong. If she is wrong, describe what conclusions you could come to about that class. [M]

6. In an experiment, Pierre was given a fixed length of time to memorise a list of items. Pierre was tested and the number of errors was noted.
This experiment was repeated for other lists, with different lengths of time. The results were as follows:

Length of time in seconds, x	5	6	7	8	9	10	11	12	13	14
Number of errors, y	9	7	8	5	6	6	4	5	3	2

These results are plotted below. The line of best fit is drawn.

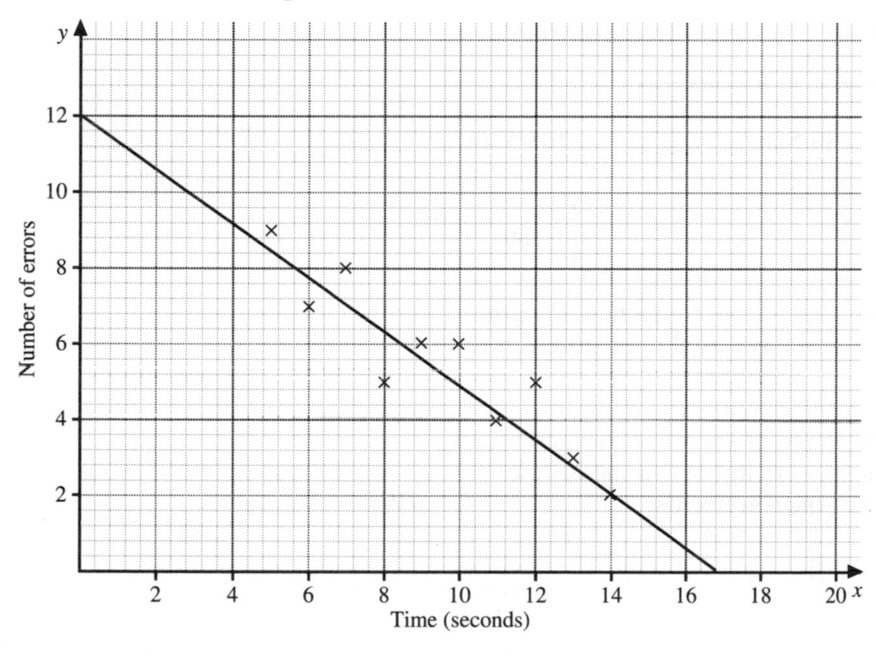

(a) Give an interpretation of the point where the line meets the x axis.
(b) Explain why there is no sensible interpretation of where the line meets the y axis.
(c) Calculate the gradient of the line, correct to one decimal place.
(d) Find the equation of the line.
(e) The experiment was repeated by Kerith. He made the same number of errors as Pierre when the time was 5 seconds, but for longer times his memory was a little worse than Pierre's. On a sketch graph draw a possible line of best fit for Kerith. Label this line K. [S]

7. Isobel is doing a survey on the use made of the school library. Write three suitable questions for a questionnaire. Each question must allow for at least three possible responses. [S]

8. The following table shows a grouped frequency distribution of the waist measurements, to the nearest centimetre, of 60 people.

Waist measurement (cm)	Number of people
66–75	7
76–80	8
81–85	12
86–90	14
91–95	10
96–105	9

(a) Using the class mid-points, calculate the mean waist measurement.

(b) Complete the following cumulative frequency table.

Waist measurement less than	75·5 cm	80·5 cm	...	...	105·5 cm
Cumulative frequency	7				60

(c) On a graph draw the cumulative frequency polygon.

(d) Showing clearly how you have used your graph in (c), find estimates for

(i) the median waist measurement,

(ii) the interquartile range of the waist measurements.

(e) State which of your answers in (d) is a measure of the dispersion of the waist measurements. [W]

8 Probability

8.1 Relative frequency

There are four different ways of estimating probabilities.

Method A Use symmetry

The probability of rolling a 3 on a fair dice is $\frac{1}{6}$. This is because all the scores, 1, 2, 3, 4, 5, 6 are equally likely.
Similarly the probability of getting a head when tossing a fair coin is $\frac{1}{2}$.

Method B Collect data from an experiment or survey

What is the estimated probability of a drawing pin landing point upwards when dropped onto a hard surface? Symmetry could not be used for obvious reasons but data could be collected from an experiment to see what happened in, say 500 trials.

Method C Look at past data

What is the estimated probability of my plane crashing as it lands at a certain airport? I would look at the accident records of the airport over the last five years or so.

Method D Make a subjective estimate

This is not really a method in the same sense as the other three. We have to use it when the event is not repeatable.

How could one estimate the probability of England beating France in a soccer match next week? Looking at past results would be of little value for many reasons. Even consulting 'experts' gives notoriously inaccurate predictions.

Exercise 1

For each question, state which method, A, B, C or D you would use to estimate the probability of the event given.

1. The probability that a person chosen at random from a class will be left-handed.
2. The probability that there will be snow in the ski resort to which a school party is going in February next year.

3. The probability of drawing an 'ace' from a pack of playing cards.
4. The probability that you hole a six-foot putt when playing golf.
5. The probability that the world record for running 1500 m will be under 3 min 20 seconds by the year 2020.
6. The probability that a person who smokes will suffer from lung cancer later in life.
7. The probability of rolling a 3 using a dice which is suspected of being biased.
8. The probability that a person selected at random would vote 'Labour' in a general election tomorrow.
9. The probability that a train will arrive within ten minutes of its scheduled arrival time.
10. The probability of winning first prize in a raffle if you bought 5 tickets and 1000 are sold.
11. The probability that the current Wimbledon Ladies Champion will successfully defend her title next year.
12. The probability that the next pupil expelled from a certain school will be a girl.

Simple probability

Probability theory is not the sole concern of people interested in betting, although it is true to say that a 'lucky' poker player is likely to be a player with a sound understanding of probability.
All major airlines regularly overbook aircraft because they can predict with accuracy the probability that a certain number of passengers will fail to arrive for the flight.
Suppose a 'trial' can have n equally likely results and suppose that a 'success' can occur in s ways (from the n). Then the probability of a 'success' $= \frac{s}{n}$.

Example 1

A fair dice is rolled 240 times. How many times would a number greater than 4 be expected?

As a 5 or a 6 is needed out of the six equally likely outcomes,

$p\text{ (number} > 4) = \frac{2}{6} = \frac{1}{3}$.

Expected number of successes
= (probability of a success) × (number of trials).

So expected number of scores greater than 4 $= \frac{1}{3} \times 240 = 80$.

Example 2

A single card is drawn from a pack of 52 playing cards. Find the probability of the following results:
(a) the card is an ace (b) the card is the ace of hearts
(c) the card is a spade (d) the card is a picture card.

We will use the notation 'p (an ace)' to represent 'the probability of selecting an ace'.

(a) p (an ace) $= \frac{4}{52} = \frac{1}{13}$. (b) p (ace of hearts) $= \frac{1}{52}$.

(c) p (a spade) $= \frac{13}{52} = \frac{1}{4}$. (d) p (a picture card) $= \frac{12}{52}$.

In each case, we have counted the number of ways in which a 'success' can occur and divided by the number of possible results of a 'trial'.

Exercise 2

1. One card is drawn at random from a pack of 52 playing cards. Find the probability of drawing
(a) a 'King',
(b) a red card,
(c) the seven of clubs,
(d) either the King, Queen or Jack of diamonds.

2. A fair die is thrown once. Find the probability of obtaining
(a) a six,
(b) an even number,
(c) a number greater than 3,
(d) a three or a five.

3. A 10p and a 5p coin are tossed at the same time. List all the possible outcomes. Find the probability of obtaining
(a) two heads, (b) a head and a tail.

4. A bag contains 6 red balls and 4 green balls.
(a) Find the probability of selecting at random:
(i) a red ball (ii) a green ball.
(b) One red ball is removed from the bag. Find the new probability of selecting at random
(i) a red ball (ii) a green ball.

5. One letter is selected at random from the word 'UNNECESSARY'. Find the probability of selecting
(a) an R (b) an E (c) an O (d) a C

6. A 'hand' of 13 cards contains the cards shown.

A card is selected at random from the 13.
Find the probability of selecting:
(a) any card of the heart suit,
(b) any card of the club suit,
(c) a 'six' of any suit,
(d) any 'picture' card [not including an ace],
(e) the 'four' of clubs,
(f) an 'eight' of any suit,
(g) any 'six' or 'four'.

7. The King, Queen and Jack of clubs are removed from a pack of 52 playing cards. One card is selected at random from the remaining cards. Find the probability that the card is:
(a) a heart (b) a King
(c) a club (d) the 10 of hearts.

8. Cards with the numbers 2 to 101 are placed in a hat. Find the probability of selecting:
(a) an even number,
(b) a number less than 14,
(c) a square number,
(d) a prime number less than 20.

9. Find the probability of the following:
(a) throwing a number less than 8 on a single die,
(b) obtaining the same numbers of heads and tails when five coins are tossed,
(c) selecting a square number from the set
A = {4, 9, 16, 25, 36, 49},
(d) selecting a prime number from the set A.

10. Louise buys five raffle tickets out of 1000 sold. She does not win first prize. What is the probability that she wins second prize?

11. Tickets numbered 1 to 1000 were sold in a raffle for which there was one prize. Mr Kahn bought all the tickets containing at least one '3' because '3' was his lucky number. What was the probability of Mr Kahn winning?

12. One ball is selected at random from a bag containing 12 balls of which x are white.

(a) What is the probability of selecting a white ball?
When a further 6 white balls are added the probability of selecting a white ball is doubled.

(b) Find x.

13. The spinner shown has four equal sectors. How many 3's would you expect in 100 spins?

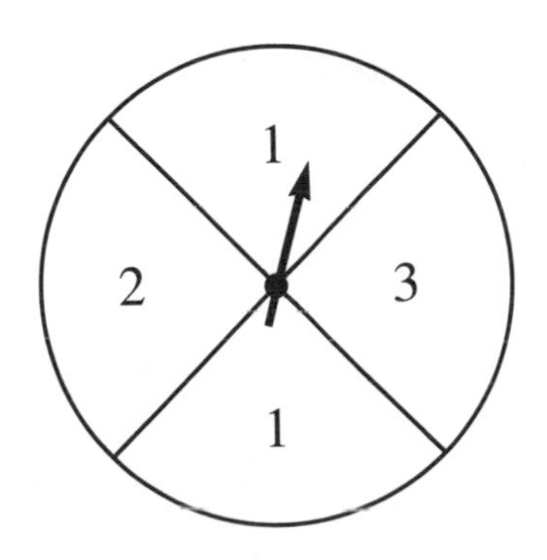

14. About one in eight of the population is left-handed. How many left-handed people would you expect to find in a firm employing 400 people?

15. A bag contains a large number of marbles of which one in five is red. If I randomly select one marble on 200 occasions, how many times would I expect to select a red marble?

16. Two cards are selected at random from the 3, 4, 5 and 6 of hearts.
Find the probability that the total of the two cards is more than 9.

17.† A point is chosen at random from inside the square.

(a) What is the probability of choosing a point which lies outside the circle?
Give your answer as a decimal to 4 s.f.

(b) In a computer simulation 5000 points are chosen using a random number generator. How many points would you expect to be chosen which lie outside the circle?

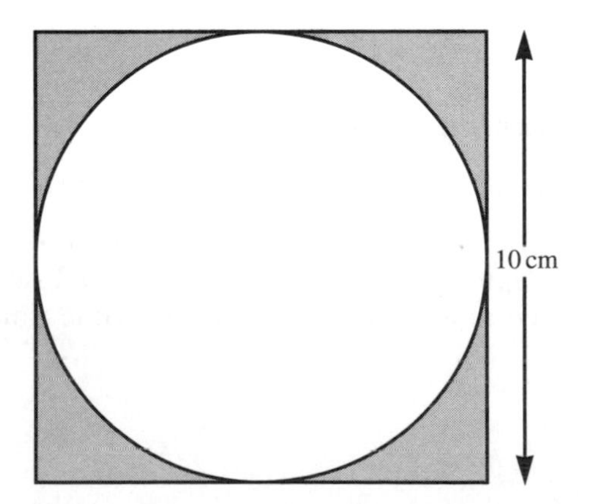

8.2 Exclusive events

Two events are *exclusive* if they cannot occur at the same time. So, selecting an Ace or selecting a Ten from a pack of cards are exclusive.

The 'OR' rule: the addition rule

For exclusive events A and B

$$p(\text{A or B}) = p(\text{A}) + p(\text{B})$$

Many questions involving exclusive events can be done *without* using the addition rule.

Example 1

A card is selected at random from a pack of 52 playing cards. What is the probability of selecting any King or Queen?

(a) Count the number of ways in which a King *or* Queen can occur.
That is 8 ways.
p (selecting a King or a Queen) $= \frac{8}{52} = \frac{2}{13}$

(b) Since 'selecting a King' and 'selecting a Queen' are exclusive events, we could use the addition law.

p (selecting a King) $= \frac{4}{52}$

p (selecting a Queen) $= \frac{4}{52}$

So, p (selecting a King or a Queen) $= \frac{4}{52} + \frac{4}{52} = \frac{8}{52}$ as before.

You can decide for yourself which method is easier in any given question.

Non-exclusive events

If the events are not exclusive, the addition rule *cannot* be used.

Here is a spinner with numbers and colours.
The events 'spinning a red' and 'spinning a 1' are *not* exclusive because a red and a 1 can occur at the same time.

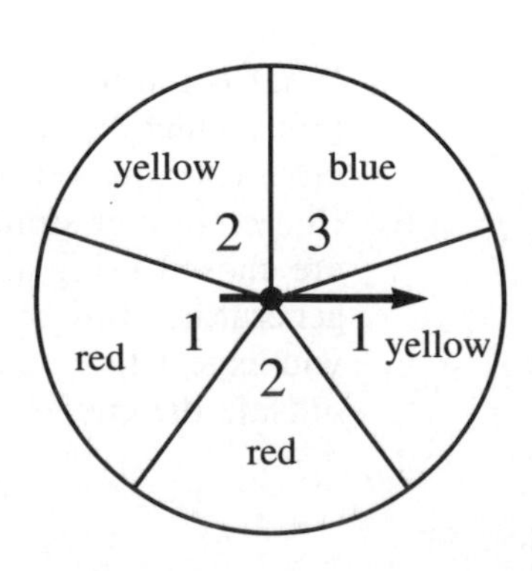

Exercise 3

1. A bag contains 10 red balls, 5 blue balls and 7 green balls. Find the probability of selecting at random:
   (a) a red ball, (b) a green ball,
   (c) a blue *or* a red ball, (d) a red *or* a green ball.

2. A roulette wheel has the numbers 1 to 36 once only. What is the probability of spinning either a 10 or a 20?

3. A fair dice is rolled. What is the probability of rolling either a 1 or a 6?

4. From a pack of cards I have already selected the King, Queen, Jack and ten of diamonds. What is the probability that on my next draw I will select either the ace or the nine of diamonds?

5. In an opinion poll people were asked to state the party for which they are going to vote.

   If one person is chosen at random from the sample, what is the probability that he or she would vote either Conservative or Labour?

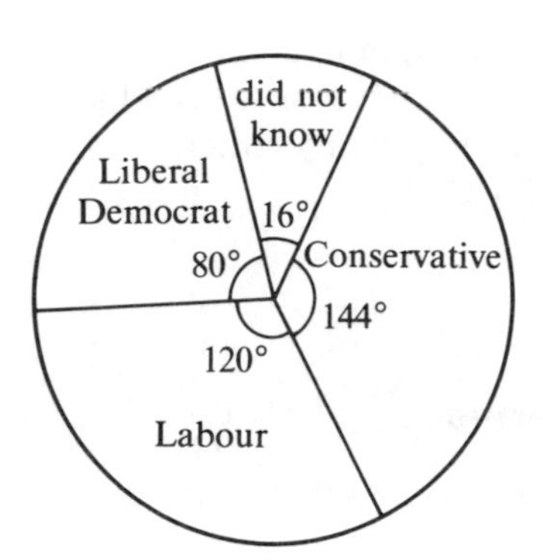

6. Thirty pupils were asked to state the activities they enjoyed from swimming (S), tennis (T) and hockey (H). The numbers in each set are shown.

   One pupil is randomly selected.

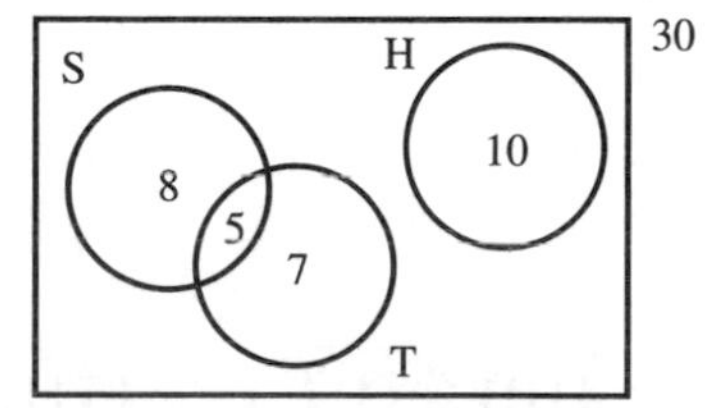

   (a) Which of the following pairs of events are exclusive?
      (i) 'selecting a pupil from S', 'selecting a pupil from H'.
      (ii) 'selecting a pupil from S', 'selecting a pupil from T'.
   (b) What is the probability of selecting a pupil who enjoyed either hockey or tennis?

7. The spinner shown has four equal sectors.
   (a) Which of the following pairs of events are exclusive?
      (i) 'spinning 1', 'spinning green',
      (ii) 'spinning 3', 'spinning 2'.
      (iii) 'spinning blue', 'spinning 1'.
   (b) What is the probability of spinning either a 1 or a green?

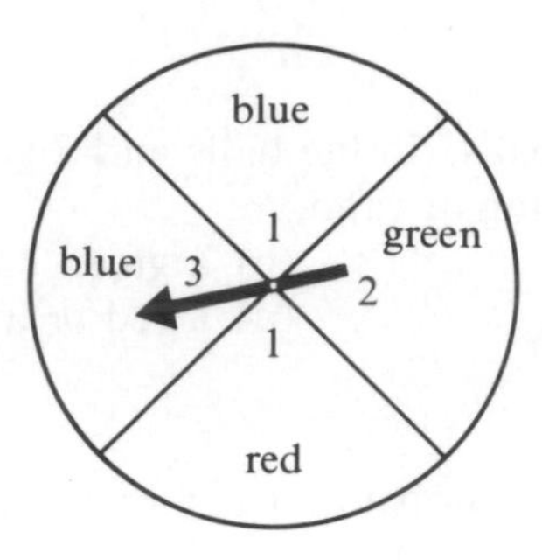

8. Here is a spinner with unequal sectors. When the pointer is spun the probability of getting each colour and number is as follows.

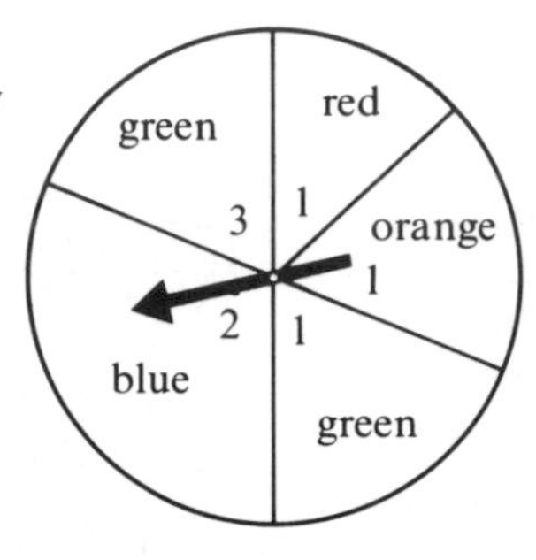

red	0·1
blue	0·3
orange	0·2
green	0·4

1	0·5
2	0·3
3	0·2

   (a) What is the probability of spinning either 1 or 2?
   (b) What is the probability of spinning either blue or green?
   (c) Why is the probability of spinning either a 1 or a green *not* $0{\cdot}5 + 0{\cdot}4$?

Listing possible outcomes

With more complicated situations, it is often helpful to list all possible outcomes in a systematic way.

Example 2

Three coins are tossed together. What is the probability of getting two tails?

List all the possible outcomes:

HHH HHT HTH (HTT) THH (THT) (TTH) TTT

There are 8 equally likely outcomes, three of which have two tails.

So, p (tossing two tails) $= \frac{3}{8}$

Example 3

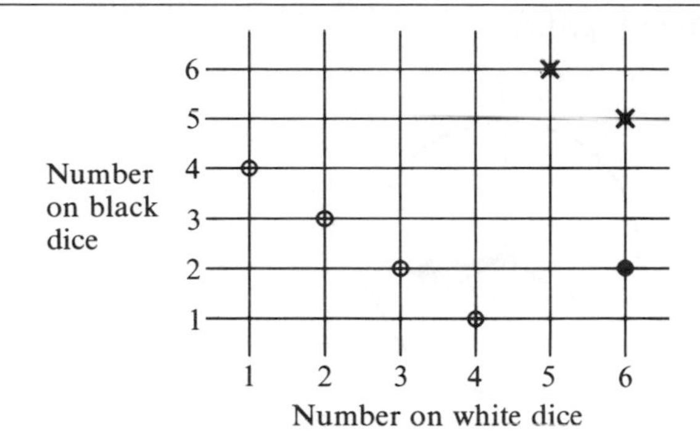

A black dice and a white dice are thrown at the same time. Display all the possible outcomes. Find the probability of obtaining:
(a) a total of 5,
(b) a total of 11,
(c) a 'two' on the black dice and a 'six' on the white dice.

It is convenient to display all the possible outcomes on a grid.

There are 36 possible outcomes, shown where the lines cross.
(a) There are four ways of obtaining a total of 5 on the two dice. They are shown circled on the diagram.

So, probability of obtaining a total of 5 $= \frac{4}{36} = \frac{1}{9}$

(b) There are two ways of obtaining a total of 11. They are shown with a cross on the diagram.

So, p (total of 11) $= \frac{2}{36} = \frac{1}{18}$

(c) There is only one way of obtaining a 'two' on the black dice and a 'six' on the white dice.

So, p (2 on black and 6 on white) $= \frac{1}{36}$

This is shown with a blob on the diagram.

Exercise 4

1. Three coins are tossed at the same time. List all the possible outcomes. Find the probability of obtaining:
(a) three heads, (b) two heads and one tail,
(c) no heads, (d) at least one head.

2. A red dice and a blue dice are thrown at the same time. List all the possible outcomes in a systematic way. Find the probability of obtaining:
(a) a total of 10, (b) a total of 12,
(c) a total less than 6, (d) the same number on both dice.

3. The two spinners shown are spun together.

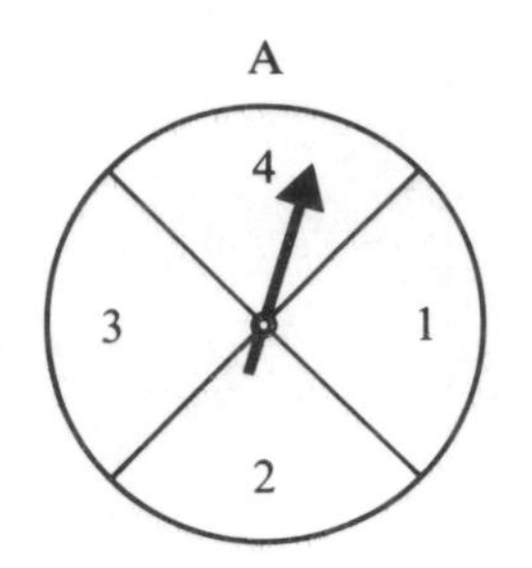

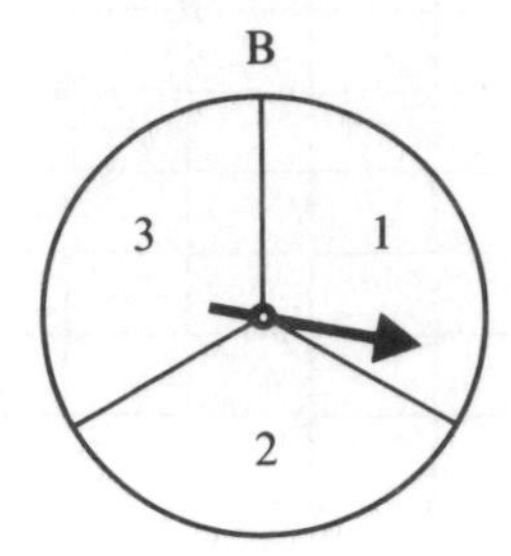

List all the possible outcomes. Find the probability of obtaining:
(a) a total of 4, (b) the same number on each spinner.

4. A dice is thrown; when the result has been recorded, the dice is thrown a second time. Display all the possible outcomes of the two throws. Find the probability of obtaining:
(a) a total of 4 from the two throws,
(b) a total of 8 from the two throws,
(c) a total between 5 and 9 inclusive from the two throws,
(d) a number on the second throw which is double the number on the first throw,
(e) a number on the second throw which is four times the number on the first throw.

5. Four coins are tossed at the same time. List all the possible outcomes in a systematic way. Find the probability of obtaining:
(a) two heads and two tails, (b) four tails,
(c) at least one tail, (d) three heads and one tail.

6. (a) How many possible outcomes are there when six coins are tossed together?
(b) What is the probability of tossing six heads?

7. The spinner is spun and the dice is thrown at the same time.

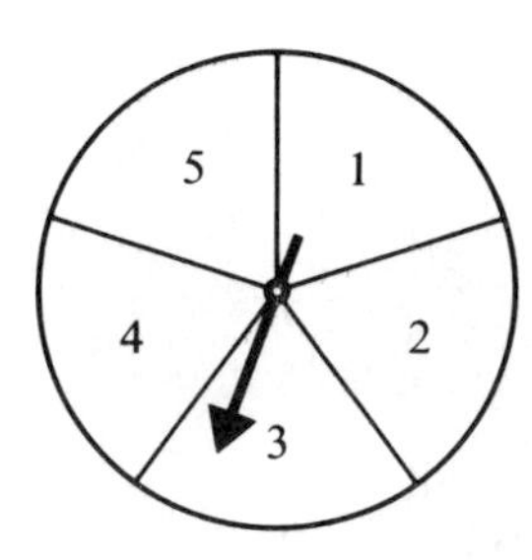

A 'win' occurs when the number on the spinner is greater than or equal to the number on the dice. What is the probability of a 'win'?

8. Four friends Wayne, Xavier, Yves and Zara each write their name on a card and the four cards are placed in a hat. Two cards are chosen to decide who does the maths homework that night. List all the possible combinations.
What is the probability that the names drawn are Xavier and Yves?

9.† Two dice and two coins are thrown at the same time. Find the probability of obtaining:
(a) two heads and a total of 12 on the dice,
(b) a head, a tail and a total of 9 on the dice,
(c) two tails and a total of 3 on the dice.

10.† A red, a blue and a white die are all thrown at the same time. Display all the possible outcomes in a suitable way. Find the probability of obtaining:
(a) a total of 18 on the three dice,
(b) a total of 4 on the three dice,
(c) a total of 10 on the three dice,
(d) a total of 15 on the three dice,
(e) a total of 7 on the three dice,
(f) the same number on each dice.

11.† The RAN # button on a calculator generates random numbers between .000 and .999. It can be used to simulate tossing three coins by letting, say, any *odd* digit be a *tail* and any *even* digit be a *head.*

So the number .568 represents THH
and .605 represents HHT

Use the RAN # button to simulate the tossing of three coins.

'Toss' the three coins 32 times and compare the relative frequencies of
(a) three heads and (b) two heads and a tail
with the value you calculated in Question **1** of this exercise.

8.3 Independent events

Two events are *independent* if the occurrence of one event is unaffected by the occurrence of the other.
So, obtaining a head on one coin and a tail on another coin when the coins are tossed at the same time are independent.

The 'AND' rule: the multiplication rule

For independent events A and B
$p(\text{A and B}) = p(\text{A}) \times p(\text{B})$

Two coins are tossed and the results listed.

HH HT TH TT

The probability of tossing two heads is $\frac{1}{4}$.

By the multiplication rule for independent events:

$p(\text{two heads}) = p(\text{head on first coin}) \times p(\text{head on second coin})$

$\therefore\ p(\text{two heads}) = \frac{1}{2} \times \frac{1}{2} = \frac{1}{4}$ as before.

Example 1

A fair coin is tossed and a fair die is rolled. Find the probability of obtaining a 'head' and a 'six'.

The two events are independent

$p(\text{head } \textit{and} \text{ six}) = p(\text{head}) \times p(\text{six}) = \frac{1}{2} \times \frac{1}{6} = \frac{1}{12}$.

Example 2

When the two spinners are spun, what is the probability of getting a B on the first and a 3 on the second?

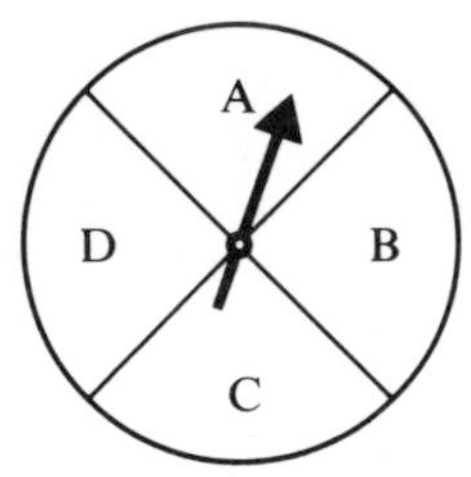

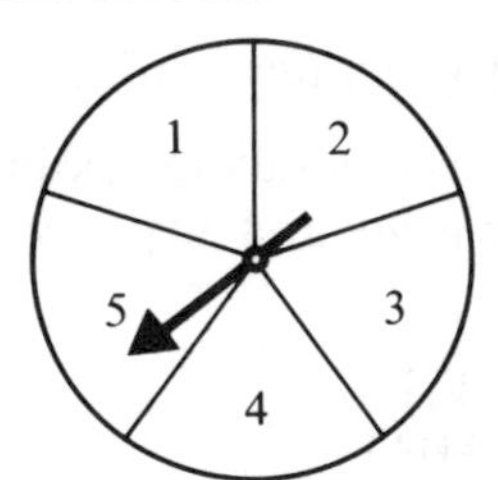

The events 'B on the first spinner' and '3 on the second spinner' are independent.

$\therefore\ p(\text{spinning B and 3}) = p(\text{B}) \times p(3) = \frac{1}{4} \times \frac{1}{5} = \frac{1}{20}$

Exercise 5

1. A card is drawn from a pack of playing cards and a dice is thrown. Events A and B are as follows:
A: 'a Jack is drawn from the pack'
B: 'a three is thrown on the dice'.
(a) Write down the values of $p(\text{A})$, $p(\text{B})$.
(b) Write down the value of $p(\text{A and B})$.

2. A coin is tossed and a dice is thrown. Write down the probability of obtaining:
(a) a 'head' on the coin,
(b) an odd number on the dice,
(c) a 'head' on the coin and an odd number on the dice.

3. Box A contains 3 red balls and 3 white balls. Box B contains 1 red and 4 white balls.

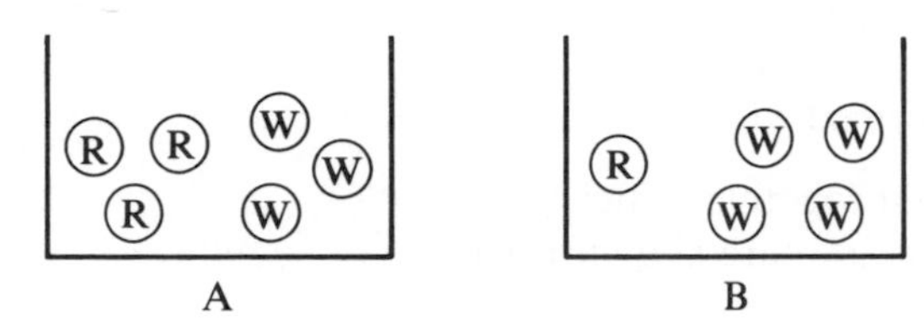

One ball is randomly selected from Box A and one from Box B. What is the probability that both balls selected are red?

4. In an experiment, a card is drawn from a pack of playing cards and a dice is thrown.
Find the probability of obtaining:
(a) a card which is an ace and a six on the dice,
(b) the king of clubs and an even number on the dice,
(c) a heart and a 'one' on the dice.

5. A card is taken at random from a pack of playing cards and replaced. After shuffling, a second card is selected. Find the probability of obtaining:
(a) two cards which are clubs,
(b) two Kings,
(c) two picture cards.

6. A ball is selected at random from a bag containing 3 red balls, 4 black balls and 5 green balls. The first ball is replaced and a second is selected. Find the probability of obtaining:
(a) two red balls, (b) two green balls.

7. The letters of the word 'INDEPENDENT' are written on individual cards and the cards are put into a box. A card is selected and then replaced and then a second card is selected. Find the probability of obtaining:
(a) the letter 'P' twice, (b) the letter 'E' twice.

8. A fruit machine has three independent reels and pays out a Jackpot of £100 when three apples are obtained.

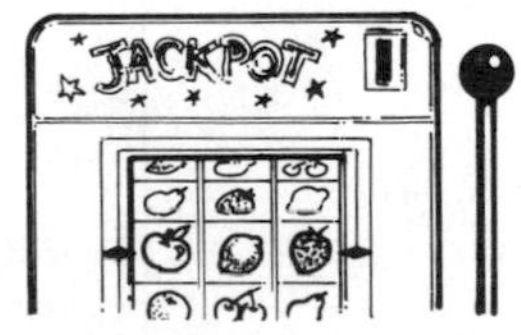

Each reel has 15 pictures. The first reel has 3 apples, the second has 4 apples and the third has 2 apples.
Find the probability of winning the Jackpot.

9. Three coins are tossed and two dice are thrown at the same time.
Find the probability of obtaining:
(a) three heads and a total of 12 on the dice,
(b) three tails and a total of 9 on the dice.

10. A coin is biased so that it shows 'Heads' with a probability of $\frac{2}{3}$. The same coin is tossed three times. Find the probability of obtaining:

(a) two tails on the first two tosses,

(b) a head, a tail and a head (in that order).

11. A fair dice and a biased dice are thrown together. The probabilities of throwing the numbers 1 to 6 are shown for the biased dice.

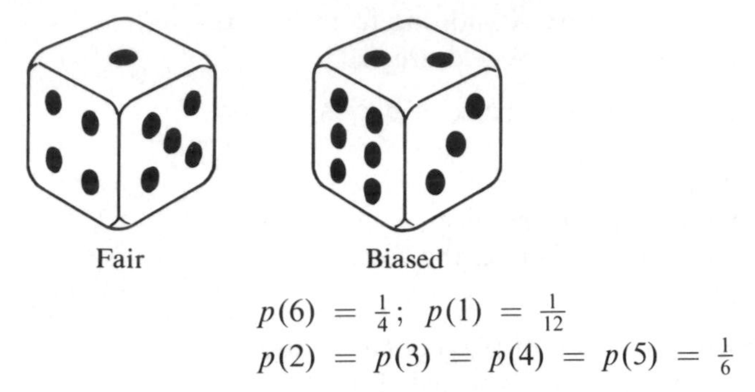

$$p(6) = \tfrac{1}{4};\ p(1) = \tfrac{1}{12}$$
$$p(2) = p(3) = p(4) = p(5) = \tfrac{1}{6}$$

Find the probability of obtaining a total of 12 on the two dice.

Tree diagrams

Example 3

A bag contains 5 red balls and 3 green balls. A ball is drawn at random and then replaced. Another ball is drawn. What is the probability that both balls are green?

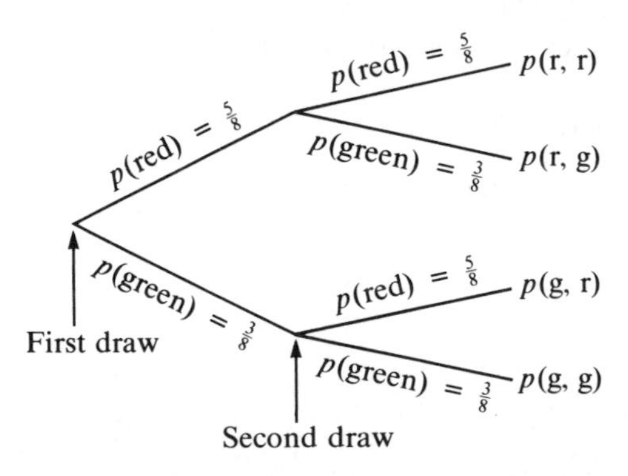

The branch marked p(g, g) involves the selection of a green ball twice.
The probability of this event is obtained by simply multiplying the fractions on the two branches.

So, $p(\text{two green balls}) = \frac{3}{8} \times \frac{3}{8} = \frac{9}{64}$

Example 4

A bag contains 5 red balls and 3 green balls. A ball is selected at random and *not* replaced. A second ball is then selected.
Find the probability of selecting:
(a) two green balls (b) one red ball and one green ball.

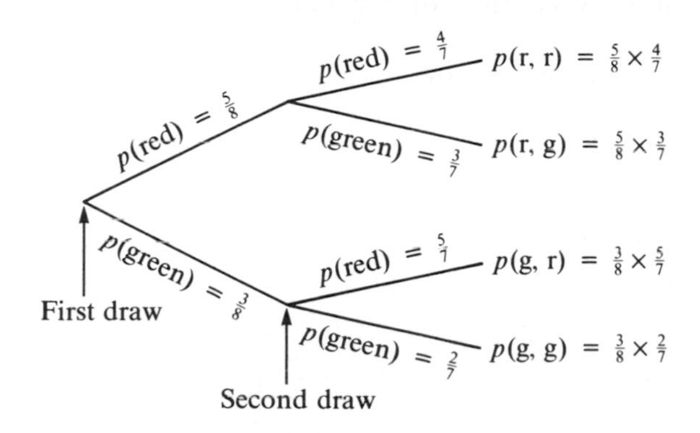

(a) $p(\text{two green balls}) = \frac{3}{8} \times \frac{2}{7} = \frac{3}{28}$

(b) $p(\text{one red, one green}) = (\frac{5}{8} \times \frac{3}{7}) + (\frac{3}{8} \times \frac{5}{7}) = \frac{15}{28}$

Note that for (b), the probabilities can be added because the events 'red then green' and 'green then red' are exclusive.
As a check, the sum of the final results at the branch ends should be one.

So $(\frac{5}{8} \times \frac{4}{7}) + (\frac{5}{8} \times \frac{3}{7}) + (\frac{3}{8} \times \frac{5}{7}) + (\frac{3}{8} \times \frac{2}{7})$

$= \frac{20}{56} + \frac{15}{56} + \frac{15}{56} + \frac{6}{56} = 1.$

Note the convention that 'not five' can be written $\overline{5}$.

Exercise 6

1. A bag contains 10 discs; 7 are black and 3 white. A disc is selected, and then replaced. A second disc is selected. Copy and complete the tree diagram showing all the probabilities and outcomes.

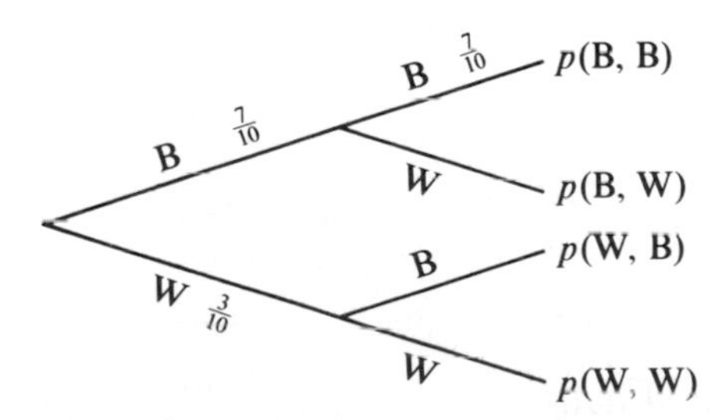

Find the probability of the following:
(a) both discs are black, (b) both discs are white.

2. A bag contains 5 red balls and 3 green balls. A ball is drawn and then replaced before a ball is drawn again. Draw a tree diagram to show all the possible outcomes. Find the probability that:
(a) two green balls are drawn,
(b) the first ball is red and the second is green.

3. A bag contains 7 green discs and 3 blue discs. A disc is drawn and *not* replaced.
A second disc is drawn. Copy and complete the tree diagram.

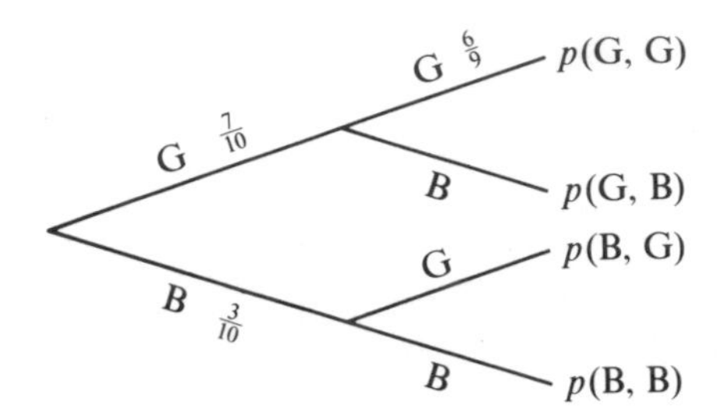

Find the probability that:
(a) both discs are green, (b) both discs are blue.

4. A bag contains 5 red balls, 3 blue balls and 2 yellow balls. A ball is drawn and not replaced. A second ball is drawn.

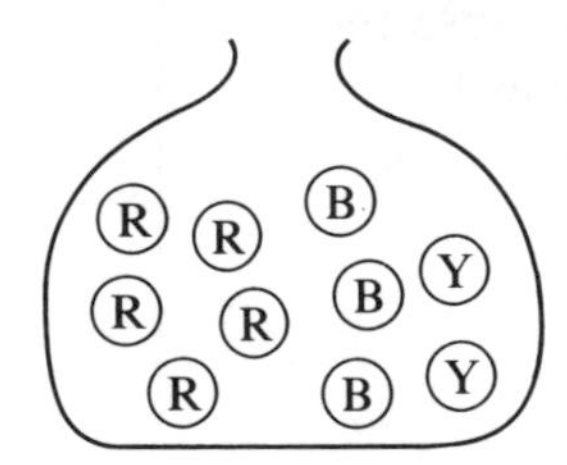

Find the probability of drawing:
(a) two red balls,
(b) one blue ball and one yellow ball,
(c) two yellow balls.

5. A bag contains 4 red balls, 2 green balls and 3 blue balls. A ball is drawn and not replaced. A second ball is drawn.

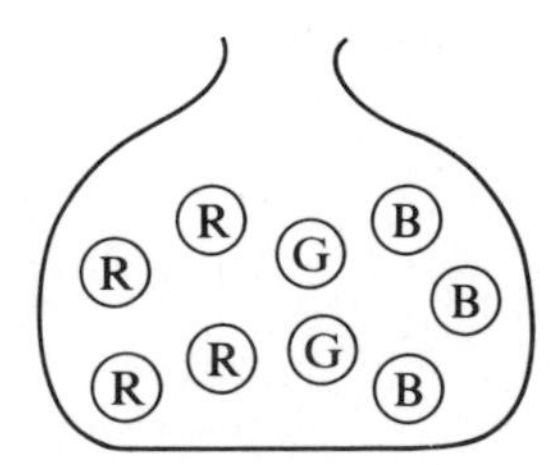

Find the probability of drawing:
(a) two blue balls, (b) two red balls
(c) one red ball and one blue ball (in any order),
(d) one green ball and one red ball (in any order).

6. A six-sided dice is thrown three times. Complete the tree diagram, showing for each branch the two events 'six' and 'not six' (written $\overline{6}$).

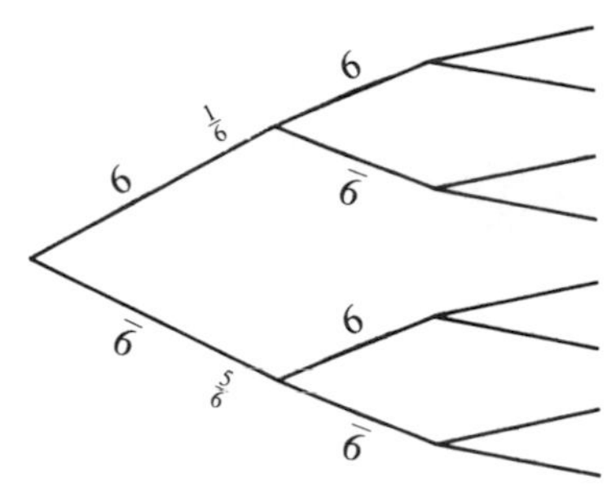

What is the probability of throwing a total of:
(a) three sixes, (b) no sixes,
(c) one six, (d) at least one six (use part (b))?

7. A card is drawn at random from a pack of 52 playing cards. The card is replaced and a second card is drawn. This card is replaced and a third card is drawn. What is the probability of drawing:
(a) three hearts, (b) at least two hearts,
(c) exactly one heart?

8. A bag contains 6 red marbles and 4 blue marbles. A marble is drawn at random and not replaced. Two further draws are made, again without replacement. Find the probability of drawing:
(a) three red marbles, (b) three blue marbles,
(c) no red marbles, (d) at least one red marble.

9. When a cutting is taken from a geranium the probability that it grows is $\frac{3}{4}$. Three cuttings are taken. What is the probability that
(a) all three grow, (b) none of them grow?

10. A dice has its six faces marked 0, 1, 1, 1, 6, 6. Two of these dice are thrown together and the total score is recorded. Draw a tree diagram.
What is the probability of obtaining a total of 7?

11. A coin is biased so that the probability of a 'head' is $\frac{3}{4}$. Find the probability that, when tossed three times, it shows:
(a) three tails, (b) two heads and one tail,
(c) one head and two tails, (d) no tails.
Write down the sum of these four probabilities.

12. A teacher decides to award exam grades A, B or C by a new fairer method. Out of 20 children, three are to receive A's, five B's and the rest C's. She writes the letters A, B and C on 20 pieces of paper and invites the pupils to draw their exam result, going through the class in alphabetical order. Find the probability that:
(a) the first three pupils all get grade 'A',
(b) the first three pupils all get grade 'B'
(c) the first four pupils all get grade 'B'.
(Do not cancel down the fractions.)

13. The probability that George, an amateur golfer, actually hits the ball is (regrettably for all concerned) only $\frac{1}{10}$.

If four separate attempts are made, find the probability that George will hit the ball:
(a) four times, (b) at least twice, (c) not at all.

Reminder

When a question says '2 balls are drawn' or '3 people are chosen', remember to draw a tree diagram *without* replacement.

Example 5

There are 10 boys and 12 girls in a class. Two pupils are chosen at random. What is the probability that one boy and one girl are chosen in either order?

Here two pupils are chosen. It is like choosing one pupil without replacement and then choosing a second. Here is the tree diagram.

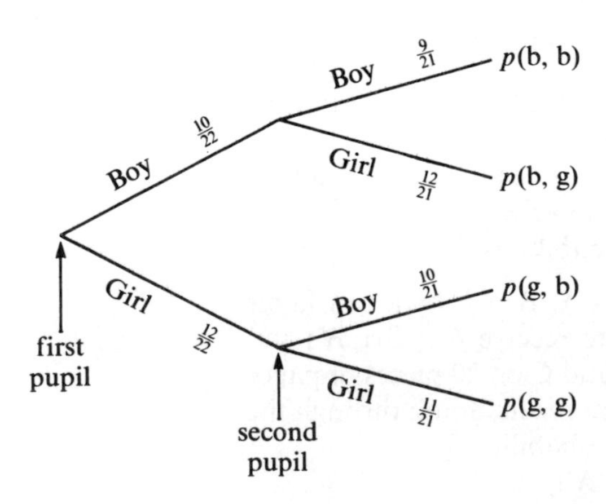

$$p(\text{b, g}) + p(\text{g, b}) = \left(\frac{10}{22} \times \frac{12}{21}\right) + \left(\frac{12}{22} \times \frac{10}{21}\right) = \frac{40}{77}$$

Exercise 7

1. There are 1000 components in a box of which 10 are known to be defective. Two components are selected at random. What is the probability that:
(a) both are defective, (b) neither are defective,
(c) just one is defective?
(Do *not* simplify your answers)

2. There are 10 boys and 15 girls in a class. Two children are chosen at random. What is the probability that:
(a) both are boys (b) both are girls,
(c) one is a boy and one is a girl?

3. Two similar spinners are spun together.
(a) What is the most likely total score on the two spinners?
(b) What is the probability of obtaining this score on three successive spins of the two spinners?

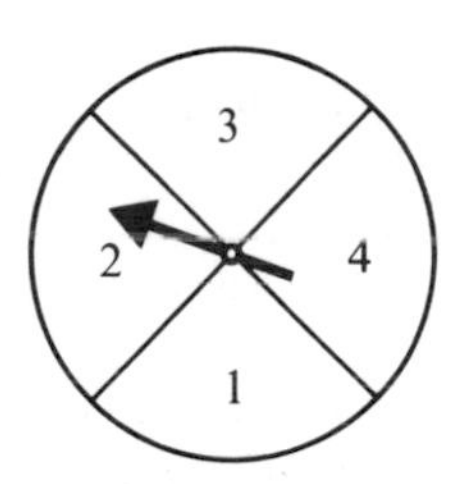

4. A bag contains 3 red, 4 white and 5 green balls. Three balls are selected without replacement.

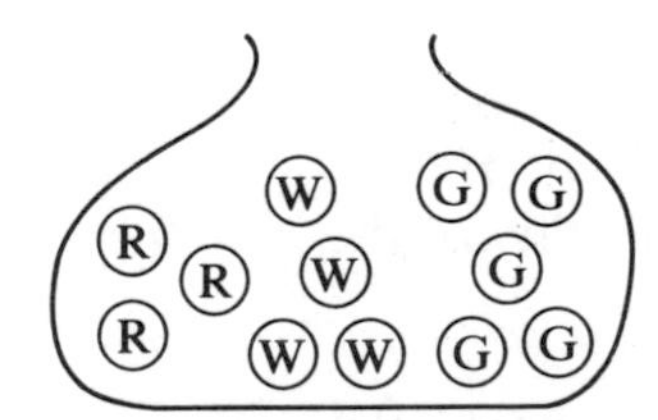

Find the probability that the three balls chosen are:
(a) all red, (b) all green, (c) one of each colour.
(d) If the selection of the three balls was carried out 1100 times, how often would you expect to choose three red balls?

5. The diagram represents 15 students in a class and shows the following.
G represents those who are girls
S represents those who are swimmers
F represents those who believe in Father Christmas.

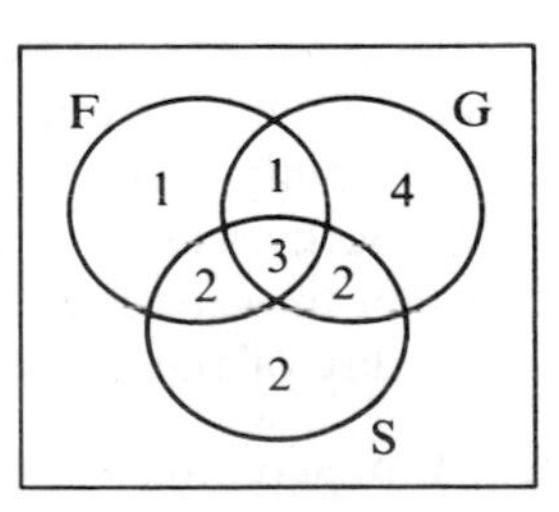

A student is chosen at random. Find the probability that the student:
(a) can swim, (b) is a girl swimmer,
(c) is a boy swimmer who believes in Father Christmas.

Two students are chosen at random. Find the probability that:
(d) both are boys, (e) neither can swim.

6. There are 500 ball bearings in a box of which 100 are known to be undersize. Three ball bearings are selected at random. What is the probability that:
(a) all three are undersize (b) none are undersize?
Give your answers as decimals correct to three significant figures.

7. There are 9 boys and 15 girls in a class. Three children are chosen at random. What is the probability that:
(a) all three are boys, (b) all three are girls,
(c) one is a boy and two are girls?
Give your answers as fractions.

8. A box contains x milk chocolates and y plain chocolates. Two chocolates are selected at random. Find, in terms of x and y, the probability of choosing:
(a) a milk chocolate on the first choice,
(b) two milk chocolates,
(c) one of each sort,
(d) two plain chocolates.

9. A pack of z cards contains x 'winning cards'. Two cards are selected at random. Find, in terms of x and z, the probability of choosing:
(a) a 'winning' card on the first choice,
(b) two 'winning cards' in the two selections,
(c) exactly one 'winning' card in the pair.

10. When a golfer plays any hole, he will take 3, 4, 5, 6, or 7 strokes with probabilities of $\frac{1}{10}$, $\frac{1}{5}$, $\frac{2}{5}$, $\frac{1}{5}$, and $\frac{1}{10}$ respectively. He never takes more than 7 strokes. Find the probability of the following events:
(a) scoring 4 on each of the first three holes,
(b) scoring 3, 4 and 5 (in that order) on the first three holes,
(c) scoring a total of 28 for the first four holes,
(d) scoring a total of 10 for the first three holes.

11. Assume that births are equally likely on each of the seven days of the week. Two people are selected at random. Find the probability that:
(a) both were born on a Sunday,
(b) both were born on the same day of the week.

12. (a) A playing card is drawn from a standard pack of 52 and then replaced. A second card is drawn. What is the probability that the second card is the same as the first?
(b) A card is drawn and *not* replaced. A second card is drawn. What is the probability that the second card is *not* the same as the first?

8.4 Conditional probability

The next two examples are more difficult.

Example 1

The performance of a climber is affected by the weather. When it rains, he falls on a mountain with a probability of $\frac{1}{10}$, whereas in dry weather he falls with a probability of only $\frac{1}{50}$. In the climbing seasons, the probability of rain on any day is $\frac{1}{4}$.

(a) Find the probability that he doesn't fall on a rainy day.
(b) What is the probability that he falls on the mountain on a day chosen at random?

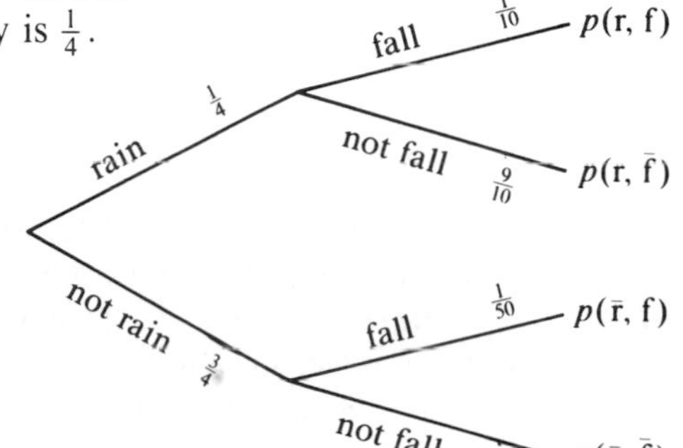

Draw a tree diagram.

(a) p (on a rainy day he doesn't fall) $= p(\text{r}, \bar{\text{f}})$

$$= \tfrac{1}{4} \times \tfrac{9}{10} = \tfrac{9}{40}$$

(b) p (he falls) $= p(\text{r, f}) + p(\bar{\text{r}}, \text{f})$

$$= (\tfrac{1}{4} \times \tfrac{1}{10}) + (\tfrac{3}{4} \times \tfrac{1}{50}) = \tfrac{1}{25}$$

Example 2

In an experiment, either Bag A or Bag B is randomly chosen and then a ball is chosen.
Given that a red ball is chosen, what is the probability that it came from Bag B?

Bag A: R R R W W　　Bag B: R W W W

From Bag A, $p(\text{red}) = \frac{3}{5}$　　From Bag B, $p(\text{red}) = \frac{1}{4}$

Suppose the experiment is repeated n times, where n is a large number. Then we would expect Bag A to be chosen $\dfrac{n}{2}$ times and Bag B also $\dfrac{n}{2}$ times.

In n trials, the total number of red balls expected $= \left(\dfrac{n}{2} \times \dfrac{3}{5}\right) + \left(\dfrac{n}{2} \times \dfrac{1}{4}\right)$

Of this total, $\left(\dfrac{n}{2} \times \dfrac{1}{4}\right)$ came from Bag B.

$$\text{So, } p(\text{ball came from Bag B}) = \frac{\left(\dfrac{n}{2} \times \dfrac{1}{4}\right)}{\left(\dfrac{n}{2} \times \dfrac{3}{5}\right) + \left(\dfrac{n}{2} \times \dfrac{1}{4}\right)} = \frac{5}{17}$$

Exercise 8

1. Bag A contains 3 red balls and 3 blue balls. Bag B contains 1 red ball and 3 blue balls.

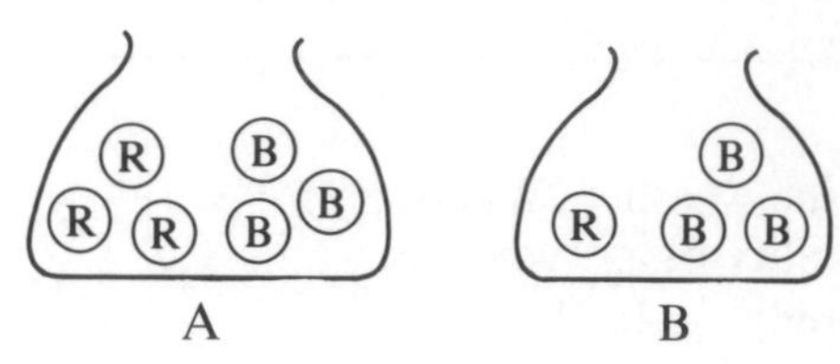

A ball is taken at random from Bag A and placed in Bag B. A ball is then chosen from Bag B. What is the probability that the ball taken from B is red?

2. On a Monday or a Thursday. Mr Gibson paints a 'masterpiece' with a probability of $\frac{1}{5}$. On any other day, the probability of producing a 'masterpiece' is $\frac{1}{100}$. In common with other great painters, Gibson never knows what day it is. Find the probability that on one day chosen at random, he will in fact paint a masterpiece.

3. If a hedgehog crosses a certain road before 7.00 a.m., the probability of being run over is $\frac{1}{10}$. After 7.00 a.m., the corresponding probability is $\frac{3}{4}$. The probability of the hedgehog waking up early enough to cross before 7.00 a.m. is $\frac{4}{5}$.

   What is the probability of the following events?
   (a) the hedgehog waking up too late to reach the road before 7.00 a.m.,
   (b) the hedgehog waking up early and crossing the road in safety,
   (c) the hedgehog waking up late and crossing the road in safety,
   (d) the hedgehog waking up early and being run over,
   (e) the hedgehog crossing the road in safety.

4. Two boxes are shown containing red and white balls.

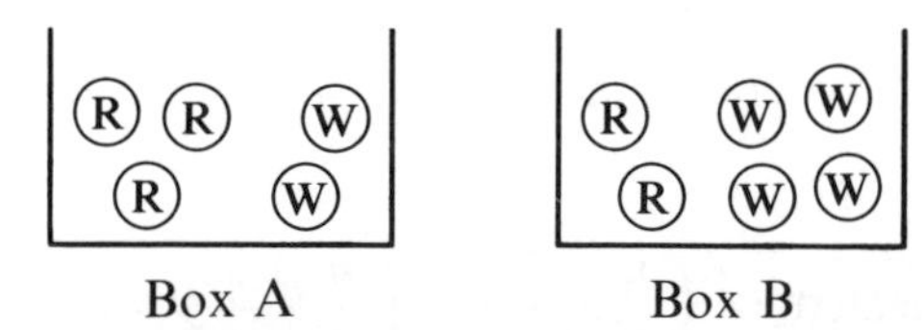

One ball is selected from Box A and placed in Box B. A ball is then selected from Box B and placed in Box A. What is the probability that Box A now contains 4 red balls?

5. Michelle has been revising for a multiple choice exam in history but she had only had time to learn 70% of the facts being tested. If there is a question on any of the facts she has revised she will get that question right. Otherwise she will simply guess one of the five possible answers.

(a) A question is chosen randomly from the paper. What is the probability that she will get it right?
(b) If the paper has 50 questions what mark do you expect her to get?

6.† Surgeons can operate to cure Pythagoratosis but the success rate at the first attempt is only 65%. If the first operation fails the operation can be repeated but this time the success rate is only 20%. After a second failure there is so little chance of success that surgeons will not operate again.

There is an outbreak of the disease at Gibson College and 38 students contract the disease.
How many of these students can we expect to be saved after both operations?

7.† In an experiment either Bag A or Bag B is randomly chosen and then a ball is chosen.

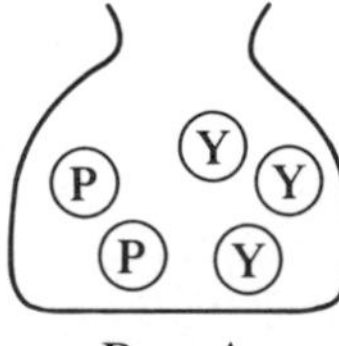

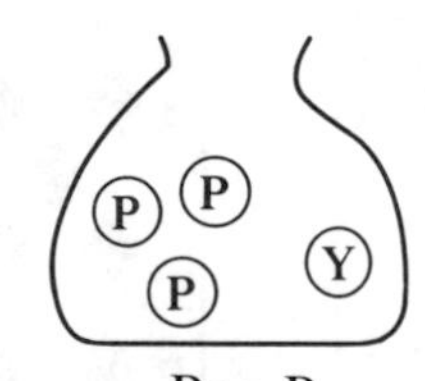

Bag A Bag B

Given that a pink ball is chosen, what is the probability that it came from Bag A?
[Hint: Suppose the experiment is repeated n times, where n is large. Work out how many pink balls you would expect from each bag.]

8.† In an experiment either Bag A, B or C is randomly chosen and then a ball is chosen.
Bag A has 3 red, 1 white
Bag B has 2 red, 2 white
Bag C has 1 red, 4 white
Given that a red ball is chosen, what is the probability that it came from Bag B?

Examination exercise 8

1. Three students A, B, C were each asked to toss a coin 30 times and record whether it came down heads (H) or tails (T) each time. They said their results were as follows:

Student A: T T T H T H T T T H T T T H T T H T T T T H
H T T T T H T T.
Student B: H T H T H T H T H T H T H T H T H T H T H
T H T H T H T H T.
Student C: H T H T H H H H T H T T H H H H T T H T H
T T T H H T H T T.

One of the students used an unbiased coin, one used a biased coin and the remaining student did not have a coin so just made up the results.

(a) (i) Which student used a biased coin?
(ii) Estimate the probability that, the next time this coin is tossed, it will come down heads.

(b) Which student just made up the results? Give a reason for your answer. [M]

2. Two dice, one red and one black, are thrown simultaneously. A score is found by adding together the numbers of dots shown on the two upper faces of the dice.
Find the probability, expressed as a fraction, of throwing
(a) a score of 2, (b) a score of 9,
(c) a double in which both dice show the same number. [L]

3.

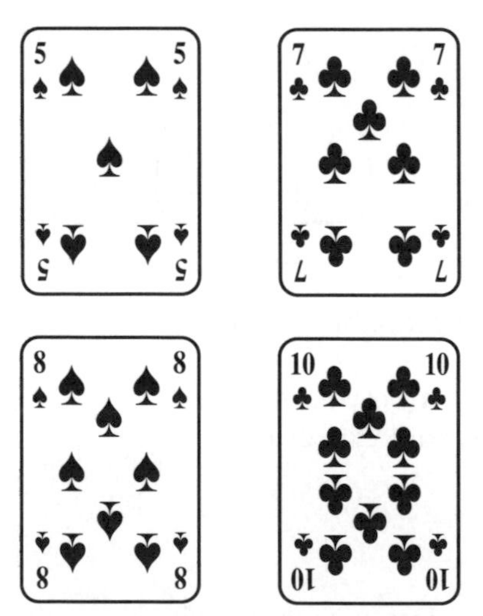

Four cards, numbered 5, 7, 8 and 10, are shuffled and laid face down. Tom offers odds of 4 to 1 against anyone choosing two of the four cards and getting a total of exactly 15.
By listing all the possibilities, or otherwise, determine whether these odds are fair, generous or unfair, to anyone playing the game. Give a reason for your answer. [N]

4. In a game to select a winner from three friends Arshad, Belinda and Connie, Arshad and Belinda both roll a normal die. If Arshad scores a number greater than 2 *and* Belinda throws an odd number, then Arshad is the winner. Otherwise Arshad is eliminated and Connie then rolls the die. If the die shows an odd number Connie is the winner, otherwise Belinda is the winner.

(a) Calculate the probability that
  (i) Arshad will be the winner,
  (ii) Connie will roll the die,
  (iii) Connie will be the winner.

(b) Is this a fair game? Give a reason for your answer. [L]

5. Some college students have to take 3 examination papers, one in English, one in Mathematics and one in Science.
In English the probability of a pass is $\frac{7}{10}$.
In Mathematics it is $\frac{5}{10}$ and in Science it is $\frac{8}{10}$.
Assume that these events are independent events.

(a) Calculate the probability that a student will
  (i) pass all three papers,
  (ii) fail all three papers,
  (iii) pass at least one of the papers.

A student needs to pass in at least two of the three papers to go on to the next year.

(b) Calculate the probability that a student will be able to go on to the next year. [L]

6. A game called "HOW'S THAT" is played between two players.
One player, A, rolls a dice which has one face labelled "HOW'S THAT" and the other five faces labelled 1, 2, 3, 4 and 6 respectively.
The numbers 1, 2, 3, 4 and 6 indicate the score obtained by Player A.
Player A continues to roll this dice, adding the scores obtained each time, until the face with "HOW'S THAT" appears uppermost on the dice.
Player B then rolls a second dice, which has four faces which show that Player A's turn is over, and two faces which show that Player A is to continue to roll his dice.

(a) What is the probability that, on the first roll of the numbered dice, Player A will score 4?

(b) What is the probability that Player A will have scored a total of 4 after two rolls of the numbered dice, assuming "HOW'S THAT" did not appear on either occasion?

(c) What is the probability that, after just one roll of the numbered dice by Player A, followed by one roll of the other dice by Player B, Player A's turn will be over? [N]

7. I played a game of cards with a friend. The pack of 52 cards was shuffled and then dealt so that we both received two cards. His cards were a ten and a nine. The first card that I received was an ace.

(a) Find the probability that my second card was a picture card (that is to say a King, Queen or Jack).

(b) In fact my second card was another ace. The rules of the game allowed me to treat the two aces separately, and so have a further two cards dealt to me, one to go alongside one ace and the other to go alongside the other ace.

Calculate the probability that these two cards were both picture cards. Give your answer as a fraction in its simplest form. [N]

8. A box of chocolates contains 9 hard centres (H) and 12 soft centres (S). One chocolate is taken at random and eaten; then a second chocolate is taken.

(a) Complete this tree diagram.

first chocolate second chocolate

$\frac{9}{21}$ H — $\frac{8}{20}$ H

H — S

S — H

S — S

(b) Giving you answers as fractions in their lowest terms, find the probability that

(i) both chocolates have soft centres,

(ii) one has a hard centre and one a soft centre. [M]

9. David and Tom each have a bag containing 5 marbles. David has 3 red and 2 green marbles, Tom has 2 red and 3 green. They play a game of "swops" where each takes a single marble at random from his bag and places it on the floor. If the colours are the same, Tom takes both marbles and puts them back into his bag. If the colours are different, David wins the marbles and puts them both into his bag.

(a) Calculate the probability

(i) that Tom wins the first swop,

(ii) that David wins the first swop.

(b) On the first swop, David takes out a red marble and Tom a green marble, so that David wins and puts the marbles into his bag. They agree to select another marble each. What is the probability that David wins again? [N]

10. Bag A contains three black and seven white balls, and Bag B contains six black and five white balls.
Each bag is chosen randomly and a ball withdrawn.

(a) If Bag A has been chosen, what is the probability of a black ball being withdrawn?

(b) The experiment is repeated n times, where n is large.

(i) How many times can you expect Bag A to be chosen?

(ii) How many times can you expect a black ball to be drawn from Bag A?

(iii) How many black balls can you expect to be drawn from both bags in n experiments?

(c) If a black ball is chosen, what is the probability that it came from Bag A? [N]

9 Handling data 2

9.1 Histograms

In a histogram, the frequency of the data is shown by the *area* of each bar. Histograms resemble bar charts but are not to be confused with them: in bar charts the frequency is shown by the height of each bar. Histograms often have bars of varying widths. Because the area of the bar represents frequency, the height must be adjusted to correspond with the width of the bar. The vertical axis is not labelled frequency but frequency density.

$$\text{frequency density} = \frac{\text{frequency}}{\text{class width}}$$

Histograms can be used to represent both discrete data and continuous data, but their main purpose is for use with continuous data.

Example 1

Draw a histogram from the table shown for the distribution of ages of passengers travelling on a flight to New York.

Ages	Frequency
$0 \leqslant x < 20$	28
$20 \leqslant x < 40$	36
$40 \leqslant x < 50$	20
$50 \leqslant x < 70$	30
$70 \leqslant x < 100$	18

Note that the data has been collected into class intervals of different widths.

To draw the histogram, the heights of the bars must be adjusted by calculating frequency density.

Ages	Frequency	Frequency density (f.d.)
$0 \leqslant x < 20$	28	$28 \div 20 = 1{\cdot}4$
$20 \leqslant x < 40$	36	$36 \div 20 = 1{\cdot}8$
$40 \leqslant x < 50$	20	$20 \div 10 = 2$
$50 \leqslant x < 70$	30	$30 \div 20 = 1{\cdot}5$
$70 \leqslant x < 100$	18	$18 \div 30 = 0{\cdot}6$

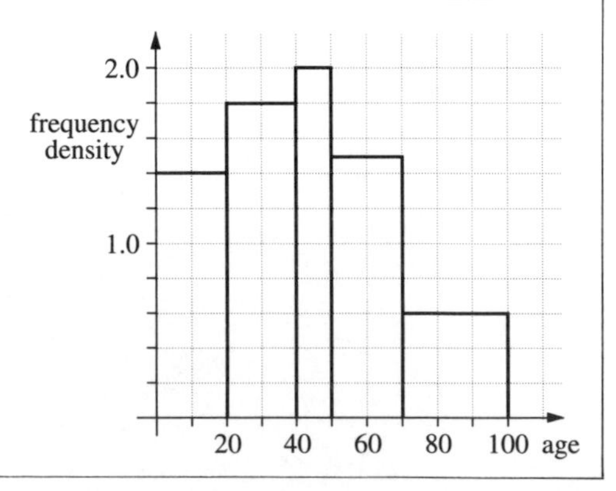

Exercise 1

1. The lengths of 20 copper nails were measured. The results are shown in the frequency table.
Calculate the frequency densities and draw the histogram as started below.

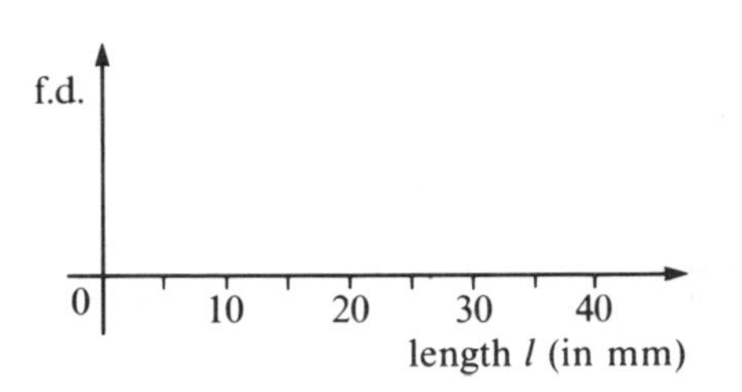

Length l (in mm)	Frequency	Frequency density (f.d.)
$0 \leqslant L < 20$	5	$5 \div 20 = 0{\cdot}25$
$20 \leqslant L < 25$	5	
$25 \leqslant L < 30$	7	
$30 \leqslant L < 40$	3	

2. The volumes of 55 containers were measured and the results presented in a frequency table as shown.
Calculate the frequency densities and draw the histogram.

Volume (mm^3)	Frequency
$0 \leqslant V < 5$	5
$5 \leqslant V < 10$	3
$10 \leqslant V < 20$	12
$20 \leqslant V < 30$	17
$30 \leqslant V < 40$	13
$40 \leqslant V < 60$	5

3. Thirty students in a class are weighed on the first day of term. Draw a histogram to represent this data.

Note that the weights do not start at zero. This can be shown on the graph as follows.

Weight (kg)	Frequency
30–40	5
40–45	7
45–50	10
50–55	5
55–70	3

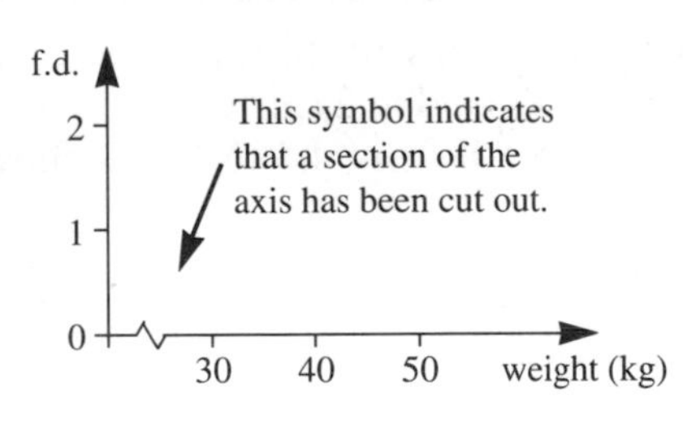

4. The ages of 120 people passing through a turnstyle were recorded and are shown in the frequency table.
The notation –10 means '$0 < \text{age} \leqslant 10$' and similarly –15 means '$10 < \text{age} \leqslant 15$'.
The class boundaries are 0, 10, 15, 20, 30, 40.
Draw the histogram for the data.

Age (yrs)	Frequency
–10	18
–15	46
–20	35
–30	13
–40	8

5. Another common notation is used here for the masses of plums picked in an orchard, shown in the table below.

Mass (g)	20–	30–	40–	60–	80–
Frequency	11	18	7	5	0

The initial 20– means $20\text{ g} \leqslant \text{mass} < 30\text{ g}$.
Draw a histogram with class boundaries at 20, 30, 40, 60, 80.

6. The heights of 50 Olympic athletes were measured as shown in the table below.

Height (cm)	170–174	175–179	180–184	185–194
Frequency	8	17	14	11

These values were rounded off to the nearest cm. For example, an athlete whose height h is 181 cm could be entered anywhere in the class $180{\cdot}5\text{ cm} \leqslant h < 181{\cdot}5\text{ cm}$. So the table is as follows.

Height	169·5–174·5	174·5–179·5	179·5–184·5	184·5–194·5
Frequency	8	17	14	11

Draw a histogram with class boundaries at 169·5, 174·5, 179·5, ...

7. The number of people travelling in 33 vehicles one day was as shown in the table below.

Number of people	1	2	3	4	5–6	7–10
Frequency	8	11	6	4	2	2

In this case, the data is discrete. To represent this information on a histogram, draw the column for the value 2, for example, from 1·5 to 2·5, and that for the values 5 – 6 from 4·5 to 6·5 as shown below.

Number of people	Frequency	Interval on histogram	Width of interval	Frequency density
1	8	0·5–1·5	1	8
2	11	1·5–2·5		
3	6			
4	4			
5–6	2	4·5–6·5	2	1
7–10	2			

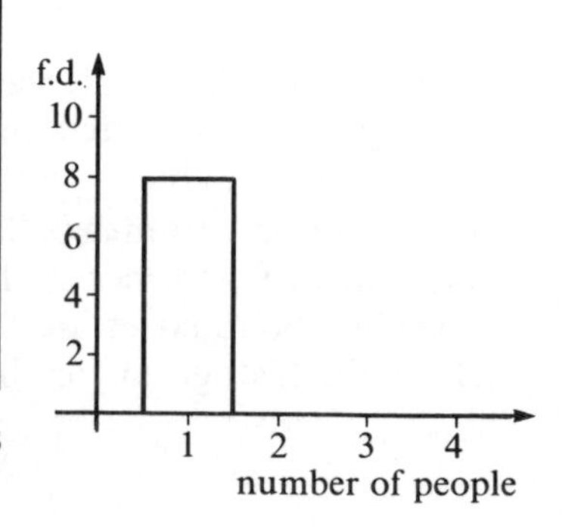

Copy and complete the above table and the histogram which has been started on the right.

Finding frequencies from histograms

If the data is already presented in a histogram, it can be useful to draw up the relevant frequency table. The actual size of the sample tested can then be found as well as the frequencies in each class. However, if a sample size is too small, the results of the analysis may not truly represent the distribution which is being examined.

Example 2

Given the histogram of children's heights shown, draw up the relevant frequency table and find the size of the sample which was tested.

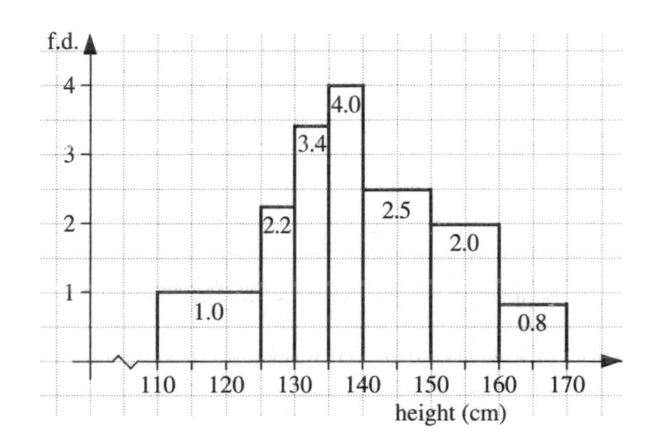

Use frequency density $= \dfrac{\text{frequency}}{\text{class width}}$

So frequency $=$ (frequency density) $\times$ (class width)

Remember:
the area of each bar represents frequency.

Height (cm)	f.d.	Frequency
$110 \leqslant x < 125$	1·0	$1{\cdot}0 \times 15 = 15$
$125 \leqslant x < 130$	2·2	$2{\cdot}2 \times 5 = 11$
$130 \leqslant x < 135$	3·4	$3{\cdot}4 \times 5 = 17$
$135 \leqslant x < 140$	4·0	$4{\cdot}0 \times 5 = 20$
$140 \leqslant x < 150$	2·5	$2{\cdot}5 \times 10 = 25$
$150 \leqslant x < 160$	2·0	$2{\cdot}0 \times 10 = 20$
$160 \leqslant x < 170$	0·8	$0{\cdot}8 \times 10 = 8$
		Total frequency $= 116$

So there were 116 children in the sample measured.

Exercise 2

1. For the histogram shown:
(a) How many of the lengths are in these intervals?
(i) 40–60 cm (ii) 30 40 cm
(b) What is the total frequency?

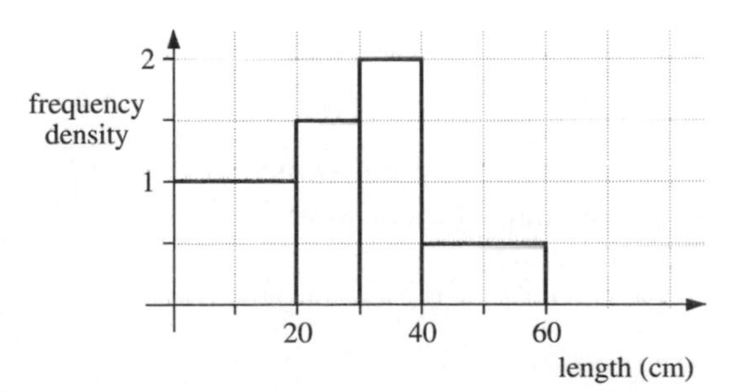

2. For the histogram shown, find the total frequency.

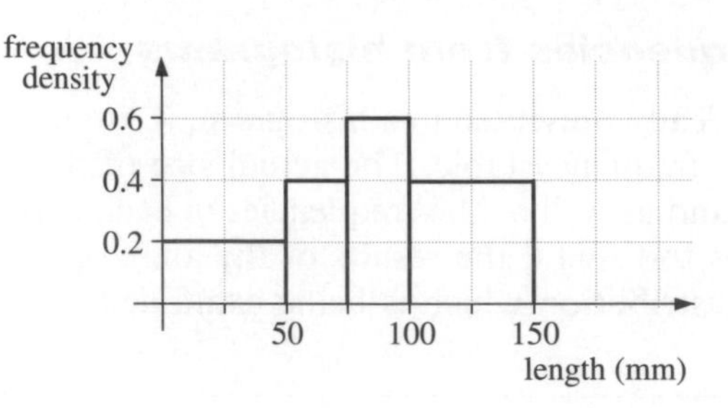

3. The histogram shows the ages of the trees in a small wood. How many trees were in the wood altogether?

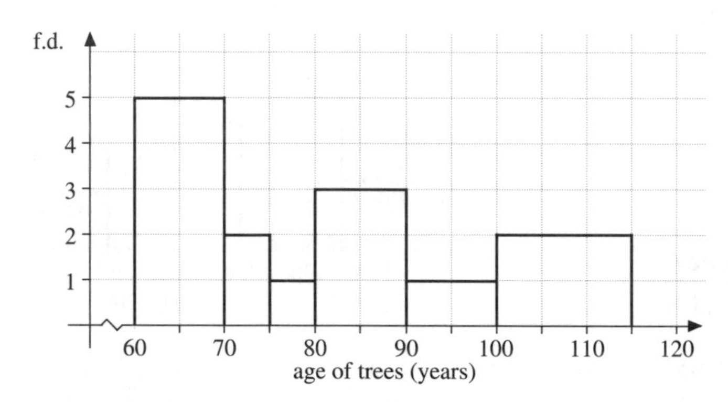

4. One day a farmer weighs all the hens' eggs which he collects. The results are shown in this histogram.
There were 8 eggs in the class 50–60 g.
Work out the numbers on the frequency density axis and hence find the total number of eggs collected.

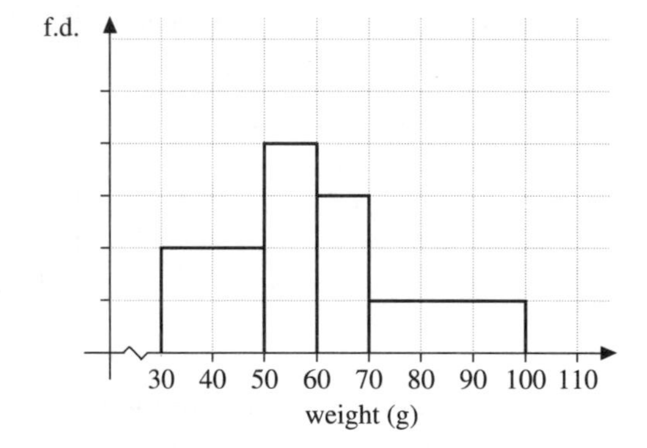

5. This histogram shows the number of letters delivered at some houses in a street. Notice that the data is discrete and that, for example, the class boundaries for 1–2 letters are 0·5–2·5.
Given that 3 letters were delivered to 8 houses, how many letters were delivered altogether?

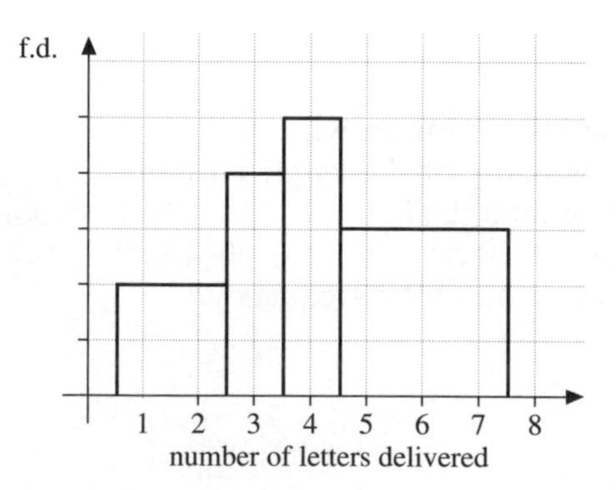

6. Here are the percentage marks obtained by 44 pupils in an exam.

41	61	11	46	51	21	63	48	25	87	47
53	52	40	55	15	70	54	74	43	85	38
47	23	52	18	75	47	73	45	80	59	88
52	46	76	49	31	57	27	58	5	70	42

Make a tally chart and a frequency table using groups of marks 0–19, 20–39, 40–49, 50–59, 60–89.
Work out the frequency densities and hence draw a histogram.

7. The manager of a small company found that the number of sick days taken by his 50 employees in 1993 was as follows.

10	31	17	22	7	1	9	36	5	13	26	2	8
6	0	24	30	11	20	28	8	18	38	3	16	12
23	15	7	21	12	8	37	22	11	19	40	18	9
6	13	30	27	3	28	4	29	41	40	14		

(a) Draw up a frequency table and construct a histogram for this set of data after grouping into classes of equal width.
(b) Repeat using more suitable class widths.
(c) In 1994, the manager introduced piped music, carpeting and better illumination in the workplace. The number of sick days taken by his employees in 1994 was as follows.

4	6	11	4	12	5	18	3	10	4	9	3	13
8	15	0	6	7	1	7	28	5	23	1	39	14
33	2	7	2	16	30	13	1	17	7	41	3	22
1	18	24	38	8	3	9	26	41	0	43		

Draw up a frequency table using the same classes as in part (b) and a histogram. Comment, in one sentence, on the two comparable histograms.

8. The histogram shows the distribution of marks obtained by students in a test.
To find the number of people with a mark from, say, 50 to 100, use $0{\cdot}3 \times 50$ to get 15. This is really the area of the rectangle with base from 50 to 100.
The number of people with a mark from 80 to 100 is given by the area of the shaded rectangle which is $0{\cdot}3 \times 20 = 6$. So 6 people have marks from 80 to 100.
Similarly, use the areas of rectangles to find the number of people with marks in the following ranges.

(a) 100–130 (b) 90–150
(c) 90–160 (d) 160–230

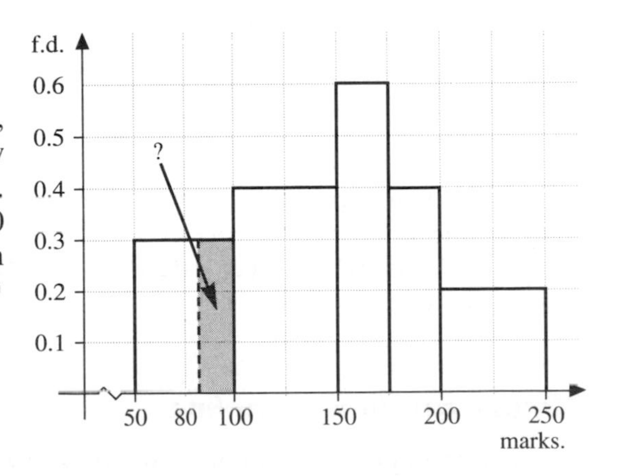

9.2 Sampling

Statisticians frequently carry out surveys to investigate a characteristic of a population. A population is the set of all possible items to be observed, not necessarily people.
Surveys are done for many reasons.

- Newspapers seek to know voting intentions before an election.
- Advertisers try to find out which features of a product appeal most to the public.
- Research workers testing new drugs need to know their effectiveness.
- Government agencies conduct tests for the public's benefit such as the percentage of first class letters arriving the next day or the percentage of '999' calls answered within 5 minutes.

Census

A census is a survey in which data is collected from every member of the population of a country. A census is done in Great Britain every ten years; the last one was in 1991. Data about age, educational qualifications, race, distance travelled to work, etc., is collected from every person in the country on the given 'Census Day'. Because no person must be omitted, it is a huge and very expensive task. Census forms take several years to prepare and many years for thorough analysis of the data collected.

Samples

If a full census is not possible or practical, then it is necessary to take a sample from the population. Two reasons for taking samples are:

- it will be much cheaper and quicker than a full census,
- the object being tested may be destroyed in the test, like testing the average lifetime of a new light bulb or the new design of a car tyre.

The choice of a sample prompts two questions.

1. Does the sample truly represent the whole population?
2. How large should the sample be?

After taking a sample, it is assumed that the result for the sample reflects the whole population. For example, if 20% of a sample of 1000 people say they will vote in a particular way, then it is assumed that 20% of the whole electorate would do so likewise, subject of course to some error.

Simple random sampling

This is a method of sampling in which every item of the population has an equal chance of selection. There are two methods of selection:

(a) Selecting *with* replacement. Here, each item selected is then replaced giving a possibility that it may be reselected.
(b) Selecting *without* replacement. Each item selected is not replaced and is therefore not available for reselection.

For large populations, there is no significant difference between these two methods, but the method chosen should always be stated clearly.

Selection techniques

Out of a school of 758 students, ten are to be selected to take part in an educational experiment in which homework is banned.
How would you select these students?
First give each student a three-digit number from 001 to 758.

Method 1

Each person's number is written on a piece of card, the cards placed in a hat and mixed up. Ten cards are then drawn.

Method 2

Use a random number table. The table is shown below.

Random number table

11 74	26 93	81 44	33 93	08 72	32 79	73 31	18 22	64 70	68 50
43 36	12 88	59 11	01 64	56 23	93 00	90 04	99 43	64 07	40 36
93 80	62 04	78 38	26 80	44 91	55 75	11 89	32 58	47 55	25 71
49 54	01 31	81 08	42 98	41 87	69 53	82 96	61 77	73 80	95 27
36 76	87 26	33 37	94 82	15 69	41 95	96 86	70 45	27 48	38 80
07 09	25 23	92 24	62 71	26 07	06 55	84 53	44 67	33 84	53 20
43 31	00 10	81 44	86 38	03 07	52 55	51 61	48 89	74 29	46 47
61 57	00 63	60 06	17 36	37 75	63 14	89 51	23 35	01 74	69 93
31 35	28 37	99 10	77 91	89 41	31 57	97 64	48 62	58 48	69 19
57 04	88 65	26 27	79 59	36 82	90 52	95 65	46 35	06 53	22 54
09 24	34 42	00 68	72 10	71 37	30 72	97 57	56 09	29 82	76 50
97 95	53 50	18 40	89 48	83 29	52 23	08 25	21 22	53 26	15 87
93 73	25 95	70 43	78 19	88 85	56 67	16 68	26 95	99 64	45 69
72 62	11 12	25 00	92 26	82 64	35 66	65 94	34 71	68 75	18 67
61 02	07 44	18 45	37 12	07 94	95 91	73 78	66 99	53 61	93 78
97 83	98 54	74 33	05 59	17 18	45 47	35 41	44 22	03 42	30 00
89 16	09 71	92 22	23 29	06 37	35 05	54 54	89 88	43 81	63 61
25 96	68 82	20 62	87 17	92 65	02 82	35 28	62 84	91 95	48 83
81 44	33 17	19 05	04 95	48 06	74 69	00 75	67 65	01 71	65 45
11 32	25 49	31 42	36 23	43 86	08 62	49 76	67 42	24 52	32 45

This table is published by kind permission of the Department of Statistics University College, London

Start anywhere in the table and go either up, down, left, or right to read numbers in groups of three digits. Starting on the bottom right-hand corner going up and then down, would give the numbers

553 108 ~~797~~ 049 370 071 605 372 ~~824~~ ~~915~~ 586 670 684

Why were 797, 824, 915 rejected?

So students with the above 10 numbers will take part in the experiment.

Method 3 Random number generator

Use a calculator or a computer.

On the CASIO calculator, random numbers are generated by pressing RAN # after using SHIFT to get this function. The numbers produced are three-digit numbers between .001 and .999. Ignore the decimal point.

Although these are not true random numbers due to their method of generation, they are adequate for school use.

Random two-digit numbers are obtained simply by omitting the last digit.

Example 1

Find 6 two-digit numbers between 01 and 60.

Ignoring the decimal points, a calculator gave the following.

~~819~~ 453 480 ~~718~~ ~~891~~ 326 050 217 ~~990~~ 437

So use: 45 48 32 05 21 43

Exercise 3

1. One hundred students took a test and their results are given in the table. Each student had a two-digit identification number between 00 and 99.

	Second number									
First number	**0**	**1**	**2**	**3**	**4**	**5**	**6**	**7**	**8**	**9**
0	53	62	48	71	41	78	64	49	82	32
1	28	66	41	83	49	62	54	67	68	33
2	75	81	47	35	26	70	85	93	36	35
3	66	42	55	58	87	38	29	15	41	68
4	84	67	39	62	47	58	62	39	72	86
5	43	32	49	61	53	80	95	26	44	33
6	77	62	41	19	26	37	56	57	40	30
7	38	75	62	41	28	33	62	39	43	26
8	64	52	53	68	36	42	39	38	65	70
9	52	45	72	67	90	47	34	55	61	47

For example, Student 31 scored 42% (Row 3, Column 1) and Student 65 scored 37%.

(a) Use your judgement to select a representative sample of ten marks from the 100 given. Find the mean mark of this sample and state its range.

(b) Select several more similar samples of ten marks. Write down the range for each sample mean and find the mean of all the sample means.

2. (a) Use the random number table (page 345) to select a random sample of ten marks from the students' marks of Question 1. Find the sample mean and the range. For example, starting at 07 at the left-hand end of the sixth row of the random number table gives these random numbers and their corresponding marks:

07	09	25	23	92	24	...
49%	32%	70%	35%	72%	26%	...

(b) Collect other sample means and their range from all the students in your class and find the mean of all the means. Compare the results from (a) and (b).

3. This example combines sampling and probability. Carefully draw the diagram below on graph paper with a scale of 1 cm to 1 unit. Points can be selected at random as follows.
From the random number table read off two pairs, like 75 48.
For the scale used here, this gives the point (7·5, 4·8).
Similarly, 83 72 gives the point (8·3, 7·2).
(a) Select 50 points in this way. Record what fraction of all the points lies inside the circle.

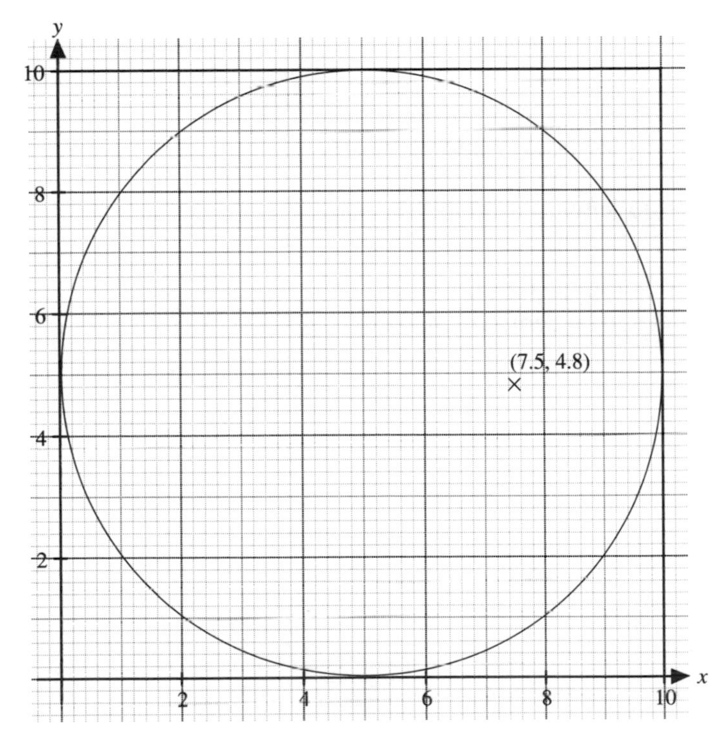

(b) The expected probability of selecting a point inside the circle can be calculated using areas.

$$\text{area of circle} = \pi . 5^2 = 25\pi$$
$$\text{area of square} = 10 \times 10 = 100\,\text{cm}^2$$

Probability of selecting a point inside the circle $= \frac{25\pi}{100} = \frac{\pi}{4}$

(c) Compare the value of the fraction obtained in (a) with the value of the fraction $\frac{\pi}{4}$ by writing the two results in decimals.

Stratified random sampling

Some populations separate naturally into a number of sub-groups or strata. Providing that the strata are quite distinct and that every member of the population belongs to one and only one stratum, then the population can be sampled using the method of stratified random sampling.
Opinion pollsters like Gallup or N.O.P. use this method when conducting polls on voting intentions. Their claim is that results are accurate to about 3 per cent.

Method

- Separate the population into suitable strata. Find what proportion of the population is in each stratum.
- Select a sample from each stratum proportional to the stratum size, by simple random sampling.

Example 2

In Year 10 of a school, the 180 students are split into three groups for games; 90 play football, 55 play hockey and 35 play badminton.
Use stratified random sample of 30 students to estimate the mean weight of the 180 students.

The sample size from each of the three groups (strata) must be proportional to the stratum size. So the 30 students are made up by selecting:

from the football group, $\frac{90}{180} \times 30 = 15$,

from the hockey group, $\frac{55}{180} \times 30 = 9{\cdot}16$, (say 9)

from the badminton group, $\frac{35}{180} \times 30 = 5{\cdot}83$, (say 6)

The three sample means were found to be:

football 53·2 kg hockey 49·4 kg badminton 46·4 kg

So the mean for the population of 180 students $= \frac{53{\cdot}2 + 49{\cdot}4 + 46{\cdot}4}{3}$

$= 49{\cdot}7$ kg (3 S.F.)

Bias

Question 1(a) on page 346 required personal judgement to select a representative sample of 10 test marks taken from 100. This provided a non-random judgmental sample. The highest mark is seldom selected in this way, so this method of sampling usually gives a mean mark which is less than the true mean.
It is *not* possible to choose a random sample using personal judgement.
Random samples may give results containing errors but these can be predicted and allowed for. The type of error introduced with judgmental sampling is unpredictable and thus corrections for it cannot be made. This type of unpredictable error is called *bias*.
Bias can come from a variety of sources including non-random sampling. For example, for data collected by questionnaire, non-response from subjects can introduce bias into the results. Qualities which these subjects have in common will not be represented in the collected data.
For data collected by a street survey, the time of day and location may well mean that there is a sector or sectors of the population who are not questioned.
Perhaps the most famous mistake in opinion sampling occurred in America in 1936. A magazine carried out an enormous poll in which over two million people were asked to state their preference for President, either the Democrat (F.D. Roosevelt) or the Republican (A.E. Landon). The sample was taken by selecting the two million people by simple random sampling from the telephone directories. The poll showed a large majority would vote for the Republican, but in the election the Democrat Roosevelt won.
A common cause of bias occurs when the questions asked in the survey are not clear or are leading questions. This was considered earlier (page 294) in the section on questionnaires.

Exercise 4

In these questions, decide whether the method of sampling is satisfactory or not. If it is not satisfactory, suggest a better way of obtaining a sample.

1. A teacher, with responsibility for school meals, wants to hear pupils' opinions on the meals currently provided. She waits next to the dinner queue and questions the first 50 pupils as they pass.

2. To find out how satisfied customers are with the service they receive from the telephone company a person telephones 200 people chosen at random from the telephone directory.

3. John, aged 10, wants to find out the average pocket money received by children of his own age. He asks 8 of his friends how much they get.

4. An opinion pollster wants to canvas opinion about our European neighbours. He questions drivers as they are waiting to board their ferry at Dover.

5. Jim has a theory that just as many men as women do the shopping in supermarkets. To test his theory he goes to Tesco one Saturday between 09:00 am and 10:00 am and counts the numbers of men and women.

6. A journalist wants to know the views of local people about a new one-way system in the town centre. She takes the electoral roll for the town and selects a random sample of 200 people.

7. A pollster working for the BBC wants to know how many people are watching a new series which is being shown. She questions 200 people as they are leaving a supermarket between 10:00 and 12:00 one Thursday.

9.3 Dispersion

Earlier we considered three averages: mean, median and mode (page 288) These are called the *measures of location* or the measures of central tendency. They locate the centre of a distribution of data.

Standard deviation

So far, we have met two measures of the way in which the data is spread. These are the range and, from a cumulative frequency curve, the interquartile range in which the middle 50% of the data lies. Standard deviation is the third measure of spread (or dispersion) about the mean.
Standard deviation gives a more detailed picture of the way in which the data is dispersed about the mean as the centre of the distribution. Its main use is to compare two sets of data.

$$\text{Standard deviation (s.d.)} = \sqrt{\frac{\Sigma\,(x - \bar{x})^2}{n}}$$

where Σ means 'the sum of'
x represents an item of data
and n is the number of items of data.
The symbol $\bar{x}$ is the mean (p. 289) where $\bar{x} = \dfrac{\Sigma x}{n}$

Sometimes the Greek symbol σ (sigma) is used for standard deviation instead of s.d.

Example 1

Find the mean and standard deviation of the set of numbers 3, 4, 7, 9, 12.

$$\text{Mean } \bar{x} = \frac{\Sigma x}{n} = \frac{3+4+7+9+12}{5} = 7$$

Make a table

x	$(x-\bar{x})$	$(x-\bar{x})^2$
3	−4	16
4	−3	9
7	0	0
9	2	4
12	5	25
		$\Sigma(x-\bar{x})^2 = 54$

$$\text{s.d.} = \sqrt{\frac{\Sigma(x-\bar{x})^2}{n}} = \sqrt{\frac{54}{5}} = 3{\cdot}286\,335\,345$$

So $\bar{x} = 7$, s.d. $= 3{\cdot}29$ (to 3 S.F.)

Example 2

Find the mean and standard deviation of the frequency distribution.

x	0	2	3	6	8	10
f	2	2	1	1	3	1

$$\bar{x} = \frac{\Sigma fx}{\Sigma f}$$

$$\bar{x} = \frac{2(0)+2(2)+1(3)+1(6)+3(8)+1(10)}{10} = 4{\cdot}7$$

x	f	$(x-\bar{x})$	$(x-\bar{x})^2$	$f(x-\bar{x})^2$
0	2	−4·7	22·09	44·18
2	2	−2·7	7·29	14·58
3	1	−1·7	2·89	2·89
6	1	1·3	1·69	1·69
8	3	3·3	10·89	32·67
10	1	5·3	28·09	28·09
	$n = \Sigma f = 10$			$\Sigma f(x-\bar{x})^2 = 124{\cdot}1$

For a frequency distribution, use the formula

$$\text{s.d.} = \sqrt{\frac{\Sigma f(x-\bar{x})^2}{n}} = \sqrt{\frac{124{\cdot}1}{10}} = 3{\cdot}522\,782\,991$$

So, $\bar{x} = 4{\cdot}7$
s.d. $= 3{\cdot}52$ (to 3 S.F.)

Example 3

Compare the results of ten students in two subjects as follows.

Mathematics	52	18	31	68	78	16	94	40	75	64
History	42	39	60	54	61	58	46	49	60	67

The mean mark for each subject is the same (53·6).
Yet the marks in Mathematics are more widely spread with a range 94–16 compared to 63–39 for History.
It can be shown that the standard deviation for the Mathematics marks was 25·2 and that for History was 8·7.

Note:
When finding the standard deviation of grouped data, the mid-point of each interval is used to represent that interval. The working then follows that of Example 2.
For example, the grouped data on the left would be replaced by the data on the right.

Interval	f
1–5	2
6–10	1
11–15	2

Mid-point	f
3	2
8	1
13	2

Exercise 5

1. Find the mean and standard deviation of the following sets of numbers.
(a) 1, 4, 8, 9, 10
(b) 3·2, 4·7, 5·1, 5·2, 6·3
(c) 103, 109, 110, 112, 125, 131
(d) −5, −2, 0, 1, 2, 3

2. A production engineer can use either of two methods to make car batteries. He takes a random sample of 8 batteries made by each method and measures the lifetime of each battery. The lifetimes in months are given below.

Made by Method A	21	24	22	23	25	23	26	20
Made by Method B	23	27	20	22	26	20	27	19

Find the mean and standard deviation for each sample. Which method would you recommend the engineer to use in future? Give your reason.

3. A professional golfer has two sets of clubs and is not sure which set gives better results. The last six scores playing with the 'Mizuno' set were 68, 71, 72, 68, 67, 74. The last six scores playing with the 'Lynx' set were 70, 65, 74, 72, 75, 64. Find the mean and standard deviation of the scores using each set of clubs.
Which set of clubs gives the more consistent results?

4. Find the mean and standard deviation of the frequency distributions.

(a)

x	4	8	12	16	20
f	4	2	3	1	3

(b)

x	3	4	5	7	9
f	1	5	4	2	3

5. Find the estimated mean and standard deviation of the frequency distributions of the following grouped data. Note that it does not matter if the intervals are of varying widths.

(a)

Interval	10–20	20–30	30–40	40–50
f	2	3	3	5

(b)

Interval	0–6	7–13	14–20	21–27
f	2	4	3	5

(c)

Interval	0–5	6–11	12–15	16–20
f	2	4	1	3

6. The weekly earnings of 30 people are shown in the histogram.

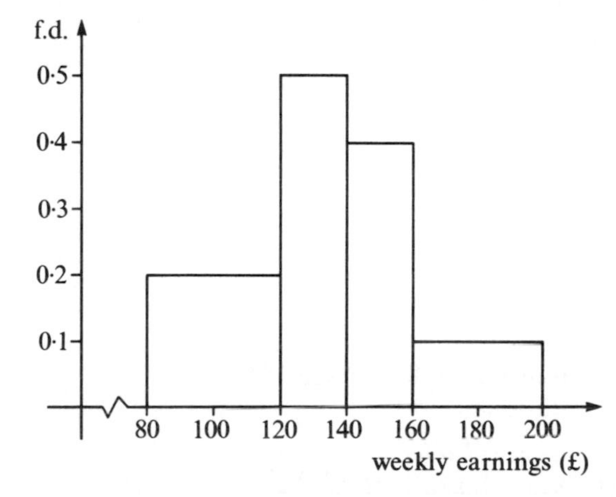

(a) Find the number of people in each interval and make a frequency distribution.
(b) Calculate the mean and standard deviation of the weekly earnings for the whole group.

Mean and standard deviation on a calculator

These instructions are for a CASIO scientific calculator with the statistical functions. Most CASIO calculators work in the same way. For another make of calculator, read the instruction handbook. The process will be similar.

[A] Set the calculator into SD mode by pressing [MODE] [3]
This gives the functions marked in blue.

[B] Clear the memories by pressing [SHIFT] [KAC]

This is essential as most calculators retain data in this mode even after switch off.

[C] Enter the data (use data from Example 1, page 351).

[3] [DATA]

[4] [DATA]

[7] [DATA]

[9] [DATA]

[12] [DATA]

[D] The answers have now been worked out!

To get:			
$\bar{x}$	press [SHIFT] [1]	$\bar{x} = 7$
s.d. ($x\sigma_n$)	press [SHIFT] [2]	s.d. $= 3{\cdot}29$
Σx	press [KOUT] [2]	$\Sigma x = 35$
Σn	press [KOUT] [3]	$n = 5$

Remember

For 'light brown' information, use [SHIFT].

For 'dark brown' information, use [KOUT].

To enter data from the frequency distribution in Example 2, page 351 (remember to clear memory first):

[0] [×] [2] [DATA]   [6] [×] [1] [DATA]

[2] [×] [2] [DATA]   [8] [×] [3] [DATA]

[3] [×] [1] [DATA]   [10] [×] [1] [DATA]

Note two things:

1. Enter $\boxed{x}$ $\boxed{\times}$ $\boxed{f}$. Do not reverse the order.
2. Do not press $\boxed{=}$.

Try some of the previous questions using the calculator in this way. Let the machine do the 'number crunching'!

The Normal distribution

By itself, a score on a test can have little value. Seen in relation to the *mean* for the class, the score has more significance.
Knowing the *spread* of the class scores gives a clearer idea of how well or how badly a single score relates to those of the class.
Like many sets of data, test results often have an approximately bell-shaped histogram.
A student scored 72% in both a History test and in a French test; the mean mark for each paper was 65%.
However, if the s.d. of the French test was 3% and of the History test was 15%, the student's two scores look very different.

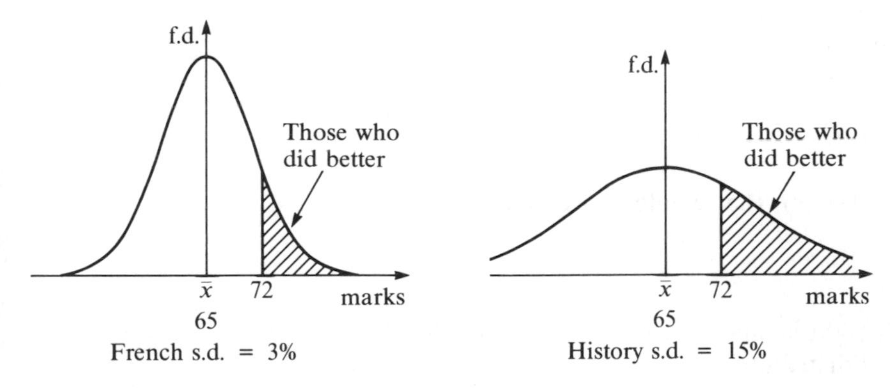

In French, very few people did better than 72% — well done!
In History, 72% is better than the mean score but the spread of marks is much greater and a lot of people did better.

If histograms with narrow bars are drawn to show the collected heights of 400 thirteen-year-old girls or the weights of 300 new-born labrador puppies for example, each curve would still resemble a bell-shaped curve. This is the usual shape of the curve if the data varies only as a result of random or accidental effects. Such data is said to form a Normal distribution; the graph forms a Normal or bell-shaped curve.

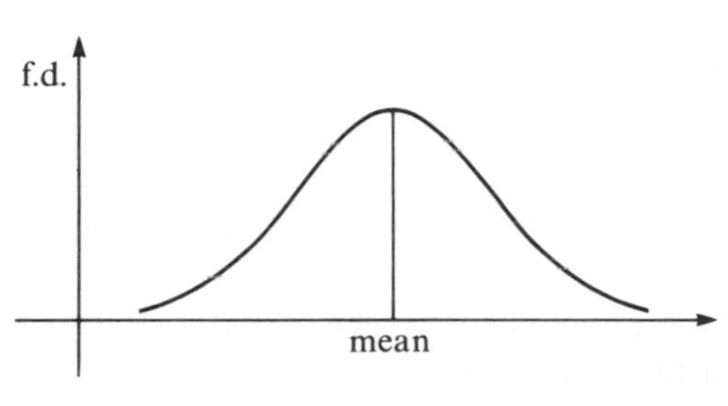

In a Normal distribution, much of the data is gathered about the mean and the rest tails off symmetrically either side.

In any Normal distribution:
approximately 68% of the data lies within one standard deviation of the mean, and 95% of the data lies within two standard deviations of the mean.

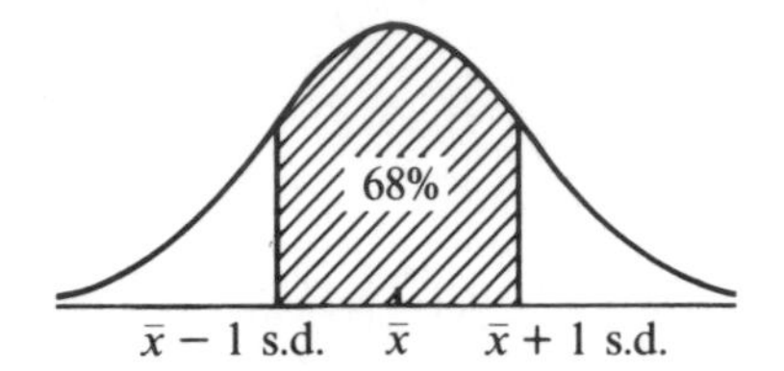

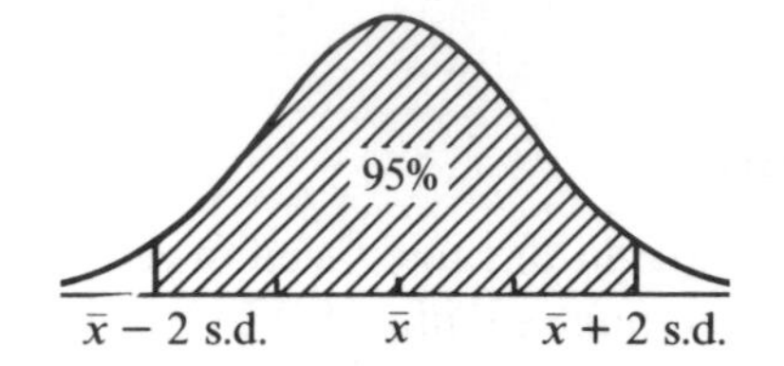

So, if the mean and the standard deviation of a set of data which fits a Normal distribution are known, then it is possible to predict with some accuracy how the members of that distribution will lie in relation to the mean.

Example 4

The mean weight of the cakes produced in a bakery is 600 g. The weights have a standard deviation of 20 g.
Find:
(a) the percentage of the cakes weighing
 (i) between 580 g and 620 g (ii) between 560 g and 640 g
 (iii) more than 640 g
(b) how many of 100 cakes bought that would be expected to weigh less than 560 g.

(a) (i) The variation in weight is random and accidental, so the data fits a Normal distribution.
$580 \text{ g} = 600 - 20 = \text{mean} - 1 \text{ s.d.}$
$620 \text{ g} = 600 + 20 = \text{mean} + 1 \text{ s.d.}$
As 68% of the data will lie within 1 s.d. of the mean, 68% of the cakes should weigh between 580 g and 620 g.

(ii) $560 \text{ g} = 600 - 40 = \text{mean} - 2 \text{ s.d.}$
$640 \text{ g} = 600 + 40 = \text{mean} + 2 \text{ s.d.}$
So, 95% of the cakes should weigh between 560 g and 640 g.

(iii) If 95% of the cakes weigh between 560 g and 640 g, then 5% are outside that range.
As the distribution is approximately symmetrical, then 2·5% should weigh more than 640 g.

(b) 2·5% of 100 = 2·5 cakes (that is 2 or 3).

Exercise 6

1. Yorkie bars have a mean weight of 62·5 g. Two hundred were weighed and the standard deviation was found to be 3·8 g.
   (a) What percentage should have a mass between 58·7 g and 66·3 g?
   (b) Find the weights between which the central 95% of the distribution should lie.

2. Bottles of orange squash contain a mean volume of 1 litre with a standard deviation of 14 ml.
   (a) What percentage should contain less than 986 ml?
   (b) What percentage should contain between 972 ml and 1028 ml?
   (c) If you bought 50 bottles, how many should contain more than 1014 ml?

3. Levi jeans with 'leg length 34' have a mean length of 34 cm and standard deviation 0·6 cm.
   (a) What percentage of the jeans should have a leg length greater than 34·6?
   (b) What is the probability that the leg length of a pair of these jeans will be greater than 34·6?
   (c) Suppose your correct leg length is 32·8. There are 50 pairs of '34 leg' jeans in your local shop. How many of them are likely to have a leg length which is 32·8 or less?

4. The scores for an I.Q. test have a Normal distribution with mean 100 and standard deviation 15. If 800 people take the test, how many would you expect
   (a) to score more than 115, (b) to score less than 70?

5. Duckhams oil cans are supposed to contain 5 litres of engine oil. The machine which fills them delivers 5030 ml on average with a standard deviation of 15 ml.
   If 15 000 cans are filled in a day, how many will be under the advertised capacity?

6. The headteacher takes 36 minutes on average for her morning journey to school, with a standard deviation of $6\frac{1}{2}$ minutes. How long should she allow for the journey to be sure of being punctual 97·5% of the time?

Exercise 7

1. A farmer records the weights of a batch of sugar beet.

   Calculate the mean and standard deviation for the weight of the sugar beet.

Weight (grams)	Frequency
204–216	7
217–229	18
230–242	30
243–255	25
256–268	10

2. The farmer in Question **1** discovered that his scales were incorrect and that each sugar beet had been weighed 10 grams less than its actual weight.
Draw up a corrected frequency distribution and calculate the mean and standard deviation of this new distribution.
(a) How does the mean compare to the old mean?
(b) How does the standard deviation compare to the old standard deviation?

3. The following are the diameters in cm of 12 flowers produced by an experimental botanist.
3·8 3·3 2·9 1·3 1·0 0·3 0·4 0·9 1·0 1·8 2·2 3·0
Find the mean and standard deviation of the measurements.

4. The botanist found a new fertiliser which doubled the diameter of the flowers in Question **3**. Write down the new set of measurements and find the new mean and standard deviation.
(a) How does the new mean compare to the original mean?
(b) How does the new standard deviation compare?

Summary of mean and standard deviation rules

The four questions of the last exercise illustrate two rules which apply to the mean and to standard deviation.

Rule 1 If a constant is added to each member of a set of data, the mean is also increased by that constant, but the standard deviation (i.e. the spread) is unchanged.

Rule 2 If each member of a distribution is multiplied by a constant, both the mean and the standard deviation are multiplied by that constant.

9.4 Critical path analysis

Networks and priority tables

The diagram shows the side view of a new prison inspired by the 'Stonehenge' principle. Its unique design means that it can be converted into a school with only minor alterations.

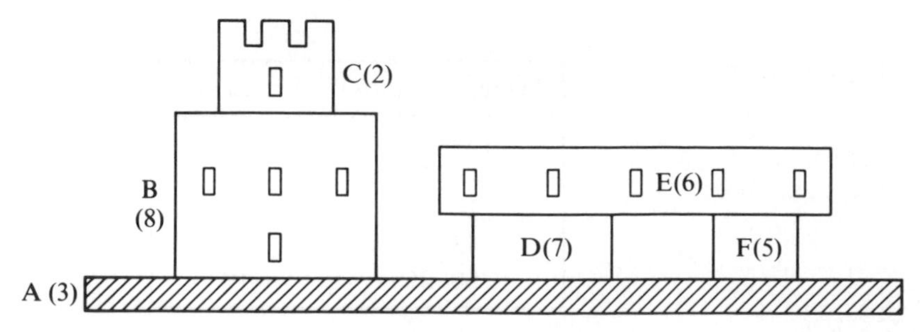

Block A contains the foundations, electric cables and sewers. Blocks B, C, D, E, F are all sections of the prison. The number of months required to build each block is shown in brackets.
Clearly A has to be built first. B has to be finished before work on C starts; D and F must be completed before E can be started.
Assume an adequate supply of men, machines and material so that work can occur on several blocks at the same time. What is the shortest possible time in which the whole prison can be built?
From the diagram, it is clear that work on D cannot begin until A is finished. E can be started as soon as D (and F) are finished. Work on F could start up to 2 months after the start of work on D.
The *critical path* is A D E and the whole prison can be built in 16 months.
The network to show the order of construction is shown below. From this network, the priority table shown on the right is obtained.

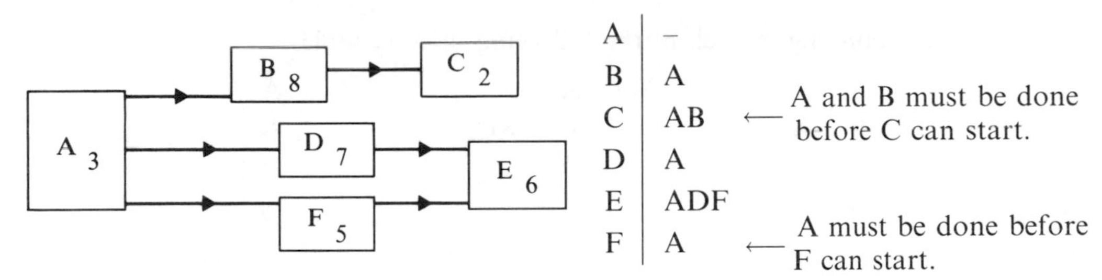

A	–
B	A
C	AB ⟵ A and B must be done before C can start.
D	A
E	ADF
F	A ⟵ A must be done before F can start.

Example 1

In the network shown below, the time that each person's job takes is given in hours.
Find the critical path and the shortest possible time for completion.

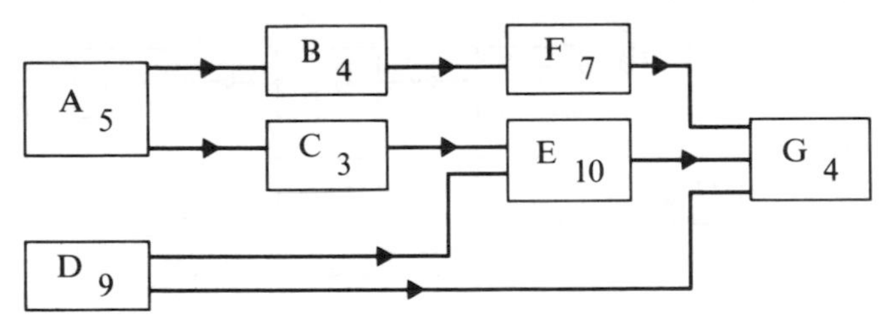

The priority table for this network is shown on the right.

By inspection, the critical path is D E G and the shortest possible time for completing the task is 23 hours.

Note. At one point *three* people need to be working at the same time.

A	–
B	A
C	A
D	–
E	ACD
F	AB
G	ABCDEF

Exercise 8

1. Draw a priority table for each of the following networks.

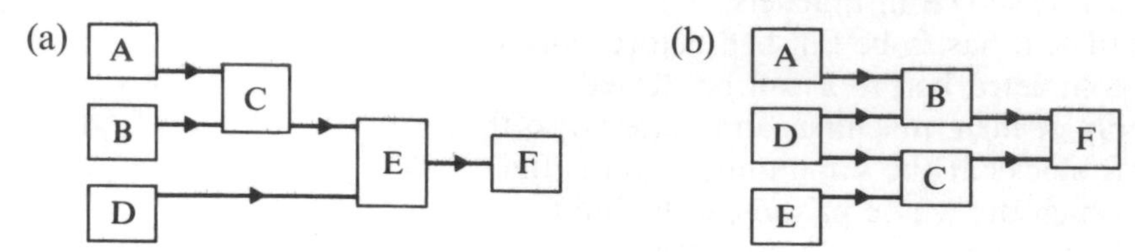

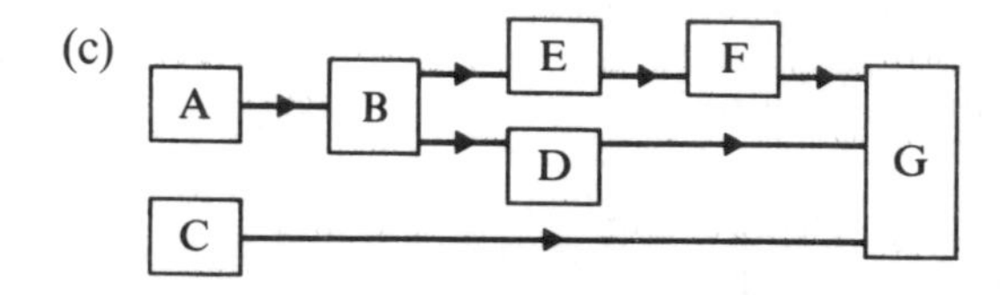

2. Draw the network corresponding to each of the following priority tables.

(a)

A	–
B	–
C	A
D	AC
E	B
F	ABCDE

(b)

A	–
B	A
C	–
D	CE
E	–
F	ABCDE
G	ABCDEF

(c)

A	–
B	AC
C	–
D	–
E	ABCD
F	ABCDEG
G	–

(d)

A	–
B	A
C	–
D	–
E	AB
F	C
G	CDF
H	ABCDEFG

3. For each of these networks find the critical path and the corresponding time required to complete the task. The number in each box gives the time taken for that job in hours.

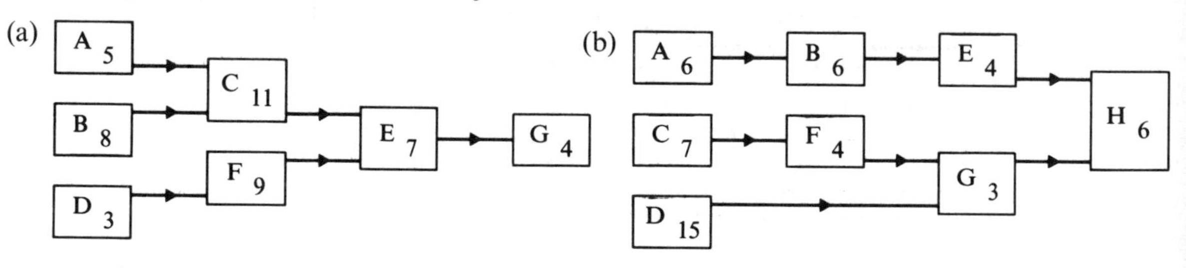

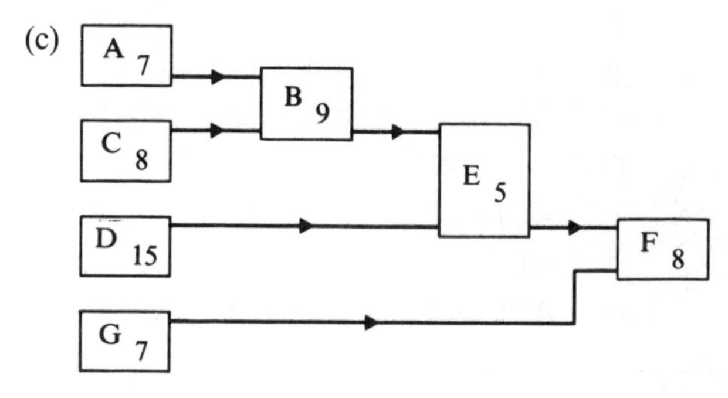

4. For Question 1, the times for each job are given below. Find the critical path and the corresponding time for each network.
   (a) A = 3, B = 7, C = 4, D = 9, E = 3, F = 2.
   (b) A = 11, B = 5, C = 6, D = 10, E = 8, F = 4.
   (c) A = 3, B = 4, C = 14, D = 9, E = 4, F = 4, G = 5.

   [You may find more than one critical path for a task. If so, give both paths.]

5. Draw the network corresponding to the priority table. Find the critical path and the corresponding time for the task.

   The time for each job is given next to the letter.

A = 3	–
B = 7	–
C = 6	AB
D = 8	B
E = 3	ABC
F = 5	ABCDE

Example 2

A room which has not previously been wall-papered, is to be completely redecorated.
Even the two windows and one door are being replaced. Three people are prepared to do some work. The jobs to be done are listed with the time required for each.
Find the time of the critical path.

A : Remove old windows and door (1 h)
B : Fit new windows and door (without glass) (4 h)
C : Paint ceiling (2 h)
D : Paint skirting board (1 h)
E : Paint windows and door (3 h)
F : Wallpaper walls (10 h)
G : Put glass in windows and door (3 h)
H : Lay carpet (3 h)

Step 1: Make a priority table, shown on the right.

Note (a) Do H last to avoid wet paint on the new carpet.
(b) Do C and D before F to avoid paint on the wallpaper.
(c) Assume the paint used for E dries quickly enough so that E and F can be done at the same time by starting in a sensible place.

A	–
B	A
C	–
D	–
E	ABG
F	CD
G	AB
H	ABCDEFG

Step 2: Draw a network for the whole job.

Step 3: Find the critical path by inspection.

The critical path is C F H and the time required is 15 hours.

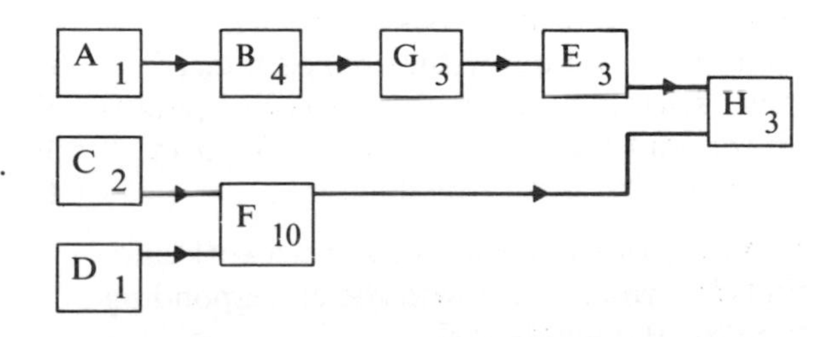

A certain amount of common sense is necessary for these questions. Do not be over-concerned about whether, for example, to paint the ceiling before or after wallpapering the walls. It is the method that is important.

Exercise 9

1. Many jobs need to be done to prepare for a meal and to clear up afterwards. Three people are willing to help, so up to three jobs can be done at the same time. Jobs A and F are reserved for Tina who likes to help but is not old enough to do anything more difficult.

(a) Make a priority table and draw a network.

(b) Find the critical path and the minimum time required.

	Job	Time
A	Put plates, cutlery etc. on table	5 min
B	Peel potatoes	10 min
C	Boil potatoes	20 min
D	Mash potatoes	5 min
E	Fry sausages	15 min
F	Warm baked beans	5 min
G	Defrost sausages in microwave	15 min
H	Eat meal	15 min
I	Clear table and wash up	15 min

2. Four people arrive starving hungry at a camp site. They need to cook a meal and get the tent up as quickly as possible. They decide to eat only when the tent and airbeds are ready. Here are the jobs to be done.

(a) Make a priority table and draw a network.

(b) Find the critical path and the minimum time required.

	Job	Time
A	Find gas cooker and matches	10 min
B	Purchase bacon, eggs from shop	15 min
C	Cook bacon and eggs	15 min
D	Pump up air beds	25 min
E	Erect main tent	30 min
F	Eat meal	15 min
G	Erect sleeping compartment inside tent (after erecting main tent)	10 min
H	Wash up	10 min

3. 'Shepherd's Pie' is made in several stages. Some minced meat is cooked and at the same time some potatoes are cooked and then mashed. Finally the cooked minced meat is put in a dish with the mashed potato on top and this dish is baked in the oven. Here are the jobs to be done:

Make a priority table, draw a network and find the critical path with the corresponding time for the whole task.

	Job	Time
A	Defrost minced meat in microwave	15 min
B	Peel potatoes	8 min
C	Cook minced meat	32 min
D	Boil potatoes	26 min
E	Mash potatoes	7 min
F	Put minced meat and mashed potatoes in dish	5 min
G	Bake Shepherd's pie in oven	15 min

9.5 Linear programming

In some business situations, the maximum or minimum value of such things as time, distance, cost, profit, number of employees are needed before decisions can be made.
Here we will solve problems with only two variables (say x and y).
Using computers, linear programming problems involving many variables can be solved.
In most linear programming problems, there are two stages:

1. to interpret the information given as a series of simultaneous inequalities and display them graphically.
2. to investigate some characteristic of the points in the unshaded solution set.

Example 1

A shopkeeper buys two types of dog food for his shop: Bruno at 40p a tin and Blaze at 60p a tin. He has £15 available and decides to buy at least 30 tins altogether. He also decides that at least one third of the tins should be Blaze.
Suppose he buys x tins of Bruno and y tins of Blaze.

(a) Write down three inequalities which correspond to the above conditions.
(b) Illustrate these inequalities on a graph, as shown below.
(c) The shopkeeper makes a profit of 10p a tin on Bruno and a profit of 20p a tin on Blaze.
Assuming all his stock can be sold, find how many tins of each type the shopkeeper should buy to maximise his profit and find that profit.

(a) Cost $40x + 60y \leqslant 1500$

$2x + 3y \leqslant 75 \ldots$ [line A on graph]

Total number

$x + y \geqslant 30 \ldots$ [line B on graph]

At least one third Blaze

$$\frac{y}{x} \geqslant \tfrac{1}{2}$$

$2y \geqslant x \ldots$ [line C on graph]

(b) The graph below shows these three inequalities.

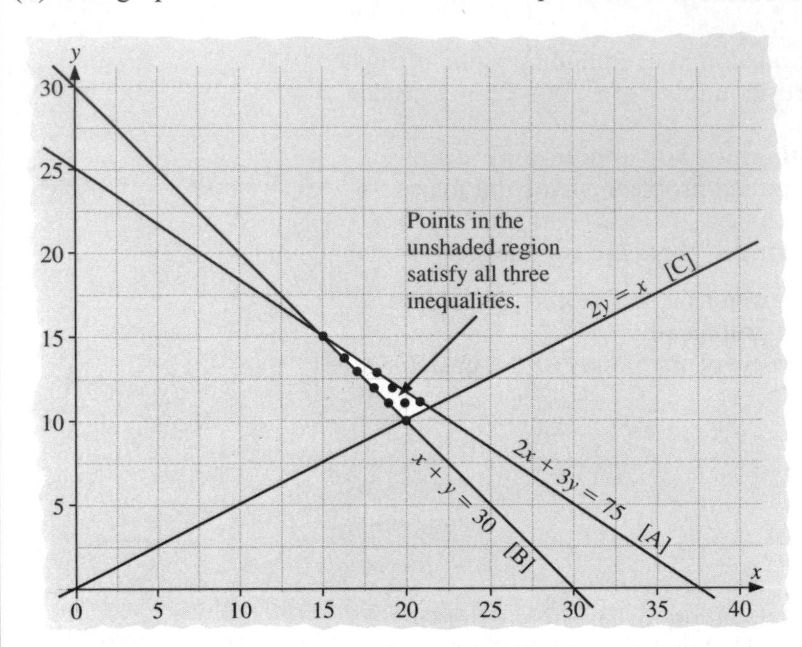

(c) The table below shows the points on the graph in the unshaded region together with the corresponding figure for the profit.
The points marked * will clearly not provide a maximum profit.

x	15	16	17*	18	19	19*	20*	20*	21
y	15	14	13	13	12	11	11	10	11
profit	150 +300 = 450p	160 +280 = 440p		180 +260 = 440p	190 +240 = 430p				210 +220 = 430p

Conclusion: the shopkeeper should buy 15 tins of Bruno and 15 tins of Blaze. The maximised profit is then 450p.

Exercise 10

For each of the Questions **1** to **3**, draw an accurate graph to represent the inequalities listed, using shading to show the unwanted regions.

1. $x + y \leqslant 11$; $y \geqslant 3$; $y \leqslant x$

Find the point having whole number coordinates and satisfying these inequalities which gives

(a) the maximum value of $x + 4y$

(b) the minimum value of $3x + y$

2. $3x + 2y > 24$; $x + y < 12$; $y < \frac{1}{2}x$; $y > 1$.
Note the use here of $>$ and $<$ instead of $\geqslant$ and $\leqslant$. So, no point actually on a line can be used. Such inequality lines should be drawn broken (see page 82).
Find the point having whole number coordinates and satisfying the above four inequalities which gives

(a) the maximum value of $2x + 3y$
(b) the minimum value of $x + y$

3. $3x + 2y \leqslant 60$; $x + 2y \leqslant 30$; $x \geqslant 10$; $y \geqslant 0$.
Find the point having whole number coordinates and satisfying these inequalities which gives

(a) the maximum value of $2x + y$
(b) the maximum value of xy

4. A girl is given £1·20 to buy some peaches and apples. Peaches cost 20p each, apples 10p each. She is told to buy at least 6 individual fruits, but not to buy more apples than peaches.
Let x be the number of peaches she buys.
Let y be the number of apples she buys.

(a) Write down three inequalities which must be satisfied.
(b) Draw a linear programming graph and use it to list the combinations of fruit that she could buy.

5. A man has a spare time job spraying cars and vans. Vans take 2 hours each and cars take 1 hour each. He has 14 hours available per week. He has an agreement with one firm to do 2 of their vans every week. Apart from that he has no fixed work.
His permission to use his back garden contains the clause that he must do at least twice as many cars as vans.
Let x be the number of vans sprayed each week.
Let y be the number of cars sprayed each week.

(a) Write down three inequalities which must be satisfied.
(b) Draw a graph and use it to list the possible combinations of vehicles which he can spray each week.

6. The manager of a football team has £1 million to spend on buying new players. He can buy defenders at £60 000 each or forwards at £80 000 each. There must be at least 6 of each sort. To cover for injuries, he must buy at least 13 players altogether. Let x represent the number of defenders he buys and y the number of forwards.

(a) In which ways can he buy players?
(b) If the wages are £1000 per week for each defender and £2000 per week for each forward, which combination of players involves the lowest wage bill?

7. A shop owner wishes to buy up to 20 televisions for stock. He can buy either type A for £150 each or type B for £300 each. A total of £4500 can be spent and at least 6 of each type must be in stock. If the shop owner buys x of type A and y of type B, write down 4 inequalities which must be satisfied, and represent the information on a graph.

(a) If he makes a profit of £40 on each of type A and £100 on each of type B, how many of each should he buy for maximum profit?

(b) If the profit is £80 on each of type A and £100 on each of type B, how many of each should he buy now?

8. The manager of a car park allows $10\,m^2$ of parking space for each car and $30\,m^2$ for each lorry. The total space available is $300\,m^2$. He decides that the maximum number of vehicles at any time must not exceed 20 and he also insists that there must be at least as many cars as lorries. If the number of cars is x and the number of lorries is y, write down three inequalities which must be satisfied.

(a) If the parking charge is £1 for a car and £5 for a lorry, find how many vehicles of each kind he should admit to maximise his income.

(b) If the charges are changed to £2 for a car and £3 for a lorry, find how many of each kind he would be advised to admit.

Examination exercise 9

1. The histogram shows the examination results of 120 candidates.

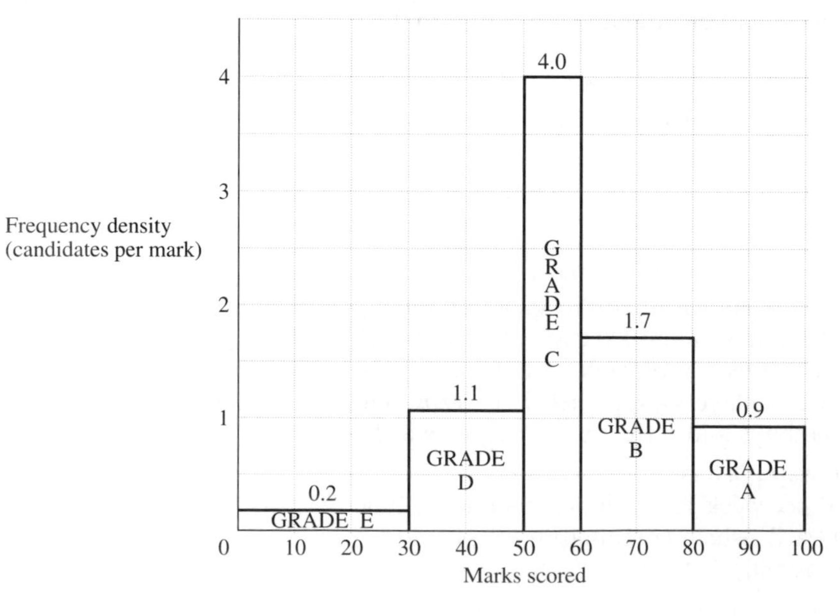

(a) Copy and complete the table below, giving the number of candidates who obtained each grade.

Grade	E	D	C	B	A
Number of candidates	6				

(b) If this information was presented on a pie chart, calculate the angle which would be used for the sector representing grade E.

(c) Complete the table below, giving the cumulative total number of candidates who obtained at the most the mark shown.

Maximum mark	30	50	60	80	100
Cumulative frequency	6				

(d) On graph paper, draw a cumulative frequency diagram to display this information.

(e) From your graph, estimate the median mark for the examination. [N]

2. The time taken for a certain job was recorded on 50 occasions and the following data obtained.

Time (nearest minute)	11–14	15–18	19–20	21–22	23–26	27–30
Frequency	6	9	5	8	12	9

Draw a histogram to illustrate the data.

3. The ages of all the 100 employees in a company A are shown.

Age (years)	16–	21–	31–	41–	51–	61–	81–
Frequency	8	12	25	30	20	5	0
Frequency Density	1·6	1·2					0

(a) (i) Complete the table of frequency densities.

(ii) On graph paper draw a histogram to illustrate the age distribution of the employees in Company A.

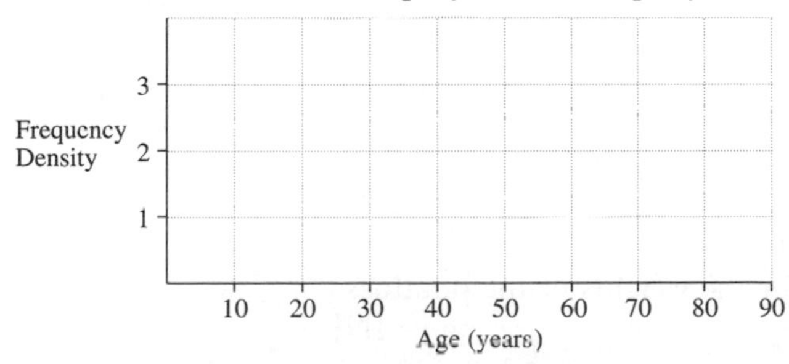

Company B has the following age distribution of its employees.

Age (years)	16–30	31–40	41–50	51–80
% of employees	8	85	7	0

(b) Write down two comparisons about the age distributions of the employees of the two companies. [S]

4. In a diving competition, each dive is marked by seven judges. To obtain the score for a dive the highest and lowest of these marks are ignored and the average (arithmetic mean) of the other five is multiplied by a figure representing the degree of difficulty of the dive.

(a) Calculate the score for a dive which the judges marked as

9·6, 8·8, 9·7, 9·2, 9·0, 8·3, 8·4,

and for which the degree of difficulty of the dive is 2·4.

(b) Calculate the standard deviation of the five marks which were used in the calculation to obtain the score for this dive. [N]

5. The data below shows the energy levels, in kilocalories per 100 g, of 10 different snack foods (such as crisps and peanuts).

440 520 480 560 572 550 620 680 545 490

(i) Calculate the mean and standard deviation of the energy levels of these snack foods.

(ii) The energy levels, in kilocalories per 100 g, of 10 different breakfast cereals had a mean of 350 kilocalories with a standard deviation of 28 kilocalories. Which of the two types of food show great variation in energy level? Give a reason for your answer. [W]

6. The lengths, in centimetres, of 10 leaves in a sample were;

5·6, 5·8, 4·9, 6·2, 6·8, 4·8, 5·4, 5·9, 5·3, 5·2

(a) Calculate the standard deviation of these lengths.

(b) The standard deviation of another sample of 10 leaves was 0·432 cm. What difference between the two samples is shown by the two standard deviations? [S]

7. The weights of a number of potatoes were recorded as follows.

Weight (grams)	Frequency
118–126	8
127–135	13
136–144	21
145–153	27
154–162	13
163–171	11
172–180	7

Calculate the standard deviation for the weights of the potatoes. [N]

8. The scores for a fitness test have a Normal distribution with mean 80 and standard deviation 12. If 1000 people take the test how many would you expect to score more than 104?
[Assume that 95% of the distribution would be contained in the interval mean ± 2 standard deviations.]

9. Your school is conducting a survey to investigate pupil attitudes to "compulsory school uniform" and you have been asked to choose a sample of pupils for the survey. State, with reasons, **two** factors which you consider to be important in your choice of sample so that the survey represents fairly the opinions of all the pupils in your school. [W]

10. It is required to investigate the television viewing preferences of pupils in a mixed comprehensive school with 1000 pupils.
(a) (i) Describe how you would obtain a simple representative sample of 100 of the pupils in order to find out their preferences.
(ii) Describe how you would obtain a stratified representative sample of 100 of the pupils in order to find out their preferences.
(iii) State, giving a reason, which of the above samples would give a more reliable picture of the television viewing preferences of the pupils in the school.
(b) Give a reason why it would be necessary to take a sample when conducting a survey of the television viewing preferences of people in Wales. [W]

11. A family decide to decorate the spare room and to have it ready in time for grandmother's visit from Australia.
The table lists the tasks that need to be done, which tasks must be done before others can be started, and the duration of each task.

	Task	Duration (hours)	Tasks that must be completed first
A	Collect tools	1	–
B	Purchase materials	3	–
C	Clear room	2	–
D	Strip walls	3	A, C
E	Repair walls	3	D
F	Redecorate	4	B, E
G	Clean tools	1	F
H	Clean floor	1	F
I	Repair furniture	8	C
J	Replace furniture	1	H

Draw a network to show the order in which the tasks need to be done (assuming that there are as many willing workers as you need, so that several tasks can be progressing simultaneously). Calculate the shortest time in which the project can be finished.
[MEI]

12. The activities involved when two workers replace a broken window pane, and the times taken for each activity, are given in the table below.

	Activity	Duration in minutes	Preceding activity
A	Remove broken pane	20	–
B	Measure size of pane	15	–
C	Purchase glass and putty	25	B
D	Put putty in frame	10	A, C
E	Put in new pane of glass	5	D
F	Putty outside and smooth	10	E
G	Sweep up broken glass	5	A
H	Clean up	5	F, G

Using a critical path diagram, or otherwise, find how the two workers should share the activities to complete the replacement of the broken window pane in the minimum possible time.
State this minimum time. [L]

13. The Wellworthy Community Centre is organising a trip to the seaside in a coach. The coach seats 61 people and costs £220. It is decided to charge £5 for an adult and £3 for a child. Adults may only go if they take at least one child.
Enough money must be collected to pay for the coach. You may use calculations or graphs.
(a) Work out the least number of adults that can go on the trip. How many children go with them?
(b) Work out the least number of children that can go on the trip. How many adults go with them?
Show clearly how you obtain your answers. [M]

14. A landscape designer has £500 to spend on planting trees and shrubs to landscape an area of $2000\,m^2$. For a tree he plans to allow $50\,m^2$ and for a shrub $5\,m^2$. Planting a tree will cost £10 and a shrub £2.
(i) If he uses x trees and y shrubs, show that two inequalities (other than $x \geqslant 0$ and $y \geqslant 0$) satisfied by x and y are $5x + y \leqslant 250$ and $10x + y \leqslant 400$.
(ii) Show these inequalities on a graph, using scales of 1 cm to 5 trees and 1 cm to 50 shrubs, shading in the unwanted region.
(iii) If he plants 80 shrubs, what is the maximum number of trees he can plant?
(iv) If he plants 10 shrubs for every tree, what is the maximum number of trees he can plant? [M]

10 Using and applying mathematics

10.1 Conjectures

A conjecture (or hypothesis) is simply a statement which may or may not be true. In mathematics people are constantly looking for rules or formulas to make calculations easier. Suppose you carry out an investigation and after looking at the results you spot what you think might be a rule.
You could make the conjecture that:

'x^2 is always greater than x'
or that 'the angle in a semicircle is always 90°'
or that 'the expression $x^2 + x + 41$ gives a prime number for all integer values of x'

There are three possibilities for any conjecture. It may be
(a) true, (b) false, (c) not proven.

(a) True

To show that a conjecture is true it is not enough to simply find lots of examples where it is true. We have to *prove* it for *all* values. We return to proof later in this section.

(b) False

To show that a conjecture is false it is only necessary to find one *counter-example*.

Example 1

Consider the conjecture 'x^2 is always greater than x'

It is true that $2^2 > 2$; $3{\cdot}1^2 > 3{\cdot}1$; $(-5)^2 > 5$

But $(\frac{1}{2})^2$ is *not* greater than $\frac{1}{2}$.

This is a counter-example so the conjecture is false.

We have *disproved* the conjecture.

Example 2

Consider the conjecture:
'the expression $x^2 + x + 41$ gives a prime number for all integer values of x'

If we try the first few values of x, we get these results.

x	1	2	3	4	5	6
x^2+x+41	43	47	53	61	71	83

The expression does give prime numbers for these values of x and indeed, it does so for all values of x up to 39.
But when $x = 40$, $\quad x^2 + x + 41 = 1681$
$= 41 \times 41$
So we do not always obtain a prime number.

The conjecture is, therefore, false.

Notice that it takes only *one* counter-example to disprove a conjecture.

(c) Not proven

Suppose you have a conjecture for which you cannot find a counter-example but which you also cannot prove. In this case the conjecture is *not proven*.
A famous example is 'Fermat's Last Theorem' which states that there are no whole numbers a, b, c for which $a^3 + b^3 = c^3$ [or further, that $a^n + b^n = c^n$ $(n > 2)$].
No one has yet proved the conjecture but also no one has found a counter-example.

Stop Press: A British mathematician has just (1993) proved Fermat's Last Theorem. The proof is extremely long and it is said that only a handful of people in the world understand it.

Exercise 1

In Questions **1** to **9**, consider the conjecture given. Some are true and some are false. Write down a counter-example where the conjecture is false.
If you cannot find a counter-example, state that the conjecture is 'not proven' as you are *not* asked to prove it.

1. $(n + 1)^2 = n^2 + 1$ for all values of n.

2. $\frac{1}{x}$ is always less than x.

3. $\sqrt{n+1}$ is always smaller than $\sqrt{n} + 1$.

4. For all values of n, $2n$ is greater than $n - 2$.

5. For any set of numbers, the median is always smaller than the mean.

6. $\sqrt{x}$ is always less than x.

7. The diagonals of a parallelogram never cut at right angles.

8. The product of two irrational numbers is always another irrational number.

9. If n is even $5n^2 + n + 1$ is odd.

10. Here is a circle with 2 points on its circumference and 2 regions.

This circle has 3 points ($n = 3$) and 4 regions ($r = 4$).

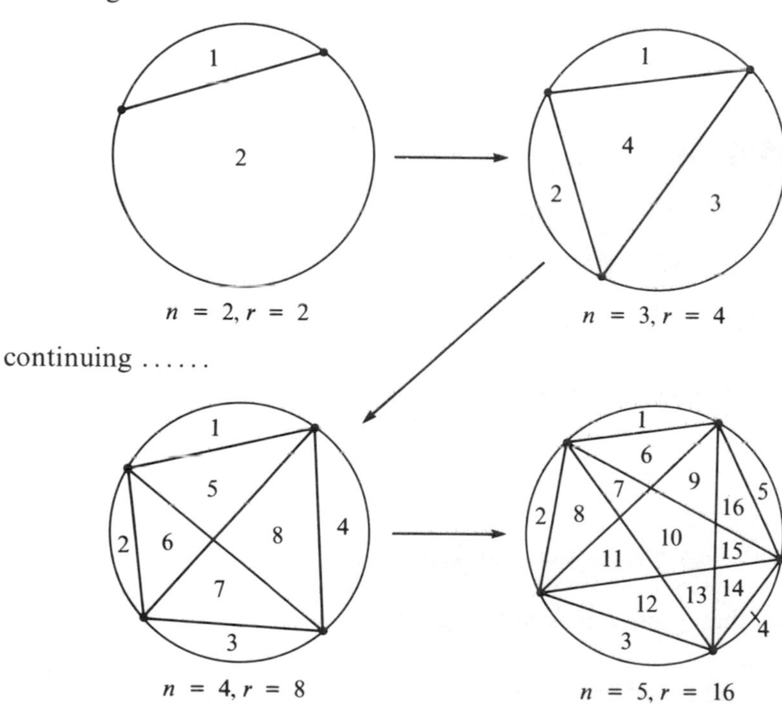

Each time, every point on the circumference is joined to all the others.

(a) Do you think a conjecture is justified at this stage?
Consider:
'The circle with 6 points on its circumference will have 32 regions'.
Or again:
'With n points on the circle the number of regions, r, is given by $r = 2^{n-1}$
Draw the next two circles in the sequence with $n = 6$ and $n = 7$. Be careful to avoid this sort of thing:

where three lines pass through a single point.

Under a magnifying glass we might see this:
and a region might be lost.
What can you say about your conjecture at
this stage?

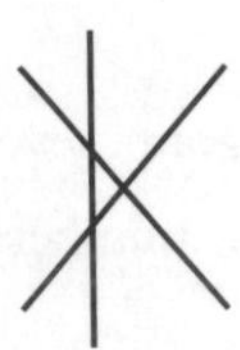

(b) When you have the results for $n = 6$ and $n = 7$ you might get somewhere (although it is quite difficult) by looking at *differences* in the results. Refer to page 56 for further details.

10.2 Implication signs

- Here is a statement

  Triangle PQR is equilateral ⇒ angle QRP = 60°.

  The statement can be read:

  If triangle PQR is equilateral, then angle QRP equals 60°.

  or

  Triangle PQR is equilateral implies that angle QRP equals 60°.

- Here is another statement.

  $a > b \Rightarrow a - b > 0$

  This can be read:

  If a is greater than b, then a less b is greater than zero

  or

  a is greater than b implies that a less b is greater than zero.

Note $x = 4 \Rightarrow x^2 = 16$ is true
but $x^2 = 16 \Rightarrow x = 4$ is not true because x could be -4.

Exercise 2

1. Write the following as sentences
(a) Today is Monday ⇒ tomorrow is Tuesday.
(b) It is raining ⇒ there are clouds in the sky.
(c) Abraham Lincoln was born in 1809 ⇒ Abraham Lincoln is dead.

2. State whether the following statements are true or false.
(a) $x - 4 = 3 \Rightarrow x = 7$
(b) n is even $\Rightarrow n^2$ is even
(c) $a = b \Rightarrow a^2 = b^2$
(d) $x > 2 \Rightarrow x = 3$
(e) $a + b$ is odd $\Rightarrow ab$ is even (a and b are whole numbers).
(f) $x^2 = 4 \Rightarrow x = 2$
(g) $x = 45° \Rightarrow \tan x = 1$

(h) $5^x = 1 \Rightarrow x = 0$
(i) $pq = 0 \Rightarrow p = 0$
(j) a is odd $\Rightarrow a$ can be written in the form $2k + 1$, where k is an integer.

3. Think of pairs of statements like those in Question **1**. Link the statements with the $\Rightarrow$ sign.

10.3 Proof

A proof starts with simple facts which are accepted. The proof then argues logically to the result which is required.

Numerical proof

Example 1

Prove that, if a and b are odd numbers, then $a + b$ is even.

If a is odd, there is remainder 1 when a is divided by 2.
$\therefore$ a may be written in the form $(2m + 1)$ where m is a whole number.
Similarly b may be written in the form $(2n + 1)$.

$$\therefore \quad a + b = 2m + 1 + 2n + 1$$
$$= 2(m + n + 1)$$

$\therefore$ $a + b$ is even, as required.

Example 2

Prove that the product of three consecutive whole numbers is divisible by 6.

e.g. $3 \times 4 \times 5 = 60 = 6 \times 10$ or $8 \times 9 \times 10 = 720 = 6 \times 120$

Proof
In any three consecutive numbers, at least one of the numbers must be even and one of the numbers must be a multiple of 3.

An even number can be written as $2a$, where a is an integer.

A multiple of 3 can be written as $3b$, where b is an integer.

Our three consecutive numbers could be x, $2a$, $3b$ in any order.

The product of these numbers is $6xab$

We deduce that the product of three consecutive numbers is divisible by 6.

Example 3 [Much harder but very interesting]

Prove that the product of 4 consecutive numbers is always one less than a square number.

e.g. $1 \times 2 \times 3 \times 4 = 24 = 25 - 1$ or $3 \times 4 \times 5 \times 6 = 360 = 19^2 - 1$

Proof
Let the first number be n.
Then the next three consecutive numbers are $(n+1)$, $(n+2)$, $(n+3)$

Find the product

$$n(n+1)(n+2)(n+3)$$
$$= (n^2+n)(n^2+5n+6)$$
$$= n^4+6n^3+11n^2+6n$$

Write the products as $(n^4+6n^3+11n^2+6n+1)-1$.
Note that this does not alter the value of the product.

The expression in the brackets factorises as
$(n^2+3n+1)(n^2+3n+1)$
$\therefore$ The product is $(n^2+3n+1)^2-1$

Since $(n^2+3n+1)^2$ is a square number, the result is proved.

Exercise 3

1. If a is odd and b is even, prove that ab is even.
2. If a and b are both odd, prove that ab is odd.
3. Prove that the sum of two odd numbers is even.
4. Prove that the square of an even number is divisible by 4.
5. Prove that the product of four consecutive numbers is divisible by 4.
6. Prove that the sum of the squares of five consecutive numbers is divisible by 5.
   e.g. $2^2+3^2+4^2+5^2+6^2 = 90 = 5 \times 18$
   Begin by writing the middle number as n, so the other numbers are $n-2$, $n-1$, $n+1$, $n+2$.
7. Here is the 'proof' that $1 = 2$.
   Let $a = b$
   $\Rightarrow ab = b^2$ [multiply by b]
   $\Rightarrow ab - a^2 = b^2 - a^2$ [subtract a^2]
   $\Rightarrow a(b-a) = (b+a)(b-a)$ [factorise]

$\Rightarrow a = b + a$ [divide by $(b - a)$]
$\Rightarrow a = a + a$ [from top line]
$\Rightarrow 1 = 2$

Which step in the argument is not allowed?

Geometric proof

Over 2000 years ago Greek mathematicians were very keen on proofs. Euclid wrote a best seller called 'The Elements' in which he first of all stated some very basic assumptions called *axioms*. He then proceeded to prove one result after another, basing each new result on a previous result which had already been proved. Examples of his axioms are 'parallel lines do not meet in either direction' and 'only one straight line can be drawn between two points'.

When you are writing a geometric proof of your own you do not need to go right back to Euclid's axioms. What you should do is state clearly after each line what facts you have used.
For example 'angles in a triangle add up to 180°'
or 'opposite angles are equal.'

Example 4

Prove that the angle at the centre of a circle is twice the angle at the circumference.

Draw the straight line COD.
Let $A\hat{C}O = y$ and $B\hat{C}O = z$.

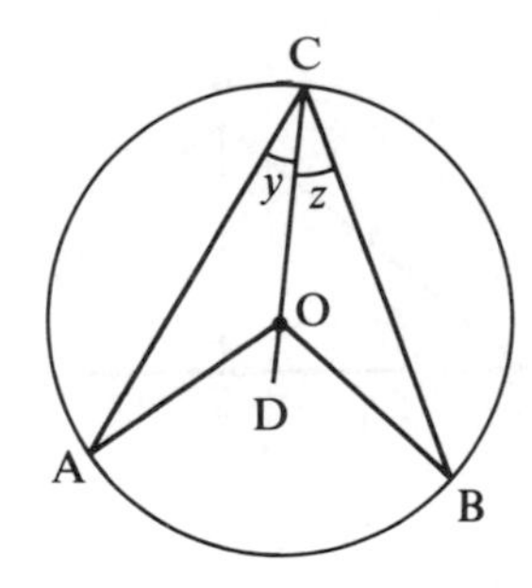

In triangle AOC,

$AO = OC$ (radii)
$\therefore\ O\hat{C}A = O\hat{A}C$ (isosceles triangle)
$\therefore\ C\hat{O}A = 180 - 2y$ (angle sum of triangle)
$\therefore\ A\hat{O}D = 2y$ (angles on a straight line)

Similarly from triangle COB, we find

$D\hat{O}B = 2z$
Now $A\hat{C}B = y + z$
and $A\hat{O}B = 2y + 2z$
$\therefore\ A\hat{O}B = 2 \times A\hat{C}B$ as required.

Example 5

Prove that opposite angles in a cyclic quadrilateral add up to 180°.

Draw radii OA and OC.
Let $A\hat{D}C = x$ and $A\hat{B}C = y$.

$$A\hat{O}C \text{ obtuse} = 2x \text{ (angle at the centre)}$$

$$A\hat{O}C \text{ reflex} = 2y \text{ (angle at the centre)}$$

$$\therefore \quad 2x + 2y = 360° \text{ (angles at a point)}$$

$$\therefore \quad x + y = 180° \text{ as required.}$$

Exercise 4

1. Triangle ABC is isosceles.

Prove that $C\hat{B}D = 2 \times C\hat{A}B$.

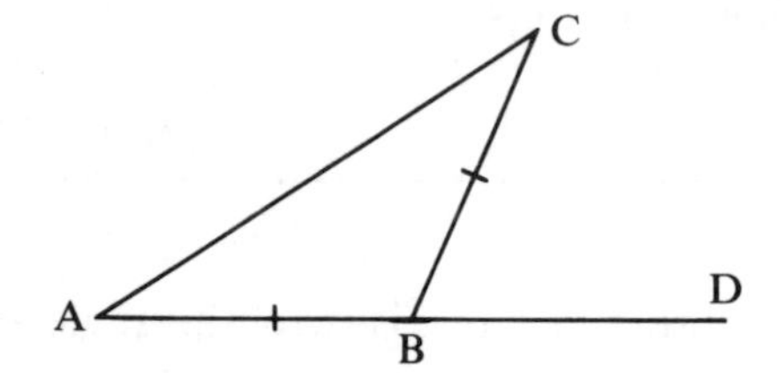

2. Triangle PQR is isosceles and $P\hat{S}Q = 90°$.

Prove that PS bisects $Q\hat{P}R$.

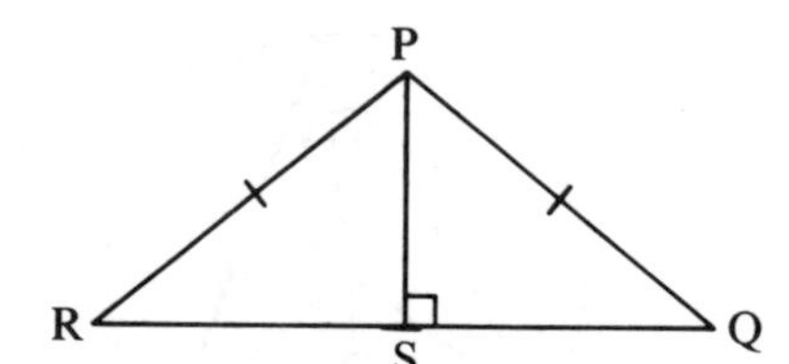

3. Two chords of a circle AB and CD intersect at X.

Use similar triangles to prove that AX.BX = CX.DX. This result is called the Intersecting Chords Theorem.

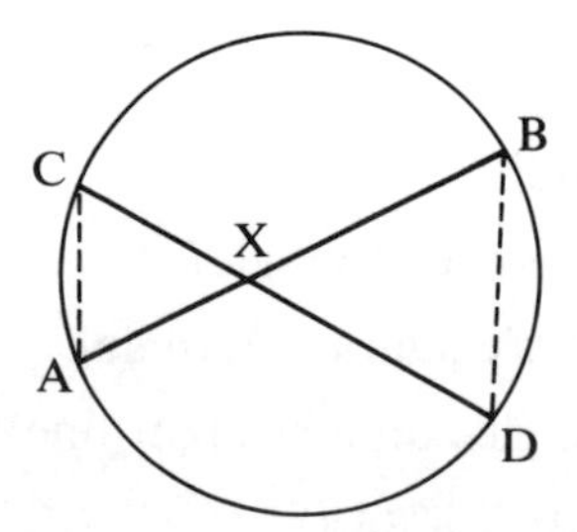

4. Line ATB touches a circle at T and TC is a diameter. AC and BC cut the circle at D and E respectively. Prove that the quadrilateral ADEB is cyclic.

5. Prove that the angle in a semicircle is a right angle.

6. TC is a tangent to a circle at C and BA produced meets this tangent at T.

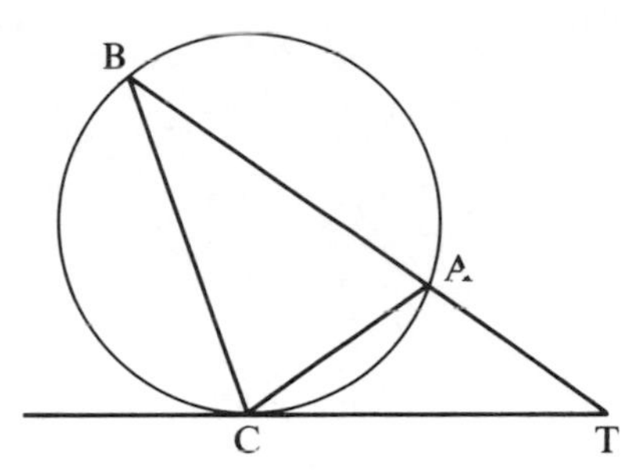

Show that triangles TCA and TBC are similar and hence prove that $TC^2 = TA.TB$.

7.

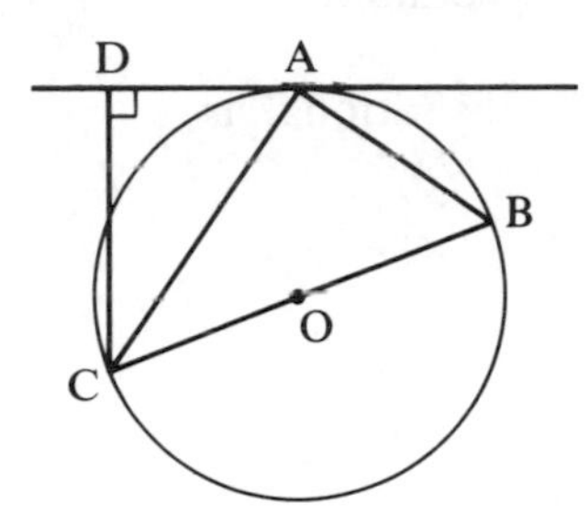

Given that BOC is a diameter and that $A\hat{D}C = 90°$, prove that AC bisects $B\hat{C}D$.

8.† Two circles touch externally at T. A chord of the first circle XY is produced and touches the other at Z. The chord ZT of the second circle, when produced, cuts the first circle at W. Prove that $X\hat{T}W = Y\hat{T}Z$.

Proof by contradiction

This is not easy to understand. Proof by contradiction is an important method in mathematics but most examples of its use are beyond the scope of Level 10. Here is a famous example.

Proof that $\sqrt{2}$ is irrational

Remember that an irrational number *cannot* be written in the form $\frac{a}{b}$ where a and b are integers.

We begin the proof by stating the negation of '$\sqrt{2}$ is irrational' which is '$\sqrt{2}$ is rational'.

$\sqrt{2}$ is rational

$\Rightarrow \sqrt{2} = \frac{a}{b}$ ... [1] where a and b are integers with no common factor.

$\Rightarrow 2 = \frac{a^2}{b^2}$

$\Rightarrow 2b^2 = a^2 \quad \ldots [2]$

$\Rightarrow a^2$ is even

$\Rightarrow a = 2n$ where n is an integer

From [2] $2b^2 = 4n^2$

$\Rightarrow b^2 = 2n^2$

$\Rightarrow b^2$ is even

$\Rightarrow b$ is even

$\Rightarrow b = 2p$

The conclusion is a contradiction since a and b have a common factor 2.
Hence, as the conclusion is false, the conjecture that '$\sqrt{2}$ is rational' is false.
$\therefore$ $\sqrt{2}$ is irrational.

Exercise 5

1. Prove that $\sqrt{3}$ is irrational by following the argument above.

Examination exercise 10

1.

$$\begin{array}{r} \text{THIS} \\ +\quad \text{IS} \\ \hline \text{HARD} \\ \hline \end{array}$$

In the sum shown, different letters stand for different digits.
(a) Explain why the letter I must be at least equal to 5.
(b) What must be the value of H?
(c) What must be the value of T? [N]

2. The sets {7, 8, 9} and {110, 111, 112} are examples of sets of three consecutive numbers. Explore the validity of the following statements.
(i) "In any set of three consecutive numbers one of them is divisible by 3."
(ii) "The product of any set of three consecutive numbers is divisible by 6."
(iii) "The product of any set of three consecutive numbers is divisible by 4." [W]

3. If you add two or more consecutive odd numbers together, you get a square number.
(a) Find two cases for which this statement is true.
(b) Prove that the statement is not always true.
(c) Modify the statement so that it is always true. [M]

4. When coming to a staircase, a man climbs either one or two steps at a time.
(a) In how many ways can he climb three steps?
(b) In how many ways can he climb four steps?
(c) Can you generalise on how many ways he could take to climb any number of steps? [N]

5. Some numbers can be written as the difference between two square numbers. (For this purpose, 0 is counted as a square number). For example,

$20 = 36 - 16 = 6^2 - 4^2$
and $15 = 16 - 1 = 4^2 - 1^2$.

(a) Write 24 as the difference between two square numbers in two different ways.
(b) Write down (in order of size) all the even numbers less than 30 which can be written as the difference between two square numbers.
(c) Write down (in order of size) the first seven numbers which cannot be written as the difference between square numbers and describe the pattern of numbers formed.
(d) By expressing 111 as the product of two whole numbers and using $x^2 - y^2 = (x - y)(x + y)$, show that 111 can be expressed as the difference between two square numbers in two, and only two ways. [M]

6. The operation $*$ is defined on the set of real numbers by

$$p * q = p.q + p$$

For example
$$2 * 3 = 2.3 + 2 = 6 + 2 = 8$$

(a) Find the value of $3 * 4$
(b) When $a = 2$, $b = 3$, $c = 5$, find the value of
 (i) $a * (b + c)$,
 (ii) $(a * b) + (a * c)$.
(c) (i) Show that $p * (q * r) = p.q.r + p.q + p$.
 (ii) Find a similar expression for $(p * q) * r$.
 (iii) Why is it not possible to find the value of $2 * 3 * 5$?
(d) Prove that $a * b \neq b * a$ if $a \neq b$.
(e) Prove that $\{a * (b + c)\} + a = (a * b) + (a * c)$.
(f) Find the values of x which satisfy the equation
$x * x + 2 * x - 20 = 0$. [W]

11 Coursework tasks, puzzles, games

11.1 Investigations

There are a large number of possible starting points for investigations here so it may be possible to allow students to choose investigations which appeal to them. On other occasions the same investigation may be set to a whole class.

Here are a few guidelines for pupils:
(a) If the set problem is too complicated try an easier case;
(b) Draw your own diagrams;
(c) Make tables of your results and be systematic;
(d) Look for patterns;
(e) Is there a rule or formula to describe the results?
(f) Can you *predict* further results?
(f) Can you *prove* any rules which you may find?
(h) Where possible extend the task further by asking questions like 'what happens if . . .'

1 Opposite corners

Here the numbers are arranged in 9 columns.

In the 2 × 2 square . . .

6	7
15	16

6 × 16 = 96
7 × 15 = 105

. . . the difference between them is 9.

In the 3 × 3 square . . .

22	23	24
31	32	33
40	41	42

22 × 42 = 924
24 × 40 = 960

. . . the difference between them is 36.

1	2	3	4	5	6	7	8	9
10	11	12	13	14	15	16	17	18
19	20	21	22	23	24	25	26	27
28	29	30	31	32	33	34	35	36
37	38	39	40	41	42	43	44	45
46	47	48	49	50	51	52	53	54
55	56	57	58	59	60	61	62	63
64	65	66	67	68	69	70	71	72
73	74	75	76	77	78	79	80	81
82	83	84	85	86	87	88	89	90

Investigate to see if you can find any rules or patterns connecting the size of square chosen and the difference.
If you find a rule, use it to *predict* the difference for larger squares.
Test your rule by looking at squares like 8 × 8 or 9 × 9.
Can you *generalise* the rule?

[What is the difference for a square of size $n \times n$?]

Can you *prove* the rule?

Hint:

In a 3 × 3 square ...

x		?
?		?

What happens if the numbers are arranged in six columns or seven columns?

1	2	3	4	5	6
7	8	9	10	11	12
13	14	15	16	17	18
19					

1	2	3	4	5	6	7
8	9	10	11	12	13	14
15	16	17	18	19	20	21
22						

2 *Hiring a car*

You are going to hire a car for one week (7 days).
Which of the firms below should you choose?

Gibson car hire	Snowdon rent-a-car	Hav-a-car
£170 per week unlimited mileage	£10 per day 6·5 p per mile	£60 per week 500 miles without charge 22p per mile over 500 miles.

Work out as detailed an answer as possible.

3 *Half-time score*

The final score in a football match was 3–2. How many different scores were possible at half-time?

Investigate for other final scores where the difference between the teams is always one goal. [1–0, 5–4, etc.]. Is there a pattern or rule which would tell you the number of possible half-time scores in a game which finished 58–57?
Suppose the game ends in a draw. Find a rule which would tell you the number of possible half-time scores if the final score was 63–63.
Investigate for other final scores [3–0, 5–1, 4–2, etc.].
Find a rule which gives the number of different half-time scores for *any* final score (say $a - b$).

4 An expanding diagram

Look at the series of diagrams.

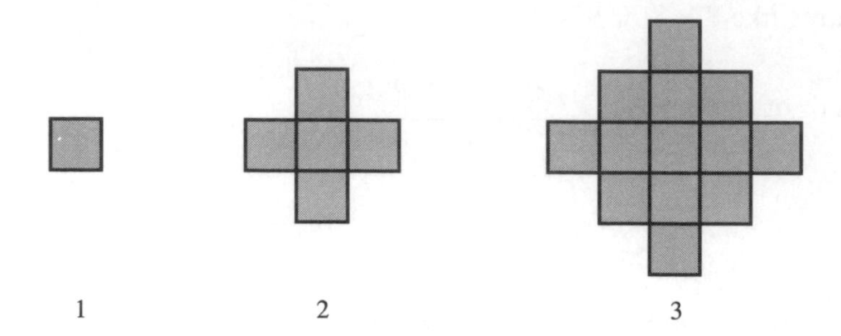

Each time new squares are added all around the outside of the previous diagram.
Draw the next few diagrams in the series and count the number of squares in each one.
How many squares are there in diagram number 15 or in diagram number 50?
What happens if we work in three dimensions? Instead of adding squares we add cubes all around the outside. How many cubes are there in the fifth member of the series or the fifteenth?

5 Maximum box

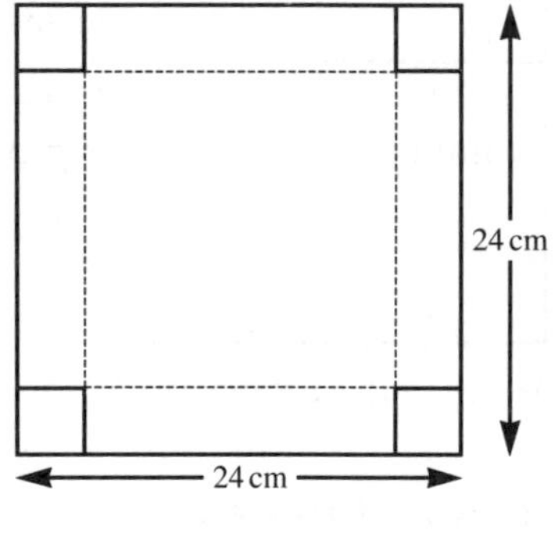

(a) You have a square sheet of card 24 cm by 24 cm.
You can make a box (without a lid) by cutting squares from the corners and folding up the sides.
What size corners should you cut out so that the volume of the box is as large as possible?
Try different sizes for the corners and record the results in a table.

Length of the side of the corner square (cm)	Dimensions of the open box (cm)	Volume of the box (cm^3)
1	$22 \times 22 \times 1$	484
2		
–		
–		

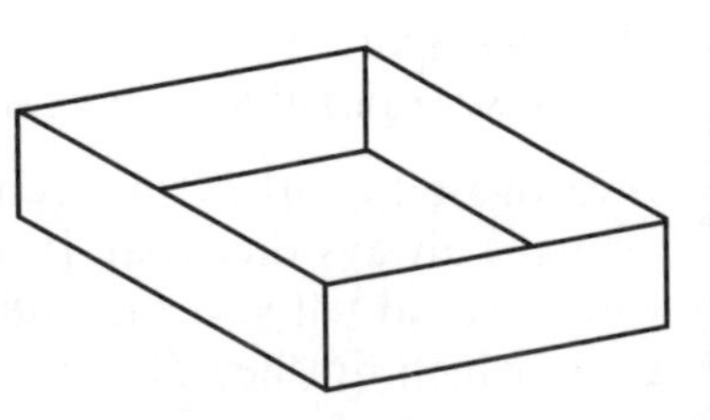

Now consider boxes made from different sized cards:
15 cm × 15 cm and 20 cm by 20 cm.
What size corners should you cut out this time so that the volume of the box is as large as possible?
Is there a connection between the size of the corners cut out and the size of the square card?

(b) Investigate the situation when the card is not square. Take rectangular cards where the length is twice the width (20×10, 12×6, 18×9 etc.)
Again, for the maximum volume is there a connection between the size of the corners cut out and the size of the original card?

6 Timetabling

(a) Every year a new timetable has to be written for the school. We will look at the problem of writing the timetable for one department (mathematics). The department allocates the teaching periods as follows:

U6	2 sets (at the same times); 8 periods in 4 doubles.
L6	2 sets (at the same times); 8 periods in 4 doubles.
Year 5	6 sets (at the same times); 5 single periods.
Year 4	6 sets (at the same times); 5 single periods.
Year 3	6 sets (at the same times); 5 single periods.
Year 2	6 sets (at the same times); 5 single periods.
Year 1	5 mixed ability forms; 5 single periods not necessarily at the same times.

Here are the teachers and the maximum number of maths periods which they can teach.

A	33	F	15	(Must be Years 5, 4, 3)
B	33	G	10	(Must be Years 2, 1)
C	33	H	10	(Must be Years 2, 1)
D	20	I	5	(Must be Year 3)
E	20			

Furthermore, to ensure some continuity of teaching, teachers B and C must teach the U6 and teachers A, B, C, D, E, F must teach year 5.

A timetable form which has been started is shown below.

M	5				U6 B, C	U6 B, C		
Tu		5	U6 B, C	U6 B, C				
W					5			
Th						5	U6 B, C	U6 B, C
F	U6 B, C	U6 B, C		5				

Your task is to write a complete timetable for the mathematics department subject to the restrictions already stated.

(b) If that was too easy, here are some changes.

U6 and L6 have 4 sets each (still 8 periods)
Two new teachers:
J 20 periods maximum
K 15 periods maximum but cannot teach on Mondays.

Because of games lessons: A cannot teach Wednesday afternoon
B cannot teach Tuesday afternoon
C cannot teach Friday afternoon

Also: A, B, C and E must teach U6
A, B, C, D, E, F must teach year 5
For the pupils, games afternoons are as follows:
Monday year 2; Tuesday year 3; Wednesday year 5 L6, U6;
Thursday year 4; Friday year 1.

7 Alphabetical order

A teacher has four names on a piece of paper which are in no particular order (say Smith, Jones, Biggs, Eaton). He wants the names in alphabetical order.
One way of doing this is to interchange each pair of names which are clearly out of order.
So he could start like this; S J B E
the order becomes: J S B E
He would then interchange S and B.

Using this method, what is the *largest* number of interchanges he could possibly have to make?
What if he had thirty names, or fifty?

8 What shape tin?

We need a cylindrical tin which will contain a volume of $600\,\text{cm}^3$ of drink.

What shape should we make the tin so that we use the minimum amount of metal?
In other words, for a volume of $600\,\text{cm}^3$, what is the smallest possible surface area?

Hint: Make a table.

r	h	A
2	?	?
3	?	?
⋮		

What shape tin should we design to contain a volume of $1000\,\text{cm}^3$?

9 Painting cubes

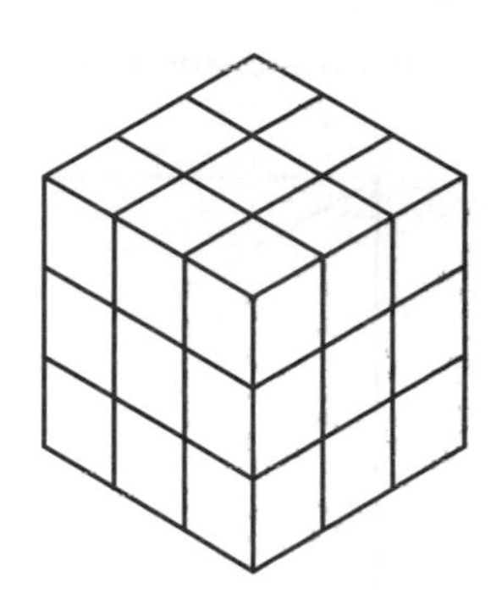

The large cube on the right consists of 27 unit cubes.

All six faces of the large cube are painted green.

- How many unit cubes have 3 green faces?
- How many unit cubes have 2 green faces?
- How many unit cubes have 1 green face?
- How many unit cubes have 0 green faces?

Answer the four questions for the cube which is $n \times n \times n$.

10 Discs

(a) You have five black discs and five white discs which are arranged in a line as shown.

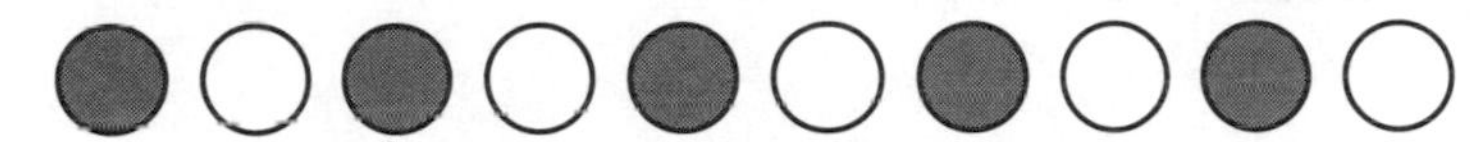

We want to get all the black discs to the right-hand end and all the white discs to the left-hand end.

The only move allowed is to interchange two neighbouring discs.

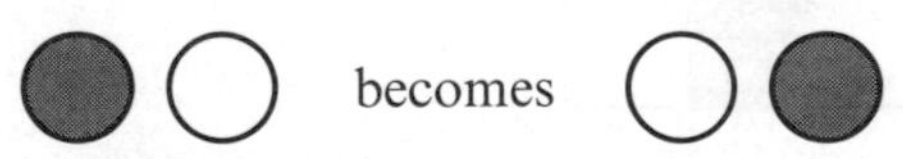

How many moves does it take?

How many moves would it take if we had fifty black discs and fifty white discs arranged alternately?

(b) Suppose the discs are arranged in pairs

How many moves would it take if we had fifty black discs and fifty white discs arranged like this?
[Hint: In both cases work with a smaller number of discs until you can see a pattern].

(c) Now suppose you have three colours black, white and green arranged alternately.

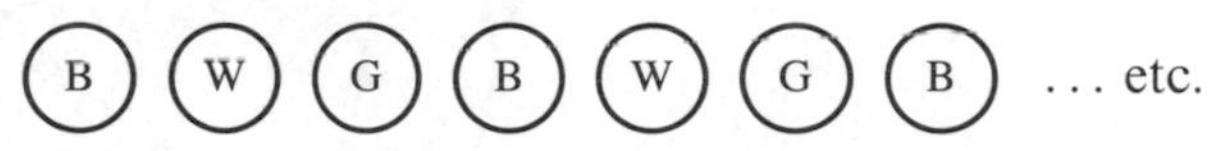

You want to get all the black discs to the right, the green discs to the left and the white discs in the middle.
How many moves would it take if you have 30 discs of each colour?

11 Diagonals

In a 4×7 rectangle, the diagonal passes through 10 squares.

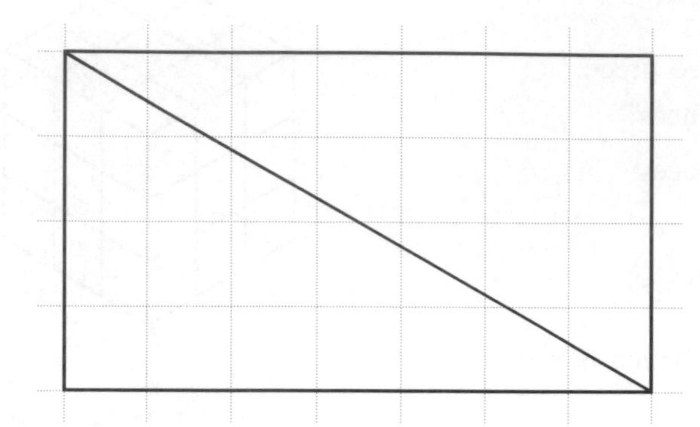

Draw rectangles of your own choice and count the number of squares through which the diagonal passes.
A rectangle is 640×250. How many squares will the diagonal pass through?

12 Chess board

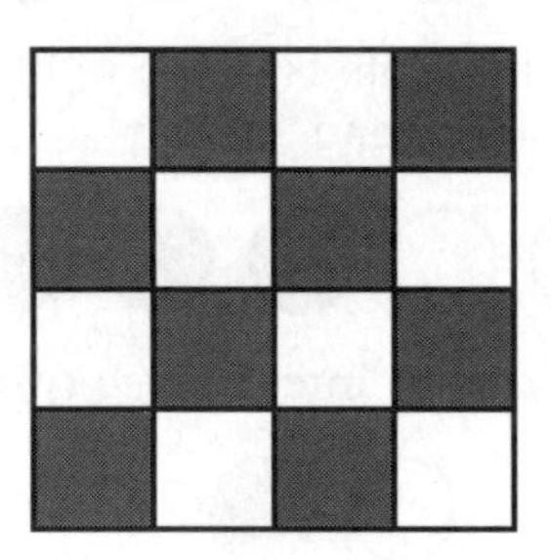

Start with a small board, just 4×4.
How many squares are there? [It is not just 16!]
How many squares are there on an 8×8 chess board?
How many squares are there on an $n \times n$ chess board?

13 Digit sum

Take the number 134.
Add the digits $1 + 3 + 4 = 8$. The digit sum of 134 is 8.

Take the number 238.
$2 + 3 + 8 = 13$ [We continue if the sum is more than 9].
$1 + 3 = 4$
The digit sum of 238 is 4.

Consider the multiples of 3:

Number	3	6	9	12	15	18	21	24	27	30	33	36
Digit sum	3	6	9	3	6	9	3	6	9	3	6	9

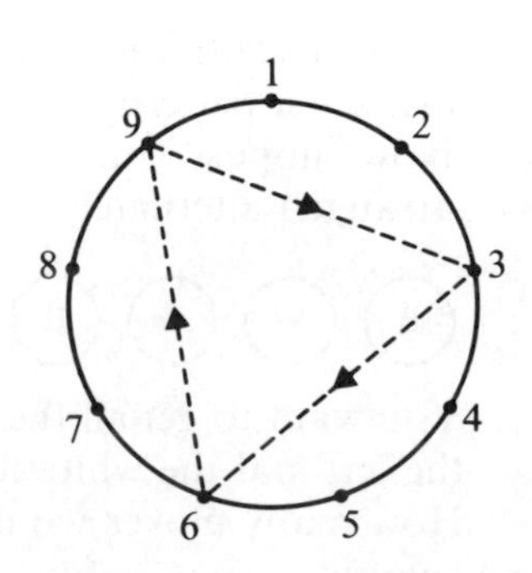

The digit sum is always 3, 6, or 9.

These numbers can be shown on the circle.

Investigate the pattern of the digit sums for multiples of:

(a) 2 (b) 4 (c) 5 (d) 6 (e) 7
(f) 8 (g) 9 (h) 11 (i) 12 (j) 13

Is there any connection between numbers where the pattern of the digit sums is the same?
Can you (without doing all the usual working) predict what the pattern would be for multiples of 43? Or 62?

14 Find the connection

Work through the flow diagram several times, using a calculator.

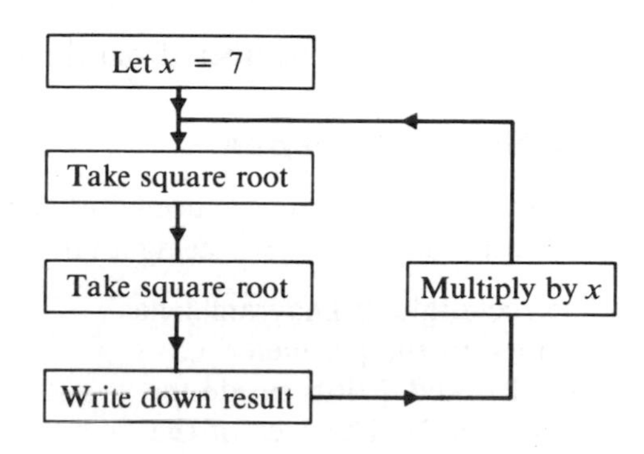

What do you notice?
Try different numbers for x (suggestions: 11, 5, 8, 27)
What do you notice?

What happens if you take the square root three times?

Suppose in the flow diagram you change 'Multiply by x' to 'Divide by x'. What happens now?

Suppose in the flow diagram you change 'Multiply by x' to 'Multiply by x^2'. What happens now?

You can vary the number of times you take the square root and whether you 'Multiply by x', 'Multiply by x^2', 'Divide by x' etc.

15 Fibonacci sequence

Fibonacci was the nickname of the Italian mathematician Leonardo de Pisa (A.D. 1170–1250). The sequence which bears his name has fascinated mathematicians for hundreds of years. You can if you like join the Fibonacci Association which was formed in 1963.
Here is the start of the sequence
1, 1, 2, 3, 5, 8, 13, 21, 34, 55, 89, 144, ...
There are no prizes for working out the next term!
The sequence has many interesting properties to investigate.
Here are a few suggestions.

(a) Add three terms.

$1 + 1 + 2$, $1 + 2 + 3$, etc.

Add four terms.

(b) Add squares of terms

$1^2 + 1^2$, $1^2 + 2^2$, $2^2 + 3^2$, ...

(c) Ratios

$\frac{1}{1} = 1$, $\frac{2}{1} = 2$, $\frac{3}{2} = 1{\cdot}5$, ...

(d) In fours | 2 3 5 8 |

$2 \times 8 = 16$, $3 \times 5 = 15$

(e) In threes | 3 5 8 |

$3 \times 8 = 24$, $5^2 = 25$

(f) Take a group of 10 consecutive terms. Compare the sum of the 10 terms with the seventh member of the group.

(g) In sixes | 1 1 2 3 5 8 |

square and add the first five numbers

$1^2 + 1^2 + 2^2 + 3^2 + 5^2 = 40$

$5 \times 8 = 40$.

Now try seven numbers from the sequence, or eight ...

16 Spotted shapes

For this investigation you need dotted paper. If you have not got any, you can make your own using a felt tip pen and squared paper.

The rectangle in Diagram 1 has 10 dots on the perimeter ($p = 10$) and 2 dots inside the shape ($i = 2$). The area of the shape is 6 square units ($A = 6$)

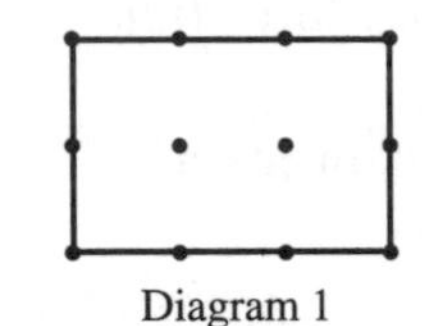

Diagram 1

The triangle in Diagram 2 has 9 dots on the perimeter ($p = 9$) and 4 dots inside the shape ($i = 4$). The area of the triangle is $7\frac{1}{2}$ square units ($A = 7\frac{1}{2}$)

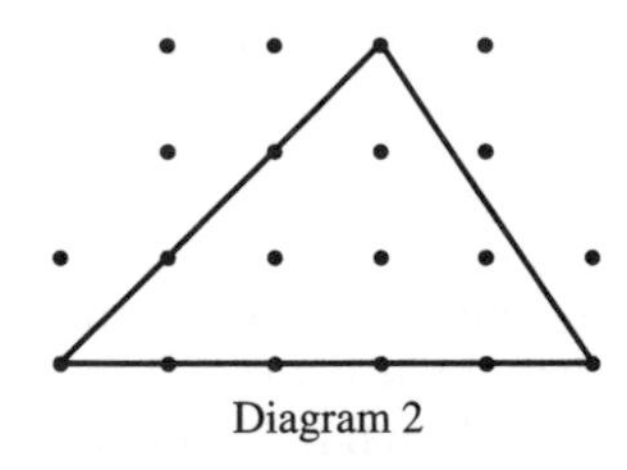

Diagram 2

Draw more shapes of your own design and record the values for p, i and A in a table. Make some of your shapes more difficult like the one in Diagram 3.

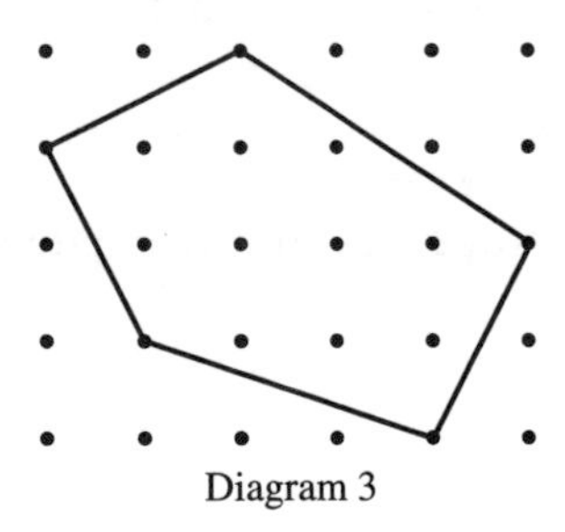

Diagram 3

This is quite difficult.
Be systematic, recording the values for i, p, A in a table. Start by drawing several shapes with $i = 0$. Find a connection between p and A. Now increase i by one at a time.
Try to find a formula connecting p, i and A.

17 Prime numbers

Write all the numbers from 1 to 104 in eight columns and draw a ring around the prime numbers 2, 3, 5 and 7.

1	②	③	4	⑤	6	⑦	8
9	10	11	12	13	14	15	16
17	18	19	20	21	22	23	24
25							

If we cross out all the multiples of 2, 3, 5 and 7, we will be left with all the prime numbers below 104. Can you see why this works?

Draw *four* lines to eliminate the multiples of 2.
Draw *six* lines to eliminate the multiples of 3.
Draw *two* lines to eliminate the multiples of 7.
Cross out all the numbers ending in 5.

Put a ring around all the prime numbers less than 104.
[Check there are 27 numbers].

Many prime numbers can be written as the sum of two squares. For example $5 = 2^2 + 1^2$, $13 = 3^2 + 2^2$. Find all the prime numbers in your table which can be written as the sum of two squares. Draw a red ring around them in the table. What do you notice?
Check any 'gaps' you may have found.

Extend the table up to 200 and see if the pattern continues. In this case you will need to eliminate the multiples of 11 and 13 as well.

11.2 Puzzles and games

1 Crossnumbers

(a) Copy out the crossnumber pattern.
(b) Fit all the given numbers into the correct spaces. Tick off the numbers from the lists as you write them in the square.

1.

2 digits	*3 digits*	*4 digits*	*5 digits*	*6 digits*
11	121	2104	14700	216841
17	147	2356	24567	588369
18	170	2456	25921	846789
19	174	3714	26759	861277
23	204	4711	30388	876452
31	247	5548	50968	
37	287	5678	51789	
58	324	6231	78967	
61	431	6789	98438	
62	450	7630		*7 digits*
62	612	9012		6645678
70	678	9921		
74	772			
81	774			
85	789			
94	870			
99				

2.

2 digits		*3 digits*	*4 digits*	*5 digits*	*6 digits*
12	47	129	2096	12641	324029
14	48	143	3966	23449	559641
16	54	298	5019	33111	956782
18	56	325	5665	33210	
20	63	331	6462	34509	
21	67	341	7809	40551	
23	81	443	8019	41503	
26	90	831	8652	44333	*7 digits*
27	91	923		69786	1788932
32	93			88058	5749306
38	98			88961	
39	99			90963	
46				94461	
				99654	

2 Estimating game

This is a game for two players. On squared paper draw an answer grid with the numbers shown.

Answer grid

891	7047	546	2262	8526	429
2548	231	1479	357	850	7938
663	1078	2058	1014	1666	3822
1300	1950	819	187	1050	3393
4350	286	3159	442	2106	550
1701	4050	1377	4900	1827	957

The players now take turns to choose two numbers from the question grid below and multiply them on a calculator.

Question grid

11	26	81
17	39	87
21	50	98

The game continues until all the numbers in the answer grid have been crossed out. The object is to get four answers in a line (horizontally, vertically or diagonally). The winner is the player with most lines of four. A line of *five* counts as *two* lines of four. A line of *six* counts as *three* lines of four.

3 The chess board problem

(a) On the 4×4 square, four objects have been placed, subject to the restriction that nowhere are there two objects on the same row, column or diagonal.

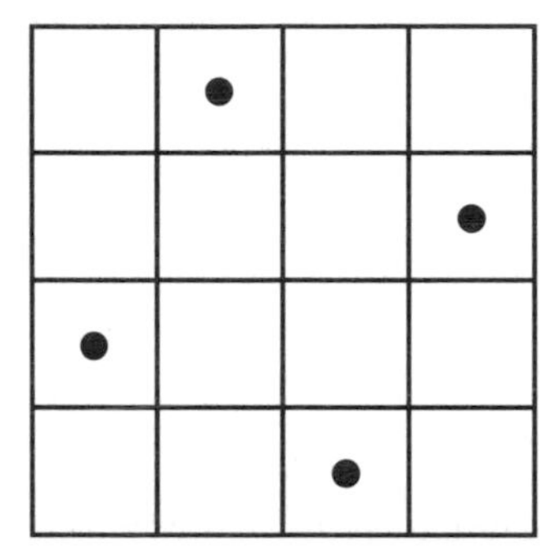

Subject to the same restrictions:
(i) find a solution for a 5×5 square, using five objects,
(ii) find a solution for a 6×6 square, using six objects,
(iii) find a solution for a 7×7 square, using seven objects,
(iv) find a solution for a 8×8 square, using eight objects.

It is called the chess board problem because the objects could be 'Queens' which can move any number of squares in any direction.

(b) Suppose we remove the restriction that no two Queens can be on the same row, column or diagonal. Is it possible to attack every square on an 8×8 chess board with less than eight Queens? Try the same problem with other pieces like knights or bishops.

4 Creating numbers

Using only the numbers 1, 2, 3 and 4 once each and the operations $+, -, \times, \div, !$ create every number from 1 to 100.
You can use the numbers as powers and you must use all of the numbers 1, 2, 3 and 4.
[4! is pronounced 'four factorial' and means $4 \times 3 \times 2 \times 1$ (i.e. 24)
similarly $3! = 3 \times 2 \times 1 = 6$
$5! = 5 \times 4 \times 3 \times 2 \times 1 = 120$]

Examples:
$1 = (4 - 3) \div (2 - 1)$
$20 = 4^2 + 3 + 1$
$68 = 34 \times 2 \times 1$
$100 = (4! + 1)(3! - 2!)$

5 Calculator words

On a calculator work out $9508^2 + 192^2 + 10^2 + 6$.
If you turn the calculator upside down and use a little imagination, you can see the word 'HEDGEHOG'.
Find the words given by the clues below.

1. $19 \times 20 \times 14 - 2{\cdot}66$ (Not an upstanding man)
2. $(84 + 17) \times 5$ (Dotty message)
3. $904^2 + 89621818$ (Prickly customer)
4. $(559 \times 6) + (21 \times 55)$ (What a surprise!)
5. $566 \times 711 - 23617$ (Bolt it down)
6. $\dfrac{9999 + 319}{8{\cdot}47 + 2{\cdot}53}$ (Sit up and plead)
7. $\dfrac{2601 \times 6}{4^2 + 1^2}$; $(401 - 78) \times 5^2$ (two words) (Not a great man)
8. $0{\cdot}4^2 - 0{\cdot}1^2$ (Little Sidney)
9. $\dfrac{(27 \times 2000 - 2)}{(0{\cdot}63 \div 0{\cdot}09)}$ (Not quite a mountain)
10. $(5^2 - 1^2)^4 - 14239$ (Just a name)
11. $48^4 + 102^2 - 4^2$ (Pursuits)
12. $615^2 + (7 \times 242)$ (Almost a goggle)
13. $(130 \times 135) + (23 \times 3 \times 11 \times 23)$ (Wobbly)
14. $164 \times 166^2 + 734$ (Almost big)
15. $8794^2 + 25 \times 342{\cdot}28 + 120 \times 25$ (Thin skin)
16. $0{\cdot}08 - (3^2 \div 10^4)$ (Ice house)
17. $235^2 - (4 \times 36{\cdot}5)$ (Shiny surface)
18. $(80^2 + 60^2) \times 3 + 81^2 + 12^2 + 3013$ (Ship gunge)
19. $3 \times 17 \times (329^2 + 2 \times 173)$ (Unlimited)
20. $230 \times 230\frac{1}{2} + 30$ (Fit feet)
21. $33 \times 34 \times 35 + 15 \times 3$ (Beleaguer)
22. $0{\cdot}32^2 + \frac{1}{1000}$ (Did he or didn't he?)
23. $(23 \times 24 \times 25 \times 26) + (3 \times 11 \times 10^3) - 20$ (Help)
24. $(16^2 + 16)^2 - (13^2 - 2)$ (Slander)
25. $(3 \times 661)^2 - (3^6 + 22)$ (Pester)
26. $(22^2 + 29{\cdot}4) \times 10$; $(3{\cdot}03^2 - 0{\cdot}02^2) \times 100^2$ (Four words) (Goliath)
27. $1{\cdot}25 \times 0{\cdot}2^6 + 0{\cdot}2^2$ (Tissue time)
28. $(710 + (1823 \times 4)) \times 4$ (Liquor)
29. $(3^3)^2 + 2^2$ (Wriggler)
30. $14 + (5 \times (83^2 + 110))$ (Bigger than a duck)
31. $2 \times 3 \times 53 \times 10^4 + 9$ (Opposite to hello, almost!)
32. $(177 \times 179 \times 182) + (85 \times 86) - 82$ (Good salesman)
33. $14^4 - 627 + 29$ (Good book, by God!)
34. $6{\cdot}2 \times 0{\cdot}987 \times 1\,000\,000 - 860^2 + 118$ (Flying ace)
35. $(426 \times 474) + (318 \times 487) + 22018$ (Close to a bubble)

36. $\dfrac{36^3}{4} - 1530$ (Foreign-sounding girl's name)
37. $(7^2 \times 100) + (7 \times 2)$ (Lofty)
38. $240^2 + 134$; $241^2 - 7^3$ (two words) (Devil of a chime)
39. $(2 \times 2 \times 2 \times 2 \times 3)^4 + 1929$ (Unhappy ending)
40. $141918 + 83^3$ (Hot stuff in France)

6 Operator squares

Each empty square contains either a number or a mathematical symbol (+, −, ×, ÷). Copy each square and fill in the missing details.

1.

0.5	−	0.01	→	
		×		
	×		→	35
↓		↓		
4	÷	0.1	→	

2.

	−	1.8	→	3.4
−		÷		
	×		→	
↓		↓		
	+	0.36	→	1

3.

	×	30	→	21
×		−		
	−		→	35
↓		↓		
	−	49	→	

4.

	×	−6	→	72
÷		+		
4	+		→	
↓		↓		
	+	1	→	−2

5.

	÷	4	→	
×		÷		
$\frac{1}{5}$	+		→	−3
↓		↓		
$-\frac{1}{25}$	−		→	1.21

6.

	+	0.21	→	1.07
		÷		
$\frac{1}{3}$			→	$-\frac{7}{24}$
↓		↓		
2.58	+		→	2.916

7.

	−	−17.2	→	10.7
÷		×		
$\frac{5}{8}$	÷		→	$\frac{1}{8}$
↓		↓		
		−86	→	75.6

8.

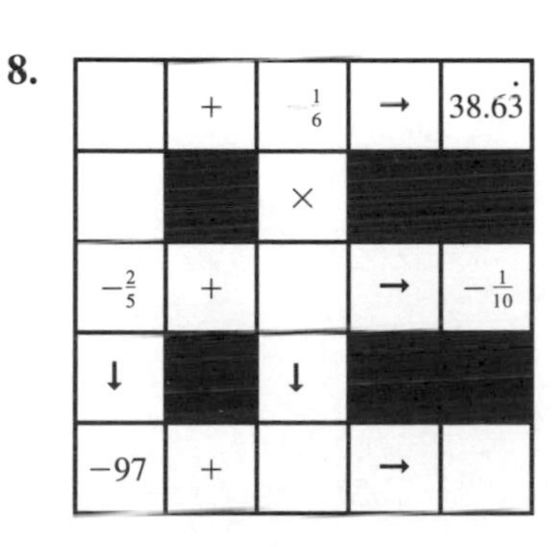

	+	$-\frac{1}{6}$	→	$38.6\dot{3}$
		×		
$-\frac{2}{5}$	+		→	$-\frac{1}{10}$
↓		↓		
−97	+		→	

7 A mad idea

In order to celebrate his mother's 100th birthday, Signor Gibsoni, the head of the largest mafia family in Sicily, decides to invite the entire population of the earth, which is about 5000 000 000 people, to a surprise party on the island.

The land area of Sicily is approximately 26 000 km^2.
Will there be room for everyone to meet on her birthday?
You may need to conduct an experiment in your class to find what area is required for say 20 or 30 pupils. What would be the effect of including babies and adults?
Make a list of some of the practical difficulties which the organisers would have to overcome.

12 Revision

12.1 Revision exercises

Revision exercise 1 Levels 7 and 8

1. $a = \frac{1}{2}$, $b = \frac{1}{4}$. Which one of the following has the greatest value?
(i) ab (ii) $a + b$ (iii) $\frac{a}{b}$
(iv) $\frac{b}{a}$ (v) $(ab)^2$

2. (a) Calculate the speed (in metres per second) of a slug which moves a distance of 30 cm in 1 minute.
(b) Calculate the time taken for a bullet to travel 8 km at a speed of 5000 m/s.
(c) Calculate the distance flown, in a time of four hours, by a pigeon which flies at a speed of 12 m/s.

3. Solve the simultaneous equations
(a) $7c + 3d = 29$, $5c - 4d = 33$ (b) $2x - 3y = 7$, $2y - 3x = -8$

4. Calculate the side or angle marked with a letter.

(a)
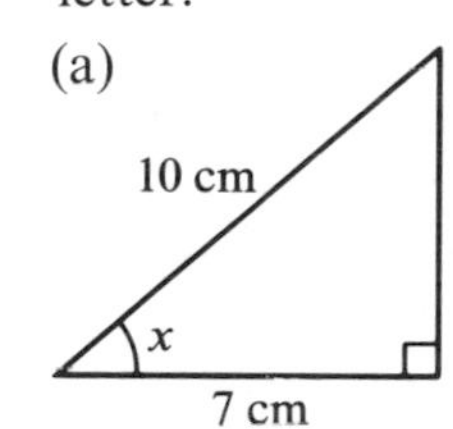

(b)
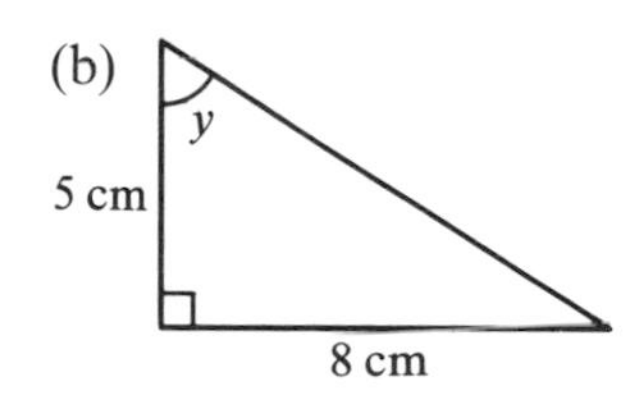

(c)
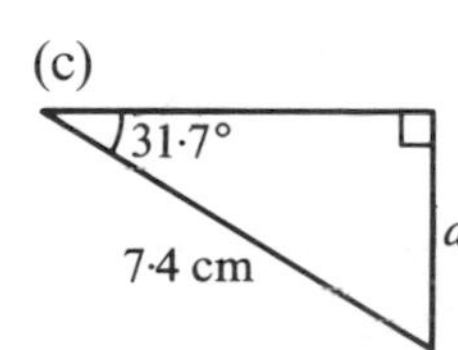

(d)
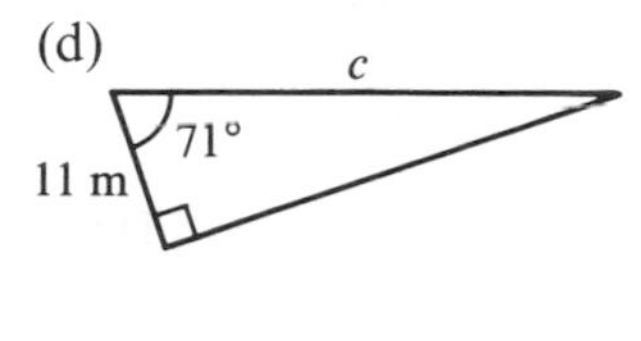

5. Given $a = 3$, $b = 4$ and $c = -2$, evaluate
(a) $2a^2 - b$ (b) $a(b - c)$ (c) $2b^2 - c^2$

6. When two dice are thrown simultaneously, what is the probability of obtaining the same number on both dice?

7. In Figure 1 a circle of radius 4 cm is inscribed in a square. In Figure 2 a square is inscribed in a circle of radius 4 cm. Calculate the shaded area in each diagram.

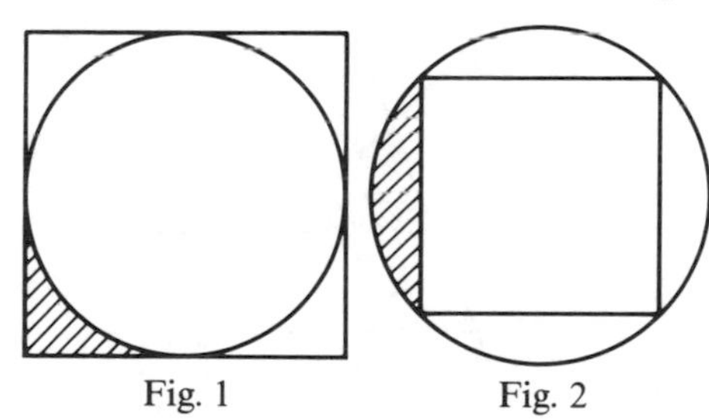
Fig. 1 Fig. 2

8. (a) A lies on a bearing of 040° from B. Calculate the bearing of B from A.
(b) The bearing of X from Y is 115°. Calculate the bearing of Y from X.

9. In the diagram, the equations of the lines are $y = 3x$, $y = 6$, $y = 10 - x$ and $y = \frac{1}{2}x - 3$.

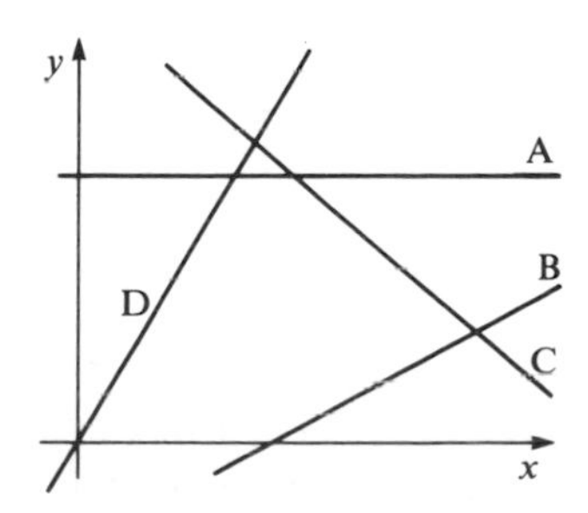

Find the equation corresponding to each line.

10. Given that $s - 3t = rt$, express
(a) s in terms of r and t
(b) r in terms of s and t

11. The mean height of 10 boys is 1·60 m and the mean height of 15 girls is 1·52 m. Find the mean height of the 25 boys and girls.

12. Find x.

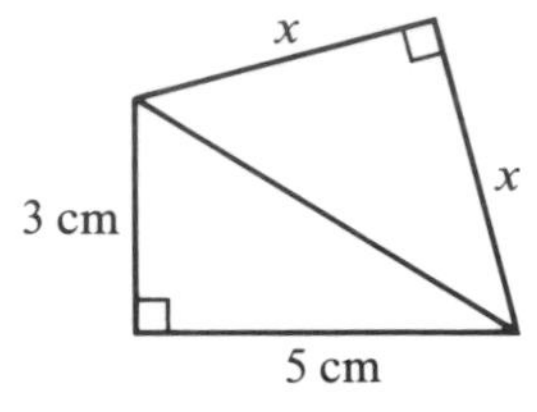

13. A cylinder of radius 8 cm has a volume of 2 litres. Calculate the cylinder height.

14. The shaded region A is formed by the lines $y = 2$, $y = 3x$ and $x + y = 6$. Write down the three inequalities which define A.

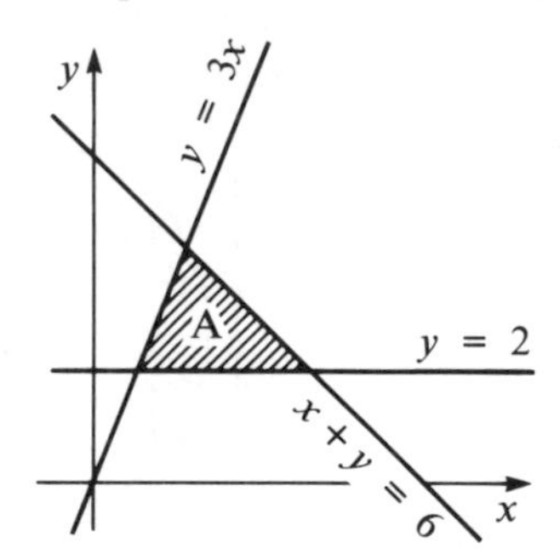

Revision exercise 2 Levels 7 and 8

1. The pump shows the price of petrol in a garage.

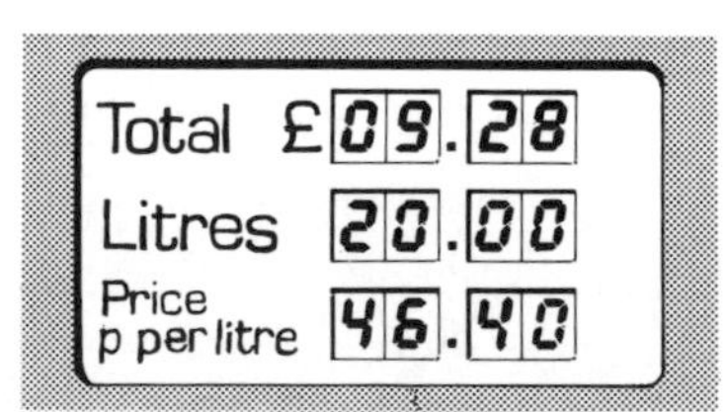

One day I buy £20 worth of petrol: How many litres do I buy?

2. Given that $x = 4$, $y = 3$, $z = -2$, evaluate
(a) $2x(y + z)$ (b) $(xy)^2 - z^2$
(c) $x^2 + y^2 + z^2$ (d) $(x + y)(x - z)$
(e) $\sqrt{[x(1 - 4z)]}$ (f) $\dfrac{xy}{z}$

3. (a) On a map, the distance between two points is 16 cm. Calculate the scale of the map if the actual distance between the points is 8 km.
(b) On another map, two points appear 1·5 cm apart and are in fact 60 km apart. Calculate the scale of the map.

4. Twenty-seven small wooden cubes fit exactly inside a cubical box without a lid. How many of the cubes are touching the sides or the bottom of the box?

5. The square has sides of length 3 cm and the arcs have centres at the corners. Find the shaded area.

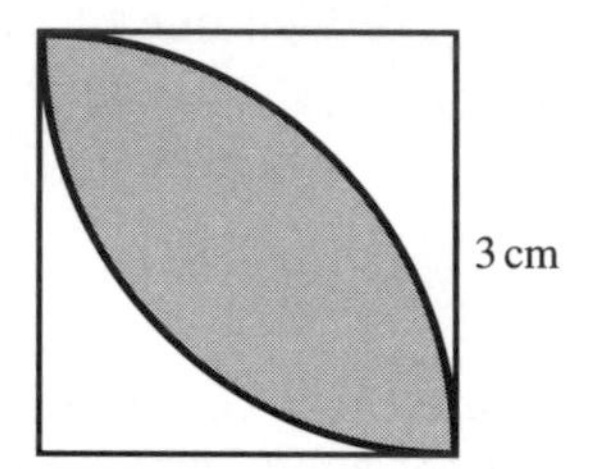

6. A coin is tossed four times. What is the probability of obtaining at least three 'heads'?

7. (a) Given that $x - z = 5y$, express z in terms of x and y.
(b) Given that $mk + 3m = 11$, express m in terms of k.
(c) For the formula $T = C\sqrt{z}$, express z in terms of T and C.

8. In a mixed school there are twice as many boys as girls and ten times as many girls as teachers. Using the letters b, g, t to represent the number of boys, girls and teachers, find an expression for the total number of boys, girls and teachers. Give your answer in terms of b only.

9. Calculate the length of AB.

10.

Marks	3	4	5	6	7	8
Number of pupils	2	3	6	4	3	2

The table shows the number of pupils in a class who scored marks 3 to 8 in a test. Find
(a) the mean mark,
(b) the modal mark,
(c) the median mark.

11. In the diagram, triangles ABC and EBD are similar but DE is *not* parallel to AC. Given that AD = 5 cm, DB = 3 cm and BE = 4 cm, calculate the length of BC.

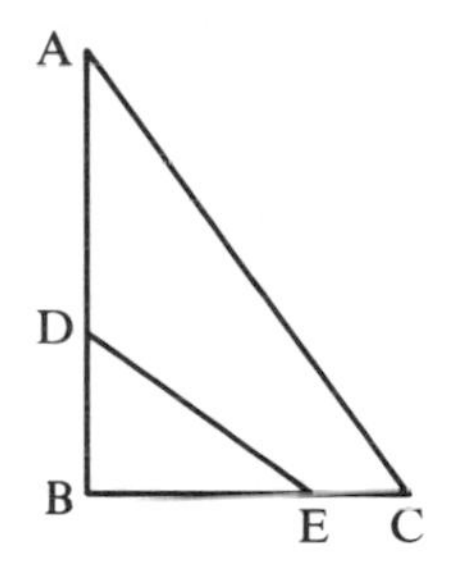

12. Draw the graph of $y = \dfrac{4000}{x} + 3x$ for $10 \leqslant x \leqslant 80$.
Find the minimum value of y.

Revision exercise 3 Levels 7 and 8

1. The mass of the planet Jupiter is about 350 times the mass of the Earth. The mass of the earth is approximately $6{\cdot}03 \times 10^{21}$ tonnes. Give an estimate correct to 2 significant figures for the mass of Jupiter.

2. Work out the difference between one ton and one tonne.

1 tonne	=	1000 kg
1 ton	=	2240 lb
1 lb	=	454 g

Give your answer to the nearest kg.

3. A target consists of concentric circles of radii 3 cm and 9 cm.

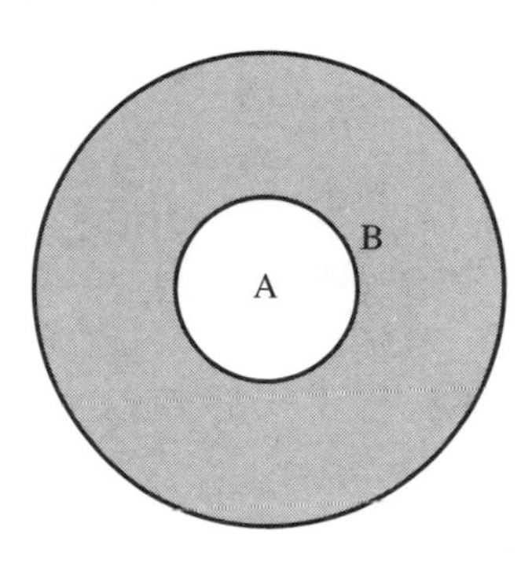

(a) Find the area of A, in terms of π

(b) Find the ratio $\dfrac{\text{area of B}}{\text{area of A}}$

4. A motorist travelled 200 miles in five hours. Her average speed for the first 100 miles was 50 m.p.h. What was her average speed for the second 100 miles?

5. Evaluate the following and give the answers to 3 significant figures:

(a) $\sqrt[3]{(9{\cdot}61 \times 0{\cdot}0041)}$

(b) $\left(\dfrac{1}{9{\cdot}5} - \dfrac{1}{11{\cdot}2}\right)^3$

(c) $\dfrac{15{\cdot}6 \times 0{\cdot}714}{0{\cdot}0143 \times 12}$ (d) $\sqrt[4]{\left(\dfrac{1}{5 \times 10^3}\right)}$

6. Throughout his life Mr Cram's heart has beat at an average rate of 72 beats per minute. Mr Cram is sixty years old. How many times has his heart beat during his life? Give the answer in standard form correct to two significant figures.

7. Two dice are thrown. What is the probability that the *product* of the numbers on top is
(a) 12, (b) 4, (c) 11?

8. The shaded region B is formed by the lines $x = 0$, $y = x - 2$ and $x + y = 7$.

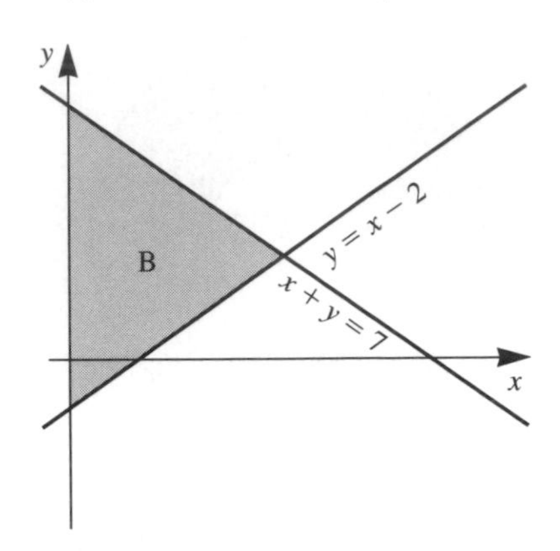

Write down the three inequalities which define B.

9. Estimate the answer correct to one significant figure. Do not use a calculator.

(a) $(612 \times 52) \div 49{\cdot}2$
(b) $(11{\cdot}7 + 997{\cdot}1) \times 9{\cdot}2$
(c) $\sqrt{\left(\dfrac{91{\cdot}3}{10{\cdot}1}\right)}$ (d) $\pi\sqrt{(5{\cdot}2^2 + 18{\cdot}2^2)}$

10. In the quadrilateral PQRS, PQ = QS = QR, PS is parallel to QR and $Q\hat{R}S = 70°$. Calculate

(a) $R\hat{Q}S$
(b) $P\hat{Q}S$.

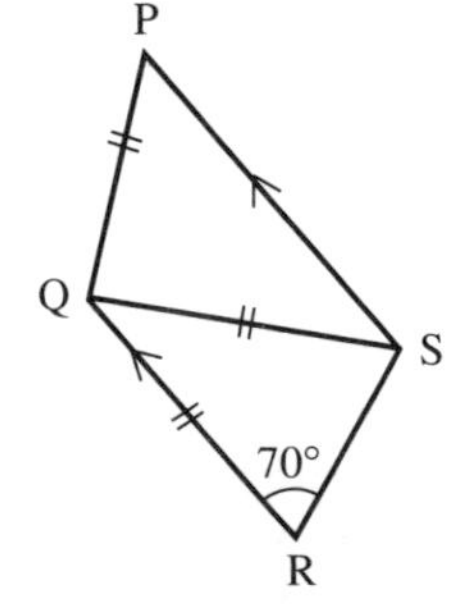

11. A bag contains x green discs and 5 blue discs. A disc is selected and replaced. A second disc is drawn. Find, in terms of x, the probability of selecting:

(a) a green disc on the first draw,
(b) a green disc on the first and second draws.

12. In the diagram, the equations of the lines are $2y = x - 8$, $2y + x = 8$, $4y = 3x - 16$ and $4y + 3x = 16$.

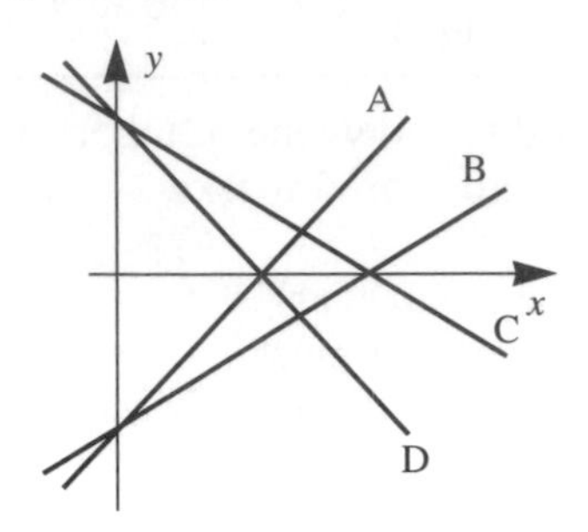

Find the equation corresponding to each line.

13. (a) A house is bought for £20 000 and sold for £24 400. What is the percentage profit?
(b) A piece of meat, initially weighing 2·4 kg, is cooked and subsequently weighs 1·9 kg. What is the percentage loss in weight?
(c) An article is sold at a 6% loss for £225·60. What was the cost price?

14. The edges of a cube are all increased by 10%. What is the percentage increase in the volume?

Revision exercise 4 Levels 7 and 8

1. Sainsburys sell their 'own-label' raspberry jam in two sizes.

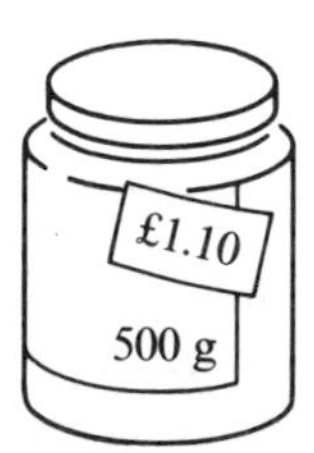

Which jar represents the better value for money? You are given that 1 kg = 2·20 lb.

2. (a) Convert into metres per second:
(i) 700 cm/s
(ii) 720 km/h
(iii) 18 km/h

(b) Convert into kilometres per hour:
(i) 40 m/s (ii) 0·6 m/s

3. Nadia said: 'I thought of a number, multiplied it by 6, then added 15. My answer was less than 200'.
   (a) Write down Nadia's statement in symbols, using x as the starting number.
   (b) Nadia actually thought of a prime number. What was the largest prime number she could have thought of?

4. In the diagram the area of the smaller square is $10\,\text{cm}^2$. Find the area of the larger square.

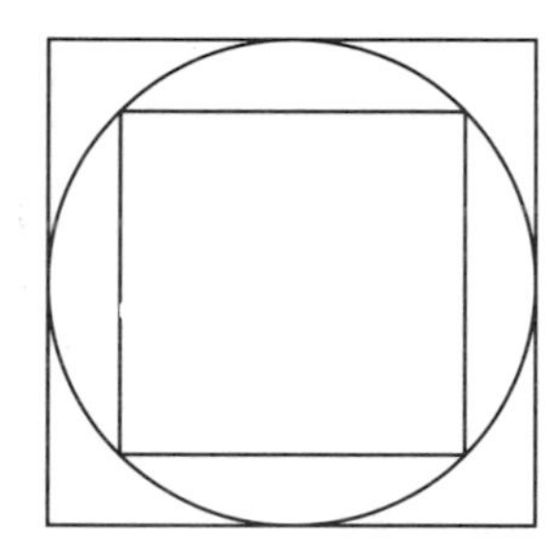

5. Given that x is an acute angle and that $3 \tan x - 2 = 4 \cos 35{\cdot}3°$ calculate
   (a) $\tan x$
   (b) the value of x in degrees correct to 1 D.P.

6. Solve the simultaneous equations
   (a) $3x + 2y = 5$, $2x - y = 8$
   (b) $2m - n = 6$, $2m + 3n = -6$

7. A regular octagon of side length 20 cm is to be cut out of a square card.

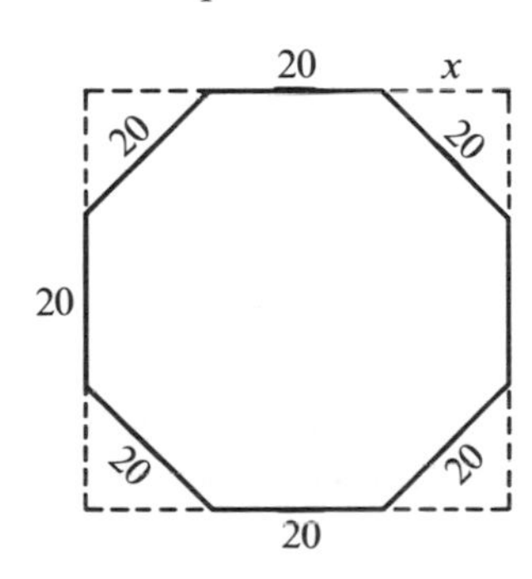

   (a) Find the length x and hence find the size of the smallest square card from which this octagon can be cut.
   (b) Calculate the area of the octagon, correct to 3 S.F.

8. Evaluate the following using a calculator: (answers to 4 sig. fig.)

   (a) $\dfrac{0{\cdot}74}{0{\cdot}81 \times 1{\cdot}631}$ (b) $\sqrt{\left(\dfrac{9{\cdot}61}{8{\cdot}34 - 7{\cdot}41}\right)}$

   (c) $\left(\dfrac{0{\cdot}741}{0{\cdot}8364}\right)^4$ (d) $\dfrac{8{\cdot}4 - 7{\cdot}642}{3{\cdot}333 - 1{\cdot}735}$

9. The mean of four numbers is 21.
   (a) Calculate the sum of the four numbers.
   Six other numbers have a mean of 18.
   (b) Calculate the mean of the ten numbers.

10. Given BD = 1 m, calculate the length AC.

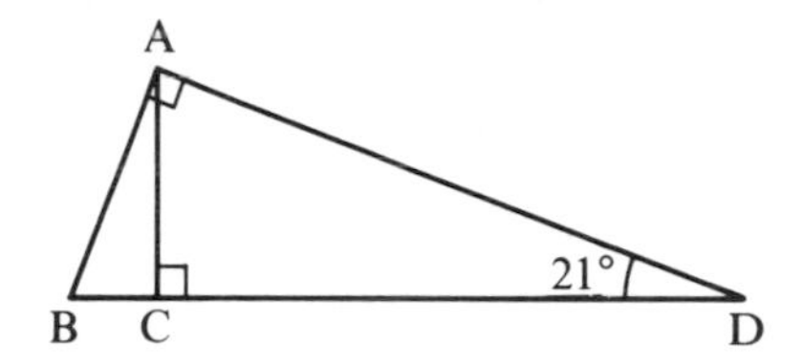

11. Use the method of trial and improvement to find a solution of the equation $x^5 = x^3 + 1$, giving your answer correct to 2 decimal places.

12. Given that $y = \dfrac{k}{k + w}$
   (a) Find the value of y when $k = \frac{1}{2}$ and $w = \frac{1}{3}$
   (b) Express w in terms of y and k.

13. It is given that $y = \dfrac{k}{x}$ and that $1 \leqslant x \leqslant 10$.
   (a) If the smallest possible value of y is 5, find the value of the constant k.
   (b) Find the largest possible value of y.

Revision exercise 5 Levels 9 and 10

1. The iteration formula for a sequence is
$$a_{n+1} = \frac{a_n}{5} + 2.$$
   (a) Take any value for u_1, and work out $a_2, a_3, a_4, a_5, \ldots$. Does the sequence appear to converge?
   (b) If so, what is the limit of the sequence?

2. Use the iteration formula $u_{n+1} = \dfrac{5}{u_n + 2}$ to find one solution of the equation $x^2 + 2x = 5$. Give your solution correct to one decimal place.

3. The letters r, h, a represent lengths. For each of the following formulas, state whether z is a length, an area, a volume or an impossible expression.
   (a) $z = r^2 + h^2$
   (b) $z = \pi h^2 a + rha$
   (c) $z = 3a^2 + h$
   (d) $z = \pi(r + h)$

4. Sketch the curve $y = \sin x$ for x from 0° to 360°. If sin 28° = 0·469, give another angle whose sine is 0·469.

5. Sketch the curve $y = \tan x$ for x from 0° to 360°. Find two solutions of the equation $\tan x = 1$.

6. The dimensions of the rectangle are correct to the nearest cm.

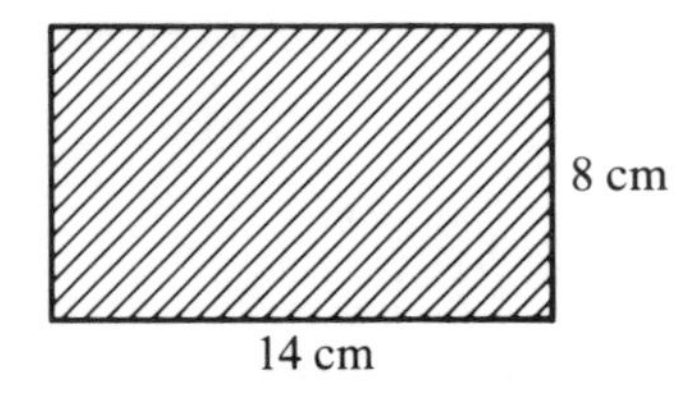

Give the maximum and minimum values for the area of the rectangle consistent with this data.

7. The fraction $\frac{139}{99}$ gives an approximate value for $\sqrt{2}$.
What is the percentage error in using this fraction? Give your answer correct to 2 S.F.

8. (a) Draw a priority table for this network of jobs.

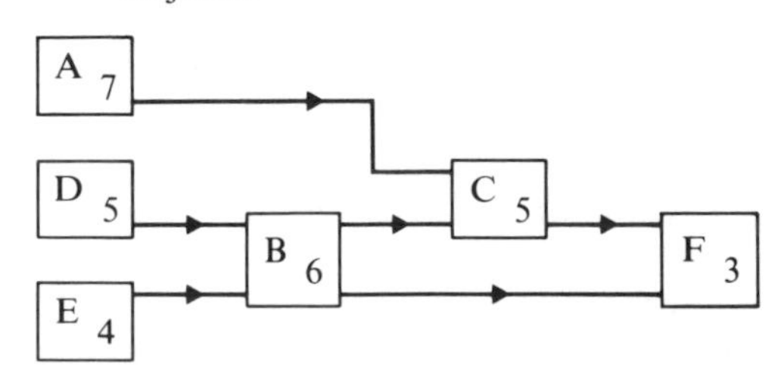

(b) The time in hours required for each job is shown inside the box. Find the critical path and the corresponding time.

9. Draw a histogram for the data below giving the ages of people at a disco.

Ages	Frequency
$14 \leqslant x < 16$	10
$16 \leqslant x < 17$	18
$17 \leqslant x < 18$	26
$18 \leqslant x < 21$	30
$21 \leqslant x < 26$	40

10. Solve these equations by factorising.
    (a) $x^2 = 2x + 15$
    (b) $x^2 + 12 = 8x$

11. Solve these equations, correct to 2 D.P.
    (a) $3x^2 - 5x - 11 = 0$
    (b) $(x - 2)^2 = 2x$

12. A formula for z is $z = ut - x^2$. The values of u, t and x are 5·2, 8·8 and 6·3 respectively, all correct to one decimal place. Work out the minimum possible value of z consistent with this data.

Revision exercise 6 Levels 9 and 10

1. A solution to the equation $x^2 - 4x - 4 = 0$ can be found by using the iteration formula $u_{n+1} = 2\sqrt{u_n + 1}$.
   Take $u_1 = 5$ and find a solution to the equation correct to one decimal place.

2. Sketch the curve $y = \cos x$ for x from 0° to 360°.
   (a) If cos 70° = 0·342, find another angle whose cosine is 0·342.
   (b) Find two values of x if $\cos x = 0{\cdot}5$.

3. Several pupils took a science test and the results are shown in the histogram below.

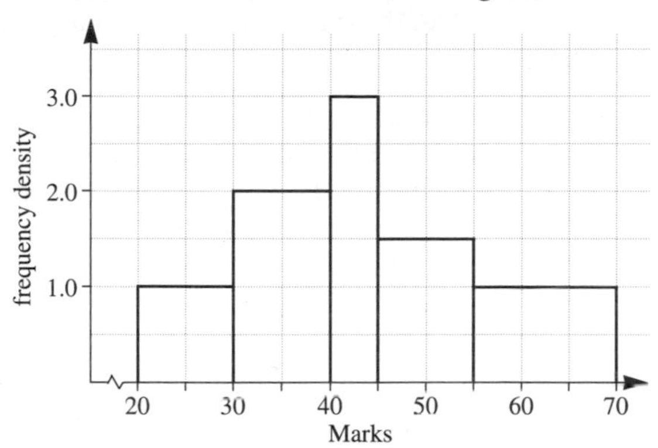

(a) If the pass mark was 45, how many pupils passed the test?
(b) How many took the test altogether?

4. Here is the priority table for a series of tasks that have to be done to complete a job.

Job	Time	Jobs which must be completed
A	4	–
B	5	–
C	6	A, B
D	7	–
E	5	D
F	8	–
G	4	A, B, C, D, E
H	3	A, B, C, D, E, F, G

(a) Draw a network to show how the jobs can be done.
(b) Find the critical path and the corresponding time for the whole job.

5. A cube of side 10 cm is melted down and made into five identical spheres. Calculate the radius of each sphere.

6. Solve the equation $4x^2 = 3x + 1$.

7. The dimensions of the cylinder are accurate to the nearest mm.

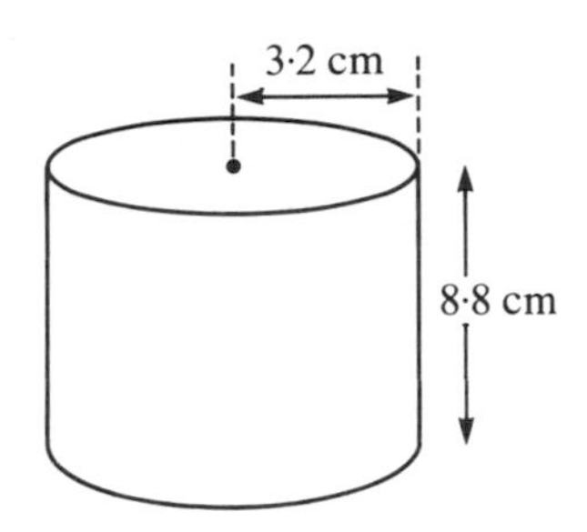

Work out the minimum possible volume of the cylinder. Give your answer to 3 S.F.

8. The figure shows a cube of side 10 cm.

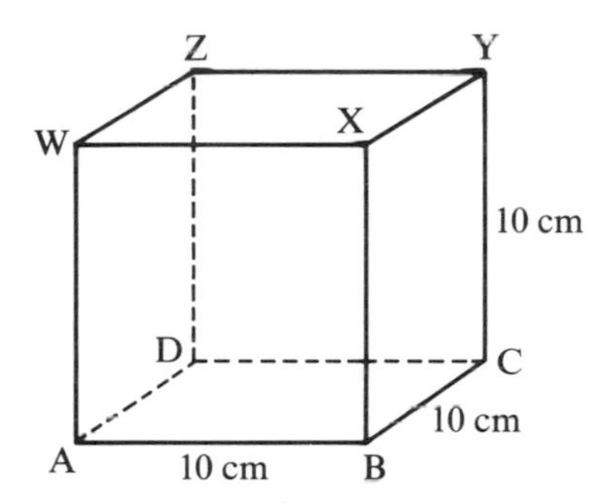

Calculate
(a) the length of AC
(b) the angle YAC

9. One solution of the equation $2x^2 - 7x + k = 0$ is $x = -\frac{1}{2}$. Find the value of k.

10. The diagram is the speed-time graph of a car.

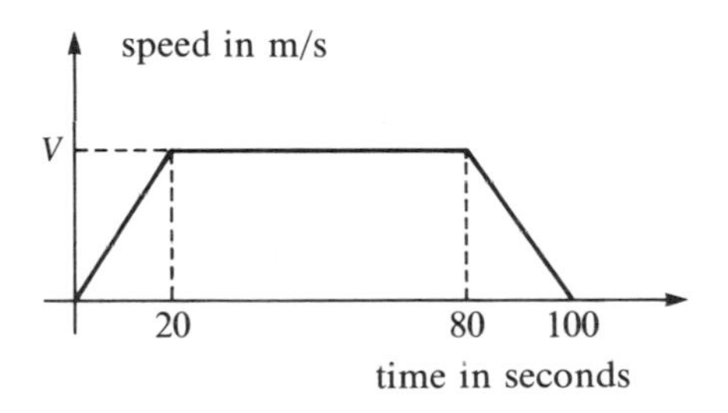

Given that the total distance travelled is 2·4 km, calculate

(a) the value of the maximum speed V,
(b) the distance travelled in the first 30 seconds of the motion.

11. A bag contains 8 balls of which 2 are red and 6 are white. A ball is selected and not replaced. A second ball is selected. Find the probability of obtaining:
    (a) two red balls, (b) two white balls,
    (c) one ball of each colour.

12. Draw the graph of $y = \frac{5}{x} + 2x - 3$, for $\frac{1}{2} \leqslant x \leqslant 7$, taking 2 cm to one unit for x and 1 cm to one unit for y.
    Use the graph to find
    (a) approximate solutions to the equation $\frac{5}{x} + 2x = 9$
    (b) the range of values of x for which $\frac{5}{x} + 2x - 3 < 6$.
    (c) the minimum value of y.

Revision exercise 7 Levels 9 and 10

1. A copper pipe has external diameter 18 mm and thickness 2 mm. The density of copper is 9 g/cm^3 and the price of copper is £150 per tonne. What is the cost of the copper in a length of 5 m of this pipe?

2. Solve the equations,
   (a) $4(y+1) = \frac{3}{1-y}$
   (b) $4(2x-1) - 3(1-x) = 0$
   (c) $\frac{x+3}{x} = 2$
   (d) $x^2 = 5x$

3. Given that OA = 10 cm and $A\hat{O}B = 70°$ (where O is the centre of the circle), calculate
   (a) the arc length AB
   (b) the area of minor sector AOB.

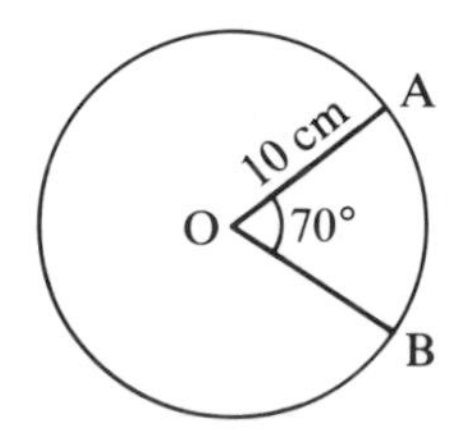

4. The radii of two spheres are in the ratio 2:5. The volume of the smaller sphere is 16 cm^3. Calculate the volume of the larger sphere.

5. The surface areas of two similar jugs are 50 cm^2 and 450 cm^2 respectively.
   (a) If the height of the larger jug is 10 cm, find the height of the smaller jug.
   (b) If the volume of the smaller jug is 60 cm^3, find the volume of the larger jug.

6. Find the angles marked with letters. (O is the centre of the circle.)

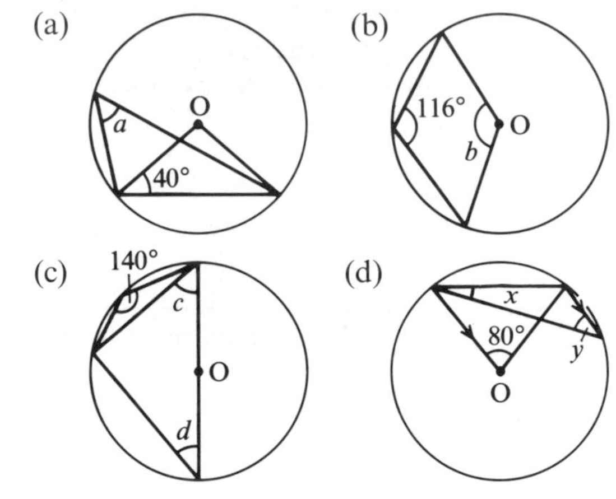

7. The matrix $\begin{pmatrix} 0 & -1 \\ 1 & 0 \end{pmatrix}$ represents the transformation **X**.
   (a) Find the image of (5, 2) under **X**.
   (b) Find the image of (−3, 4) under **X**.
   (c) Describe the transformation **X**.

8. The probability that it will be wet today is $\frac{1}{6}$. If it is dry today, the probability that it will be wet tomorrow is $\frac{1}{8}$. What is the probability that both today and tomorrow will be dry?

9. Given $\cos A\hat{C}B = 0{\cdot}6$, AC = 4 cm, BC = 5 cm and CD = 7 cm, find the length of AB and AD.

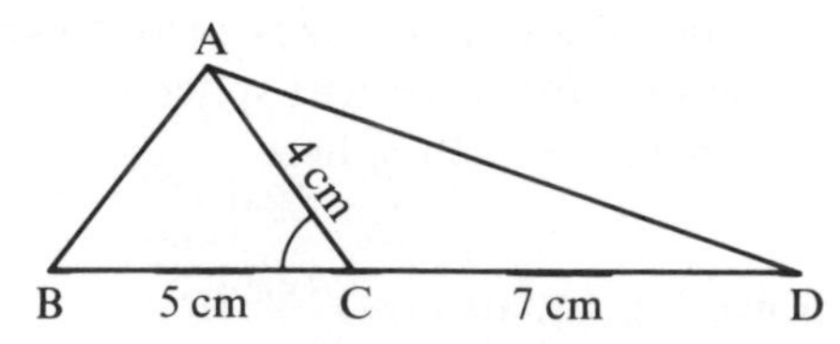

10. A sphere of radius 5 cm is melted down and made into a solid cube. Find the length of a side of the cube.

11. The sides of a right-angled triangle have lengths $(x-3)$ cm, $(x+11)$ cm and $2x$ cm, where $2x$ is the hypotenuse. Find x.

12. In the parallelogram OABC, M is the mid-point of AB and N is the mid-point of BC. If $\overrightarrow{OA} = \mathbf{a}$ and $\overrightarrow{OC} = \mathbf{c}$, express in terms of **a** and **c**.

(a) $\overrightarrow{CA}$ (b) $\overrightarrow{ON}$ (c) $\overrightarrow{NM}$

Describe the relationship between CA and NM.

Revision exercise 8 Levels 9 and 10

1. The diagram is the speed-time graph of a bus.

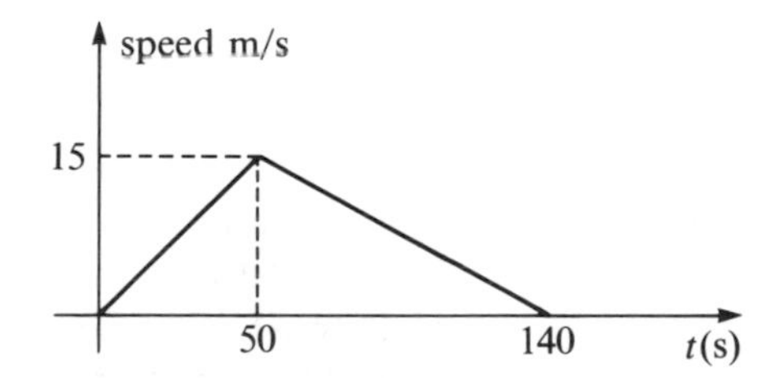

Calculate

(a) the acceleration during the first 50 seconds,

(b) the total distance travelled,

(c) how long it takes before it is moving at 12 m/s for the first time.

2. A car is an enlargement of a model, the scale factor being 10.

(a) If the windscreen of the model has an area of 100 cm^2, find the area of the windscreen on the actual car (answer in m^2).

(b) If the capacity of the boot of the car is 1 m^3, find the capacity of the boot on the model (answer in cm^3).

3. ABCD is a parallelogram and AE bisects angle A. Prove that DE = BC.

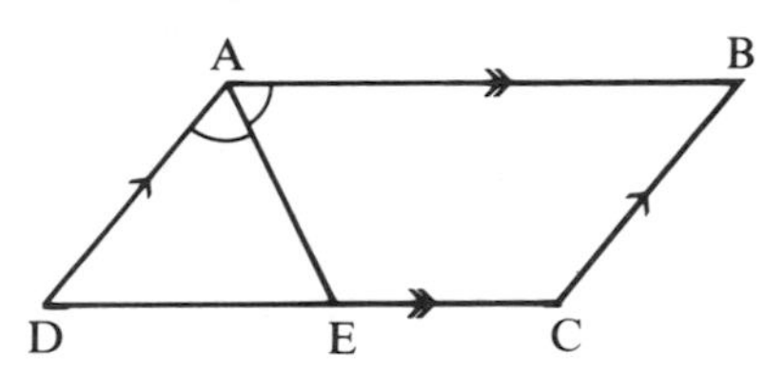

4. Without using a calculator, work out

(a) $9^{-\frac{1}{2}} + (\frac{1}{8})^{\frac{1}{3}} + (-3)^0$

(b) $(1000)^{-\frac{1}{3}} - (0{\cdot}1)^2$

5. Two lighthouses A and B are 25 km apart and A is due West of B. A submarine S is on a bearing of 137° from A and on a bearing of 170° from B. Find the distance of S from A and the distance of S from B.

6. The diagram shows a rectangular block. AY = 12 cm, AB = 8 cm, BC = 6 cm.

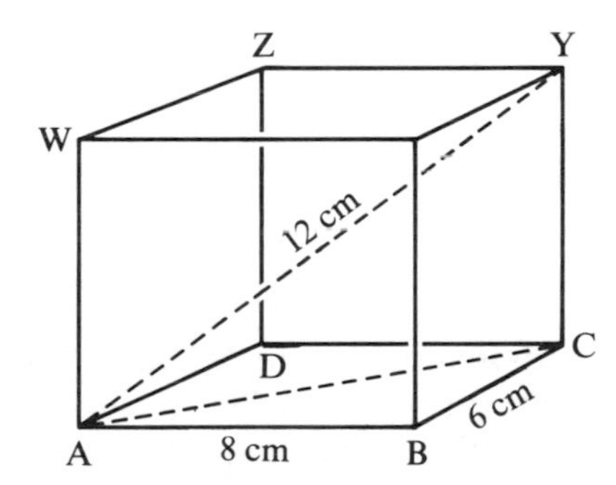

Calculate

(a) the length YC

(b) the angle $Y\hat{A}Z$

7. Here is the graph of $y = f(x)$.

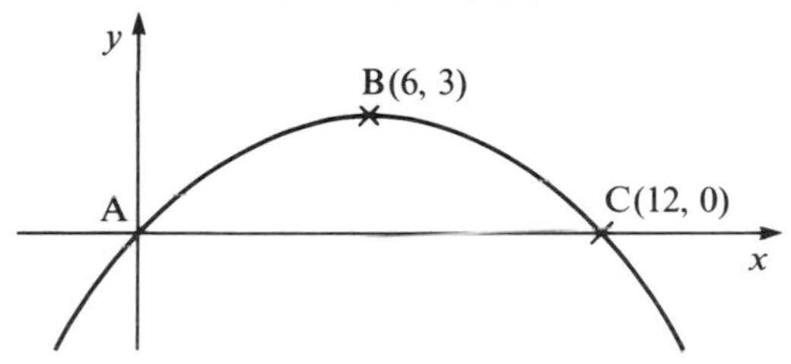

Draw three sketches to show

(a) $y = f(x) - 3$

(b) $y = f(x-3)$

(c) $y = f(3x)$

Give the new coordinates of A, B and C on each sketch.

8. A school has 958 pupils. You are going to conduct a survey about pupils' reactions to a proposed change to the length of the lunch break.
Describe how you would choose a sample of pupils to question.

9. Describe the single transformation which maps
(a) $\triangle$ABC onto $\triangle$DEF
(b) $\triangle$ABC onto $\triangle$PQR
(c) $\triangle$ABC onto $\triangle$XYZ

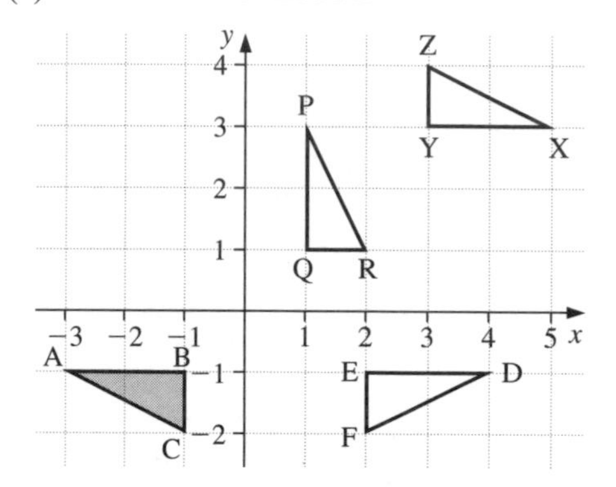

10. Transformation **N**, which is given by

$$\begin{pmatrix} x' \\ y' \end{pmatrix} = \begin{pmatrix} 2 & 0 \\ 0 & 2 \end{pmatrix} \begin{pmatrix} x \\ y \end{pmatrix} + \begin{pmatrix} 5 \\ -2 \end{pmatrix},$$

is composed of two single transformations.
(a) Describe each of the transformations,
(b) Find the image of the point $(3, -1)$ under **N**,
(c) Find the image of the point $(-1, \frac{1}{2})$ under **N**,
(d) Find the point which is mapped by **N** onto the point $(7, 4)$.

11. ABCD is a rectangle, where AB $= x$ cm and BC is 1·5 cm less than AB.

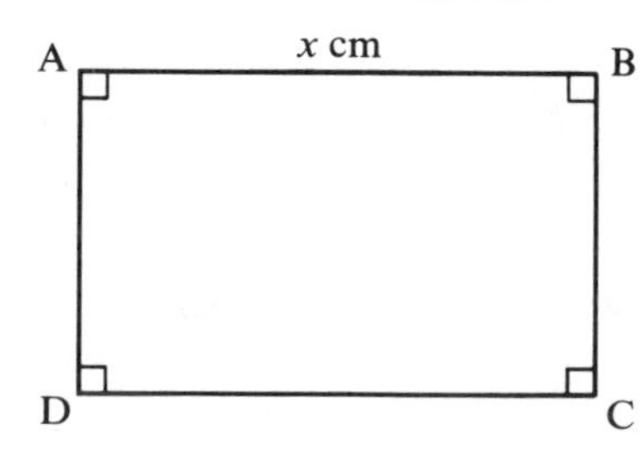

If the area of the rectangle is $52\,\text{cm}^2$, form an equation in x and solve it to find the dimensions of the rectangle.

12. Find the mean and standard deviation of these numbers.
5, 17, 23, 11, 4, 8, 15, 32, 9, 12.

Revision exercise 9

Which of the following statements are true?

1. All prime numbers are odd numbers.

2. Every positive integer greater than 10 has an even number of factors.

3. If $x^2 = x$, then x must be the number 1.

4. The translation $\begin{pmatrix} 0 \\ 0 \end{pmatrix}$ is the only transformation that leaves *any* shape completely unchanged.

5. All non-negative numbers are positive.

6. If we add a given number to both the numerator and denominator of a fraction, then the new fraction is equivalent to the original fraction.

7. If both the numerator and denominator of a fraction are squared, then the new fraction is equivalent to the original fraction.

8. If we multiply both the numerator and denominator of a fraction by a given non-zero number, then the new fraction is equivalent to the original fraction.

9. $a \times (b \times c) = (a \times b) \times c$.

10. $a \div (b \div c) = (a \div b) \div c$.

11. All mathematical curves cross the x-axis or the y-axis or both.

12. If a quadrilateral has exactly 2 lines of symmetry, then it must be a rectangle.

13. Except for 1, no cube number is also a square number.

14. x^2 is never equal to $5x + 14$.

15. If n is a positive integer, then $n^2 + n + 5$ is a prime number.

16. If m and n are positive integers, then $6m + 4n + 13$ is an odd number.

17. No square number differs from a cube number by exactly 2.

18. A polygon having all its sides equal is a *regular* polygon.

19. If $a^2 = 7^2$, then a must be 7.

20. $\dfrac{a+b}{c} = \dfrac{a}{c} + \dfrac{b}{c}$.

21. $\dfrac{a}{b+c} = \dfrac{a}{b} + \dfrac{a}{c}$.

22. An enlargement always changes the area of a shape, unless the scale factor of the enlargement is 1.

23. $a \times (b + c) = ab + ac$.

24. $a \times (b \times c) = ab \times ac$.

25. $(a + b)^3 = a^3 + b^3$.

26. $\sqrt{x+y} = \sqrt{x} + \sqrt{y}$.

27. x^2 is never less than x.

28. $(a + b)(c + d) = ac + bd$.

29. 2^x is always positive.

30.

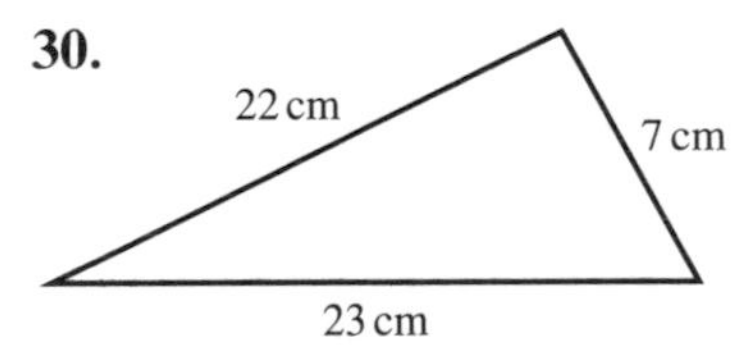

This triangle (not drawn to scale) is a *right-angled* triangle.

31. Suppose that we accurately draw the straight line $y = x$ on a set of axes. Then if we take a protractor and measure the acute angle between the line and the x-axis, it will be found to be 45°.

32. If we have an unlimited supply of 5p, 7p and 11p stamps, then we can make up any amount above 13p from just these stamps.

33. If n is a positive integer greater than 1, then $2^n < n^3$.

34. There is exactly one point which lies on *both* of the straight lines $y = \frac{1}{2}x + 5$ and $x - 2y = 3$.

35. If n is a positive integer, then $n^4 - 10n^3 + 35n^2 - 48n + 24$ is equal to $2n$.

36. If the number x is multiplied by 78·39, and the result is then divided by 78·39, then the final answer is x.

37. If the number x is increased by 8·3%, and the result is then decreased by 8·3%, then the final answer is x.

38. Written as a fraction, π is $\frac{22}{7}$.

39. If triangle ABC is isosceles, then $A\hat{B}C = A\hat{C}B$.

40. If the product of two numbers is 8, then one of the numbers must be 8.

41. If the product of two numbers is 0, then one of the numbers must be 0.

42. A cuboid has 6 faces, 12 edges, and 8 vertices.

43. A pyramid has 5 faces, 8 edges, and 5 vertices.

44. The number 133! (i.e. the number obtained by working out the product $1 \times 2 \times 3 \times 4 \times 5 \times \ldots\ldots \times 132 \times 133$) ends in exactly 26 noughts.

12.2 Multiple choice tests

Test 1

1. How many mm are there in 1 m 1 cm?

A 1001
B 1110
C 1010
D 1100

2. The circumference of a circle is 16π cm. The radius, in cm, of the circle is:

A 2
B 4
C $\frac{4}{\pi}$
D 8

3. In the triangle below the value of $\cos x$ is:

A 0·8
B 1·333
C 0·75
D 0·6

4. The line $y = 2x - 1$ cuts the x-axis at P. The coordinates of P are:

A $(0, -1)$
B $(\frac{1}{2}, 0)$
C $(-\frac{1}{2}, 0)$
D $(-1, 0)$

5. The formula $b + \frac{x}{a} = c$ is rearranged to make x the subject. What is x?

A $a(c - b)$
B $ac - b$
C $\frac{c - b}{a}$
D $ac + ab$

6. The mean weight of a group of 11 men is 70 kg. What is the mean weight of the remaining group when a man of weight 90 kg leaves?

A 80 kg
B 72 kg
C 68 kg
D 62 kg

7. Find x if $2^{x+2} = 16^{x-7}$

A 8
B 9
C 10
D 11

8. In standard form the value of $2000 \times 80\,000$ is:

A 16×10^6
B $1{\cdot}6 \times 10^9$
C $1{\cdot}6 \times 10^7$
D $1{\cdot}6 \times 10^8$

9. The solutions of the equation $(x - 3)(2x + 1) = 0$ are:

A $-3, \frac{1}{2}$
B $3, -2$
C $3, -\frac{1}{2}$
D $-3, -2$

10. In the triangle the size of angle x is:

A 35°
B 70°
C 110°
D 40°

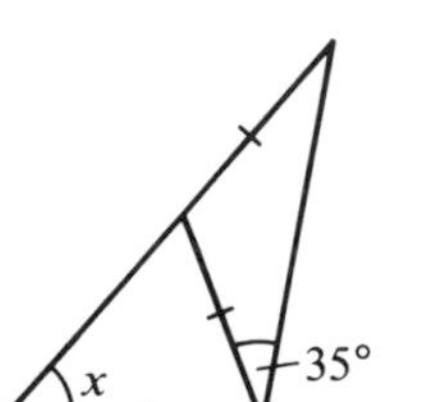

11. A man paid tax on £9000 at 30%. He paid the tax in 12 equal payments. Each payment was:

A £2·25
B £22·50
C £225
D £250

12. The approximate value of $\frac{3{\cdot}96 \times (0{\cdot}5)^2}{97{\cdot}1}$ is:

A 0·01
B 0·02
C 0·04
D 0·1

13. Given that $\frac{3}{n} = 5$, then $n =$

A 2
B -2
C $1\frac{2}{3}$
D 0·6

14. Cube A has side 2 cm. Cube B has side 4 cm. $\left(\frac{\text{Volume of B}}{\text{Volume of A}}\right) =$

A 2
B 4
C 8
D 16

15. How many tiles of side 50 cm will be needed to cover the floor shown?

A 16
B 32
C 64
D 84

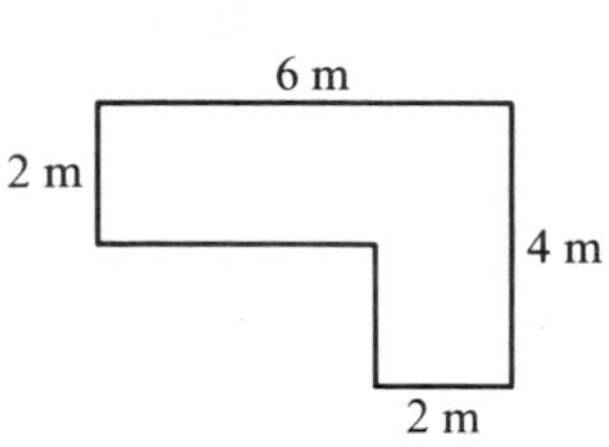

16. The equation $ax^2 + x - 6 = 0$ has a solution $x = -2$ What is a?

A 1
B -2
C $\sqrt{2}$
D 2

17. Which of the following is/are correct?
1. $\sqrt{0{\cdot}16} = 0{\cdot}4$
2. $0{\cdot}2 \div 0{\cdot}1 = 0{\cdot}2$
3. $\frac{4}{7} > \frac{3}{5}$

A **1** only
B **2** only
C **3** only
D **1** and **2**

18. How many prime numbers are there between 30 and 40?

A 0
B 1
C 2
D 3

19. A man is paid £180 per week after a pay rise of 20%. What was he paid before?

A £144
B £150
C £160
D £164

20. A car travels for 20 minutes at 45 m.p.h. and then for 40 minutes at 60 m.p.h. The average speed for the whole journey is:

A $52\frac{1}{2}$ m.p.h.
B 50 m.p.h
C 54 m.p.h.
D 55 m.p.h.

21. The point $(3, -1)$ is reflected in the line $y = 2$. The new coordinates are:

A $(3, 5)$
B $(1, -1)$
C $(3, 4)$
D $(0, -1)$

22. Two discs are randomly taken from a bag containing 3 red discs and 2 blue discs. What is the probability of taking 2 red discs?

A $\frac{9}{25}$
B $\frac{1}{10}$
C $\frac{3}{10}$
D $\frac{2}{5}$

23. The shaded area, in cm^2, is:

A $16 - 2\pi$
B $16 - 4\pi$
C $\frac{4}{\pi}$
D $64 - 8\pi$

24. Given the equation $5^x = 120$, the best approximate solution is $x =$

A 2
B 3
C 4
D 25

25. What is the sine of 45°?

A 1
B $\frac{1}{2}$
C $\frac{1}{\sqrt{2}}$
D $\sqrt{2}$

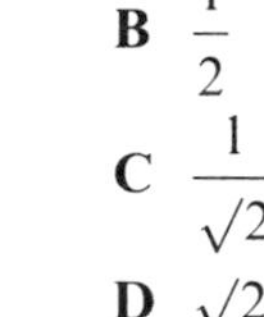

Test 2

1. What is the value of the expression $(x - 2)(x + 4)$ when $x = -1$?

A 9
B -9
C 5
D -5

2. The perimeter of a square is 36 cm. What is its area?

A 36 cm^2
B 324 cm^2
C 81 cm^2
D 9 cm^2

3. AB is a diameter of the circle. Find the angle BCO.

A 70°
B 20°
C 60°
D 50°

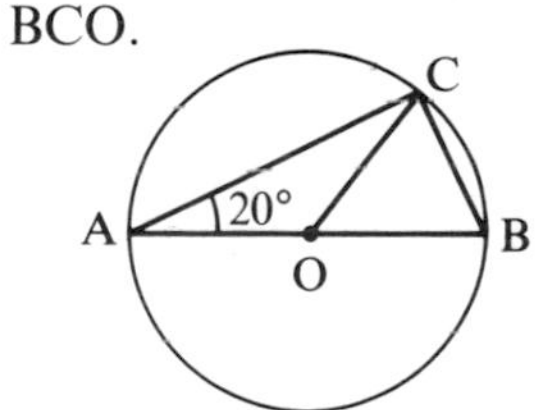

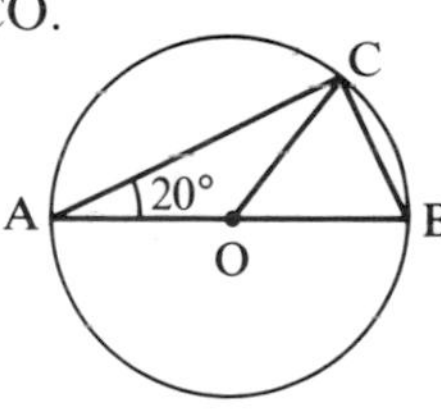

4. The gradient of the line $2x + y = 3$ is:

A 3
B -2
C $\frac{1}{2}$
D $-\frac{1}{2}$

5. A firm employs 1200 people, of whom 240 are men. The percentage of employees who are men is:

A 40%
B 10%
C 15%
D 20%

6. A car is travelling at a constant speed of 30 m.p.h. How far will the car travel in 10 minutes?

A $\frac{1}{3}$ mile
B 3 miles
C 5 miles
D 6 miles

7. What are the coordinates of the point $(1, -1)$ after reflection in the line $y = x$?

A $(-1, 1)$
B $(1, 1)$
C $(-1, -1)$
D $(1, -1)$

8. $\frac{1}{3} + \frac{2}{5} =$

A $\frac{2}{8}$
B $\frac{3}{8}$
C $\frac{3}{15}$
D $\frac{11}{15}$

9. In the triangle the size of the largest angle is:

A 30°
B 90°
C 120°
D 80°

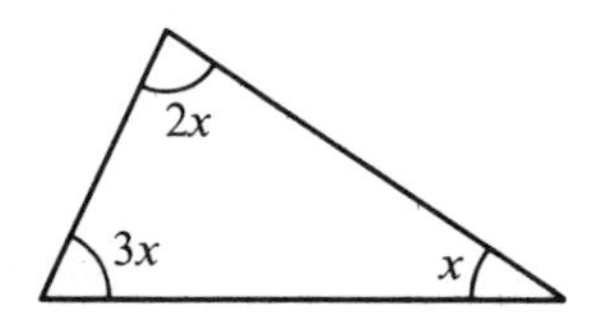

10. 800 decreased by 5% is:

A 795
B 640
C 760
D 400

11. Which of the statements is (are) true?
1. $\tan 60° = 2$
2. $\sin 60° = \cos 30°$
3. $\sin 30° > \cos 30°$

A **1** only
B **2** only
C **3** only
D **2** and **3**

12. Given $a = \frac{3}{5}$, $b = \frac{1}{3}$, $c = \frac{1}{2}$ then

A $a < b < c$
B $a < c < b$
C $a > b > c$
D $a > c > b$

13. The *larger* angle between South-West and East is:

A 225°
B 240°
C 135°
D 315°

14. In a triangle PQR, $P\hat{Q}R = 50°$ and point X lies on PQ such that QX = XR. Calculate $Q\hat{X}R$.

A 100°
B 50°
C 80°
D 65°

15. What is the value of $1 - 0{\cdot}05$ as a fraction?

A $\frac{1}{20}$
B $\frac{9}{10}$
C $\frac{19}{20}$
D $\frac{5}{100}$

16. Find the length x.

A 5
B 6
C 8
D $\sqrt{50}$

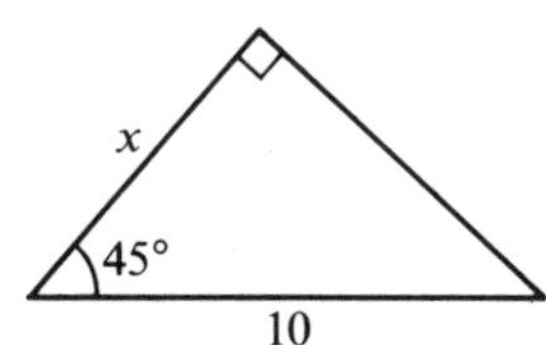

17. Given that $m = 2$ and $n = -3$, what is mn^2?

A -18
B 18
C -36
D 36

18. The graph of $y = (x - 3)(x - 2)$ cuts the y-axis at P. The coordinates of P are:

A (0, 6)
B (6, 0)
C (2, 0)
D (3, 0)

19. £240 is shared in the ratio 2 : 3 : 7. The largest share is:

A £130
B £140
C £150
D £160

20. Adjacent angles in a parallelogram are $x°$ and $3x°$. The smallest angles in the parallelogram are each:

A 30° **B** 45° **C** 60° **D** 120°

21. When the sides of a square are increased by 10% the area is increased by:

A 10% **B** 20% **C** 21% **D** 15%

22. The volume, in cm^3, of the cylinder is:

A 9π **B** 12π **C** 600π **D** 900π

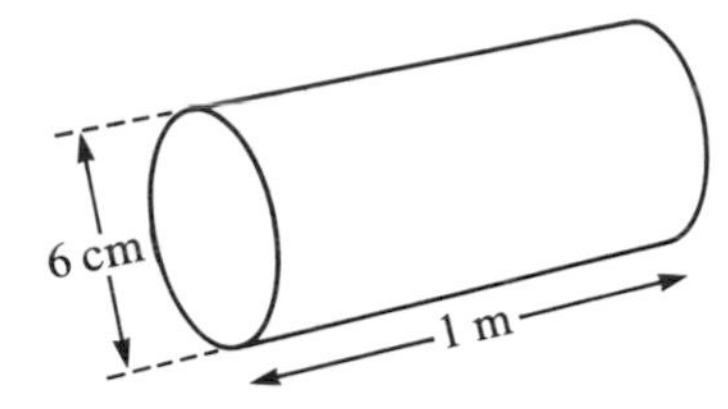

23. A car travels for 10 minutes at 30 m.p.h. and then for 20 minutes at 45 m.p.h. The average speed for the whole journey is:

A 40 m.p.h. **B** $37\frac{1}{2}$ m.p.h. **C** 20 m.p.h. **D** 35 m.p.h.

24. Four people each toss a coin. What is the probability that the fourth person will toss a 'tail'?

A $\frac{1}{2}$ **B** $\frac{1}{4}$ **C** $\frac{1}{8}$ **D** $\frac{1}{16}$

25. A rectangle 8 cm by 6 cm is inscribed inside a circle. What is the area, in cm^2, of the circle?

A 10π **B** 25π **C** 49π **D** 100π

Test 3

1. The price of a T.V. changed from £240 to £300. What is the percentage increase?

A 15% **B** 20% **C** 60% **D** 25%

2. Find the length x.

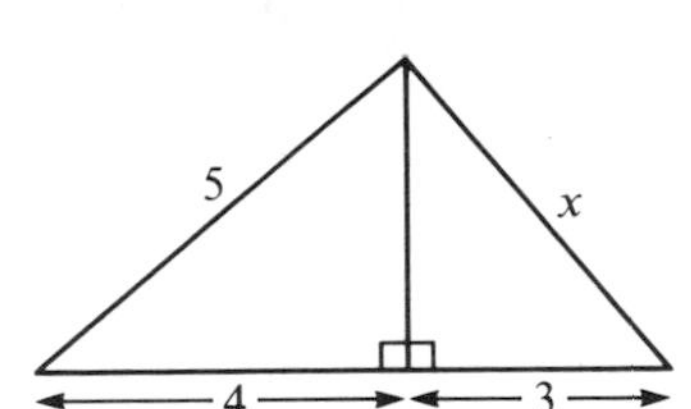

A 6 **B** 5 **C** $\sqrt{44}$ **D** $\sqrt{18}$

3. The bearing of A from B is 120°. What is the bearing of B from A?

A 060° **B** 120° **C** 240° **D** 300°

4. Numbers m, x and y satisfy the equation $y = mx^2$. When $m = \frac{1}{2}$ and $x = 4$ the value of y is:

A 4 **B** 8 **C** 1 **D** 2

5. A school has 400 pupils, of whom 250 are boys. The ratio of boys to girls is:

A 5 : 3 **B** 3 : 2 **C** 3 : 5 **D** 8 : 5

6. A train is travelling at a speed of 30 km per hour. How long will it take to travel 500 m?

A 2 minutes **B** $\frac{3}{50}$ hour **C** 1 minute **D** $\frac{1}{2}$ hour

7. The approximate value of $\dfrac{9{\cdot}65 \times 0{\cdot}203}{0{\cdot}0198}$ is:

A 99 **B** 9·9 **C** 0·99 **D** 180

8. Which point does *not* lie on the curve $y = \dfrac{12}{x}$?

A (6, 2) **B** $(\frac{1}{2}, 24)$ **C** (−3, −4) **D** (3, −4)

9. $t = \dfrac{c^3}{y}$, $y =$

A $\dfrac{t}{c^3}$ **B** c^3t **C** $c^3 - t$ **D** $\dfrac{c^3}{t}$

10. The largest number of 1 cm cubes which will fit inside a cubical box of side 1 m is:

A 10^3
B 10^6
C 10^8
D 10^{12}

11. The nth term of a sequence is $u_n = n(n-2)$. Find the largest value of n for which $u_n < 2000$.

A 50
B 2001
C 44
D 45

12. Which of the following has the largest value?

A $\sqrt{100}$
B $\sqrt{\dfrac{1}{0{\cdot}1}}$
C $\sqrt{1000}$
D $\dfrac{1}{0{\cdot}01}$

13. Two dice numbered 1 to 6 are thrown together and their scores are added. The probability that the sum will be 12 is:

A $\frac{1}{6}$
B $\frac{1}{12}$
C $\frac{1}{18}$
D $\frac{1}{36}$

14. The length, in cm, of the minor arc is:

A 2π
B 3π
C 6π
D $13\frac{1}{2}\pi$

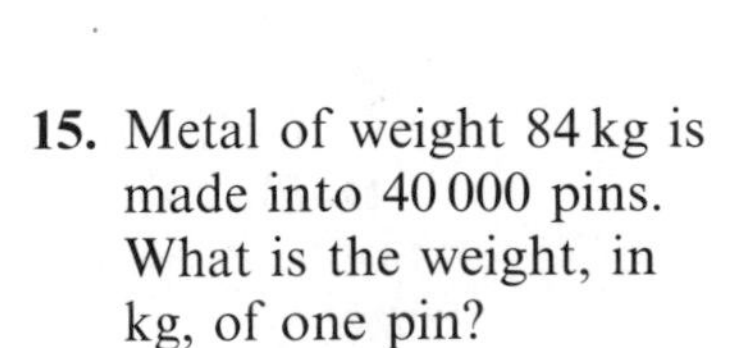

15. Metal of weight 84 kg is made into 40 000 pins. What is the weight, in kg, of one pin?

A 0·0021
B 0·0036
C 0·021
D 0·21

16. What is the value of x which satisfies both equations?
$3x + y = 1$
$x - 2y = 5$

A -1
B 1
C -2
D 2

17. What is the new fare when the old fare of £250 is increased by 8%?

A £258
B £260
C £270
D £281·25

18. What is the area of this triangle?

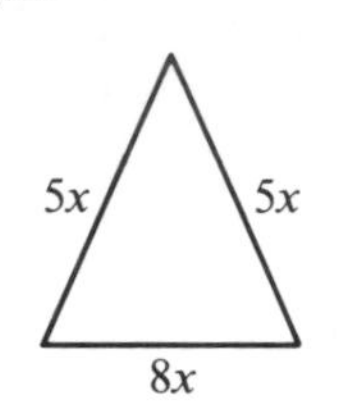

A $12x^2$
B $15x^2$
C $16x^2$
D $30x^2$

19. What values of x satisfy the inequality $2 - 3x > 1$?

A $x < -\frac{1}{3}$
B $x > -\frac{1}{3}$
C $x > \frac{1}{3}$
D $x < \frac{1}{3}$

20. A right-angled triangle has sides in the ratio 5 : 12 : 13. The tangent of the smallest angle is:

A $\frac{12}{5}$
B $\frac{12}{13}$
C $\frac{5}{13}$
D $\frac{5}{12}$

21. The area of $\triangle$ABE is 4 cm^2. The area of $\triangle$ACD is:

A 10 cm^2
B 6 cm^2
C 25 cm^2
D 16 cm^2

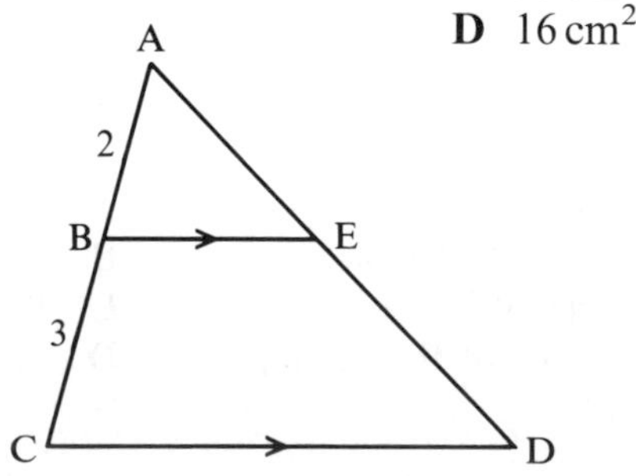

22. Given $2^x = 3$ and $2^y = 5$, the value of 2^{x+y} is:

A 15
B 8
C 4
D 125

23. The probability of an event occurring is 0·35. The probability of the event *not* occurring is:

A $\frac{1}{0 \cdot 35}$
B 0·65
C 0·35
D 0

24. What fraction of the area of the rectangle is the area of the triangle?

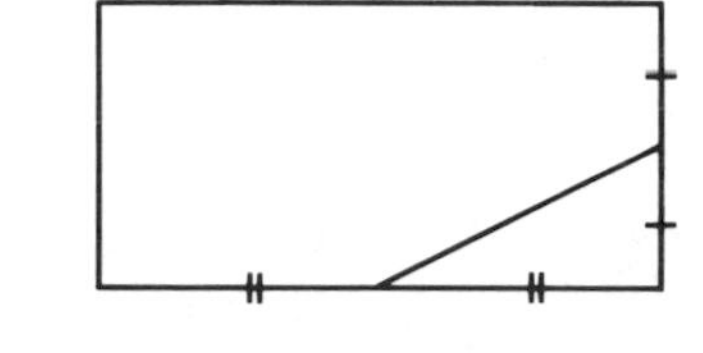

A $\frac{1}{4}$
B $\frac{1}{8}$
C $\frac{1}{16}$
D $\frac{1}{32}$

25. On a map a distance of 36 km is represented by a line of 1·8 cm. What is the scale of the map?

A 1 : 2000
B 1 : 20 000
C 1 : 200 000
D 1 : 2000 000

Test 4

1. What is the value of x satisfying the simultaneous equations
$3x + 2y = 13$
$x - 2y = -1$?

A 7
B 3
C $3\frac{1}{2}$
D 2

2. A straight line is 4·5 cm long. $\frac{2}{5}$ of the line is:

A 0·4 cm
B 1·8 cm
C 2 cm
D 0·18 cm

3. The mean of four numbers is 12. The mean of three of the numbers is 13. What is the fourth number?

A 9
B 12·5
C 7
D 1

4. How many cubes of edge 3 cm are needed to fill a box with internal dimensions 12 cm by 6 cm by 6 cm?

A 8
B 18
C 16
D 24

For Questions **5** to **7** use the diagram below.

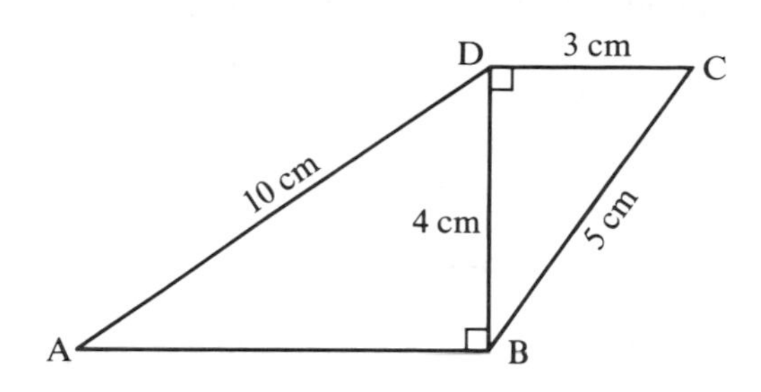

5. The length of AB, in cm, is:

A 6
B $\sqrt{116}$
C 8
D $\sqrt{84}$

6. The sine of angle DCB is:

A 0·8
B 1·25
C 0·6
D 0·75

7. The tangent of angle CBD is:

A 0·6
B 0·75
C 1·333
D 1·6

8. The value of 4865·355 correct to 2 significant figures is:

A 4865·36
B 4865·35
C 4900
D 49

9. What values of y satisfy the inequality $4y - 1 < 0$?

A $y < 4$
B $y < -\frac{1}{4}$
C $y > \frac{1}{4}$
D $y < \frac{1}{4}$

10. The area of a circle is $100\pi\,cm^2$. The radius, in cm, of the circle is:

A 50
B 10
C $\sqrt{50}$
D 5

11. If $f(x) = x^2 - 3$, then $f(3) - f(-1) =$

A 5
B 10
C 8
D 9

12. In the triangle BE is parallel to CD. What is x?

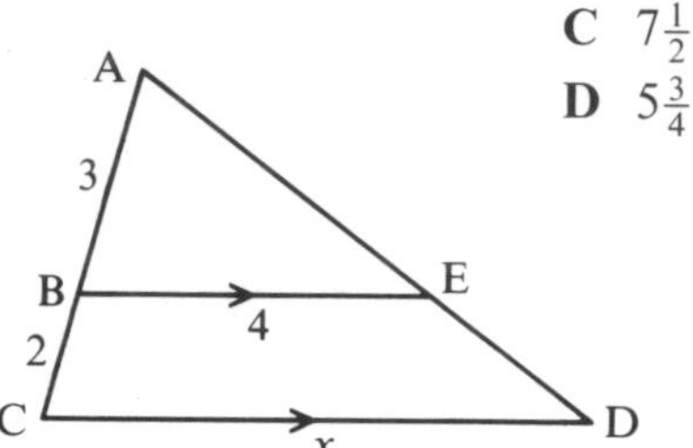

A $6\frac{2}{3}$
B 6
C $7\frac{1}{2}$
D $5\frac{3}{4}$

13. The cube root of 64 is:

A 2
B 4
C 8
D 16

14. In a group of 20 people, 5 cannot swim. If two people are selected at random, what is the probability that neither of them can swim?

A $\frac{1}{16}$
B $\frac{1}{19}$
C $\frac{35}{76}$
D $\frac{12}{19}$

15. Given $16^x = 4^4$, what is x?

A -2
B $-\frac{1}{2}$
C $\frac{1}{2}$
D 2

16. What is the area, in m^2, of a square with each side 0·02 m long?

A 0·0004
B 0·004
C 0·04
D 0·4

17. I start with x, then square it, multiply by 3 and finally subtract 4. The final result is:

A $(3x)^2 - 4$
B $(3x - 4)^2$
C $3x^2 - 4$
D $3(x - 4)^2$

18. How many prime numbers are there between 50 and 60?

A 1
B 2
C 3
D 4

19. What are the coordinates of the point $(2, -2)$ after reflection in the line $y = -x$?

A $(-2, 2)$
B $(2, -2)$
C $(-2, -2)$
D $(2, 2)$

20. The area of a circle is $36\pi\ cm^2$. The circumference, in cm, is:

A 6π
B 18π
C $12\sqrt{\pi}$
D 12π

21. The gradient of the line $2x - 3y = 4$ is:

A $\frac{2}{3}$
B $1\frac{1}{2}$
C $-\frac{4}{3}$
D $-\frac{3}{4}$

22. When all three sides of a triangle are trebled in length, the area is increased by a factor of:

A 3
B 6
C 9
D 27

23. $a = \sqrt{\left(\frac{m}{x}\right)}$

$x =$

A a^2m
B $a^2 - m$
C $\frac{m}{a^2}$
D $\frac{a^2}{m}$

24. A coin is tossed three times. The probability of getting three 'heads' is:

A $\frac{1}{3}$
B $\frac{1}{6}$
C $\frac{1}{8}$
D $\frac{1}{16}$

25. A triangle has sides of length 5 cm, 5 cm and 6 cm. What is the area, in cm^2?

A 12
B 15
C 18
D 20

ANSWERS

Part 1

page 1 ***Exercise 1***

1. 7·91 **2.** 22·22 **3.** 7·372 **4.** 0·066 **5.** 466·2
6. 1·22 **7.** 1·67 **8.** 1·61 **9.** 16·63 **10.** 24·1
11. 26·7 **12.** 3·86 **13.** 0·001 **14.** 1·56 **15.** 0·0288
16. 2·176 **17.** 0·02 **18.** 0·0001 **19.** 7·56 **20.** 0·7854
21. 360 **22.** 34 000 **23.** 18 **24.** 0·74 **25.** 2·34
26. 1620 **27.** 8·8 **28.** 1200 **29.** 0·001 75 **30.** 13·2
31. 200 **32.** 0·804 **33.** 0·8 **34.** 0·077 **35.** 0·0009
36. 0·01 **37.** 184 **38.** 20 **39.** 0·099 **40.** 3
41. 31 r 14 **42.** 41 r 13 **43.** 349 r 10 **44.** 62 r 2 **45.** 83 r 27
46. 2687 r 10

page 1 ***Exercise 2***

1. 20 **2.** 256; 65536 **3.** $a = 100$, $b = 1$ or $a = 35$, $b = 19$
4. (a) $54 \times 9 = 486$ (b) $57 \times 8 = 456$ (c) Three possible answers **5.** 8
6. (a) $3r + 2b$ (b) $4b - 3r$ (c) $2b - 3r$
7. 37 **10.** (a) 66 666 (b) 82 (c) 455 551 **11.** 500

page 3 ***Exercise 3***

1. 3, 11, 19, 23, 29, 31, 37, 47, 59, 61, 67, 73
2. (a) $2^3 \times 3 \times 5^2$
(b) $3^2 \times 7 \times 11$
(c) $2^5 \times 7 \times 11$
(d) $2 \times 3^3 \times 5 \times 13$
(e) $2^5 \times 5^3$
(f) $2^2 \times 5 \times 7^2 \times 23$
3. 8 **4.** (b) 168 (c) 7 **5.** (b) 990 (c) 22
6. Divide by prime numbers less than the square root of the number.

page 4 ***Exercise 4***

1. 70·56 **2.** 118·958 **3.** 451·62 **4.** 33678·8 **5.** 0·6174
6. 1068 **7.** 19·53 **8.** 18914·4 **9.** 38·72 **10.** 0·00979
11. 2·4 **12.** 11 **13.** 41 **14.** 8·9 **15.** 4·7
16. 56 **17.** 0·0201 **18.** 30·1 **19.** 1·3 **20.** 0·31
21. 210·21 **22.** 294 **23.** 282·131 **24.** 35 **25.** 242
26. Yes **27.** Yes **28.** He got it wrong. Correct answer = £10·45

page 6 ***Exercise 5***

1. 1 : 3 **2.** 1 : 6 **3.** 1 : 50 **4.** 1 : 1·6 **5.** 1 : 0·75
6. 1 : 0·375 **7.** 2·4 : 1 **8.** 2·5 : 1 **9.** 0·8 : 1 **10.** £15, £25
11. £36, £84 **12.** 15 kg, 75 kg, 90 kg **13.** 46 min, 69 min, 69 min
14. £39 **15.** 5 : 3 **16.** £200 **17.** 3 : 7 **18.** $\frac{1}{7}x$
19. 6 **20.** £120 **21.** 300 g **22.** 625

page 7 ***Exercise 6***

1. (a) £15 (b) £900 (c) \$2·80 (d) 125 **2.** £32 **3.** 13·2p
4. 52·8 kg **5.** (a) £1·02 (b) £21·58 (c) £2·22 (d) £0·53
6. £248·57 **7.** £26 182 **8.** 96·8% **9.** 77·5% **10.** £73·03
11. 200 kg **12.** 29 000 **13.** 500 cm **14.** £6·30 **15.** 400 kg
16. 325 **17.** £56 **18.** £8425·60

page 9 ***Exercise 7***

1. £49 **2.** £136 **3.** 14 000 **4.** 5500 **5.** 22 000
6. 12 000 **7.** 62p **8.** 72p

page 10 ***Exercise 8***

1. (a) 25%, profit (b) 25%, profit (c) 10%, loss (d) 20%, profit
(e) 30%, profit (f) 7·5%, profit (g) 12%, loss (h) 54%, loss
2. 28% **3.** $44\frac{4}{9}$% **4.** 46·9% **5.** 12% **6.** $5\frac{1}{3}$%
7. (a) £50 (b) £450 (c) £800 (d) £12·40
8. £500 **9.** £12 **10.** £5 **11.** 60p **12.** £220
13. 14·3% **14.** 20% **15.** 350 g **16.** 21%

page 12 ***Exercise 9***

1. (a) £2180 (b) £2376·20 (c) £2590·06
2. (a) £5550 (b) £6838·16 (c) £8425·29 **3.** £13 108 **4.** £5657
5. No. Should be > £193. **6.** (a) £14 033 (b) £734 (c) £107 946
7. 8 **8.** 11 **9.** (b) $x = 9$ (c) 9 years

page 13 ***Exercise 10***

1. 3·041 **2.** 1460 **3.** 0·030 83 **4.** 47·98 **5.** 130·6
6. 0·4771 **7.** 0·3658 **8.** 37·54 **9.** 8·000 **10.** 0·6537
11. 0·037 16 **12.** 34·31 **13.** 0·7195 **14.** 3·598 **15.** 0·2445
16. 2·043 **17.** 0·3798 **18.** 0·7683 **19.** −0·5407 **20.** 0·070 40
21. 2·526 **22.** 0·094 78 **23.** 0·2110 **24.** 3·123 **25.** 2·230
26. 128·8 **27.** 4·268 **28.** 3·893 **29.** 0·6290 **30.** 0·4069
31. 9·298 **32.** 0·1010 **33.** 0·3692 **34.** 1·125 **35.** 1·677
36. 0·9767 **37.** 0·8035 **38.** 0·3528 **39.** 2·423 **40.** 1·639
41. 0·000 465 9 **42.** 0·3934 **43.** −0·7526 **44.** 2·454 **45.** 40 000
46. 0·070 49 **47.** 405 400 **48.** 471·3 **49.** 20 810 **50.** $2{\cdot}218 \times 10^6$
51. $1{\cdot}237 \times 10^{-24}$ **52.** 3·003 **53.** 0·035 81 **54.** 47·40 **55.** −1748
56. 0·011 38 **57.** 1757 **58.** 0·026 35 **59.** 0·1651 **60.** 5447
61. 0·006 562 **62.** 0·1330 **63.** 0·4451 **64.** 0·036 16 **65.** 19·43

page 15 ***Exercise 11***

1. 4×10^3 **2.** 5×10^2 **3.** 7×10^4 **4.** 6×10 **5.** $2{\cdot}4 \times 10^3$
6. $3{\cdot}8 \times 10^2$ **7.** $4{\cdot}6 \times 10^4$ **8.** $4{\cdot}6 \times 10$ **9.** 9×10^5 **10.** $2{\cdot}56 \times 10^3$
11. 7×10^{-3} **12.** 4×10^{-4} **13.** $3{\cdot}5 \times 10^{-3}$ **14.** $4{\cdot}21 \times 10^{-1}$ **15.** $5{\cdot}5 \times 10^{-5}$
16. 1×10^{-2} **17.** $5{\cdot}64 \times 10^5$ **18.** $1{\cdot}9 \times 10^7$ **19.** $1{\cdot}1 \times 10^9$ **20.** $1{\cdot}67 \times 10^{-24}$
21. $5{\cdot}1 \times 10^8$ **22.** $2{\cdot}5 \times 10^{-10}$ **23.** $6{\cdot}023 \times 10^{23}$ **24.** 3×10^{10} **25.** £$3{\cdot}6 \times 10^6$

page 17 ***Exercise 12***

1. $1{\cdot}5 \times 10^7$ **2.** 3×10^8 **3.** $2{\cdot}8 \times 10^{-2}$ **4.** 7×10^{-9} **5.** 2×10^6
6. 4×10^{-6} **7.** 9×10^{-2} **8.** $6{\cdot}6 \times 10^{-8}$ **9.** $3{\cdot}5 \times 10^{-7}$ **10.** 10^{-16}
11. 8×10^9 **12.** $7{\cdot}4 \times 10^{-7}$ **13.** *c, a, b* **14.** 13 **15.** 16
16. (i) $8{\cdot}75 \times 10^2$, $3{\cdot}75 \times 10^2$ (ii) 10^8, $4{\cdot}29 \times 10^7$
17. 50 min **18.** 6×10^2 **19.** $2{\cdot}4 \times 10^9$ kg **20.** 25 000
21. (a) $9{\cdot}46 \times 10^{12}$ km (b) 144 million km **22.** (a) 20·5 s (b) $6{\cdot}3 \times 10^{91}$ years

page 19 ***Exercise 13***

1. 13 **2.** 211 **3.** −12 **4.** −31 **5.** −66
6. 6·1 **7.** 9·1 **8.** −35 **9.** 18·7 **10.** −9
11. −3 **12.** 3 **13.** −2 **14.** −14 **15.** −7
16. 3 **17.** 181 **18.** −2·2 **19.** 8·2 **20.** 17
21. 2 **22.** −6 **23.** −15 **24.** −14 **25.** −2
26. −12 **27.** −80 **28.** −13·1 **29.** −4·2 **30.** 12·4
31. −7 **32.** 8 **33.** 4 **34.** −10 **35.** 11
36. 4 **37.** −20 **38.** 8 **39.** −5

page 19 ***Exercise 14***

1. −8 **2.** 28 **3.** 12 **4.** 24 **5.** 18
6. −35 **7.** 49 **8.** −12 **9.** −2 **10.** 9
11. −4 **12.** 4 **13.** −4 **14.** 8 **15.** 70
16. −7 **17.** $\frac{1}{4}$ **18.** $-\frac{3}{5}$ **19.** −0·01 **20.** 0·0002
21. 121 **22.** 6 **23.** −600 **24.** −1 **25.** −20
26. −2·6 **27.** −700 **28.** 18 **29.** −1000 **30.** 640
31. −6 **32.** −42 **33.** −0·4 **34.** −0·4 **35.** −200
36. −35 **37.** −2 **38.** $\frac{1}{2}$ **39.** $-\frac{1}{4}$ **40.** −90

page 20 ***Exercise 15***

1. −10 **2.** 1 **3.** 12 **4.** −28 **5.** −2
6. 16 **7.** −3 **8.** 14 **9.** −28 **10.** 4
11. $-\frac{1}{6}$ **12.** 9 **13.** −30 **14.** 24 **15.** −1
16. −2 **17.** −30 **18.** 7 **19.** 3 **20.** 16
21. 93 **22.** 2400 **23.** 10 **24.** 1

page 20 ***Exercise 16***

1. $1\frac{11}{20}$ **2.** $\frac{11}{24}$ **3.** $1\frac{1}{2}$ **4.** $\frac{5}{12}$ **5.** $\frac{4}{15}$
6. $\frac{1}{10}$ **7.** $\frac{8}{15}$ **8.** $\frac{5}{42}$ **9.** $\frac{15}{26}$ **10.** $\frac{5}{12}$
11. $4\frac{1}{2}$ **12.** $1\frac{2}{3}$ **13.** $\frac{23}{40}$ **14.** $\frac{3}{40}$ **15.** $1\frac{7}{8}$
16. $1\frac{1}{12}$ **17.** $1\frac{1}{6}$ **18.** $2\frac{5}{8}$ **19.** $6\frac{1}{10}$ **20.** $9\frac{1}{10}$
21. $1\frac{9}{26}$ **22.** $\frac{1}{9}$ **23.** $\frac{2}{3}$ **24.** $5\frac{1}{4}$ **25.** $2\frac{2}{25}$
26. (a) $\frac{1}{2}, \frac{7}{12}, \frac{2}{3}$ (b) $\frac{2}{3}, \frac{3}{4}, \frac{5}{6}$ (c) $\frac{1}{3}, \frac{5}{8}, \frac{17}{24}, \frac{3}{4}$ (d) $\frac{5}{6}, \frac{8}{9}, \frac{11}{12}$
27. (a) $\frac{1}{2}$ (b) $\frac{3}{4}$ (c) $\frac{17}{24}$ (d) $\frac{7}{18}$ (e) $\frac{3}{10}$ (f) $\frac{5}{12}$
28. 5 **29.** £39 **30.** 3 **31.** 123 cm **32.** $\frac{1}{5}$
33. $1\frac{16}{17}$ **34.** 9 **35.** $\frac{5}{24}$ **36.** same

page 22 ***Exercise 17***

1. (a) 0·25 (b) 0·4 (c) 0·375 (d) $0{\cdot}41\dot{6}$ (e) $0{\cdot}1\dot{6}$ (f) $0{\cdot}\dot{2}8571\dot{4}$
2. (a) $\frac{1}{5}$ (b) $\frac{9}{20}$ (c) $\frac{9}{25}$ (d) $\frac{1}{8}$ (e) $1\frac{1}{20}$ (f) $\frac{7}{1000}$
3. (a) 25% (b) 10% (c) 72% (d) 7·5% (e) 2% (f) $33\frac{1}{3}$%
4. (a) 45%; $\frac{1}{2}$; 0·6 (b) 4%; $\frac{6}{16}$; 0·38 (c) 11%; 0·111; $\frac{1}{9}$ (d) 0·3; 32%; $\frac{1}{3}$
5. 0·58 **6.** 1·42 **7.** 0·65 **8.** 1·61 **9.** 0·07
10. 0·16 **11.** 3·64 **12.** 0·60 **13.** $\frac{4}{15}$; 0·33; $\frac{1}{3}$ **14.** $\frac{2}{7}$; 0·3 ; $\frac{4}{9}$
15. $\frac{7}{11}$; 0·705; 0·71 **16.** $\frac{5}{18}$; 0·3; $\frac{4}{13}$

page 23 ***Exercise 18***

1. 21 **2.** 1·62 **3.** 396 **4.** 650 **5.** 63·8
6. 9×10^{12} **7.** $10\frac{1}{2}$ **8.** 800
9. (a) 1245 km/h (b) 5°C (c) 1008 km/h **10.** $ac + ab - a^2$ **11.** $r - p + q$
12. 802; $5n + 2$ **13.** $w = 2n + 6$

page 25 ***Exercise 19***

1. 7 **2.** 13 **3.** 13 **4.** 22 **5.** 1
6. −1 **7.** 18 **8.** −4 **9.** −3 **10.** 37
11. 0 **12.** −4 **13.** −7 **14.** −2 **15.** −3
16. −8 **17.** −30 **18.** 16 **19.** −10 **20.** 0
21. 7 **22.** −6 **23.** −2 **24.** −7 **25.** −5
26. 3 **27.** 4 **28.** −8 **29.** −2 **30.** 2
31. 0 **32.** 4 **33.** −4 **34.** −3 **35.** −9
36. 4

page 26 ***Exercise 20***

1. 9 **2.** 27 **3.** 4 **4.** 16 **5.** 36
6. 18 **7.** 1 **8.** 6 **9.** 2 **10.** 8
11. −7 **12.** 15 **13.** −23 **14.** 3 **15.** 32
16. 36 **17.** 144 **18.** −8 **19.** −7 **20.** 13
21. 5 **22.** −16 **23.** 84 **24.** 17 **25.** 6
26. 0 **27.** −25 **28.** −5

page 26 ***Exercise 21***

1. −20 **2.** 16 **3.** −42 **4.** −4 **5.** −90
6. −160 **7.** −2 **8.** −81 **9.** 4 **10.** 22
11. 14 **12.** 5 **13.** 1 **14.** $\sqrt{5}$ **15.** 4
16. $-6\frac{1}{2}$ **17.** 54 **18.** 25 **19.** 4 **20.** 312
21. 45 **22.** 22 **23.** 14 **24.** −36 **25.** −7
26. 1 **27.** 901 **28.** −30 **29.** −5 **30.** $7\frac{1}{2}$
31. −7 **32.** $-\frac{3}{13}$ **33.** $1\frac{1}{3}$ **34.** $-\frac{5}{36}$

page 27 ***Exercise 22***

1. (a) Fr 218 (b) $121·80 (c) Ptas 36 400 (d) DM 6·3 (e) Lire 5244 (f) $1·57
2. (a) £45·87 (b) £1436·78 (c) £1·79 (d) £10·34 (e) £219·30 (f) £5·22
3. £1·80 **4.** Spain by £1·20 **5.** £20 000
6. Germany £12 400; USA £14 300; Britain £15 000; Belgium £17 200, France £17 800
7. 3·25 Swiss francs = £1 **8.** DM 1690 **9.** £2197·46 **10.** 63p

page 29 ***Exercise 23***

1. 12·3 km **2.** 4·71 km **3.** 50 cm **4.** 64 cm **5.** 5·25 cm
6. 40 m by 30 m; 12 cm^2; 1200 m^2 **7.** 1 m^2, 6 m^2 **8.** 0·32 km^2

page 30 ***Exercise 24***

1. (a) $2\frac{1}{2}$ h (b) $3\frac{1}{8}$ h (c) 75 s (d) 4 h
2. (a) 20 m/s (b) 108 km/h (c) 1·2 cm/s (d) 90 m/s
3. (a) 75 km/h (b) 4×10^6 m/s (c) 3 km/h
4. (a) 10 000 m (b) 56 400 m (c) 4500 m (d) 50 400 m
5. (a) 3·125 h (b) 76·8 km/h **6.** (a) 4·45 h (b) 23·6 km/h
7. (a) 8 m/s (b) 7·6 m/s (c) 102·63 s (d) 7·79 m/s
8. 1230 km/h **9.** 3 h **10.** 100 s **11.** $1\frac{1}{2}$ minutes **12.** 600 m
13. $53\frac{1}{3}$ s **14.** 5 cm/s **15.** 60 s **16.** 120 mph

page 32 ***Exercise 25***

1. 400 **3.** 225 mm **4.** 1
5. (a) 323 g (b) 23 (c) 67p (d) 29 **6.** 6 m **7.** ≈ 190
8. 1105 **9.** July by 0·1% **11.** 13 **12.** 10 **13.** £10 485·76
14. 200 g **15.** (a) £150 (b) £0 (c) £8 **16.** (a) $\frac{1}{66}$ (b) 16
17. £118

page 34 ***Exercise 26***

1. 6 **2.** 17 **3.** $6^2 + 7^2 + 42^2 = 43^2$; $x^2 + (x+1)^2 + [x(x+1)]^2 = (x^2 + x + 1)^2$
4. (a) 87p (b) 240p (c) 72p (d) 15p (e) 9p
5. 32 **8.** 21 **10.** (a) 1 (b) 15 **11.** 50
12. 106 **13.** £$6·3 \times 10^{10}$ (2 s.f.) **15.** (a) $\frac{1}{994}, \frac{1}{949}, \frac{1}{499}$ (b) $\frac{11}{94}$

page 36 ***Exercise 27***

1. Rational: $(\sqrt{17})^2$; 3·14; $\frac{\sqrt{12}}{\sqrt{3}}$; $3^{-1} + 3^{-2}$; $\frac{22}{7}$; $\sqrt{2·25}$ **3.** (a) $\frac{2}{9}$ (b) $\frac{29}{99}$ (c) $\frac{541}{999}$
6. (a) No (b) Yes e.g. $\sqrt{8} \times \sqrt{2} = 4$ **7.** (a) $5\sqrt{5}$ (b) $\frac{4}{3}$ (c) $\sqrt{2}$ (d) 6

page 39 ***Exercise 28***

1. 16·5, 17·5 **2.** 255·5, 256·5 **3.** 2·35, 2·45 **4.** 0·335, 0·345 **5.** 2·035, 2·045
6. 11·95, 12·05 **7.** 81·35, 81·45 **8.** 0·25, 0·35 **9.** 3·995, 4·005 **10.** 0·065, 0·075
11. 0·05, 0·15 **12.** 614·5, 615·5 **13.** 7·125, 7·135 **14.** 51·5 million, 52·5 million
15.–28. For discussion

page 41 ***Exercise 29***

1. (i) 10·5 (ii) 4·3 **2.** (i) 11 (ii) 1 (iii) 0·6
3. 56 cm^2 **4.** 58·848 039, 55·706 604 **5.** 17·20 m/s, 17·09 m/s
6. 3·298 8372, 2·872 2222 **7.** 4·101 6355 **8.** 7·163 6234
9. (a) \$35 390 → \$38 250 (b) \$35 640 → \$35 660 (c) Only one figure to approximate

page 42 ***Exercise 30***

1. 7·15% **2.** 4·23% **3.** 12·9% **4.** 0·177%
5. 6·15% **6.** 0·0402% **7.** 1·55% **8.** 2·1%; 5·5%; Katy
9. 8% **10.** 0·28% **11.** sin 45°, 1·01% **12.** Second
13. 1·02 **14.** (a) 1256·6371 (b) (i) 1307·4052 (ii) 50·7681 (iii) 4·0%

page 44 ***Examination Exercise 1***

1. (a) D. Express (b) D. Mail, D. Express (c) Today, Independent; Guardian, Times
(d) 3·17% (e) $4{\cdot}15 \times 10^6$ **2.** 15p
3. (a) $1{\cdot}35085 \times 10^8$ (b) 22·3% **4.** (a) 2·750 (b) 57·14 (c) 7·698 (d) 107·6
5. 1 : 4000 **6.** (a) $\frac{1}{9}$ (b) 24 **7.** (a) 10% (b) £900 **8.** (a) 4·1
9. (a) $\frac{11}{80}$ (b) 0·1375 **10.** £2593·74 **11.** (a) $3 \times 7 \times 13$ (b) $2 \times 7 \times 11 \times 13$, 91
12. (a) Saved by 47 votes (b) 843 **13.** $1{\cdot}1 \times 10^{34}$ **14.** (a) 480 000 (b) $4{\cdot}8 \times 10^5$
15. 85 s **16.** 2·7 min **17.** (a) 36·6 km/h (b) 9·2%
18. (a) (i) 146·8 m (ii) 145·5 m (b) 150 m **19.** (a) $16\frac{2}{3}$ m/s (b) 19 m/s

Part 2

page 49 ***Exercise 1***

1. $w = b + 4$ **2.** $w = 2b + 6$ **3.** $w = 2b - 12$ **4.** $m = 2t + 1$ **5.** $m = 3t + 2$
6. $s = t + 2$ **7.** (a) $p = 5n - 2$ (b) $k = 7n + 3$ (c) $w = 2n + 11$ **8.** $m = 8c + 4$
9. (a) $y = 3n + 1$ (b) $h = 4n - 3$ (c) $k = 3n + 5$
10. (a) $t = 2n + 4$ (b) $e = 3n + 11$ (c) $e = \dfrac{3t + 10}{2}$
11. (a) £7800 (b) $R = 5000 + 400N$

page 52 ***Exercise 2***

1. $3n$ **2.** $5n$ **3.** 3^n **4.** $2n - 1$ **5.** $n(n + 2)$
6. $2n + 1$ **7.** $3n - 1$ **8.** $n^2 - 1$ **9.** $(n + 1)(n + 2)$ **10.** $\dfrac{n}{n + 1}$

11. $\dfrac{n}{n+2}$ **12.** $\dfrac{1}{n^2}$ **13.** $n(n+1)$ **14.** $n(n+3)$ **15.** 3×2^n
16. $2n^2$ **17.** $2n(n+1)$ **18.** $(10^n - 1)$

page 54 ***Exercise 3***

1. 60, $m = 2n(n+1)$ **2.** $c = \frac{1}{2}s(s-1)$
3. (a) $r = n(n+3)$ (b) $t = \dfrac{n(n+1)}{2}$ (c) $p = (n+1)^2$ or $p = n(n+2)+1$
4. $s = 2n(n-1)+1$ **5.** (a) $t = n(n+2)$ (b) $t = n^2 + 2n$ **6.** $m = 2^d - 1$

page 57 ***Exercise 4***

1. $x^2 + 2x + 4$ **2.** $x^2 + 4x + 3$ **3.** $2x^2 + 7x + 3$ **4.** $3x^2 - x$
5. $\dfrac{x(x+1)}{2}$ **6.** $x^3 + 5$ **7.** $x(x+1)(x+2)$ **8.** $x^3 + x - 1$

page 59 ***Exercise 5***

1. 8 **2.** 9 **3.** 7 **4.** 10 **5.** $\frac{1}{3}$
6. 10 **7.** $1\frac{1}{2}$ **8.** -1 **9.** $-1\frac{1}{2}$ **10.** $\frac{1}{3}$
11. $\frac{99}{100}$ **12.** 0 **13.** 1000 **14.** $-\frac{1}{1000}$ **15.** 1
16. -7 **17.** -5 **18.** $1\frac{1}{6}$ **19.** 1 **20.** 2
21. -5 **22.** -3 **23.** $-1\frac{1}{2}$ **24.** 2 **25.** 1
26. $3\frac{1}{2}$ **27.** 2 **28.** -1 **29.** $10\frac{2}{3}$ **30.** 1·1
31. -1 **32.** 2

page 59 ***Exercise 6***

1. 35 **2.** 130 **3.** 14 **4.** $\frac{2}{3}$ **5.** $3\frac{1}{3}$
6. $-2\frac{1}{2}$ **7.** 3 **8.** $1\frac{1}{8}$ **9.** $\frac{3}{10}$ **10.** $-1\frac{1}{4}$
11. 10 **12.** 27 **13.** 20 **14.** 18 **15.** 28
16. -15 **17.** $2\frac{1}{2}$ **18.** $1\frac{1}{3}$

page 60 ***Exercise 7***

1. $-1\frac{1}{2}$ **2.** 2 **3.** $-\frac{2}{5}$ **4.** $-\frac{1}{3}$ **5.** $1\frac{2}{3}$
6. 6 **7.** $-\frac{2}{5}$ **8.** $-3\frac{1}{5}$ **9.** $\frac{1}{2}$ **10.** -4
11. 18 **12.** 5 **13.** 4 **14.** 3 **15.** $2\frac{3}{4}$
16. $-\frac{7}{22}$ **17.** $\frac{1}{4}$ **18.** 1 **19.** 4 **20.** -11

page 60 ***Exercise 8***

1. $\frac{1}{3}$ **2.** $\frac{1}{5}$ **3.** $1\frac{2}{3}$ **4.** -3 **5.** $\frac{5}{11}$
6. -2 **7.** -7 **8.** $-7\frac{2}{3}$ **9.** 2 **10.** 3
11. 2 **12.** 3 **13.** 5 **14.** -4 **15.** 4
16. $\frac{3}{5}$ **17.** $1\frac{1}{8}$ **18.** -1 **19.** 1 **20.** 6
21. $1\frac{5}{7}$

page 61 ***Exercise 9***

1. $3\frac{1}{3}$ cm **2.** 12 cm **3.** 91, 92, 93
4. 21, 22, 23, 24 **5.** 57, 59, 61 **6.** 506, 508, 510
7. $12\frac{1}{2}$ **8.** 20 **9.** $18\frac{1}{2}$, $27\frac{1}{2}$
10. 20°, 60°, 100° **11.** 5, 15, 8 **12.** 5 cm
13. $59\frac{2}{3}$ kg, $64\frac{2}{3}$ kg, $72\frac{2}{3}$ kg **14.** 7 cm **15.** (b) $14 + 8x$ (c) 0·75 m

page 63 ***Exercise 10***

1. 40 cm **2.** £3700 **3.** 3 **4.** 26, 58 **5.** 2 km
6. 8 km **7.** 400 m **8.** 21 **9.** 23 **10.** £3600
11. 15 **12.** 2 km **13.** 6, 7, 8, 9 **14.** 2, 3, 4, 5 **15.** 26 cm

page 65 ***Exercise 11***

1. 12 cm **2.** 15×5
3. (a) 26×13 (b) 16×8 (c) 32×16 (d) $9 \times 4{\cdot}5$ (e) $6{\cdot}5 \times 3{\cdot}25$
4. (a) 6×5 (b) 12×11 (c) 20×19 (d) $6{\cdot}5 \times 5{\cdot}5$ (e) $8{\cdot}7 \times 7{\cdot}7$
5. not given

page 67 ***Exercise 12***

1. (a) 9·51/9·52 (b) 7·57/7·58 **2.** (a) 5·14/5·15 (b) 3·82/3·83 (c) 6·69/6·70
(d) 3·37/3·38 (e) 5·82/5·83 or 0·17/0·18 (f) 3·15/3·16 (g) 3·59/3·60
3. 8·07/8·08 **4.** 9·55/9·56 **5.** (a) $x + 10$ (c) $34 \times 44 \times 6$
6. $x = 3{\cdot}8$ **7.** $x = 1{\cdot}62$

page 69 ***Exercise 13***

10. (a) 180 (b) $C = 0{\cdot}2x + 35$ **11.** (a) $2{\cdot}5h$ (b) $C = 15h + 18$
12. (a) (i) £560 (ii) 2400 miles **13.** (a) £188 (b) 158 km/h

page 71 ***Exercise 14***

1. (a) (3, 7) (b) (1, 3) (c) (11, −1)
2. (2, 4) **3.** (2, 3) **4.** (3, 1) **5.** (1, 5) **6.** (5, 3)
7. (a) (4, 0) (b) (1, 6) (c) (−2, −3) (d) (8, −1) (e) (−0·6, 1·2)

page 73 ***Exercise 15***

1. $x = 2, y = 1$ **2.** $x = 4, y = 2$ **3.** $x = 3, y = 1$ **4.** $x = -2, y = 1$
5. $x = 3, y = 2$ **6.** $x = 5, y = -2$ **7.** $x = 2, y = 1$ **8.** $x = 5, y = 3$
9. $x = 3, y = -1$ **10.** $a = 2, b = -3$ **11.** $a = 5, b = \frac{1}{4}$ **12.** $a = 1, b = 3$
13. $m = \frac{1}{2}, n = 4$ **14.** $w = 2, x = 3$ **15.** $x = 6, y = 3$ **16.** $x = \frac{1}{2}, z = -3$
17. $m = 1\frac{15}{17}, n = \frac{11}{17}$ **18.** $c = 1\frac{16}{23}, d = -2\frac{12}{23}$

page 74 ***Exercise 16***

1. $x = 2, y = 4$ **2.** $x = 1, y = 4$ **3.** $x = 2, y = 5$ **4.** $x = 3, y = 7$
5. $x = 5, y = 2$ **6.** $a = 3, b = 1$ **7.** $x = -2, y = 3$ **8.** $x = 4, y = 1$

9. $x = \frac{5}{7}$, $y = 4\frac{3}{7}$ **10.** $x = 1$, $y = 2$ **11.** $x = 2$, $y = 3$ **12.** $x = 4$, $y = -1$
13. $x = 1$, $y = 2$ **14.** $a = 4$, $b = 3$ **15.** $x = 4$, $y = 3$ **16.** $x = 5$, $y = -2$
17. $x = 3$, $y = -1$ **18.** $x = 5$, $y = 0{\cdot}2$

page 75 ***Exercise 17***

1. $5\frac{1}{2}$, $9\frac{1}{2}$ **2.** 6, 3 or $2\frac{2}{5}$, $5\frac{2}{5}$ **3.** 4, 10
4. 10·5, 7·5 **5.** $a = 2$, $c = 7$ **6.** $m = 4$, $c = -3$
7. $a = 30$, $b = 5$ **8.** TV £200, video £450 **9.** white 2 oz, brown $3\frac{1}{2}$ oz
10. 12 m, 24 m **11.** 150 m, 350 m **12.** 2p × 15, 5p × 25
13. 10p × 14, 50p × 7 **14.** 20 **15.** current 4 m/s, kipper 10 m/s
16. $\frac{5}{7}$ **17.** boy 10, mouse 3 **18.** (4, −3), (−2, 3)
19. walks 4 m/s, runs 5 m/s **20.** £1 × 15, £5 × 5 **21.** $a = 1$, $b = 2$, $c = 5$
22. $y = x^2 + 3x + 4$

page 79 ***Exercise 18***

1. 2, 3 **2.** 1, 2, 3, 4, 5 **3.** 1, 2 **4.** 1, 2, 3 **5.** $x > 13$
6. $x < -1$ **7.** $x < 12$ **8.** $x \leqslant 2\frac{1}{2}$ **9.** $x > 3$ **10.** $x \geqslant 8$
11. $x < \frac{1}{4}$ **12.** $x \geqslant -3$ **13.** $x < -8$ **14.** $x < 4$ **15.** $x > -9$
16. $x < 8$ **17.** $x > 3$ **18.** $x \geqslant 1$ **19.** $x < 1$ **20.** $x > 2\frac{1}{3}$
21. $x < -3$ **22.** $x > 7\frac{1}{2}$ **23.** $x > 0$ **24.** $x < 0$ **25.** $5 \leqslant x \leqslant 9$
26. $-1 < x < 4$ **27.** $\frac{11}{2} \leqslant x \leqslant 6$ **28.** $\frac{1}{2} < x < 8$ **29.** $-8 < x < 2$

page 80 ***Exercise 19***

1. $-5 < x < 5$ **2.** $-4 \leqslant x \leqslant 4$ **3.** $x > 1$, $x < -1$
4. $x \geqslant 6$, $x \leqslant -6$ **5.** all values except zero **6.** $-2 < x < 2$
7. 1, 2, 3, 4, 5, 6 **8.** 7, 11, 13, 17, 19 **9.** 4, 9, 16, 25, 36, 49
10. −4, −3, −2, −1 **11.** 2, 3, 4, ... 12 **12.** 2, 3, 5, 7, 11
13. 2, 4, 6, ... 18 **14.** 1, 2, 3, 4 **15.** 5
16. 16, −16, 20, −5 **17.** > **18.** $\frac{1}{2}$ (or others)
19. 19 **20.** (a) $-3 \leqslant x < 6$ (b) $-2 < x < 2$ (c) $-3 \leqslant x \leqslant 2$ (d) $-3 \leqslant x < 7$
21. 17 **22.** $x > 3\frac{2}{3}$ **23.** 7 **24.** 5
25. 6 **26.** 3, 4, 5 **27.** $30 < x < 90$ **28.** $0 < x < 75{\cdot}5$
29. $y < \dfrac{d+c}{a}$

page 82 ***Exercise 20***

1. $x \leqslant 3$ **2.** $y \geqslant 2\frac{1}{2}$ **3.** $1 \leqslant x \leqslant 6$ **4.** $x < 7$, $y < 5$ **5.** $y \geqslant x$
6. $x + y < 10$ **7.** $2x - y \leqslant 3$ **8.** $y \leqslant x$, $x \leqslant 8$, $y \geqslant -2$
9. (a) $x + y \leqslant 7$, $x \geqslant 0$, $y \geqslant x - 1$ (b) $x + y \leqslant 6$, $y \geqslant 0$, $y \leqslant x + 2$
28. A: $x + y < 5$, $y > x + 1$ B: $x + y > 5$, $y > x + 1$
C: $x + y > 5$, $y < x + 1$ D: $x + y < 5$, $y < x + 1$
29. $-2 < x < 1$ **30.** (2, 6), (3, 5), (3, 4), (4, 4), (4, 3), (5, 3), (6, 2)

page 85 ***Exercise 21***

1. x^2+4x+3 **2.** x^2+5x+6 **3.** $y^2+9y+20$ **4.** x^2+x-12
5. $x^2+3x-10$ **6.** x^2-5x+6 **7.** $a^2-2a-35$ **8.** $z^2+7z-18$
9. x^2-9 **10.** k^2-121 **11.** $2x^2-5x-3$ **12.** $3x^2-2x-8$
13. $2y^2-y-3$ **14.** $49y^2-1$ **15.** $x^2+8x+16$ **16.** x^2+4x+4
17. x^2-4x+4 **18.** $4x^2+4x+1$ **19.** $2x^2+6x+5$ **20.** $2x^2+2x+13$
21. $5x^2+8x+5$ **22.** $2y^2-14y+25$ **23.** $10x-5$ **24.** $-8x+8$
25. (d) Yes. $n=11$, 121, 22 etc.

page 86 ***Exercise 22***

1. $x(x+5)$ **2.** $x(x-6)$ **3.** $x(7-x)$ **4.** $y(y+8)$
5. $y(2y+3)$ **6.** $2y(3y-2)$ **7.** $3x(x-7)$ **8.** $2a(8-a)$
9. $3c(2c-7)$ **10.** $3x(5-3x)$ **11.** $7y(8-3y)$ **12.** $x(a+b+2c)$
13. $x(x+y+3z)$ **14.** $y(x^2+y^2+z^2)$ **15.** $ab(3a+2b)$ **16.** $xy(x+y)$
17. $2a(3a+2b+c)$ **18.** $m(a+2b+m)$ **19.** $2k(x+3y+2z)$ **20.** $a(x^2+y+2b)$
21. $xk(x+k)$ **22.** $ab(a^2+2b)$ **23.** $bc(a-3b)$ **24.** $ae(2a-5e)$
25. $ab(a^2+b^2)$ **26.** $x^2y(x+y)$ **27.** $2xy(3y-2x)$ **28.** $3ab(b^2-a^2)$
29. $a^2b(2a+5b)$ **30.** $ax^2(y-2z)$ **31.** $2ab(x+b+a)$ **32.** $yx(a+x^2-2yx)$

page 87 ***Exercise 23***

1. $\frac{1}{4}$ **2.** -3 **3.** 4 **4.** $-7\frac{2}{3}$ **5.** -43
6. 11 **7.** $-\frac{1}{2}$ **8.** 0 **9.** 1 **10.** $-1\frac{2}{3}$
11. 10, 8, 6 **12.** 13, 12, 5 **13.** 4 cm **14.** 5 m

page 88 ***Examination Exercise 2***

1. (a) -1, $1\frac{4}{5}$ (b) $3z(2z-3)$ (c) $6x^2+x-35$ **2.** -5 **3.** $a=8$, $b=3$
4. (a) $300+4n$, $9n$ (b) $9n>300+4n$ (c) $n>60$
5. (a) 0, 0 (b) sum of cubes = square of the sum of integers
6. (a) $2n+6$ (b) 37 **7.** (a) 2 (b) 105 (c) (ii) $5x+30$ (iii) 65 (d) 5
8. (a) 189 225 tonnes (b) 29·2 s **9.** 34
10. (b) C: 18, 1; D: 14, 3; E: 10, 5. (c) $P=20-2N$
11. (b) 36 (c) 28 (d) 60 (e) 4, 392, 9604, 200
12. 0·6 (or −1·6) **13.** 3 **14.** (b) $h=\frac{1}{2}n(n+1)$ (c) $a=1\frac{1}{2}$, $b=4\frac{1}{2}$
15. (a) 8, 9 (b) 8·46 **16.** $x=34\frac{2}{7}$ mph

Part 3

page 93 ***Exercise 1***

1. A (2, 4, 0) B (0, 4, 3) C (2, 4, 3) D (2, 0, 3)
2. (a) B (3, 0, 0) C (3, 4, 0) Q (3, 0, 2) R (3, 4, 2)
(b) (i) (0, 2, 0) (ii) (0, 4, 1) (iii) ($1\frac{1}{2}$, 4, 0)
(c) (i) ($1\frac{1}{2}$, 2, 0) (ii) ($1\frac{1}{2}$, 2, 2) (iii) ($1\frac{1}{2}$, 4, 1) (d) ($1\frac{1}{2}$, 2, 1)
3. (a) C (2, 2, 0) R (2, 2, 3) B (2, −2, 0) P (0, −2, 3) Q (2, 3)
(b) (i) (2, −2, $1\frac{1}{2}$) (ii) (1, −2, 3)

4. (i) 5 (ii) 5·83 (iii) 6·40
5. (i) $(2, 3\frac{1}{2}, 5)$ (ii) $(2, 7, 2\frac{1}{2})$ (iii) $(2, 3\frac{1}{2}, 0)$
6. (20, 45, 5) **7.** (a) 4 (b) 5 (c) $5\sqrt{2}$ **8.** Square-based pyramid
9. (a) 10 (b) 45° **10.** 29·9 m

page 97 ***Exercise 2***

8. 10 cm, 40 cm **9.** (a) anticlockwise (b) 15 (c) 200 r.p.m. **11.** a plane

page 101 ***Exercise 3***

1. 10 cm **2.** 4·12 cm **3.** 4·24 cm **4.** 9·90 cm
5. 9·85 cm **6.** 7·07 cm **7.** 9·49 cm **8.** 3·46 m
9. 40·3 km **10.** 32·6 cm **11.** 5·39 units **12.** Yes
13. 8·72 **14.** 5·66 **15.** 6·63 **16.** 2·24

page 102 ***Exercise 4***

2. (a) 6·40 cm (b) 13·6 cm **3.** 6·34 m **4.** 4·58 cm
5. (a) 7·55 (b) 12·5 (c) 14·9 **6.** 24 cm
7. 10 feet **8.** $x = 4$ m, 20·6 m **9.** 18·5 km

page 105 ***Exercise 5***

1. 42 cm^2 **2.** 22 cm^2 **3.** 103 cm^2 **4.** 60·5 cm^2
5. 143 cm^2 **6.** 9 cm^2 **7.** 24 cm^2 **8.** 35 cm^2
9. 32 cm^2 **10.** 46 cm^2 **11.** 47 cm^2 **12.** $81\frac{3}{4}$ cm^2
13. 13 m **14.** 15 cm **15.** 2500 **16.** 2·4 cm
17. (a) $\frac{1}{3}$ (b) $\frac{4}{9}$ (c) 25 cm^2 **19.** 1100 m **20.** 6 square units
21. 14 square units **22.** 1849 **23.** 10 cm

page 108 ***Exercise 6***

1. 31·4 cm, 78·5 cm^2 **2.** 18·8 cm, 28·3 cm^2 **3.** 129 m, 982 m^2 **4.** 56·6 cm, 190 cm^2
5. 53·7 m, 198 m^2 **6.** 28·1 m, 54·7 m^2 **7.** 20·6 cm, 24·6 cm^2 **8.** 25·1 cm, 43·4 cm^2
9. 20·3 cm, 24·6 cm^2 **10.** 35·1 cm, 84·1 cm^2 **11.** 25·1 cm, 13·7 cm^2 **12.** 25·1 cm, 25·1 cm^2
13. 18·8 cm, 12·6 cm^2

page 109 ***Exercise 7***

1. 2·19 cm **2.** 31·8 cm **3.** 2·65 km **4.** 9·33 cm
5. 17·8 cm **6.** 14·2 mm **7.** 497 000 km^2 **8.** 21·5 cm^2
9. 30; (a) 1508 cm^2 (b) 508 cm^2 **10.** 5305 **11.** 29
12. (a) 40·8 m^2 (b) 6 **13.** (a) 80 (b) 7 **14.** 5·39 cm
15. 118 m^2 **16.** (a) 33·0 cm (b) 70·9 cm^2
17. (a) 98 cm^2 (b) 14·0 cm^2 **18.** 796 m^2 **19.** 57·5°
20. Yes **21.** 1·716 cm

page 112 ***Exercise 8***

1. 2 : 1 **2.** 112 cm^2 **3.** 0·586 m **4.** $\frac{7}{16}$
5. 20% **6.** $t = -1$ **7.** 7·172 cm **8.** $112\frac{1}{2}°$
9. 60° **10.** 55·4 cm^2 **11.** $27\frac{1}{2}°$ **12.** 8 cm
13. 4 : 1 **14.** (a) 2·41 (b) 1·85 (c) $1 + \sqrt{2}$

page 115 ***Exercise 9***

1. (a) 30 cm^3 (b) 168 cm^3 (c) 110 cm^3 (d) 94·5 cm^3 (e) 754 cm^3 (f) 283 cm^3
2. (a) 503 cm^3 (b) 760 m^3 **3.** 358 m^3 **4.** 3·98 cm
5. 6·37 cm **6.** 1·89 cm **7.** 9·77 cm **8.** 7·38 cm
9. 1270 cm (to 3 s.f.) **10.** 4·24 litres **11.** 106 cm/s **12.** 1570 cm^3, 12·57 kg
13. No **14.** 1·19 cm **15.** 53 times **16.** 191 cm

page 118 ***Exercise 10***

1. (a), (b), (d) **2.** (a) 6 (b) 4 (c) $a = 10$, $b = 6$, $c = 10$, $d = 10$ (d) 64 cm^3
3. (a) 168 mm^2 (b) 16 800 mm^3 **4.** $a = \sqrt{2}$, $b = \sqrt{2}$, $c = \sqrt{3}$, $x = \sqrt{2}$, $y = \sqrt{3}$

page 120 ***Exercise 11***

1. C only **2.** $m = 10$, $a = 16\frac{2}{3}$ **3.** $x = 12$, $y = 8$ **4.** $a = 2\frac{1}{2}$, $e = 3$
5. $x = 6$, $y = 10$ **6.** $y = 6$ **7.** $x = 4$, $w = 1\frac{1}{2}$ **8.** $e = 9$, $f = 4\frac{1}{2}$
9. $x = 13\frac{1}{3}$, $y = 9$ **10.** $m = 6$, $n = 6$ **11.** $m = 5\frac{1}{3}$, $z = 4\frac{4}{5}$ **12.** $v = 5\frac{1}{3}$, $w = 6\frac{2}{3}$
13. No **14.** 2, 6 **15.** 16 m **16.** 5
17. (a) Yes (b) No (c) No (d) Yes (e) Yes (f) No (g) No (h) Yes
20. $3\frac{2}{3}$, $1\frac{1}{11}$ **21.** 0·618; 1·618 : 1

page 124 ***Exercise 12***

1. 4·54 **2.** 3·50 **3.** 3·71 **4.** 6·62 **5.** 8·01
6. 31·9 **7.** 45·4 **8.** 4·34 **9.** 17·1 **10.** 13·2
11. 38·1 **12.** 3·15 **13.** 516 **14.** 79·1 **15.** 5·84
16. 2·56 **17.** 18·3 **18.** 8·65 **19.** 11·9 **20.** 10·6
21. 5, 5·55 **22.** 13·1, 27·8 **23.** 41·3 **24.** 8·82 **25.** 20·4
26. 26·2

page 126 ***Exercise 13***

1. 36·9° **2.** 44·4° **3.** 48·2° **4.** 60° **5.** 36·9°
6. 50·2° **7.** 29·0° **8.** 56·4° **9.** 38·9° **10.** 43·9°
11. 41·8° **12.** 39·3° **13.** 60·3° **14.** 50·5° **15.** 13·6°
16. 34·8° **17.** 60·0° **18.** 42·0° **19.** 36·9° **20.** 51·3°
21. 19·6° **22.** 17·9° **23.** 32·5° **24.** 59·6° **25.** 17·8°

page 128 ***Exercise 14***

1. 19·5° **2.** 4·12 m **3.** 3·91 m **4.** (a) 26·0 km (b) 23·4 km
5. (a) 88·6 km (b) 179·3 km **6.** 8·60 m
7. (a) 484 km (b) 858 km (c) 986 km, 060·6° **8.** 954 km, 133° **9.** 56·3°

10. 35·5° **11.** 71·6° **12.** 91·8° **13.** 180 m **14.** 36·4°
15. 10·3 cm **16.** 72°, 8·23 cm **17.** 71·1° **18.** 67·1 m **19.** 76·5 m/s
20. 83·2 km **21.** 60° **22.** Yes **23.** 11·1 m; 11·1 s; 222 m
24. 2·90 m **25.** 4·41 m **26.** 3·13 m **27.** 25·6 cm **28.** 27·1 cm

page 133 ***Exercise 15***

1. (a) 13 cm (b) 13·6 cm (c) 17·1°
2. (a) 4·04 m (b) 38·9° (c) 11·2 m (d) 19·9°
3. (a) 8·49 cm (b) 8·49 cm (c) 10·4 cm (d) 35·3°
4. (a) 14·1 cm (b) 18·7 cm (c) 69·3°
5. (a) 4·47 m (b) 7·48 m (c) 63·4° (d) 74·5°
6. 10·8 cm; 21·8° **7.** (a) $h \tan 65°$ (b) $h \tan 57°$ (c) 22·7 m
8. 22·6 m **9.** 55·0 m

page 137 ***Exercise 16***

1. (a) 11·3 cm, 12·4 cm (b) 23·8° (c) 32·0°
2. (a) 10 cm, 9·43 cm (b) 26·6° (c) 32·5° **3.** (a) 29·5° (b) 38·7°
4. (a) 58·0° (b) 66·1° **5.** (a) 57·5° (b) 61·0° (c) 61·0°
6. (a) 16·0° (b) 19·3° (c) 19·3° **7.** 43·3°

page 139 ***Exercise 17***

1. Yes, S.S.S. **2.** Yes, S.A.S. **3.** No **4.** Yes, A.A.S.
5. No **6.** Yes, A.A.S

page 140 ***Examination Exercise 3***

1. (a) 154 m^2 (b) 42 m^2 (c) 3360 g **2.** (a) 198 cm^3 (b) 1360 mm^3 (c) 145
3. $b = 2·9$ cm, $a = 5·8$ cm, $p = 11·6$ cm **4.** 900 **5.** (5, 3, 4)
6. (a) (3, 0, 3), (3, 4, 0), (3, 4, 3), (1, 4, 3) (b) 13 **7.** 65 cm **8.** 36 km/h
9. (a) 108 m^2 (b) 14° **10.** (a) 700 (b) 26·7 inches (c) 0·30 miles
11. (a) 6 m (b) 16 m^2 **12.** (a) (i) $\frac{\pi r^2}{8}$ (ii) $\frac{\pi r^2}{8}$ (b) 90°
13. (a) 31·6° (b) 67·4° (c) 12 cm (d) 33·7°
14. (a) 139 m (b) 19·1° (c) (i) 333 m (ii) 22·7° **15.** (i) No (ii) No (iii) Yes

Part 4

page 145 ***Exercise 1***

1. $\frac{B}{A}$ **2.** $\frac{T}{N}$ **3.** $\frac{K}{M}$ **4.** $\frac{4}{y}$ **5.** $\frac{T+N}{9}$
6. $\frac{B-R}{A}$ **7.** $\frac{R+T}{C}$ **8.** $\frac{N-R^2}{L}$ **9.** $\frac{R-S^2}{N}$ **10.** 2
11. $S - B$ **12.** $N - D$ **13.** $T - N^2$ **14.** $N + M - L$ **15.** $A + R$
16. $E + A$ **17.** $F + B$ **18.** $F^2 + B^2$ **19.** $L + B$ **20.** $N + T$
21. 2 **22.** $4\frac{1}{2}$ **23.** $\frac{N-C}{A}$ **24.** $\frac{L-D}{B}$ **25.** $\frac{F-E}{D}$

26. $\frac{H+F}{N}$ **27.** $\frac{Q-m}{V}$ **28.** $\frac{n+a+m}{t}$ **29.** $\frac{c-b}{V^2}$ **30.** $\frac{r+6}{n}$
31. $2\frac{2}{3}$ **32.** $\frac{C-AB}{A}$ **33.** $\frac{F-DE}{D}$ **34.** $\frac{a-hn}{h}$ **35.** $\frac{q+bd}{b}$
36. $\frac{n-rt}{r}$

page 146 **Exercise 2**

1. 12 **2.** 10 **3.** BD **4.** TB **5.** RN
6. bm **7.** 26 **8.** $BT+A$ **9.** $AN+D$ **10.** B^2N-Q
11. $ge+r$ **12.** $4\frac{1}{2}$ **13.** $\frac{DC-B}{A}$ **14.** $\frac{pq-m}{n}$ **15.** $\frac{vS+t}{r}$
16. $\frac{qt+m}{z}$ **17.** $\frac{bc-m}{A}$ **18.** $\frac{AE-D}{B}$ **19.** $\frac{nh+f}{e}$ **20.** $\frac{qr-b}{g}$
21. 4 **22.** -2 **23.** 2 **24.** $A-B$ **25.** $C-E$
26. $D-H$ **27.** $n-m$ **28.** $q-t$ **29.** $s-b$ **30.** $r-v$
31. $m-t$ **32.** 2 **33.** $\frac{T-B}{X}$ **34.** $\frac{M-Q}{N}$ **35.** $\frac{V-T}{M}$
36. $\frac{N-L}{R}$ **37.** $\frac{v^2-r}{r}$ **38.** $\frac{w-t^2}{n}$ **39.** $\frac{n-2}{q}$ **40.** $\frac{1}{4}$
41. $-\frac{1}{7}$ **42.** $\frac{B-DE}{A}$ **43.** $\frac{D-NB}{E}$ **44.** $\frac{h-bx}{f}$ **45.** $\frac{v^2-Cd}{h}$
46. $\frac{NT-MB}{M}$ **47.** $\frac{mB+ef}{fN}$ **48.** $\frac{TM-EF}{T}$ **49.** $\frac{yx-zt}{y}$ **50.** $\frac{k^2m-x^2}{k^2}$

page 147 **Exercise 3**

1. $\frac{1}{2}$ **2.** $1\frac{2}{3}$ **3.** $\frac{B}{C}$ **4.** $\frac{T}{X}$ **5.** $\frac{v}{t}$
6. $\frac{n}{\sin 20^\circ}$ **7.** $\frac{7}{\cos 30^\circ}$ **8.** $\frac{B}{x}$ **9.** $\frac{vs}{m}$ **10.** $\frac{mb}{t}$
11. $\frac{B-DC}{C}$ **12.** $\frac{Q+TC}{T}$ **13.** $\frac{V+TD}{D}$ **14.** $\frac{L}{MB}$ **15.** $\frac{N}{BC}$
16. $\frac{m}{cd}$ **17.** $\frac{tc-b}{t}$ **18.** $\frac{xy-z}{x}$ **19.** 1 **20.** $\frac{5}{6}$
21. $\frac{A}{C-B}$ **22.** $\frac{V}{H-G}$ **23.** $\frac{r}{n+t}$ **24.** $\frac{b}{q-d}$ **25.** $\frac{m}{t+n}$
26. $\frac{b}{d-h}$ **27.** $\frac{d}{C-e}$ **28.** $\frac{m}{r-e^2}$ **29.** $\frac{n}{b-t^2}$ **30.** $\frac{d}{mn-b}$
31. $\frac{N-2MP}{2M}$ **32.** $\frac{B-6Ac}{6A}$ **33.** $\frac{m^2}{n-p}$ **34.** $\frac{q}{w-t}$

page 148 ***Exercise 4***

1. 4
2. 11
3. $D^2 - C$
4. $\frac{c^2 - b}{a}$
5. $\frac{h^2 + t}{g}$
6. $d - t^2$
7. $n - c^2$
8. $c - g^2$
9. $\frac{D - B}{A}$
10. $\pm\sqrt{g}$
11. $\pm\sqrt{B}$
12. $\pm\sqrt{(M + A)}$
13. $\pm\sqrt{(C - m)}$
14. $\pm\sqrt{\frac{n}{m}}$
15. $\frac{at}{z}$
16. $\pm\sqrt{(a - n)}$
17. $\pm\sqrt{(B^2 + A)}$
18. $\pm\sqrt{(t^2 - m)}$
19. $\frac{M^2 - A^2 B}{A^2}$
20. $\frac{N}{B^2}$
21. $\pm\sqrt{(a^2 - t^2)}$
22. $\frac{4}{\pi^2} - t$
23. $\pm\sqrt{\left(\frac{C^2 + b}{a}\right)}$
24. $\pm\sqrt{(x^2 - b)}$

page 148 ***Exercise 5***

1. $3\frac{2}{3}$
2. 3
3. $\frac{D - B}{2N}$
4. $\frac{E + D}{3M}$
5. $\frac{2b}{a - b}$
6. $\frac{e + c}{m + n}$
7. $\frac{3}{x + k}$
8. $\frac{C - D}{R - T}$
9. $\frac{z + x}{a - b}$
10. $\frac{nb - ma}{m - n}$
11. $\frac{d + xb}{x - 1}$
12. $\frac{a - ab}{b + 1}$
13. $\frac{d - c}{d + c}$
14. $\frac{M(b - a)}{b + a}$
15. $\frac{x - n - mn}{m}$
16. $\frac{m^2 + 5}{2 - m}$
17. $\frac{2 + n^2}{n - 1}$
18. $\frac{e - b^2}{b - a}$
19. $\frac{3x}{a + x}$
20. $\frac{e - c}{a - d}$
21. $\frac{d}{a - b - c}$
22. $\frac{ab}{m + n - a}$
23. $\frac{s - t}{b - a}$
24. $2x$
25. $\frac{v}{3}$
26. $\frac{a(b + c)}{b - 2a}$
27. $\frac{5x}{3}$
28. $-\frac{4z}{5}$
29. $\frac{mn}{p^2 - m}$
30. $\frac{mn + n}{4 + m}$

page 149 ***Exercise 6***

1. $-\left(\frac{by + c}{a}\right)$
2. $\pm\sqrt{\left(\frac{e^2 + ab}{a}\right)}$
3. $\frac{n^2}{m^2} + m$
4. $\frac{a - b}{1 + b}$
5. $3y$
6. $\frac{a}{e^2 + c}$
7. $-\left(\frac{a + lm}{m}\right)$
8. $\frac{t^2 g}{4\pi^2}$
9. $\frac{4\pi^2 d}{t^2}$
10. $\pm\sqrt{\frac{a}{3}}$
11. $\pm\sqrt{\left(\frac{t^2 e - ba}{b}\right)}$
12. $\frac{1}{a^2 - 1}$
13. $\frac{a + b}{x}$
14. $\pm\sqrt{(x^4 - b^2)}$
15. $\frac{c - a}{b}$
16. $\frac{a^2 - b}{a + 1}$
17. $\pm\sqrt{\left(\frac{G^2}{16\pi^2} - T^2\right)}$
18. $-\left(\frac{ax + c}{b}\right)$
19. $\frac{1 + x^2}{1 - x^2}$
20. $\pm\sqrt{\left(\frac{a^2 m}{b^2} + n\right)}$
21. $\frac{P - M}{E}$
22. $\frac{RP - Q}{R}$
23. $\frac{z - t^2}{x}$
24. $(g - e)^2 - f$

page 151 **Exercise 7**

1. (a) $S = ke$ (b) $v = kt$ (c) $x = kz^2$ (d) $y = k\sqrt{x}$ (e) $T = k\sqrt{L}$
2. (a) 9 (b) $2\frac{2}{3}$ **3.** (a) 35 (b) 11 **4.** (a) 75 (b) 4

5.

x	1	3	4	$5\frac{1}{2}$
z	4	12	16	22

6.

r	1	2	4	$1\frac{1}{2}$
V	4	32	256	$13\frac{1}{2}$

7. 333 N/cm^2 **8.** 180 m; 2 s **9.** 675 J; $\sqrt{\frac{4}{3}}$ cm **10.** 9000 N; 25 m/s
11. $p \propto w^3$ **12.** $15^4 : 1$ (50625 : 1)

page 153 **Exercise 8**

1. (a) $x = \frac{k}{y}$ (b) $s = \frac{k}{t^2}$ (c) $t = \frac{k}{\sqrt{q}}$ (d) $m = \frac{k}{w}$ (e) $z = \frac{k}{t^2}$

2. (a) 1 (b) 4
3. (a) 36 (b) ± 4 **4.** (a) 6 (b) 16

5.

y	2	4	1	$\frac{1}{4}$
z	8	4	16	64

6.

t	2	5	20	10
v	25	4	$\frac{1}{4}$	1

7. (a) 6 (b) 50 **8.** 2·5 m^3; 200 N/m^2 **9.** 3 h; 48 men **10.** 6 cm
11. 2 days; 200 days **12.** (a) 20 min (b) 2 min **13.** $k = 100$, $n = 3$ **14.** $k = 12$, $n = 2$

x	1	2	4	10
z	100	$12\frac{1}{2}$	1·5625	$\frac{1}{10}$

v	1	4	36	10000
y	12	6	2	$\frac{3}{25}$

page 156 **Exercise 9**

1. AB : $\frac{1}{5}$; BC : $\frac{5}{2}$; AC : $-\frac{4}{3}$
2. PQ : $\frac{4}{5}$; PR : $-\frac{1}{6}$; QR : -5
3. (a) 3 (b) $\frac{3}{2}$ (c) 4 (d) 5
4. $3\frac{1}{2}$ **5.** (a) $\frac{n+4}{2m-3}$ (b) $n = -4$ (c) $m = 1\frac{1}{2}$

page 158 **Exercise 10**

21. A : $y = 3x - 4$; B : $y = x + 2$ **22.** C : $y = \frac{2}{3}x - 2$ or $3y = 2x - 6$; D : $y = -2x + 4$
23. (a) A (0, −8), B (4, 0) (b) 2 (c) $y = 2x - 8$

page 159 **Exercise 11**

1. $y = 3x + 7$ **2.** $y = 2x - 9$ **3.** $y = -x + 5$ **4.** $y = 2x - 1$ **5.** $y = 3x + 5$
6. $y = -x + 7$ **7.** $y = \frac{1}{2}x - 3$ **8.** $y = 2x - 3$ **9.** $y = 3x - 11$ **10.** $y = -x + 5$

page 160 **Exercise 12**

1. $a = 2{\cdot}5$, $c = 1{\cdot}5$; $Z = 2{\cdot}5X + 1{\cdot}5$ **2.** $m = 2$, $c = 3$ **3.** $n = -2{\cdot}5$, $k = 15$
4. $m = 0{\cdot}2$, $c = 3$ **5.** $a = 12$, $b = 3{\cdot}5$; $V = \frac{12}{T} + 3{\cdot}5$ **6.** $k = 0{\cdot}092$

page 162 **Exercise 13**

1. (a) 45 min (b) 0915 (c) 60 km/h (d) 100 km/h (e) 57·1 km/h
2. (a) 0915 (b) 64 km/h (c) 37·6 km/h (d) 47 km (e) 80 km/h
3. (b) 1105 **4.** (b) 1242 **5.** (b) 1235 **6.** $1\frac{1}{8}$ h **7.** 1 h

page 164 **Exercise 14**

2. (a) £2000 (b) £22 000 (c) £3000, 75%
3. (a) 6 gallons (b) 40 mpg; 30 mpg (c) $33\frac{1}{3}$ mpg; $5\frac{1}{2}$ gallons
4. (a) 740p (b) £280 (c) £14 000 (d) £11 000
5. (a) 2000 (b) 270 (c) $1{\cdot}6 \leqslant x \leqslant 2{\cdot}4$
6. (a) Yes (b) No (c) About £250–£270

page 167 **Exercise 15**

1. B **2.** (a) C (b) A (c) D (d) B **3.** D
7. (a) (i) B (ii) A (b) 8 s to 18 s (c) About 15 s (d) About 9 s (e) B (f) A

page 170 **Exercise 16**

1. (a) quadratic, x^2 term is negative (b) cubic, x^3 term is positive
(c) reciprocal (d) cubic, x^3 term is negative
(e) quadratic, x^2 term is positive (f) reciprocal
3. (a): (iii) (b): (ii) (c): (i) (d): (vi) (e): (iv) (f): (v)

page 173 **Exercise 18**

12. (a) 7·25 (b) −0·8, 3·8 **13.** (a) 0·75 (b) 1·23
14. (a) 3·13 (b) 3·35 **15.** (a) −2·45 (b) 1·9
17. (a) 245 (b) 41 (c) $25 < x < 67$

page 174 **Exercise 19**

1. (a) 10·7 cm^2 (b) 1·7 cm × 5·3 cm (c) 12·25 cm^2 (d) 3·5 cm × 3·5 cm (e) square
2. 15 m × 30 m **3.** (a) 2·5 s (b) 31·3 m (c) $2 < t < 3$
4. (a) 108 m/s (b) 1·4 s (c) $2{\cdot}3 < t < 3{\cdot}6$

page 177 **Exercise 20**

1. (a) 0·4, 2·4 (b) −0·8, 3·8 (c) −1, 3 (d) −0·4, 2·4
2. −0·3, 3·3 **3.** 0·6, 3·4 **4.** 0·3, 3·7
5. (a) $y = 3$ (b) $y = -2$ (c) $y = x + 4$ (d) $y = x$ (e) $y = 6$
6. (a) $y = 6$ (b) $y = 0$ (c) $y = 4$ (d) $y = 2x$ (e) $y = 2x + 4$
7. (a) $y = -4$ (b) $y = 2x$ (c) $y = x - 2$ (d) $y = -3$ (e) $y = 2$
8. (a) −1·65, 3·65 (b) −1·3, 2·3 (c) −1·45, 3·45
9. (a) 1·7, 5·3 (b) 0·2, 4·8 **10.** (a) −2·35, 0·85 (b) −2·7, 1·7
11. (a) 3·35 (b) 2·4, 7·6 (c) 4·25 **12.** (a) ±3·74 (c) ±2·83
13. (a) 1·75 (b) 0, ±1·4 **14.** (a) 2·6 (b) 0·45, 3·3 (c) 0·64; 5·66
15. (a) 0, 2·9 (b) −0·65, 1·35, 5·3

page 181 ***Exercise 21***

1. 3^4 **2.** $4^2 \times 5^3$ **3.** 3×7^3 **4.** $2^3 \times 7$ **5.** 10^{-3}
6. $2^{-2} \times 3^{-3}$ **7.** $15^{\frac{1}{2}}$ **8.** $3^{\frac{1}{3}}$ **9.** $10^{\frac{1}{5}}$ **10.** $5^{\frac{3}{2}}$
11. x^7 **12.** y^{13} **13.** z^4 **14.** z^{100} **15.** m
16. e^{-5} **17.** y^2 **18.** w^6 **19.** y **20.** x^{10}
21. 1 **22.** w^{-5} **23.** w^{-5} **24.** x^7 **25.** a^8
26. k^3 **27.** 1 **28.** x^{29} **29.** y^2 **30.** x^6
31. z^4 **32.** t^{-4} **33.** $4x^6$ **34.** $16y^{10}$ **35.** $6x^4$
36. $10y^5$ **37.** $15a^4$ **38.** $8a^3$ **39.** 3 **40.** $4y^2$
41. $\frac{5}{2}y$ **42.** $32a^4$ **43.** $108x^5$ **44.** $4z^{-3}$ **45.** $2x^{-4}$
46. $\frac{5}{2}y^5$ **47.** 1 **48.** $21w^{-3}$ **49.** $2n^4$ **50.** $2x$

page 182 ***Exercise 22***

1. 27 **2.** 1 **3.** $\frac{1}{9}$ **4.** 25 **5.** 2
6. 4 **7.** 9 **8.** 2 **9.** 27 **10.** 3
11. $\frac{1}{3}$ **12.** $\frac{1}{2}$ **13.** 1 **14.** $\frac{1}{5}$ **15.** 10
16. 8 **17.** 32 **18.** 4 **19.** $\frac{1}{9}$ **20.** $\frac{1}{8}$
21. 18 **22.** 10 **23.** 1000 **24.** $\frac{1}{1000}$ **25.** $\frac{1}{9}$
26. 1 **27.** $1\frac{1}{2}$ **28.** $\frac{1}{25}$ **29.** $\frac{1}{10}$ **30.** $\frac{1}{4}$
31. $\frac{1}{4}$ **32.** 100 000 **33.** 1 **34.** $\frac{1}{32}$ **35.** 0·1
36. 0·2 **37.** 1·5 **38.** 1 **39.** 9 **40.** $1\frac{1}{2}$
41. $\frac{3}{10}$ **42.** 64 **43.** $\frac{1}{100}$ **44.** $1\frac{2}{3}$ **45.** $\frac{1}{100}$

page 182 ***Exercise 23***

1. 3 **2.** 4 **3.** -1 **4.** -2 **5.** 3
6. 3 **7.** 1 **8.** $\frac{1}{5}$ **9.** 0 **10.** -4
11. 2 **12.** -5 **13.** 1 **14.** $\frac{1}{18}$ **15.** 2, 4
16. (a) 3·60 (b) 5·44 **17.** (b) 1, 7, 9 **18.** 21
19. (a) 512 (b) 6 h (c) 2^{21} **20.** 1, 2; 1
21. (a) 2·37, 2·44 (b) 2·59, 2·72, 2·718 (d) As $x \to \infty$, $f(x) \to e^1$

page 184 ***Examination Exercise 4***

1. (a) $c = d + x(a + b)$ (b) $x = \dfrac{c - d}{a + b}$ (c) 12
2. (b) (i) £76 (ii) 47 (iii) £25 (iv) £1·50 (c) (i) 36 (ii) £9
3. (a) $x = 2$, $y = -1$ (b) $v = \dfrac{uf}{u - f}$ **4.** (a) 9·6 (b) $\dfrac{Vr}{12 - V}$
5. $R = \dfrac{12}{I} - 2$ **6.** (a) $h = \dfrac{A}{2\pi r} - r$ (b) 3·8 **7.** 56·25 m
8. (a) 3, -1 (b) $11x^8$ **9.** (a) 1 (b) 2 (c) 0·889 (3 s.f.)
10. (e) 55 mph **11.** (a) $x(6 - x)$ (d) 1·4 cm
12. (a) (iii) -8 (b) (ii) $x = 1·5$ or $5·1$ **13.** $y = 1·3x - 1·5$

Part 5

page 189 ***Exercise 1***

1. 16 cm^2 **2.** 27 cm^2 **3.** $11\frac{1}{4}$ cm^2 **4.** $14\frac{1}{2}$ cm^2 **5.** 128 cm^2
6. 12 cm^2 **7.** 8 cm **8.** 18 cm **9.** $4\frac{1}{2}$ cm **10.** $7\frac{1}{2}$ cm
11. (a) $16\frac{2}{3}$ cm^2 (b) $10\frac{2}{3}$ cm^2 **12.** (a) 25 cm^2 (b) 21 cm^2 **13.** 24 cm^2
14. 150 **15.** 360 **16.** Less (for the same weight) **17.** 6·29 cm
18. $\frac{4}{9}$

page 192 ***Exercise 2***

1. 480 cm^3 **2.** 540 cm^3 **3.** 160 cm^3 **4.** 4500 cm^3 **5.** 81 cm^3
6. 11 cm^3 **7.** 16 cm^3 **8.** $85\frac{1}{3}$ cm^3 **9.** 4 cm **10.** 21 cm
11. 4·6 cm **12.** 9 cm **13.** 6·6 cm **14.** $4\frac{1}{2}$ cm **15.** $168\frac{3}{4}$ cm^3
16. 106·3 cm^3 **17.** 12 cm **18.** (a) 2 : 3 (b) 8 : 27 **19.** 8 : 125
20. 108 **21.** 21 m, 62·5 m^3, 700 cm^2, 12, 92·5 m^2 **22.** 54 kg
23. 240 cm^2 **24.** $9\frac{3}{8}$ litres **25.** $2812\frac{1}{2}$ cm^2

page 196 ***Exercise 3***

1. (a) 2·09 cm; 4·19 cm^2 (b) 7·85 cm; 39·3 cm^2 (c) 8·20 cm; 8·20 cm^2 **2.** 31·9 cm^2
3. 31·2 cm^2 **4.** (a) 7·07 cm^2 (b) 19·5 cm^2 **5.** (a) 3·98 cm (b) 74·9°
6. (a) 12 cm (b) 30° **7.** (a) 30° (b) 10·5 cm
8. (a) 85·9° (b) 57·3° (c) 6·25 cm **9.** 30·6 cm^2 **10.** 15·1 km^2
11. 36° **12.** (a) 6·14 cm (b) 27·6 m (c) 28·6 cm^2 **13.** (a) 18 cm (b) 38·2°
14. (a) 10 cm (b) 43·0° **15.** (b) 66·8° **16.** $x = 38{\cdot}1°$

page 199 ***Exercise 4***

1. (a) 14·5 cm (b) 72·6 cm^2 (c) 24·5 cm^2 (d) 48·1 cm^2
2. (a) 5·08 cm^2 (b) 82·8 m^2 (c) 5·14 cm^2
3. (a) 60°, 9·06 cm^2 (b) 106·3°, 11·2 cm^2 **4.** 3 cm **5.** 3·97 cm
6. (a) 13·5 cm^2 (b) 405 cm^3 **7.** (a) 130 cm^2 (b) 184 cm^2 **8.** 19·6 cm^2
9. 0·313 r^2 **10.** (a) 8·37 cm (b) 54·5 cm (c) 10·4 cm **11.** 81·2 cm^2

page 202 ***Exercise 5***

1. 20·9 cm^3 **2.** 524 cm^3 **3.** 4190 cm^3 **4.** 101 cm^3 **5.** 268 cm^3
6. 550 cm^3 **7.** 776 cm^3 **8.** 262 cm^3 **9.** 235 cm^3 **10.** 415 cm^3
11. £3870 **12.** £331 **13.** 303 cm^2 **14.** 2·43 cm **15.** 23·9 cm
16. 3·46 cm **17.** 3·72 cm **18.** 1930 g **19.** 106 s
20. (a) 2·9 m^3 (b) 1·7 m **21.** 488 cm^3 **22.** (a) 125 (b) $2{\cdot}7 \times 10^7$
23. (a) 0·36 cm (b) 0·427 cm **24.** (a) 6·69 cm (b) 39·1 cm **25.** 4·19 cm^3
26. (a) 16π cm (b) 8 cm (c) 6 cm **27.** 2720 cm^3

page 206 ***Exercise 6***

1. (a) 2 (b) 2 (c) 3 (d) 1 (e) 2 (f) 3
2. (a) 2 (b) 3 (c) 2 (d) 1 (e) 3 (f) 1 (g) 0 (h) 2
3. (a) A (b) L (c) V (d) A (e) F (f) A (g) V (h) A
4. 2 **5.** (a) 2 (b) 3 (c) 1 (d) 2, 1 (e) 3, 3 (f) 2

page 210 ***Exercise 7***

1. $\mathbf{d}$ **2.** $2\mathbf{c}$ **3.** $3\mathbf{c}$ **4.** $3\mathbf{d}$ **5.** $5\mathbf{d}$
6. $3\mathbf{c}$ **7.** $-2\mathbf{d}$ **8.** $-2\mathbf{c}$ **9.** $-3\mathbf{c}$ **10.** $-\mathbf{c}$
11. $\mathbf{c}+\mathbf{d}$ **12.** $\mathbf{c}+2\mathbf{d}$ **13.** $2\mathbf{c}+\mathbf{d}$ **14.** $3\mathbf{c}+\mathbf{d}$ **15.** $2\mathbf{c}+2\mathbf{d}$
16. $\overrightarrow{QI}$ **17.** $\overrightarrow{QU}$ **18.** $\overrightarrow{QH}$ **19.** $\overrightarrow{QB}$ **20.** $\overrightarrow{QF}$
21. $\overrightarrow{QJ}$ **22.** (a) $-\mathbf{a}$ (b) $\mathbf{a}+\mathbf{b}$ (c) $2\mathbf{a}-\mathbf{b}$ (d) $-\mathbf{a}+\mathbf{b}$
23. (a) $\mathbf{a}+\mathbf{b}$ (b) $\mathbf{a}-2\mathbf{b}$ (c) $-\mathbf{a}+\mathbf{b}$ (d) $-\mathbf{a}-\mathbf{b}$
24. (a) $-\mathbf{a}-\mathbf{b}$ (b) $3\mathbf{a}-\mathbf{b}$ (c) $2\mathbf{a}-\mathbf{b}$ (d) $-2\mathbf{a}+\mathbf{b}$
25. (a) $\mathbf{a}-2\mathbf{b}$ (b) $\mathbf{a}-\mathbf{b}$ (c) $2\mathbf{a}$ (d) $-2\mathbf{a}+3\mathbf{b}$

page 211 ***Exercise 8***

1. (a) $\mathbf{a}$ (b) $-\mathbf{a}+\mathbf{b}$ (c) $2\mathbf{b}$ (d) $-2\mathbf{a}$ (e) $-2\mathbf{a}+2\mathbf{b}$ (f) $-\mathbf{a}+\mathbf{b}$
(g) $\mathbf{a}+\mathbf{b}$ (h) $\mathbf{b}$ (i) $-\mathbf{b}+2\mathbf{a}$ (j) $-2\mathbf{b}+\mathbf{a}$
2. (a) $\mathbf{a}$ (b) $-\mathbf{a}+\mathbf{b}$ (c) $3\mathbf{b}$ (d) $-2\mathbf{a}$ (e) $-2\mathbf{a}+3\mathbf{b}$ (f) $-\mathbf{a}+\frac{3}{2}\mathbf{b}$
(g) $\mathbf{a}+\frac{3}{2}\mathbf{b}$ (h) $\frac{3}{2}\mathbf{b}$ (i) $-\mathbf{b}+2\mathbf{a}$ (j) $-3\mathbf{b}+\mathbf{a}$
3. (a) $2\mathbf{a}$ (b) $-\mathbf{a}+\mathbf{b}$ (c) $2\mathbf{b}$ (d) $-3\mathbf{a}$ (e) $-3\mathbf{a}+2\mathbf{b}$ (f) $-\frac{3}{2}\mathbf{a}+\mathbf{b}$
(g) $\frac{3}{2}\mathbf{a}+\mathbf{b}$ (h) $\frac{1}{2}\mathbf{a}+\mathbf{b}$ (i) $-\mathbf{b}+3\mathbf{a}$ (j) $-2\mathbf{b}+\mathbf{a}$
4. (a) $\frac{1}{2}\mathbf{a}$ (b) $-\mathbf{a}+\mathbf{b}$ (c) $4\mathbf{b}$ (d) $-\frac{3}{2}\mathbf{a}$ (e) $-\frac{3}{2}\mathbf{a}+4\mathbf{b}$ (f) $-\mathbf{a}+\frac{8}{3}\mathbf{b}$
(g) $\frac{1}{2}\mathbf{a}+\frac{8}{3}\mathbf{b}$ (h) $-\frac{1}{2}\mathbf{a}+\frac{8}{3}\mathbf{b}$ (i) $\frac{3}{2}\mathbf{a}-\mathbf{b}$ (j) $\mathbf{a}-4\mathbf{b}$
5. $\frac{1}{2}\mathbf{s}-\frac{1}{2}\mathbf{t}$ **6.** $\frac{1}{3}\mathbf{a}+\frac{2}{3}\mathbf{b}$ **7.** $\mathbf{a}+\mathbf{c}-\mathbf{b}$ **8.** $2\mathbf{m}+2\mathbf{n}$
9. (a) $\mathbf{b}-\mathbf{a}$ (b) $\mathbf{b}-\mathbf{a}$ (c) $2\mathbf{b}-2\mathbf{a}$ (d) $\mathbf{b}-2\mathbf{a}$ (e) $\mathbf{b}-2\mathbf{a}$ (f) $2\mathbf{b}-3\mathbf{a}$
10. (a) $\mathbf{y}-\mathbf{z}$ (b) $\frac{1}{2}\mathbf{y}-\frac{1}{2}\mathbf{z}$ (c) $\frac{1}{2}\mathbf{y}+\frac{1}{2}\mathbf{z}$ (d) $-\mathbf{x}+\frac{1}{2}\mathbf{y}+\frac{1}{2}\mathbf{z}$
(e) $-\frac{2}{3}\mathbf{x}+\frac{1}{3}\mathbf{y}+\frac{1}{3}\mathbf{z}$ (f) $\frac{1}{3}\mathbf{x}+\frac{1}{3}\mathbf{y}+\frac{1}{3}\mathbf{z}$

page 214 ***Exercise 9***

1. (a) 13·0 N, 032·5° (b) 34 N, 208·1° (c) 8·94 N, 026·6° (d) 12·0 N, 048·4°
2. (a) 15·6 N, 26·3° (b) 5·79 N, 89·0°
3. 11·6 N at 46·7° to vertical **4.** 10·3 N, 127·4°

page 216 ***Exercise 10***

1. 4·47 m/s, 026·6° **2.** 286 km/h, 102·1° **3.** 8·16 knots, 164·9° **4.** 133 m
5. bearing 113·6° **6.** 043·5° **7.** 009·6° **8.** 151 km/h

page 218 ***Exercise 11***

4. 162° **5.** 153° **6.** (a) 140° (b) 110° (c) 50°
7. 290° **8.** 315° **9.** (a) 350° (b) 304° (c) 60°
10. 220° **11.** 160° **12.** 82° **13.** 315° **14.** 240°
15. 250° **16.** 58°, 122° **17.** 20·5°, 159·5° **18.** 53·1°, 306·9°

19. (a) 46·1°, 133·9° (b) 72·5°, 287·5° (c) 78·7°, 258·7° (d) 220·5°, 319·5°
20. 30°, 150°, 210°, 330° **21.** (a) 40°, 140° (b) 30°, 150°
22. (a) (i) 16°, 111° (ii) 153° (b) 2·24 (c) 63° **23.** (a) 48°, 205° (b) 37°, 217°

page 220 ***Exercise 12***

1. 6·38 m **2.** 12·5 m **3.** 5·17 cm **4.** 40·4 cm **5.** 7·81 m, 7·10 m
6. 3·55 m, 6·68 m **7.** 8·61 cm **8.** 9·97 cm **9.** 8·52 cm **10.** 15·2 cm
11. 35·8° **12.** 42·9° **13.** 32·3° **14.** 37·8° **15.** 35·5°, 48·5°
16. 68·8°, 80·0° **17.** 64·6° **18.** 34·2° **19.** 50·6° **20.** 39·1°
21. 39·5° **22.** 21·6°

page 222 ***Exercise 13***

1. 6·24 **2.** 6·05 **3.** 5·47 **4.** 9·27 **5.** 10·1
6. 8·99 **7.** 5·87 **8.** 4·24 **9.** 11·9 **10.** 154
11. 25·2° **12.** 78·5° **13.** 115·0° **14.** 111·1° **15.** 24·0°
16. 92·5° **17.** 99·9° **18.** 38·2° **19.** 137·8° **20.** 34·0°

page 224 ***Exercise 14***

1. (a) 50·2 km (b) 054·7° **2.** 35·6 km **3.** 25·2 m **4.** 101·5°
5. 92·9° **6.** 40·4 m **7.** (a) 9·8 km (b) 085·7°
8. (a) 29·6 km (b) 050·5° **9.** (a) 10·8 m (b) 72·6° (c) 32·6°
10. 378 km, 048·4° **11.** (a) 62·2° (b) 2·33 km **12.** 9·64 m **13.** 8·6°
14. (a) 5·66 cm (b) 4·47 cm (c) 3·66 cm **15.** 70·2°

page 227 ***Exercise 15***

1. $a = 27°$, $b = 30°$ **2.** $c = 20°$, $d = 45°$ **3.** $c = 58°$, $d = 41°$, $e = 30°$
4. $f = 40°$, $g = 55°$, $h = 55°$ **5.** $a = 32°$, $b = 80°$, $c = 43°$ **6.** $x = 34°$, $y = 34°$, $z = 56°$
7. 43° **8.** 92° **9.** 42°
10. $c = 46°$, $d = 44°$ **11.** $e = 49°$, $f = 41°$ **12.** $g = 76°$, $h = 52°$
13. 48° **14.** 32° **15.** 22°
16. $a = 36°$, $x = 36°$

page 229 ***Exercise 16***

1. $a = 94°$, $b = 75°$ **2.** $c = 101°$, $d = 84°$ **3.** $x = 92°$, $y = 116°$
4. $c = 60°$, $d = 45°$ **5.** 37° **6.** 118°
7. $e = 36°$, $f = 72°$ **8.** 35° **9.** 18°
10. 90° **11.** 30° **12.** $22\frac{1}{2}°$
13. $n = 58°$, $t = 64°$, $w = 45°$ **14.** $a = 32°$, $b = 40°$, $c = 40°$ **15.** $a = 18°$, $c = 72°$
16. 55° **17.** $e = 41°$, $f = 41°$, $g = 41°$ **18.** 8°
19. $x = 30°$, $y = 115°$ **20.** $x = 80°$, $z = 10°$

page 230 ***Exercise 17***

1. $a = 18°$ **2.** $x = 40°$, $y = 65°$, $z = 25°$ **3.** $c = 30°$, $e = 15°$
4. $f = 50°$, $g = 40°$ **5.** $h = 70°$, $k = 40°$, $i = 40°$ **6.** $m = 108°$, $n = 36°$
7. $x = 50°$, $y = 68°$ **8.** $a = 74°$, $b = 32°$ **9.** $e = 36°$
10. $k = 63°$, $m = 54°$ **11.** $k = 50°$, $m = 50°$, $n = 80°$, $p = 80°$
12. (a) p (b) $2p$ (c) $90 - 2p$ **13.** $x = 70°$, $y = 20°$, $z = 55°$

page 232 ***Exercise 18***

3. (c) (8, 8) (8, −6) (−8, 6)
4. (f) (1, −1) (−3, −1) (−3, −3)
5. (f) (8, −2) (6, −6) (8, −6)
6. (c) (−2, 1) (−2, −1) (1, −2)
7. (e) (−5, 2) (−5, 6) (−3, 5)
8. (b) (i) 90° ACW (0, 0) (ii) 180° (2, 1) (iii) 90° CW (2, 0)
(iv) 180° $(3\frac{1}{2}, 2\frac{1}{2})$ (v) 90° ACW (6, 1) (vi) 90° CW (1, 3)

page 233 ***Exercise 19***

1. (a) $\begin{pmatrix} 7 \\ 3 \end{pmatrix}$ (b) $\begin{pmatrix} 0 \\ -9 \end{pmatrix}$ (c) $\begin{pmatrix} 9 \\ 10 \end{pmatrix}$ (d) $\begin{pmatrix} -10 \\ 3 \end{pmatrix}$ (e) $\begin{pmatrix} -1 \\ 13 \end{pmatrix}$
(f) $\begin{pmatrix} 10 \\ 0 \end{pmatrix}$ (g) $\begin{pmatrix} -9 \\ -4 \end{pmatrix}$ (h) $\begin{pmatrix} -10 \\ 0 \end{pmatrix}$

12. (a) Rotation 90° clockwise, centre (0, −2)
(b) Reflection in $y = x$
(c) Translation $\begin{pmatrix} 3 \\ 7 \end{pmatrix}$
(d) Enlargement, scale factor 2, centre (−5, 5)
(e) Translation $\begin{pmatrix} -7 \\ -3 \end{pmatrix}$
(f) Reflection in $y = x$

13. (a) Rotation 90° clockwise, centre (4, −2)
(b) Translation $\begin{pmatrix} 8 \\ 2 \end{pmatrix}$
(c) Reflection in $y = x$
(d) Enlargement, scale factor $\frac{1}{2}$, centre (7, −7)
(e) Rotation 90° anticlockwise, centre (−8, 0)
(f) Enlargement, scale factor 2, centre (−1, −9)
(g) Rotation 90° anticlockwise, centre (7, 3)

14. (a) Enlargement, scale factor $1\frac{1}{2}$, centre (1, −4)
(b) Rotation 90° clockwise, centre (0, −4)
(c) Reflection in $y = -x$
(d) Translation $\begin{pmatrix} 11 \\ 10 \end{pmatrix}$
(e) Enlargement, scale factor $\frac{1}{2}$, centre (−3, 8)
(f) Rotation 90° anticlockwise, centre $(\frac{1}{2}, 6\frac{1}{2})$
(g) Enlargement, scale factor 3, centre (−2, 5)

page 235 ***Exercise 20***

1. (a) (−4, 4) (b) (2, −2) (c) (0, 0) (d) (0, 4) (e) (0, 0)
2. (a) (−2, 5) (b) (−4, 0) (c) (2, −2) (d) (1, −1)
3. (a) reflection in y-axis
(b) rotation 180°, centre (−2, 2)
(c) rotation 90° clockwise, centre (2, 2)

4. (a) rotation 90° clockwise, centre (0, 0)
(b) translation $\begin{pmatrix} -2 \\ 5 \end{pmatrix}$
(c) rotation 90° anticlockwise, centre (2, −4)
(d) rotation 90° anticlockwise, centre $(-\frac{1}{2}, 3\frac{1}{2})$

5. (a) rotation 90° anticlockwise, centre (2, 2)
(b) enlargement, scale factor $\frac{1}{2}$, centre (8, 6)
(c) rotation 90° clockwise, centre $(-\frac{1}{2}, -3\frac{1}{2})$

6. $\mathbf{A}^{-1}$: reflection in $x = 2$
$\mathbf{B}^{-1}$: B
$\mathbf{C}^{-1}$: translation $\begin{pmatrix} 6 \\ -2 \end{pmatrix}$
$\mathbf{D}^{-1}$: D
$\mathbf{E}^{-1}$: E
$\mathbf{F}^{-1}$: translation $\begin{pmatrix} -4 \\ -3 \end{pmatrix}$
$\mathbf{G}^{-1}$: 90° rotation anticlockwise, centre (0, 0)
$\mathbf{H}^{-1}$: enlargement, scale factor 2, centre (0, 0)

7. (a) (4, 0) (b) (−6, −1) (c) (−2, −2) (d) (2, −2) (e) (6, 2)
8. (a) (1, −6) (b) (4, −2) (c) (2, 7) (d) (4, −6) (e) (2, −4)
9. (a) (−1, −2) (b) (8, 2) (c) (4, −6) (d) (0, −2)
10. (b) rotation, 180°, centre (4, 0) (c) translation $\begin{pmatrix} 12 \\ -4 \end{pmatrix}$

page 237 ***Exercise 21***

1. $\begin{pmatrix} -1 & 12 \\ 4 & 7 \end{pmatrix}$ **2.** $\begin{pmatrix} 15 & 20 \\ -4 & -9 \end{pmatrix}$ **3.** $\begin{pmatrix} 5 & -10 \\ 2 & 7 \end{pmatrix}$ **4.** $\begin{pmatrix} 3 & 14 \\ -2 & 9 \end{pmatrix}$

5. $\begin{pmatrix} 5 \\ 13 \end{pmatrix}$ **6.** $\begin{pmatrix} 5 \\ 1 \end{pmatrix}$ **7.** $\begin{pmatrix} -2 & 16 & 1 \\ 19 & -9 & 7 \end{pmatrix}$ **8.** $\begin{pmatrix} 16 & 2 & 7 \\ -7 & 17 & -1 \end{pmatrix}$

9. $\begin{pmatrix} 8 & -27 \\ 23 & -2 \end{pmatrix}$ **10.** $\begin{pmatrix} 8 & -27 \\ 23 & -2 \end{pmatrix}$

11. $x = 1, y = 4$ **12.** $m = 5, n = -\frac{1}{3}$ **13.** $x = 1, y = 2, z = -1, w = -2$
14. $x = 2\frac{2}{3}$ **15.** $k = \pm 1$ **16.** (a) $k = 2$ (b) $m = 4$
17. (a) $n = 3$ (b) $q = 9$

page 238 ***Exercise 22***

2. A: reflection in x-axis B: reflection in y-axis C: reflection in $y = x$
D: rotation, −90°, centre (0, 0) E: reflection in $y = -x$ F: rotation, 180°, centre (0, 0)
G: rotation, +90°, centre (0, 0) H: identity (no change)

3. (d) rotation 45° anticlockwise; enlargement scale factor $\sqrt{2}$ (1·41)
4. (d) rotation 26·6° clockwise; enlargement scale factor $\sqrt{5}$ (2·24)
5. (d) rotation 90° clockwise, centre (0, 0)
6. (c) $OB = \sqrt{20}$, $OB' = 3\sqrt{20}$ (d) 36·9° (e) rotation 36·9°; enlargement scale factor 3
7. (a) reflection in $y = x - 1$ (b) reflection in $y = 1$
(c) rotation −90°, centre (2, −2) (d) enlargement, scale factor 3, centre (2, −1)
8. $\mathbf{BA} \equiv \mathbf{A}$ then $\mathbf{B}$

page 240 ***Examination Exercise 5***

1. (a) $3{\cdot}14\ m^2$ (b) $0{\cdot}157\ m^3$ (c) 13·6 m (d) $177\ m^2$
2. (a) $8000\ cm^3$ (b) $268\ mm^3$ (c) 29 800 **3.** (a) 49°, $7{\cdot}34\ m^2$ (b) $21\,200\ m^3$
4. (a) 35 (b) 175 (c) 6·8 (d) 823 200
5. (a) $180\ cm^3$ (b) 55 (c) $92\ cm^3$ **6.** (a) $\frac{\pi}{6}d^3 + \frac{\pi}{4}d^2h$ (b) $\pi d^2 + \pi dh$
7. 12·4 m/s, bearing 104° **8.** 60°, 300°, 420° (and others) **9.** 16·97
10. (a) 57° (b) 39 cm **11.** (a) 4850 m (b) 1070 m
12. (a) 42° (b) 53° (c) 21° (d) 106°
13. (c) (i) 5·8 m (ii) 3:40, 8:20 (d) 0·78 m/h (e) 9:00 am
14. (a) (i) $2a + x$ (ii) $\pi a - x$ (c) 32·7° (d) 073·6°

Part 6

page 245 ***Exercise 1***

1. (b) 96 min **2.** (a) B (b) A
4. (b) after 76/77 years i.e. 2056/57 **6.** (b) about 130 years

page 248 ***Exercise 2***

1. (a) 3 (b) −5 (c) 1·5 **2.** 3·3 **3.** (a) (i) 4 (ii) 8

page 249 ***Exercise 3***

1. 7·47 **2.** 13·4 **3.** 23·5 **4.** (b) 38·5 (c) less
5. (a) 58 (b) greater **6.** (a) $x = 1$, $x = 4$ (b) about $7\frac{1}{2}$ square units

page 252 ***Exercise 4***

1. (a) $1\frac{1}{2}\ m/s^2$ (b) 675 m **2.** (a) 600 m (b) 225 m (c) $-2\ m/s^2$
3. (a) 600 m (b) $387\frac{1}{2}$ m (c) $0\ m/s^2$ **4.** (a) 20 m/s (b) 750 m
5. (a) 8 s (b) 496 m (c) 12·4 m/s
6. (a) $0{\cdot}75\ m/s^2$ (b) 680 m (both approximate)
7. (a) $0{\cdot}35\ m/s^2$ (b) 260 m (both approximate) **8.** $1{\cdot}0 < \text{speed} < 1{\cdot}1$ m/s
9. (a) 50 m/s (b) 20 s **10.** (a) 20 m/s (b) 20 s

page 254 ***Exercise 5***

1. 225 m **2.** 120 m **3.** 1 km **4.** 10 s **5.** 60 s **6.** 88 yards
7. 55 yards **8.** 1·39 km **9.** 250 m **10.** Yes, Stopping distance = 46·5 m
11. 94 375 m

page 255 ***Exercise 6***

1. (a) Yes (b) 10·5 **2.** 2·5
3. (a) 10 (b) 0·742 (3 sf) (c) No (d) 3·37 (e) 1·93
4. 2·14 **6.** −5·140 **7.** 2·1926
8. (a) diverges (b) oscillates between −1 and 2
9. (c) limit of sequence which starts a, b is $a + \frac{2}{3}(b - a)$ or $\frac{1}{3}a + \frac{2}{3}b$ **10.** (c) 2·828

page 259 ***Exercise 7***

1. (a) 5·19, −0·19 (b) 1·56, −2·56 (c) 2·30, −1·30 **2.** 2·69, −0·19
3. $x = 3$, equal roots **4.** no roots **5.** 3·28, −1·32 **6.** 1·79

page 261 ***Exercise 8***

1. $\frac{5}{7}$ **2.** $\frac{7}{8}$ **3.** $5y$ **4.** $\frac{1}{2}$ **5.** 4
6. $\frac{x}{2y}$ **7.** 2 **8.** $\frac{a}{2}$ **9.** $\frac{2b}{3}$ **10.** $\frac{a}{5b}$
11. a **12.** $\frac{7}{8}$ **13.** $\frac{5+2x}{3}$ **14.** $\frac{3x+1}{x}$ **15.** $\frac{32}{25}$
16. $\frac{4+5a}{5}$ **17.** $\frac{3}{4-x}$ **18.** $\frac{b}{3+2a}$ **19.** $\frac{5x+4}{8x}$ **20.** $\frac{2x+1}{y}$
21. $\frac{x+2y}{3xy}$ **22.** $\frac{6-b}{2a}$ **23.** $\frac{2b+4a}{b}$ **24.** $x-2$ **25.** $\frac{x+2}{x-3}$
26. $\frac{x}{x+1}$ **27.** $\frac{x+4}{2(x-5)}$ **28.** $\frac{x+5}{x-2}$ **29.** $\frac{x+3}{x+2}$ **30.** $\frac{x+5}{x-2}$

page 262 ***Exercise 9***

1. $\frac{3}{5}$ **2.** $\frac{3x}{5}$ **3.** $\frac{3}{x}$ **4.** $\frac{4}{7}$ **5.** $\frac{4x}{7}$
6. $\frac{4}{7x}$ **7.** $\frac{7}{8}$ **8.** $\frac{7x}{8}$ **9.** $\frac{7}{8x}$ **10.** $\frac{5}{6}$
11. $\frac{5x}{6}$ **12.** $\frac{5}{6x}$ **13.** $\frac{23}{20}$ **14.** $\frac{23x}{20}$ **15.** $\frac{23}{20x}$
16. $\frac{1}{12}$ **17.** $\frac{x}{12}$ **18.** $\frac{1}{12x}$ **19.** $\frac{5x+2}{6}$ **20.** $\frac{7x+2}{12}$
21. $\frac{9x+13}{10}$ **22.** $\frac{1-2x}{12}$ **23.** $\frac{2x-9}{15}$ **24.** $\frac{-3x-12}{14}$ **25.** $\frac{3x+1}{x(x+1)}$
26. $\frac{7x-8}{x(x-2)}$ **27.** $\frac{8x+9}{(x-2)(x+3)}$ **28.** $\frac{4x+11}{(x+1)(x+2)}$ **29.** $\frac{-3x-17}{(x+3)(x-1)}$ **30.** $\frac{11-x}{(x+1)(x-2)}$

page 263 ***Exercise 10***

1. $(x+2)(x+5)$ **2.** $(x+3)(x+4)$ **3.** $(x+3)(x+5)$ **4.** $(x+3)(x+7)$
5. $(x+2)(x+6)$ **6.** $(y+5)(y+7)$ **7.** $(y+3)(y+8)$ **8.** $(y+5)(y+5)$
9. $(y+3)(y+12)$ **10.** $(a+2)(a-5)$ **11.** $(a+3)(a-4)$ **12.** $(z+3)(z-2)$
13. $(x+5)(x-7)$ **14.** $(x+3)(x-8)$ **15.** $(x-2)(x-4)$ **16.** $(y-2)(y-3)$
17. $(x-3)(x-5)$ **18.** $(a+2)(a-3)$ **19.** $(a+5)(a+9)$ **20.** $(b+3)(b-7)$
21. $(x-4)(x-4)$ **22.** $(y+1)(y+1)$ **23.** $(y-7)(y+4)$ **24.** $(x-5)(x+4)$
25. $(x-20)(x+12)$ **26.** $(x-15)(x-11)$ **27.** $(y+12)(y-9)$ **28.** $(x-7)(x+7)$
29. $(x-3)(x+3)$ **30.** $(x-4)(x+4)$

page 264 **Exercise 11**

1. $(2x+3)(x+1)$ **2.** $(2x+1)(x+3)$ **3.** $(3x+1)(x+2)$ **4.** $(2x+3)(x+4)$
5. $(3x+2)(x+2)$ **6.** $(2x+5)(x+1)$ **7.** $(3x+1)(x-2)$ **8.** $(2x+5)(x-3)$
9. $(2x+7)(x-3)$ **10.** $(3x+4)(x-7)$ **11.** $(2x+1)(3x+2)$ **12.** $(3x-2)(x-3)$
13. $(y-2)(3y-5)$ **14.** $(2y+3)(3y-1)$ **15.** $(5x+2)(2x+1)$ **16.** $(6x-1)(x-3)$
17. $(4x+1)(2x-3)$ **18.** $(3x+2)(4x+5)$ **19.** $(4y-3)(y-5)$ **20.** $(2x-5)(3x-6)$

page 264 **Exercise 12**

1. $(y-a)(y+a)$ **2.** $(m-n)(m+n)$ **3.** $(x-t)(x+t)$ **4.** $(y-1)(y+1)$
5. $(x-3)(x+3)$ **6.** $(a-5)(a+5)$ **7.** $(x-\frac{1}{2})(x+\frac{1}{2})$ **8.** $(x-\frac{1}{3})(x+\frac{1}{3})$
9. $(2x-y)(2x+y)$ **10.** $(a-2b)(a+2b)$ **11.** $(5x-2y)(5x+2y)$ **12.** $(3x-4y)(3x+4y)$
13. $\left(x-\frac{y}{2}\right)\left(x+\frac{y}{2}\right)$ **14.** $(3m-\frac{2}{3}n)(3m+\frac{2}{3}n)$ **15.** $(4t-\frac{2}{5}s)(4t+\frac{2}{5}s)$ **16.** $\left(2x-\frac{z}{10}\right)\left(2x+\frac{z}{10}\right)$
17. $x(x-1)(x+1)$ **18.** $a(a-b)(a+b)$ **19.** $x(2x-1)(2x+1)$ **20.** $2x(2x-y)(2x+y)$

page 265 **Exercise 13**

1. $-3, -4$ **2.** $-2, -5$ **3.** $3, -5$ **4.** $2, -3$ **5.** $2, 6$
6. $-3, -7$ **7.** $2, 3$ **8.** $5, -1$ **9.** $-7, 2$ **10.** $-\frac{1}{2}, 2$
11. $\frac{2}{3}, -4$ **12.** $1\frac{1}{2}, -5$ **13.** $\frac{2}{3}, 1\frac{1}{2}$ **14.** $\frac{1}{4}, 7$ **15.** $\frac{3}{5}, -\frac{1}{2}$
16. $7, 8$ **17.** $\frac{5}{6}, \frac{1}{2}$ **18.** $7, -9$ **19.** $-1, -1$ **20.** $3, 3$
21. $-5, -5$ **22.** $7, 7$ **23.** $-\frac{1}{3}, \frac{1}{2}$ **24.** $-1\frac{1}{4}, 2$ **25.** $13, -5$
26. $-3, \frac{1}{6}$ **27.** $\frac{1}{10}, -2$ **28.** $1, 1$ **29.** $\frac{2}{9}, -\frac{1}{4}$ **30.** $-\frac{1}{4}, \frac{3}{5}$

page 266 **Exercise 14**

1. $0, 3$ **2.** $0, -7$ **3.** $0, 1$ **4.** $0, \frac{1}{3}$ **5.** $4, -4$
6. $7, -7$ **7.** $\frac{1}{2}, -\frac{1}{2}$ **8.** $\frac{2}{3}, -\frac{2}{3}$ **9.** $0, -1\frac{1}{2}$ **10.** $0, 1\frac{1}{2}$
11. $0, 5\frac{1}{2}$ **12.** $\frac{1}{4}, -\frac{1}{4}$ **13.** $\frac{1}{2}, -\frac{1}{2}$ **14.** $0, \frac{5}{8}$ **15.** $0, \frac{1}{12}$
16. $0, 6$ **17.** $0, 11$ **18.** $0, 1\frac{1}{2}$ **19.** $0, 1$ **20.** $0, 4$

page 267 **Exercise 15**

1. $-\frac{1}{2}, -5$ **2.** $-\frac{2}{3}, -3$ **3.** $-\frac{1}{2}, -\frac{2}{3}$ **4.** $\frac{1}{3}, 3$ **5.** $\frac{2}{5}, 1$
6. $\frac{1}{3}, 1\frac{1}{2}$ **7.** $-0{\cdot}63, -2{\cdot}37$ **8.** $-0{\cdot}27, -3{\cdot}73$ **9.** $0{\cdot}72, 0{\cdot}28$ **10.** $6{\cdot}70, 0{\cdot}30$
11. $0{\cdot}19, -2{\cdot}69$ **12.** $0{\cdot}85, -1{\cdot}18$ **13.** $0{\cdot}61, -3{\cdot}28$ **14.** $-1\frac{2}{3}, 4$ **15.** $-1\frac{1}{2}, 5$
16. $3{\cdot}56, -0{\cdot}56$ **17.** $0{\cdot}16, -3{\cdot}16$ **18.** $-\frac{1}{2}, 2\frac{1}{3}$ **19.** $-\frac{1}{3}, -8$ **20.** $1\frac{2}{3}, -1$
21. $2{\cdot}28, 0{\cdot}22$ **22.** $-0{\cdot}35, -5{\cdot}65$ **23.** $-\frac{2}{3}, \frac{1}{2}$ **24.** $-0{\cdot}58, 2{\cdot}58$

page 267 **Exercise 16**

1. $-3, 2$ **2.** $-3, -7$ **3.** $-\frac{1}{2}, 2$ **4.** $1, 4$ **5.** $-1\frac{2}{3}, \frac{1}{2}$
6. $-0{\cdot}39, -4{\cdot}28$ **7.** $-0{\cdot}16, 6{\cdot}16$ **8.** 3 **9.** $2, -1\frac{1}{3}$ **10.** $-3, -1$
11. $0{\cdot}66, -22{\cdot}66$ **12.** $-7, 2$ **13.** $\frac{1}{4}, 7$ **14.** $-\frac{1}{2}, \frac{3}{5}$ **15.** $0, 3\frac{1}{2}$
16. $-\frac{1}{4}, \frac{1}{4}$ **17.** $-2{\cdot}77, 1{\cdot}27$ **18.** $-\frac{2}{3}, 1$ **19.** $-\frac{1}{2}, 2$ **20.** $0, 3$
21. (a) -1 (b) $0{\cdot}6258$ (c) $0{\cdot}5961$ (d) $0{\cdot}2210$

page 269 ***Exercise 17***

1. $(x+4)^2-16$ **2.** $(x-6)^2-36$ **3.** $(x+\frac{1}{2})^2-\frac{1}{4}$ **4.** $(x+2)^2-3$ **5.** $(x-3)^2$
6. $(x+1)^2-16$ **7.** $2[(x+4)^2-\frac{27}{2}]$ **8.** $2[(x-\frac{5}{2})^2-\frac{25}{4}]$ **9.** $10-(x-2)^2$ **10.** $4-(x+1)^2$
11. (a) 0·65, −4·65 (b) 3·56, −0·56 (c) 0·08, −12·08
15. (a) 3 (b) −2 (c) $\frac{1}{3}$ **16.** (a) $\frac{3}{4}$ (b) $\frac{1}{2}$ (c) $\frac{4}{3}$

page 270 ***Exercise 18***

1. 8, 11 or −8, −11 **2.** 11, 13 or −11, −13 **3.** 12 cm **4.** 6 cm **5.** $x = 11$
6. 10 cm × 24 cm **7.** 8 km north, 15 km east **8.** 12 eggs **9.** 1 cm
10. 4 **11.** 2, 5 **12.** $\frac{40}{x}$ h, $\frac{40}{x-2}$ h, 10 km/h **13.** 4 km/h
14. 20 mph **15.** 6 cm **16.** 157 km **17.** $x = 2$ **18.** $x = 3$
19. $\frac{3}{4}$ **20.** 9 cm or 13 cm **21.** (c) $x = 595$ mm, $y = 841$ mm (d) 297 mm

page 277 ***Exercise 19***

2. Stretch parallel to the x-axis, scale factor 2 **4.** (0, −2), (7, 0)
5. (a) A′(−2, −1), B′(0, −3), C′(2, 0) (b) A′(0, 1), B′(2, 3), C′(4, 0)
(c) A′(−1, 1), B′(0, 3), C′(1, 0)
6. (a) A′(2, 4) (b) B′(0, $\frac{1}{5}$), C′(3, 1), D′(−3, 1)
7. (a) (1, 5) (b) (5, −4)
8. (a) $y = x^2+3x+5$ (b) $y = x^2-x-2$ (c) $y = -(x^2+3x)$
9. $a = 2$, $b = 3$

page 279 ***Examination Exercise 6***

1. (a) $21-x$ (b) 12 cm, 9 cm **2.** (b) 9·28, 3·72 **3.** (c) $n = 48$; 47, 48, 49
4. (a) 10 g (b) 56·6 g (c) 80 g
5. (a) $x = 0{\cdot}3$ (b) (i) 10% (ii) 6·97 kg (iii) about 7 years **6.** 1·653, 1·706
7. (a) −1, 5 (b) root between $x = 2$ and $x = 3$ (d) $x_2 = 2{\cdot}2$, $x_3 = 2{\cdot}227$ (e) $x = -0{\cdot}225$
8. (a) 6·4 (b) 6 (c) $x^2-4x-12=0$, $x=-2$
9. (a) 0·3 ms^{-2} (b) 1200 m (c) 10 m/s **10.** (b) 350 m
11. (a) $(x+\frac{5}{2})^2-\frac{53}{4}$ (b) $x = 1{\cdot}14, -6{\cdot}14$
12. (c) (i) $t = 1{\cdot}2, 3{\cdot}3$ (ii) $t = 2{\cdot}25$ s (d) 27·5 m
13. (c) −9 **14.** (a) A′(1, 4), B′(2, 0)

Part 7

page 286 ***Exercise 1***

1. (a) £3000 (b) £4000 (c) £6000 (d) £11 000
2. eggs 270°; milk 12°; butter 23·4°; cheese 54°; salt/pepper 0·6° **3.** 18°, 54°, 54°, 234°
4. (a) 22·5% (b) $x = 45°$, $y = 114°$ **5.** $x = 8$

page 289 ***Exercise 2***

1. (a) 6, 5, 4 (b) 9, 7, 7 (c) 6·5, 8, 9 (d) 3·5, 3·5, 4
2. (a) mean = £33 920, median = £22 500, mode = £22 500
(b) the mean is skewed by one large figure
3. (a) mean = 157·1 kg, median = 91 kg
(b) mean; No: Over three quarters of the cattle are below the mean weight
4. (a) mean = 74·5 cm, median = 91 cm (b) Yes
5. 78 kg **6.** 35·2 cm **7.** (a) 2 (b) 9
8. (a) 20·4 m (b) 12·8 m (c) 1·66 m **9.** 55 kg

page 292 ***Exercise 3***

1. 3·38 **2.** (a) mean = 6·62; median = 8; mode = 3 (b) mode
3. (a) mean = 3·025; median = 3; mode = 3 (b) mean = 17·75; median = 17; mode = 17
4. (a) 68·25 (b) median in 55-69; modal group 55-69
5. 3·8 **6.** (a) 9 (b) 9 (c) 15 **7.** (a) 5 (b) 10 (c) 10
8. 12 **9.** $3\frac{2}{3}$ **10.** 4·68
11. (a) N (b) mean = $N^2 + 2$; median = N^2 (c) 2 **12.** $\dfrac{ax + by + cz}{a + b + c}$

page 298 ***Exercise 5***

1. 44 **2.** 4·2 km **3.** (b) Bogota; High altitude (c) about 73°C

page 300 ***Exercise 6***

1. 6, 24, 12, 42, 18, 60, 24 **2.** 54, 62, 71, 38, 42, 46, 50
3. Tends to 4 **4.** Integral part of $\sqrt[3]{N}$
5. (a) 4, 4, 6, 7 (b) 5 (c) How many times 5 goes into N (d) 56

page 303 ***Exercise 7***

1. (a) 46 (b) 28; 62 (c) 34 (d) 35 [all approximate]
2. (a) 100 (b) 250 (c) 2250 h (d) 750 h
3. (a) 28 (b) 25; 37 (c) 12 (d) about 27 (e) 60 [all approx]
4. (b) (i) 45 (ii) 17 **5.** (b) France 18·5; Britain 24·5 (c) 8·5
(d) Results for Britain bunched together more closely with a higher median.
6. (b) (i) 80·5 g (ii) 22 g (c) Half the population within 22 g (d) 71 g
7. (b) 10·2 cm (c) Many failed to germinate
(d) Median is better as it discounts those which do not germinate
8. (a) 36·5 g (b) 20 g (c) about 25 or 26 people
9. (c) 44 million (d) A: 21, B: 59 (e) A has much younger population

page 307 ***Examination Exercise 7***

1. (a) £4500 (b) £9000 (c) £8646 **2.** 169°, 83°, 108°
3. (a) (i) £52 000 (ii) £57 000 **4.** (a) 1·6-1·8 (b) 1·2-1·4 (c) 1·44 kg
5. Second year: wrong, no correlation; fourth year: wrong, do better at back
6. (c) −0·7 (d) $y = -0{\cdot}7x + 12$
8. (a) 86·3 cm (d) (i) 86·5 cm (ii) 12 cm (e) part (ii)

Part 8

page 311 ***Exercise 1***

1. B **2.** C **3.** A **4.** B or C **5.** C or D **6.** C
7. B **8.** B **9.** C **10.** A **11.** D **12.** C

page 313 ***Exercise 2***

1. (a) $\frac{1}{13}$ (b) $\frac{1}{2}$ (c) $\frac{1}{52}$ (d) $\frac{3}{52}$ **2.** (a) $\frac{1}{6}$ (b) $\frac{1}{2}$ (c) $\frac{1}{2}$ (d) $\frac{1}{3}$
3. (a) $\frac{1}{4}$ (b) $\frac{1}{2}$ **4.** (a) (i) $\frac{3}{5}$ (ii) $\frac{2}{5}$ (b) (i) $\frac{5}{9}$ (ii) $\frac{4}{9}$
5. (a) $\frac{1}{11}$ (b) $\frac{2}{11}$ (c) 0 (d) $\frac{1}{11}$
6. (a) $\frac{5}{13}$ (b) $\frac{4}{13}$ (c) $\frac{2}{13}$ (d) $\frac{3}{13}$ (e) $\frac{1}{13}$ (f) 0 (g) $\frac{3}{13}$
7. (a) $\frac{13}{49}$ (b) $\frac{3}{49}$ (c) $\frac{10}{49}$ (d) $\frac{1}{49}$ **8.** (a) $\frac{1}{2}$ (b) $\frac{3}{25}$ (c) $\frac{9}{100}$ (d) $\frac{2}{25}$
9. (a) 1 (b) 0 (c) 1 (d) 0 **10.** $\frac{5}{999}$ **11.** $\frac{271}{1000}$ **12.** $\frac{x}{12}$, 3 **13.** 25
14. 50 **15.** 40 **16.** $\frac{1}{3}$ **17.** (a) 0·2146 (b) 1073

page 317 ***Exercise 3***

1. (a) $\frac{5}{11}$ (b) $\frac{7}{22}$ (c) $\frac{15}{22}$ (d) $\frac{17}{22}$ **2.** $\frac{1}{18}$ **3.** $\frac{1}{3}$ **4.** $\frac{1}{24}$ **5.** $\frac{264}{360} = \frac{11}{15}$
6. (a) (i) exclusive (ii) not exclusive (b) $\frac{11}{15}$
7. (a) (i) exclusive (ii) exclusive (iii) not exclusive (b) $\frac{3}{4}$
8. (a) 0·8 (b) 0·7 (c) not exclusive

page 319 ***Exercise 4***

1. (a) $\frac{1}{8}$ (b) $\frac{3}{8}$ (c) $\frac{1}{8}$ (d) $\frac{7}{8}$ **2.** (a) $\frac{1}{12}$ (b) $\frac{1}{36}$ (c) $\frac{5}{18}$ (d) $\frac{1}{6}$
3. (a) $\frac{1}{4}$ (b) $\frac{1}{4}$ **4.** (a) $\frac{1}{12}$ (b) $\frac{5}{36}$ (c) $\frac{2}{3}$ (d) $\frac{1}{12}$ (e) $\frac{1}{36}$
5. (a) $\frac{3}{8}$ (b) $\frac{1}{16}$ (c) $\frac{15}{16}$ (d) $\frac{1}{4}$ **6.** (a) 64 (b) $\frac{1}{64}$ **7.** $\frac{1}{2}$ **8.** $\frac{1}{6}$
9. (a) $\frac{1}{144}$ (b) $\frac{1}{18}$ (c) $\frac{1}{72}$ **10.** (a) $\frac{1}{216}$ (b) $\frac{1}{72}$ (c) $\frac{1}{8}$ (d) $\frac{5}{108}$ (e) $\frac{5}{72}$ (f) $\frac{1}{36}$

page 322 ***Exercise 5***

1. (a) $\frac{1}{13}, \frac{1}{6}$ (b) $\frac{1}{78}$ **2.** (a) $\frac{1}{2}$ (b) $\frac{1}{2}$ (c) $\frac{1}{4}$ **3.** $\frac{1}{10}$
4. (a) $\frac{1}{78}$ (b) $\frac{1}{104}$ (c) $\frac{1}{24}$ **5.** (a) $\frac{1}{16}$ (b) $\frac{1}{169}$ (c) $\frac{9}{169}$
6. (a) $\frac{1}{16}$ (b) $\frac{25}{144}$ **7.** (a) $\frac{1}{121}$ (b) $\frac{9}{121}$ **8.** $\frac{8}{1125}$
9. (a) $\frac{1}{288}$ (b) $\frac{1}{72}$ **10.** (a) $\frac{1}{9}$ (b) $\frac{4}{27}$ **11.** $\frac{1}{24}$

page 325 ***Exercise 6***

1. (a) $\frac{49}{100}$ (b) $\frac{9}{100}$ **2.** (a) $\frac{9}{64}$ (b) $\frac{15}{64}$ **3.** (a) $\frac{7}{15}$ (b) $\frac{1}{15}$
4. (a) $\frac{2}{9}$ (b) $\frac{2}{15}$ (c) $\frac{1}{45}$ **5.** (a) $\frac{1}{12}$ (b) $\frac{1}{6}$ (c) $\frac{1}{3}$ (d) $\frac{2}{9}$
6. (a) $\frac{1}{216}$ (b) $\frac{125}{216}$ (c) $\frac{25}{72}$ (d) $\frac{91}{216}$ **7.** (a) $\frac{1}{64}$ (b) $\frac{5}{32}$ (c) $\frac{27}{64}$
8. (a) $\frac{1}{6}$ (b) $\frac{1}{30}$ (c) $\frac{1}{30}$ (d) $\frac{29}{30}$ **9.** (a) $\frac{27}{64}$ (b) $\frac{1}{64}$ **10.** $\frac{1}{3}$
11. (a) $\frac{1}{64}$ (b) $\frac{27}{64}$ (c) $\frac{9}{64}$ (d) $\frac{27}{64}$; Sum = 1
12. (a) $\frac{3}{20} \times \frac{2}{19} \times \frac{1}{18} (= \frac{1}{1140})$ (b) $\frac{1}{4} \times \frac{4}{19} \times \frac{1}{16} (= \frac{1}{114})$ (c) $\frac{5}{20} \times \frac{4}{19} \times \frac{3}{18} \times \frac{2}{17}$
13. (a) $\frac{1}{10\,000}$ (b) $\frac{523}{10\,000}$ (c) $\frac{9^4}{10^4}$

page 329 ***Exercise 7***

1. (a) $\frac{10\times9}{1000\times999}$ (b) $\frac{990\times989}{1000\times999}$ (c) $\frac{2\times10\times990}{1000\times999}$ **2.** (a) $\frac{3}{20}$ (b) $\frac{7}{20}$ (c) $\frac{1}{2}$
3. (a) 5 (b) $\frac{1}{64}$ **4.** (a) $\frac{1}{220}$ (b) $\frac{1}{22}$ (c) $\frac{3}{11}$ (d) 5
5. (a) $\frac{3}{5}$ (b) $\frac{1}{3}$ (c) $\frac{2}{15}$ (d) $\frac{2}{21}$ (e) $\frac{1}{7}$ **6.** (a) 0·007 81 (b) 0·511
7. (a) $\frac{21}{506}$ (b) $\frac{455}{2024}$ (c) $\frac{945}{2024}$
8. (a) $\frac{x}{x+y}$ (b) $\frac{x(x-1)}{(x+y)(x+y-1)}$ (c) $\frac{2xy}{(x+y)(x+y-1)}$ (d) $\frac{y(y-1)}{(x+y)(x+y-1)}$
9. (a) $\frac{x}{z}$ (b) $\frac{x(x-1)}{z(z-1)}$ (c) $\frac{2x(z-x)}{z(z-1)}$ **10.** (a) $\frac{1}{125}$ (b) $\frac{1}{125}$ (c) $\frac{1}{10\,000}$ (d) $\frac{3}{500}$
11. (a) $\frac{1}{49}$ (b) $\frac{1}{7}$ **12.** (a) $\frac{1}{52}$ (b) 1

page 332 ***Exercise 8***

1. $\frac{3}{10}$ **2.** $\frac{9}{140}$ **3.** (a) $\frac{1}{5}$ (b) $\frac{18}{25}$ (c) $\frac{1}{20}$ (d) $\frac{2}{25}$ (e) $\frac{77}{100}$
4. $\frac{4}{35}$ **5.** (a) $\frac{19}{25}$ (b) 38 out of 50 **6.** 27 **7.** $\frac{8}{23}$ **8.** $\frac{10}{29}$

page 334 ***Examination Exercise 8***

1. (a) (i) A (ii) $\frac{4}{15}$ (b) B **2.** (a) $\frac{1}{36}$ (b) $\frac{1}{9}$ (c) $\frac{1}{6}$
3. Generous, $p(\text{total} = 15) = \frac{1}{3}$ **4.** (a) (i) $\frac{1}{3}$ (ii) $\frac{2}{3}$ (iii) $\frac{1}{3}$ (b) Yes
5. (a) (i) $\frac{7}{25}$ (ii) $\frac{3}{100}$ (iii) $\frac{97}{100}$ (b) $\frac{3}{4}$ **6.** (a) $\frac{1}{6}$ (b) $\frac{1}{12}$ (c) $\frac{1}{9}$
7. (a) $\frac{12}{49}$ (b) $\frac{11}{188}$ **8.** (b) (i) $\frac{11}{35}$ (ii) $\frac{18}{35}$ **9.** (a) (i) $\frac{12}{25}$ (ii) $\frac{13}{25}$ (b) $\frac{1}{2}$
10. (a) $\frac{3}{10}$ (b) (i) $\frac{n}{2}$ (ii) $\frac{3}{10}\times\frac{n}{2}$ (iii) $\left(\frac{3}{10}\times\frac{n}{2}+\frac{6}{11}\times\frac{n}{2}\right)$ (c) $\frac{11}{31}$

Part 9

page 341 ***Exercise 2***

1. (a) (i) 10 (ii) 20 (b) 65 **2.** 55 **3.** 135 **4.** 28 **5.** 44
8. (a) 12 (b) 23 (c) 29 (d) 25

page 349 ***Exercise 4***

1. Pupils who do not eat school meals will not be questioned.
2. OK.
3. His friends may not be typical.
4. People going on ferry may not be representative of the population.
5. Might not be typical of the whole week.
6. OK.
7. People are unlikely to include those at work/school etc.

page 352 ***Exercise 5***

1. (a) $\overline{x} = 6{\cdot}4$; s.d. = 3·38 (b) $\overline{x} = 4{\cdot}9$; s.d. = 1·00
(c) $\overline{x} = 115$; s.d. = 9·75 (d) $\overline{x} = 0{\cdot}16$; s.d. = 2·67

2. A: $\bar{x} = 23$; s.d. = 1·87 B: $\bar{x} = 23$; s.d. = 3·08 Method A involves less variation in lifetimes.
3. Mizuno: $\bar{x} = 70$; s.d. = 2·52. Lynx: $\bar{x} = 70$; s.d. = 4·20.
Use Mizuno for consistent results as s.d. is lower.
4. (a) $\bar{x} = 11{\cdot}08$, s.d. = 6·11 (b) $\bar{x} = 5{\cdot}6$, s.d. = 1·99
5. (a) mean ≈ 33·5; s.d. ≈ 11 (b) mean ≈ 15·5; s.d. ≈ 7·6 (c) mean ≈ 10·65; s.d. ≈ 5·7
6. (a) 8, 10, 8, 4 (b) mean = £134; s.d. = £25·8

page 357 ***Exercise 6***

1. (a) 68% (b) 54·9 g ⩽ wt ⩽ 70·1 g **2.** (a) 16% (b) 95% (c) 8
3. (a) 16% (b) 0·16 (c) 1 pair **4.** (a) 128 (b) 20
5. 375 **6.** 49 mins

page 357 ***Exercise 7***

1. mean = 237·9 g; s.d. = 14·3 g **2.** mean = 247·9 g; s.d. = 14·3 g
3. mean = 1·825; s.d. = 1·14 **4.** mean = 3·65; s.d. = 2·28

page 360 ***Exercise 8***

1. (a)

A	–
B	–
C	AB
D	–
E	ABCD
F	ABCDE

(b)

A	–
B	AD
C	DE
D	–
E	–
F	ABCDE

(c)

A	–
B	A
C	–
D	AB
E	AB
F	ABE
G	ABCDEF

3. (a) BCEG, 30 h (b) DGH, 24 h (c) CBEF, 30 h
4. (a) BCEF, 16 h (b) ABDG, 21 h (c) DCF or ABF, 20 h **5.** BCEF, 21 h

page 362 ***Exercise 9***

1. BCDHI, 65 min **2.** EGFH, 65 min **3.** ACFG, 67 min

page 364 ***Exercise 10***

1. (a) 26 [at (6, 5)] (b) 12 [at (3, 3)] **2.** (a) 25 [at (8, 3)] (b) 9 [at (7, 2)]
3. (a) 40 [at (20, 0)] (b) 112 [at (14, 8)] **4.** (3, 3), (4, 2), (4, 3), (4, 4), (5, 1), (5, 2), (6,0)
5. (2, 4), (2, 5), (2, 6), (2, 7), (2, 8), (2, 9), (2, 10), (3, 6), (3, 7), (3, 8)
6. (a) (6, 7), (7, 7), (8, 6), (7, 6), (6, 8) (b) 7 defenders, 6 forwards – £19 000
7. (a) (6, 12), £1440 (b) (10, 10), £1800 **8.** (a) (9, 7), £44 (b) (15, 5), £45

page 366 ***Examination Exercise 9***

1. (a) 22, 40, 34, 18 (b) 18° (e) 58 marks **3.** (a) (i) 2·5, 3, 2, 0·25
(b) e.g. more elderly employees in A, more young employees in A, smaller spread in B
4. (a) 21·6 (b) 0·4
5. (i) mean = 545·7, s.d. = 66·1 (ii) snack foods show greater variation as s.d. is higher
6. (a) 0·579 cm (b) second sample less widely spread
7. 14·6 g **8.** 25 **11.** 14 hours **12.** 70 minutes
13. (a) 19 adults, 42 children (b) 28 children, 28 adults **14.** (iii) 32 (iv) 16

Part 10

page 374 Exercise 2

1. (a) If today is Monday, then tomorrow is Tuesday.
(b) If it is raining, then there are clouds in the sky.
(c) If Abraham Lincoln was born in 1809, then Abraham Lincoln is dead.

2. (a) T (b) T (c) T (d) F (e) T (f) F (g) T (h) T (i) F (j) T

Part 11

page 382 ***11.1 Investigations***

Note. It must be emphasised that the *process* of obtaining reliable results is far more important than these few results. It is not suggested that 'obtaining a formula' is the only aim of these coursework tasks. The results are given here merely as a check for teachers or students working on their own.

It is not possible to summarise the enormous number of variations which students might think of for themselves. Obviously some original thoughts will be productive while many others will soon 'dry up'.

1. With the numbers written in c columns, the difference for a $(n \times n)$ square is $(n-1)^2 \times c$.

2. From 0 to 774 miles, choose 'Hav-a-car'.
From 775 to 1538 miles, choose 'Snowdon'.
Over 1538 miles, choose 'Gibson'.

3. For a final score $a - b$, number of possible half-time scores $= (a+1)(b+1)$.
e.g. For 7–5, number of half-time scores $= (7+1)(5+1) = 48$.

4. For diagram number n, number of squares $= 2n^2 - 2n + 1$.

5. For a square card, corner cut out $= \frac{1}{6}$ (size of card).

For a rectangle $a \times 2a$, corner cut out $\cong \dfrac{a}{4{\cdot}732}$.

7. For n names, maximum possible number of interchanges $= \dfrac{n(n-1)}{2}$.

8. For smallest surface area, height of cylinder $= 2 \times$ radius.

9. For $n \times n \times n$: 3 green faces $= 8$
2 green faces $= 12(n-2)$
1 green face $= 6(n-2)^2$
0 green face $= (n-2)^3$

10. (a) For n blacks and n whites, number of moves $= \dfrac{n(n+1)}{2}$.

(b) For n blacks and n whites, number of moves $= \dfrac{n(n+2)}{2}$.

(c) For n of each colour, number of moves $= \frac{3}{2}n(n+1)$.

11. Consider three cases of rectangles $m \times n$
(a) m and n have no common factor. Number of squares $= m + n - 1$.
e.g. 3×7, number $= 3 + 7 - 1 = 9$
(b) n is a multiple of m. Number of squares $= n$ e.g. 3×12, number $= 12$
(c) m and n share a common factor a
so $m \times n = a(m' + n')$
number of squares $= a(m' + n' - 1)$
e.g. $640 \times 250 = 10(64 \times 25)$, number of squares $= 10(64 + 25 - 1) = 880$.

12. 4×4: There are 30 squares (i.e. $16 + 9 + 4 + 1$)
8×8: There are 204 squares ($64 + 49 + 36 + 25 + 16 + 9 + 4 + 1$)
$n \times n$: Number of squares $= 1^2 + 2^2 + 3^2 + \ldots + n^2$

$$= \frac{n}{6}(n + 1)(2n + 1)$$

This could be found using method of differences or from the standard result for $\sum_1^n r^2$.

13. Pattern of digit sums for multiples of 43 = pattern for multiples of 7. $(4 + 3 = 7)$
Pattern of digit sums for multiples of 62 = pattern for multiples of 8. $(6 + 2 = 8)$

14. (a) square root twice, 'multiply by x' gives cube root of x.
(b) square root three times, 'multiply by x' gives seventh root of x.
(c) square root twice, 'divide by x' gives $1/(\sqrt[3]{x})$
(d) square root twice, 'multiply by x^2' gives $(\sqrt[3]{x})^2$

15. (a) Another Fibonacci sequence.
(b) Terms are alternate terms of original sequence.
(c) Ratio tends towards 1·618 (to 4 s.f.), the 'Golden Ratio'
(d) (first $\times$ fourth) = (second $\times$ third) + 1
(e) (first $\times$ third) $\pm$ 1 = (second)^2, alternating + and −
(f) sum of 10 terms = 11 $\times$ seventh term.
(g) For six terms $a\ b\ c\ d\ e\ f$ let $x = e \times f - (a^2 + b^2 + c^2 + d^2 + e^2)$

	x	first difference
first six	0	
		1
second six	1	
		1
third six	2	
		4
fourth six	6	
		9
fifth six	15	
		25
sixth six	40	
		64
seventh six	104	
		169
eighth six	279	

The numbers in the first difference column are the squares of the terms in the original Fibonacci sequence.

16. Pick's theorem: $A = i + \frac{1}{2}p - 1$

17. The prime numbers in columns 1 and 5 can all be written as the sum of two squares.

Part 12

page 397 **Revision exercise 1**

1. $\frac{a}{b}$ **2.** (a) 0·005 m/s (b) 1·6 s (c) 173 km
3. (a) $c = 5, d = -2$ (b) $x = 2, y = -1$
4. (a) 45·6° (b) 58·0° (c) 3·89 cm (d) 33·8 m **5.** (a) 14 (b) 18 (c) 28
6. $\frac{1}{6}$ **7.** 3·43 cm^2, 4·57 cm^2 **8.** (a) 220° (b) 295°
9. A : $y = 6$; B : $y = \frac{1}{2}x - 3$; C : $y = 10 - x$; D : $y = 3x$
10. (a) $s = t(r + 3)$ (b) $r = \frac{s - 3t}{t}$ **11.** 1·552 m **12.** 4·12 cm
13. 9·95 cm **14.** $y \geqslant 2$, $x + y \leqslant 6$; $y \leqslant 3x$

page 398 **Revision exercise 2**

1. 43·1 litres **2.** (a) 8 (b) 140 (c) 29 (d) 42 (e) 6 (f) −6
3. (a) 1 : 50 000 (b) 1 : 4 000 000 **4.** 25 **5.** 5·14 cm^2 **6.** $\frac{5}{16}$
7. (a) $z = x - 5y$ (b) $m = \frac{11}{k + 3}$ (c) $z = \frac{T^2}{C^2}$ **8.** $\frac{31}{20}b$ **9.** 5·39 cm
10. (a) 5·45 (b) 5 (c) 5 **11.** 6 cm **12.** 219

page 399 **Revision exercise 3**

1. $2{\cdot}1 \times 10^{24}$ **2.** 17 kg **3.** (a) 9π cm^2 (b) 8 : 1 **4.** $33\frac{1}{3}$ mph
5. (a) 0·340 (b) $4{\cdot}08 \times 10^{-6}$ (c) 64·9 (d) 0·119 **6.** $2{\cdot}3 \times 10^9$
7. (a) $\frac{1}{9}$ (b) $\frac{1}{12}$ (c) 0 **8.** $x \geqslant 0$, $y \geqslant x - 2$, $x + y \leqslant 7$
9. (a) 600 (b) 9000 or 10 000 (c) 3 (d) 60
10. 40° (b) 100° **11.** (a) $\frac{x}{x + 5}$ (b) $\left(\frac{x}{x + 5}\right)^2$
12. A : $4y = 3x - 16$; B : $2y = x - 8$; C : $2y + x = 8$; D : $4y + 3x = 16$
13. (a) 22% (b) 20·8% (c) £240 **14.** 33·1%

page 400 **Revision exercise 4**

1. 95p for 1 lb **2.** (a) (i) 7 m/s (ii) 200 m/s (iii) 5 m/s (b) (i) 144 km/h (ii) 2·16 km/h
3. (a) $6x + 15 < 200$ (b) 29 **4.** 20 cm^2 **5.** (a) 1·75 (b) 60·3°
6. (a) (3, −2) (b) $(1\frac{1}{2}, -3)$ **7.** (a) $x = 14{\cdot}1$ cm, size of card = 48·3 cm (b) 1930 cm^2
8. (a) 0·5601 (b) 3·215 (c) 0·6161 (d) 0·4743 **9.** (a) 84 (b) 19·2
10. 0·335 m **11.** 1·24 **12.** (a) $\frac{3}{5}$ (b) $w = \frac{k(1 - y)}{y}$ **13.** (a) 50 (b) 50

page 401 **Revision exercise 5**

1. (a) Yes (b) 2·5 **2.** 1·4
3. (a) A (b) V (c) Impossible (d) L **4.** 152° **5.** 45°, 225°
6. 123·25 cm^2, 101·25 cm^2 **7.** 0·72% **8.** (b) DBCF, 19 h
10. (a) −3, 5 (b) 2, 6 **11.** 2·92, −1·25 (b) 5·24, 0·76 **12.** 4·74

page 402 ***Revision exercise 6***

1. 4·8 **2.** (a) 290° (b) 60°, 300° **3.** (a) 30 (b) 75
4. DEGH, 19 h **5.** 3·63 cm **6.** 1, $-\frac{1}{4}$ **7.** 273 cm^3
8. (a) 14·1 cm (b) 35·3° **9.** -4 **10.** (a) 30 m/s (b) 600 m
11. (a) $\frac{1}{28}$ (b) $\frac{15}{28}$ (c) $\frac{3}{7}$ **12.** (a) 0·65, 3·85 (b) $0{\cdot}65 < x < 3{\cdot}85$ (c) 3·3

page 404 ***Revision exercise 7***

1. 68p **2.** (a) $\frac{1}{2}, -\frac{1}{2}$ (b) $\frac{7}{11}$ (c) 3 (d) 0, 5
3. (a) 12·2 cm (b) 61·1 cm^2 **4.** 250 cm^3 **5.** (a) $3\frac{1}{3}$ cm (b) 1620 cm^3
6. (a) 50° (b) 128° (c) $c = 50°$, $d = 40°$ (d) $x = 10°$, $y - 40°$
7. (a) $(-2, 5)$ (b) $(-4, -3)$ (c) Rotation 90° anticlockwise about (0, 0)
8. $\frac{35}{48}$ **9.** 4·12 cm, 9·93 cm **10.** 8·06 cm **11.** 13
12. (a) $\mathbf{a} - \mathbf{c}$ (b) $\frac{1}{2}\mathbf{a} + \mathbf{c}$ (c) $\frac{1}{2}\mathbf{a} - \frac{1}{2}\mathbf{c}$
CA is parallel to NM and CA = 2NM.

page 405 ***Revision exercise 8***

1. (a) 0·3 m/s^2 (b) 1050 m (c) 40 s **2.** (a) 1 m^2 (b) 1000 cm^3
4. (a) $1\frac{5}{6}$ (b) 0·09 **5.** 45·2 km, 33·6 km **6.** (a) 6·63 cm (b) 41·8°
7. (a) A′ (0, −3), B′ (6, 0), C′ (12, −3) (b) A′ (3, 0), B′ (9, 3), C′ (15, 0)
(c) A′ (0, 0), B′ (2, 3), C′ (4, 0)
9. (a) reflection in $x = \frac{1}{2}$ (b) reflection in $y = -x$ (c) rotation 180° about (1, 1)
10. (a) enlargement, scale factor 2, centre (0, 0); translation $\begin{pmatrix} 5 \\ -2 \end{pmatrix}$
(b) (11, −4) (c) (3, −1) (d) (1, 3)
11. $x = 8$; dimensions 8 cm × 6·5 cm **12.** mean = 13·6, s.d. = 8·2

page 408 ***Test 1***

1. C **2.** D **3.** D **4.** B **5.** A
6. C **7.** C **8.** D **9.** C **10.** B
11. C **12.** A **13.** D **14.** C **15.** C
16. D **17.** A **18.** C **19.** B **20.** D
21. A **22.** C **23.** B **24.** B **25.** C

page 409 ***Test 2***

1. B **2.** C **3.** A **4.** B **5.** D
6. C **7.** A **8.** D **9.** B **10.** C
11. B **12.** D **13.** A **14.** C **15.** C
16. D **17.** B **18.** A **19.** B **20.** B
21. C **22.** D **23.** A **24.** A **25.** B

page 411 ***Test 3***

1. D **2.** D **3.** D **4.** B **5.** A
6. C **7.** A **8.** D **9.** D **10.** B
11. D **12.** D **13.** D **14.** B **15.** A

16. B	**17.** C	**18.** A	**19.** D	**20.** D
21. C	**22.** A	**23.** B	**24.** B	**25.** D

page 413 ***Test 4***

1. B	**2.** B	**3.** A	**4.** C	**5.** D
6. A	**7.** B	**8.** C	**9.** D	**10.** B
11. C	**12.** A	**13.** B	**14.** B	**15.** D
16. A	**17.** C	**18.** B	**19.** B	**20.** D
21. A	**22.** C	**23.** C	**24.** C	**25.** A

INDEX